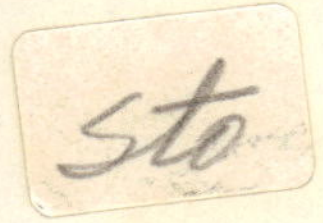

Semiconductors

Semiconductors

R. A. SMITH
CBE, FRS, FRSE
PRINCIPAL EMERITUS AND HONORARY PROFESSOR
HERIOT-WATT UNIVERSITY, EDINBURGH

SECOND EDITION

CAMBRIDGE UNIVERSITY PRESS
CAMBRIDGE
LONDON · NEW YORK · MELBOURNE

Published by the Syndics of the Cambridge University Press
The Pitt Building, Trumpington Street, Cambridge CB2 1RP
Bentley House, 200 Euston Road, London NW1 2DB
32 East 57th Street, New York, NY 10022, USA
296 Beaconsfield Parade, Middle Park, Melbourne 3206, Australia

First published 1959
Reprinted 1961, 1964, 1968
Second edition 1978

Printed in Great Britain by
J. W. Arrowsmith Ltd., Bristol BS3 2NT

Library of Congress cataloguing in publication data

Smith, Robert Allan, 1909–
Semiconductors

Bibliography: p. 501 Includes indexes
1. Semiconductors I. Title
QC611.S59 1978 537.6'22 77–28181
ISBN 0 521 21824 1 hard covers
ISBN 0 521 29314 6 paperback
(ISBN 0 521 06500 3 first edition)

Contents

Preface

When the first edition of *Semiconductors* was first published there was already an extensive literature on the subject, but mainly in the form of papers in the scientific and technical journals, so there was a clear need for a book giving a comprehensive account of the basic physics of semiconductors and indicating their role in technology. Since then a vast literature has grown up, consisting not only of published papers but of review articles and books dealing with special aspects of the subject. Indeed, the most notable feature of recent publications has been the number of specialist books both in the form of monographs and as collections of articles by a number of authors, each expert in some aspect of semiconductor research or development.

It is perhaps somewhat surprising that *Semiconductors* has so long held its place as one of the main textbooks for advanced undergraduate and postgraduate courses. But the book was clearly in need of drastic revision if it was to continue to serve this purpose. While the basic physics of semiconductors has remained virtually unchanged, so much that is new in the field has happened in the last decade. For example the striking developments in magneto-optics associated with magnetic quantization (except for cyclotron resonance) had not taken place when the first edition was being written. Hot electrons were known and had been studied, but the importance of transferred-electron effects had not been appreciated. New developments through the use of very high magnetic fields and high pressures were still to come and new phenomena involving interaction of light with various other excitations had yet to be discovered. The laser, which has played such a vital part in modern optics had not yet become available and the importance of light emission from semiconductors was just beginning to be appreciated. Although some studies of amorphous semiconductors had been made, the phenomenal growth of interest in this subject has only taken place over the last few years.

While it was necessary to retain much of the basic physics treated in the first edition it was clearly desirable to include a great deal of new material. A boundary condition set by the publishers was that a new edition should not be substantially larger than the former one. This presented a very difficult task in selection of material and it was only the pressure and persuasion of colleagues and friends that induced me to undertake it, mainly with the argument that a book like *Semiconductors* is *still* needed in spite of the very large literature available, there being few overall treatments of the subject at the appropriate level. Clearly a lot of material had to go. This was mainly in the chapters on applications, on methods of measurement and on detailed listing of the values of the physical characteristics of many semiconductors; there are, however, many books on electronic and optical applications and tables of physical constants available, so these can be found elsewhere. Also, application tends to lag behind new discoveries in time. With the publication of my book *Wave Mechanics of Crystalline Solids* (2nd edition), written partly as a complementary theoretical treatment to go along with *Semiconductors*, a certain amount of theoretical work could be left out of the new edition and reference made to *W.M.C.S.* This has been done throughout the text.

In preparing the new edition my task has been made much easier by the availability of the proceedings of the biennial series of conferences on the physics of semiconductors up to the 13th held in Rome in 1976. These are goldmines of information and give a vivid historical picture of the development of semiconductor physics. The 'series' publications, such as *Progress in Semiconductors* (which sadly ceased publication in 1967), *Semiconductors and Semimetals* and *Solid State Physics*, have also been invaluable and again I should like to express my thanks to the authors of these and of other review articles. The availability of several Russian journals devoted specifically to semiconductors, in English translation, has also been a great help.

With so extensive a literature the problem of referencing has been difficult. To give full references would have taken up much too large a proportion of the book. I have tried to give only some of the early key references and some of the more important later ones and have chosen these so that the reader can readily compile a much more complete set for any particular topic if he so desires. I have also included a short and very incomplete bibliography, indicating some of the publications I have found most helpful.

The subject matter which I have included covers most of the main developments. Some of the treatment is a good deal shorter than I should have liked, but lack of space prevented more detailed exposition. Although most of the book deals with crystalline semiconductors I have included a final chapter on amorphous semiconductors, not with any hope of dealing adequately with this subject for which the publication rate now may even exceed that for crystalline materials, but in order to contrast the ideas appropriate to this rapidly growing field with those which have been so successful in enabling us to give a coherent account of this remarkable class of substances in crystalline form.

To the students I have taught and to many others who have written to me I want to express my grateful thanks for helping me to clarify the exposition, and also to my colleagues at M.I.T. and more recently at Heriot-Watt University with whom I have had many stimulating discussions both on basic physics and on presentation.

Finally I should like to thank the staff of the Cambridge University Press for their ready help, and in particular for a number of valuable suggestions for clarification of the text.

Edinburgh 1977 R. A. Smith

From preface to the first edition

The phenomenal growth in the research effort devoted to the study of semiconductors during the past decade has resulted in a very large literature on the subject. As is usual with a rapidly developing field of research, this literature has been mainly in the form of original papers published in the scientific and technical journals, supplemented by review articles dealing more comprehensively with special aspects of the subject. Fortunately, within the last few years, a considerable number of excellent review articles has become available; some have been published in the scientific and technical journals, some in books giving annual series of review articles on wider aspects of the physics of solids, such as *Solid State Physics*,[1] and at least one annual publication is devoted entirely to review articles on semiconductors.[2]

This book came to be written as a result of the author's being invited to give a course of lectures on the physics of semiconductors as a guest lecturer in the Department of Engineering at Edinburgh University in the spring of 1955. The material used for the lecture course has now been considerably expanded and brought up-to-date, and the aim of the book is rather wider than that of the original lecture course, which was mainly intended for graduate engineers. The book is intended principally for physicists and aims at giving an account of the main physical properties of semiconductors; it is hoped, however, that it will also be of use to the large number of engineers now employed in the development of electronic devices based on semiconductors, in helping them to understand the basic physics of the semiconducting materials which they use. Without such an understanding the best use of the materials cannot

[1] Ed. F. Seitz and D. Turnbull (Academic Press, 1955 *et seq.*).

[2] *Progress in Semiconductors*, ed. A. F. Gibson, P. Aigrain and R. E. Burgess (Heywood, 1956 *et seq.*).

be made, since semiconductors are by no means simple and have some rather subtle properties.

The treatment is based on the 'effective mass' concept and a brief and elementary account only is given of the quantum theory underlying the motion of electrons through a crystal. The form of the electron energy levels in a crystalline solid is discussed briefly but without a full quantum-mechanical treatment. Several advanced treatments of the quantum theory of solids are available and a more elementary treatment by the author, intended as a supplement to the present volume, will be published elsewhere.[3] Although mathematical analysis is used, where necessary, in the present text, the mathematics involved is in no case of an advanced character.

In the section of the book devoted to methods of measurement, the aim has been to give the reader the principles of the methods used rather than to include a lot of detail on particular pieces of apparatus; for these reference is generally made to the original papers. As regards references, no attempt has been made to compile a comprehensive list; this would indeed be a formidable task. Fortunately, there exist many long lists in the review articles referred to; in addition, there is an annual publication by the Battelle Memorial Institute giving a classified set of abstracts of papers on semiconductors.[4] An impression of the number of papers being published on semiconductor research may be obtained from the fact that the 1957 issue contains abstracts of 1258 papers published in 1955. Here our aim has been to refer to one or two of the principal papers initiating some new development and also to a few of the later 'key' papers; from these the reader will generally be able to compile a more comprehensive list of references.

I am indebted to a number of my colleagues at the Royal Radar Establishment for a great deal of help during the preparation of the book; in particular, I have benefited from many valuable discussions with Dr A. F. Gibson, Dr E. G. S. Paige, Dr C. A. Hogarth and Dr E. H. Putley on a large number of technical matters and also on methods of presentation. I am also greatly indebted to Mr W. E. J. Farvis for encouragement to write the book and for arranging the lectures out of which it grew: also to Dr D. Shoenberg for helpful criticism and advice on presentation. A number of authors have kindly allowed use to be made of figures adapted from their original papers; to these acknow-

[3] *Wave Mechanics of Crystalline Solids* (Chapman and Hall, and John Wiley and Sons, 1961).
[4] *Semiconductor Abstracts* (John Wiley and Sons, 1955 *et seq.*).

ledgment has been made in the text. To those who have written review articles, particularly those listed in the bibliography, I should like to acknowledge my indebtedness – without these articles the task of writing this book would have been immeasurably harder.

1

The elementary properties of semiconductors

1.1 Early work on semiconductors

The history of research on the substances known as semiconductors is a long one, extending over more than a century. Much of the early work was carried out under very great difficulties, which are only appreciated now with the fuller understanding which we have of the subject. The purity of the materials available to the early workers fell far below the very high standards which we now know to be necessary if unambiguous results are to be obtained. It is, nevertheless, a high tribute to the skill and care of many experimenters that, in spite of this, semiconductors had been recognized as a distinct class of substances and their main properties appreciated long before a comprehensive theory was available to account for them. That mistakes were made is not surprising. A few substances included once in the class have now been shown to be metals, and a number of substances thought to show metallic behaviour have now been shown, when pure, to be semiconductors.

The first feature used to distinguish this class of electrical conductors from metals and other poor conductors was their negative temperature coefficient of resistance – i.e. their resistance generally falls as the temperature is raised, while that of a metal rises. Michael Faraday would appear to have been the first to notice this effect, when carrying out experiments on silver sulphide.[1] This criterion is now known to be inadequate, and over a certain range of temperature the resistance of a semiconductor may increase as the temperature is raised, particularly if it contains a fair amount of impurity. At high temperatures, however, a point is reached where a rapid decrease in resistance sets in as the

[1] *Experimental Researches in Electricity*, series IV (1833), §§ 433–9; also *Beibl. Ann. Phys.* (1834) **31**, 25.

temperature is further increased. Again, certain metallic films show a negative temperature coefficient of resistance as may also polycrystalline ingots of some metals. These effects are now known to be due to oxide films or actual gaps separating the individual crystals but led to the metals titanium and zirconium once being listed as semiconductors. With these exceptions in mind, however, it is generally true to say that pure semiconductors have a negative temperature coefficient of resistance. They are generally associated with a number of other properties which have been used to distinguish them, and which we shall now discuss. It must be admitted, however, that no infallible criterion was available till the quantum theory of solids gave an understanding of the reasons for the various properties observed.

Except at temperatures not much below their melting point, semiconductors have resistivities considerably higher than good metallic conductors, and also much less than good insulators have. The range of resistivities in solids is indeed enormous. A good metallic conductor has a resistivity of the order of $10^{-6}\ \Omega$ cm at room temperature, whereas semiconductors generally have room temperature resistivities in the range 10^{-3} to $10^{6}\ \Omega$ cm. There are, moreover, a great many substances with resistivities in this range which are not semiconductors as we now define them. Good insulators, on the other hand, have resistivities of the order of $10^{12}\ \Omega$ cm.

The progress made in the next forty years after Faraday's observations was not very great, though it was noted by various workers that substances belonging to the class of 'poor conductors' had exceptionally high values of their thermo-electric power, e.g. tellurium. Two important advances were made in 1873 and 1874. The phenomenon of rectification was observed by F. Braun[2] using substances like lead sulphide and iron pyrites, and photo-conductivity was observed in selenium by W. Smith.[3] Other substances were soon found to show some of these effects but not necessarily all of them. These were mainly metallic sulphides and oxides and the element silicon.

After this, a great deal of work was carried out and a class of substances, called semiconductors (*Halbleiter*), with these properties began to emerge. A review of this early work has been given by K. Lark-Horowitz,[4] together with a very extensive bibliography containing

[2] *Ann. Phys. Chem.* (1874) **153**, 556.

[3] *J. Soc. Telegraph Engrs.* (1873) **2**, 31.

[4] 'The New Electronics', in *The Present State of Physics* (American Assn for the Advancement of Science, Washington, 1954).

over 350 references. Earlier reviews by B. Gudden[5] also deal extensively with this phase and discuss in some detail the problems of identifying semiconductors. An excellent account of the earlier Russian work has been given by A. F. Ioffe.[6]

The main properties of these substances which emerged were:

(*a*) negative temperature coefficient of resistance;

(*b*) resistivity in range roughly 10^6 to 10^{-3} Ω cm;

(*c*) generally high thermo-electric power, both positive and negative relative to a given metal;

(*d*) rectifying effects or at least non-ohmic behaviour;

(*e*) sensitivity to light – either producing a photo-voltage or change of resistance.

It was not long before the important part played by impurities came to be appreciated. It was noted that some of the properties, in particular the negative temperature coefficient of resistance at high temperatures, were always the same for a given substance. Others varied considerably from sample to sample. The former were called 'intrinsic' properties. This variation hampered the study of the 'intrinsic' properties as one could never be quite sure that an observed effect was not due to impurities. This doubt has now been removed, since impurities in many semiconductors can be accurately controlled.

A most important event, as it turned out later, took place in 1879 with the discovery of the Hall effect,[7] the transverse voltage developed across a conductor carrying a current in a magnetic field. Although at this time the electron had not been discovered, so that the idea of number of current carriers had not arisen, this effect turned out to be the key to understanding electrical conduction in semiconductors, and in distinguishing them from other poorly conducting substances. The reason for this is, as we shall see, that a measurement of the Hall voltage enables us to determine directly the number of current carriers per unit volume, and also whether they are positively or negatively charged. It also enables us readily to distinguish ionic conduction from electronic conduction, a distinction which is very necessary, as the conductivity due to the former increases rapidly with temperature, and may lead to a false conclusion.

The various methods of distinguishing between ionic and electronic conduction have been discussed by K. Lark-Horowitz (*loc. cit.*) in some

[5] *Ergebn. exakt. Naturw.* (1924) **3**, 143; (1934) **13**, 223.
[6] *Physics of Semiconductors* (Infosearch, 1960).
[7] E. H. Hall, *Amer. J. Math.* (1879) **2**, 287.

detail and also the difficulties of interpretation caused by failure to make the distinction. Before the use of the Hall effect the distinction was usually made by noting whether electrolytic transport took place or not. The Hall effect due to an ionic current is, as we shall see, negligible compared with that due to the electronic current and this allows the latter to be identified.

The electronic conductivity of a substance depends on two factors, the number of current carriers per unit volume and the ease with which the carriers move through the substance under an applied electric field. The latter, which is generally known as the carrier mobility, and which we shall define more precisely later (see § 5.1, equation (14)) generally tends to decrease as the temperature is raised, especially at the higher temperatures, and this accounts for the decrease in conductivity of metals with increasing temperature. For these, the number of current carriers remains constant as shown by Hall effect measurements on metal foils. For a semiconductor, however, the number of current carriers increases rapidly with temperature, especially at higher temperatures, and this accounts for the rapid decrease in resistance. Here then we have the essential distinction between a metal and a *pure* semiconductor. We emphasize the word pure, since too much impurity may mask the distinction. A metal is a conductor with essentially a constant number of current carriers – a pure semiconductor is one in which the number increases as the temperature is raised.

The first systematic use of the Hall effect to study semiconductors appears to be due to K. Baedeker[8] using CuI. Detailed studies of a large number of substances were made by J. Königsberger,[9] using the Hall effect and other properties listed above. It was found that the number of current carriers in semiconductors was very much less than in metals but that their mobility, in general, was somewhat higher. One very important result of this work was that the elements silicon, selenium and tellurium were classed as semiconductors. It was much later that germanium was added.[10] Königsberger fully appreciated the distinction outlined above.

A very interesting observation was made regarding the sign of the current carriers. This was sometimes found to be negative as expected for electrons but was also found in many instances to be positive, and even to change sign from positive to negative as the temperature was raised. Positive Hall coefficients had also been found for certain metals

[8] *Phys. Z.* (1909) **29**, 506. [9] *Jb. Radioakt.* (1907) **4**, 158; *ibid.* (1914) **11**, 84.
[10] E. Merrit, *Proc. Nat. Acad. Sci., Wash.* (1925) **11**, 743.

and this gave rise to a theoretical puzzle which was not solved till the advent of the quantum theory.

1.1.1 'Excess' and 'defect' semiconductors

A considerable amount of work on a large variety of substances thought to be semiconductors was carried out between 1910 and 1930 but not a great deal of fundamental progress was made. Increased interest in these substances was aroused about 1930 largely due to the stimulus of technological applications. Hall effect, conductivity and thermo-electric power measurements were mainly used for their study and it was shown that the sign of the Hall effect, at low temperatures, and of the thermo-electric power are generally the same. A study of the chemistry of a number of semiconducting compounds led C. Wagner[11] to identify two distinct types of semiconductor – 'defect' and 'excess' semiconductors. The substances concerned were mainly metallic oxides and sulphides and the 'defect' semiconductors were those with a metallic content less than that corresponding to stoichiometric composition, i.e. oxidized compounds. They generally showed a positive Hall coefficient at low temperatures and a positive thermo-electric power. The 'excess' semiconductors were 'reduced' compounds and had an excess of metal. They had generally a negative Hall coefficient at all temperatures. For the 'defect' semiconductors the Hall coefficient sometimes became negative at high temperatures. These are what we now call respectively p-type (p for positive) and n-type (n for negative) semiconductors. The importance of this work was in showing the vital part played by small deviations from stoichiometric composition in determining the properties of compound semiconductors.

1.1.2 The alkali halides

Although they are not strictly semiconductors but insulators, mention must be made of the large amount of research carried out by R. W. Pohl[12] and his collaborators on the alkali halides, since this helped greatly to clarify many of the properties of semiconductors. One of the main reasons for choosing this group of substances was that they may be

[11] *Z. Chem. Phys.* B (1930) **11**, 163; *ibid.* (1933) **22**, 195.

[12] This work was started about 1920. A good account of some of it is given by Pohl in *Proc. Phys. Soc.* (1937) **43**, 3. See also N. F. Mott and R. W. Gurney, *Electronic Processes in Ionic Crystals* (Oxford University Press, 1940).

readily obtained in the form of large single crystals of high purity. Much of the work on the metallic oxides and sulphides was carried out with compressed powders and evaporated or chemically deposited thin films. These are now known frequently to give very misleading results. It should be noted that the conductivity in the alkali halides is mainly ionic in character but electronic conductivity may be produced by illumination with ultra-violet light. Study of this photo-conductivity paved the way for understanding the similar effect produced by visible light and infra-red radiation in semiconductors.

1.1.3 Surface and bulk effects

Much of the uncertainty of the early work on semiconductors arose through a failure to differentiate between effects which arise in the bulk of the material and those which are characteristic of the surface or of the interface between two different materials. Extensive use of compressed powder samples accentuated the surface effects. It was later thought that a negative temperature coefficient of resistance is always a bulk effect but it is now known that this is by no means so. Rectification was rightly classed as a surface or interface effect but a great deal of confusion arose over photo-voltaic and photo-conductive effects.

For a few semiconductors, natural crystals were available for study, for example lead sulphide (galena), but these were usually of doubtful purity. A new phase in the study of these substances began with the growth of synthetic crystals of a number of semiconductors. It is mainly with this phase of the work that we shall be concerned in this book. The use of single crystals has enabled not only the separation of the bulk and surface properties but has also enabled the surface and the interface between two types of semiconductor, or between a semiconductor and a metal, to be studied in much greater detail. Such studies have shown in no uncertain way how very misleading results may be obtained when such surfaces are not taken into account, as for example in polycrystalline materials.

More recently the study of amorphous semiconductors has led to a fuller appreciation of those properties, such as high carrier mobility, that depend principally on the quality of the crystals being studied and those more fundamental properties which do not depend on long-range order.

1.2 Applications of semiconductors

The first important application of semiconductors was to provide rectifiers for low-frequency alternating currents. Such rectifiers, using selenium, were made as early as 1886 by C. E. Fritts[13] although they were not used to any extent in power engineering or in electronic equipment till much later. The copper oxide rectifier was introduced by L. O. Grondahl and P. H. Geiger[14] in 1927 and came to be used extensively as a low-power rectifier in battery chargers, wireless sets, etc. The development of selenium rectifiers on a commercial scale also began to take place at about the same time and these largely replaced the copper oxide type in electronic equipment. The development of these rectifiers has led to a great deal of work on both copper oxide and selenium. In spite of this the amount of detailed fundamental knowledge of their properties as semiconductors is still rather scanty compared with what we now know of element semiconductors such as germanium and silicon. In the case of selenium this is largely due to its complex crystal structure and to the existence of different forms, together with the difficulty of obtaining the substance in a pure state. The photo-conductive properties of selenium, and of copper oxide, have been used to provide exposure meters for photography and photo-cells which are used in the film industry for transforming the markings on the sound track into electric currents for amplification and reproduction by loud-speakers. Such cells are also used in a number of automatic devices such as burglar alarms, train counters, etc. Rectifiers handling very large powers in the electricity supply industry are now generally made of the element semiconductor silicon which is also the main material used to make diode rectifiers handling small quantities of high-frequency power in such things as radio receivers.

Although radio waves had been demonstrated by Hertz in 1888, it was not until about 1904[15] that it was appreciated that the rectifying properties of semiconductors could be used to provide a detector of the high-frequency currents set up in an electric circuit by these waves. In the following few years a great variety of substances were tried. The discovery that a fine wire, or 'cat's whisker', in contact with a crystal of semiconducting material made an excellent rectifier for high-frequency currents led to a great increase in the sensitivity of radio receivers, and this type of device was widely used in the early days of broadcasting. The two substances generally preferred were silicon and lead sulphide, the latter being in the form of natural crystals of galena. An interesting controversy arose as to whether the rectification effects were electrical or thermal in origin. This was settled by the extensive work of G. W.

[13] *Amer. J. Sci.* (1883) **26**, 465. [14] *Trans. Amer. Inst. Elect. Engrs.* (1927) **46**, 357.
[15] J. C. Bose, U.S. Patent, 755 840 (1904).

Pierce[16] who showed that there could be no doubt that they are electrical.

The crystal detector was soon replaced by the thermionic valve and had largely become obsolete by 1939. With the development of microwaves (of the order of 10 cm and less) for radar it came into its own once again, since no other device was found to act as an efficient rectifier or frequency changer at these extremely high frequencies. Silicon proved to be the best substance for this application. The British work in this field has been described by B. Bleaney, J. W. Ryde and T. H. Kinsman[17] and a full account of this and the American work has been given by H. C. Torrey and C. A. Whitmer.[18] The stimulus of this application led to a great deal of fundamental work on silicon, particularly at the Bell Telephone Laboratories and at Purdue University in the U.S.A. Germanium, being the next element to silicon in the same column of the periodic table, was also extensively studied in both of the above laboratories. Since it has a much lower melting point and is easier to purify, it was a more suitable substance than silicon for fundamental studies and soon became the semiconductor about which far more was known than about any other. Although it was not effective when first tried as a radar detector it found a very important use in the development of small compact rectifiers for use at low frequencies – the so-called germanium high back-voltage diodes. As it turned out, its study as a semiconductor was most fortunate since it led to the discovery of transistor action by J. Bardeen and W. H. Brattain[19] and to the invention of the transistor[20] in the Bell Telephone Laboratories.

The detailed study of transistor action and of the controlled injection of current carriers into semiconductors by W. Shockley, G. L. Pearson and J. R. Haynes,[21] led to an enormous increase in research on the properties of germanium, and later of silicon, as it was realized that the transistor would revolutionize the future development of electronics, and would largely replace the vast number of vacuum tubes used by the electronics industry. The large research effort employed in transistor development has led to a far better understanding of the fundamental properties of semiconductors. The transistor effect has itself proved to

[16] *Phys. Rev.* (1907) **25**, 31; *ibid.* (1909) **28**, 153; *ibid.* (1909) **29**, 478.

[17] *J. Inst. Elect. Engrs.* IIIA (1946) **93**, 847.

[18] *Crystal Rectifiers* (Radiation Laboratory Series) (McGraw Hill, 1948).

[19] *Phys. Rev.* (1949) **74**, 1208.

[20] Accounts of the research work which led to the discovery of transistor action are given by W. Shockley, *Electrons and Holes in Semiconductors* (Van Nostrand, 1950), ch. 2, and by G. L. Pearson and W. H. Brattain, *Proc. I.R.E.* (1955), **43**, 1794.

[21] *Bell Syst. Tech. J.* (1949) **28**, 344.

be a most powerful tool for investigating the fundamental properties of these substances so that the interaction between pure and applied research has been most fruitful.

Another group of semiconductors which has found important applications is that known as the infra-red photo-conductors, particularly the sulphide, selenide and telluride of lead. These have been used to make sensitive infra-red detectors. This in turn has led to a great deal of work on these substances, started in Germany during the years 1940–45 and greatly extended in recent years in a number of laboratories. Work on these substances and the development of infra-red detectors has been reviewed by R. A. Smith.[22] The properties of these detectors and their use in infra-red spectroscopy have been fully described by R. A. Smith, F. E. Jones and R. P. Chasmar.[23]

Many other applications of semiconductors have been made in recent years and the electrical and electronics industry based on these substances is now one of the largest and most rapidly developing. In power engineering, for example, they find applications in rectifiers, switchgear and power-monitoring devices as well as in a great variety of control equipment. In the telecommunications industry most mechanical switching systems have now given way to solid-state devices on which automatic telephone exchanges are largely based. The use of solar-cells to provide power for satellites and even their possibilities for widescale power generation, now being actively examined, are based on the photo-electric effect in semiconductors.

The modern successor to the transistor is the integrated circuit (I.C.) in which many transistors and their associated components such as resistors, capacitors etc. are produced by controlled diffusion of impurities into a small 'chip' of silicon. This has revolutionized the electronics industry for the second time in two decades and has made possible incredibly complex systems such as the modern electronic digital computer.

One other application requires special mention, namely the solid-state laser based on the optical properties of semiconductors. This has provided spectroscopists with a tunable light (or infra-red) source of very narrow bandwidth, and is making possible new advances in spectroscopy. As a source which can be modulated and readily linked to optical fibres it is likely to play a vital part in optical communications. A

[22] *Advanc. Phys.* (1953) **2**, 321; *Sci. Mon.* (1956) **82**, 3.

[23] *The Detection and Measurement of Infra-red Radiation* (Oxford University Press, 2nd edition, 1968).

less complex but related device, the light-emitting diode (L.E.D.) is finding widespread use to provide displays for calculators, computers and signalling systems and may also have a key role in optical communications. For these optical applications the compound semiconductors, for reasons which we shall later see, are most suitable. Compounds such as GaAs and alloys like $Ga_xAl_{1-x}As$ are at present the most widely used types of semiconductor for these purposes.

1.3 Elementary theory of semiconductors

As we have seen, a great deal of experimental work had been carried out on semiconductors before any satisfactory theory had been put forward to account for their properties. This is no longer surprising when we realize that it requires the quantum theory, in the form of wave-mechanics, to account for even the most elementary properties of semiconductors. It was not till after this theory had been applied to the motion of electrons in crystalline solids that a satisfactory theory of semiconductors emerged. This was given by A. H. Wilson in 1931.[24] It was not until this theory was available that a satisfactory definition of a semiconductor could be given. It is interesting to note that in Wilson's original papers doubt is expressed as to whether silicon is a semiconductor or a metal. This is indicative of the uncertainty that still existed in the early 1930s and which Wilson's theory was largely responsible for removing.

The first application of quantum mechanics to the motion of electrons in solids was the treatment of conduction of electricity in metals by A. Sommerfeld.[25] The only essential difference between this and earlier theories, based on the classical theory of electrons, was that the energy levels which the various electrons can occupy are determined by the wave-mechanics. It was assumed that, in a metal, the outer atomic electrons, known as the valence electrons since they are the ones which take part in chemical binding, are not tied to individual atoms but are free to move through the solid. The electrons are assumed to move in a field-free space, the fields of force due to the atomic cores and other electrons being smoothed out except at the boundary of the solid (see Fig. 1.1(a)). Here these forces are supposed to attract the electron strongly if it moves outside the boundary, and in the simplest form of the theory it is assumed that they set up an impenetrable potential barrier

[24] *Proc. Roy. Soc.* A (1931) **133**, 458; *ibid.* (1931) **134**, 277.
[25] *Z. Phys.* (1928) **47**, 1.

which holds the electrons in the solid. Later refinements enabled such phenomena as field emission and thermionic emission to be discussed in terms of potential barriers of finite height. Sommerfeld's theory did not change appreciably the ideas on conduction of electricity in metals, but did give an explanation of the very small contribution of the electrons to the specific heat. The assumption that the electrons are quite free and do not make frequent collisions with the atoms of the solid is clearly a very crude one. Moreover, this theory gives no explanation of the vast differences in the properties of metals, semiconductors and insulators.

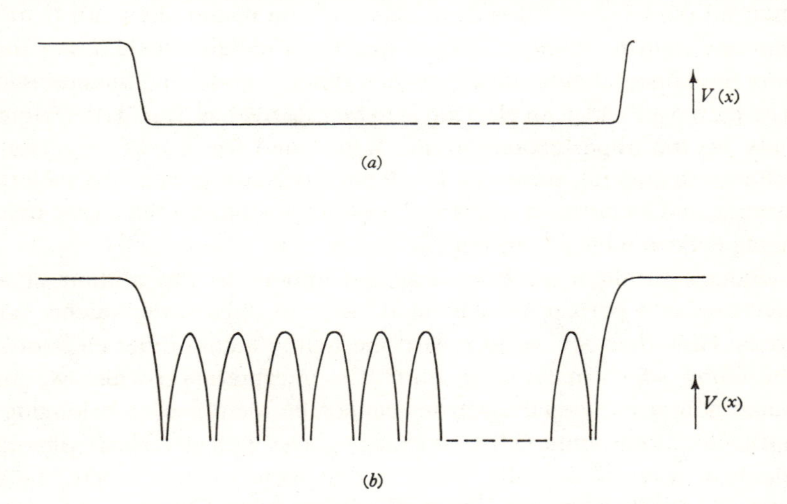

Fig. 1.1. Potential energy of electron in crystalline solid. (*a*) Sommerfeld model. (*b*) Periodic potential due to atomic cores.

It was generally believed that in a metal the valence electrons are easily detached from the individual atoms, whereas in an insulator they are tightly bound and cannot be detached. This is somewhat difficult to understand, since, in the isolated atoms, the energy required for detachment is not so very different and may even be greater for a metal than for a non-metal (e.g. 9.20 eV for Au and 8.09 eV for Ge). It soon became clear that something much less crude than Sommerfeld's free-electron approximation was required. The next step was to take into account the interaction of the valence electrons with the atomic cores, assuming these to be placed at the lattice points of the crystal. These electrons are still assumed to move independently but the smoothed out potential used by Sommerfeld is replaced by one of the form shown in Fig. 1.1(*b*) which has been drawn for a direction passing through a line of atoms in the crystal. The particular feature of a potential of this form is that it is periodic having the same periodicity as the lattice. Motion of electrons in such a potential was discussed by F. Bloch[26] by means of quantum mechanics and a most fundamental result came to light which changed completely the whole approach to the problem of electronic motion in crystalline solids. This result is independent of the particular form of periodic potential. It turns out that in a perfect periodic lattice an electron may move freely and is not 'scattered' by the individual atoms of the lattice but only by *deviations* from perfection. The Sommerfeld model would therefore appear to be nearer the truth than we had any right to suppose. This is true for a metal, but there are some very fundamental differences between Bloch's model and Sommerfeld's. The path over which an electron is to be regarded as 'free' is determined only by the imperfections of the lattice and for a perfect crystal is infinite. In general, such a path will be very much greater than a lattice spacing and an electron will travel past many atoms of the lattice before being deflected by a 'collision'.

Bloch's result is quite general and applies to the motion of any electrons in a perfect periodic field, and not only to the valence electrons. How then are we to regard the tightly bound inner electrons of the atoms which make up a solid? The interpretation which we must make is that individual electrons cannot be identified as belonging to particular atoms, but are free to change places with electrons from other identical atoms of the solid. What then happens when an electric field is applied? Do these electrons take part in conduction? It was in considering this problem that A. H. Wilson (*loc. cit*) was led to his theory of semiconductors.

In Sommerfeld's theory, the allowed energy levels for the valence electrons of a crystal of macroscopic dimensions lie very close together and their values extend from nearly the bottom of the potential trough in which the electrons move to indefinitely high values. For the other electrons the energy levels are assumed to be undisturbed and are just the atomic energy levels. These are the same for all the atoms of the crystal and are therefore at least N-fold degenerate, corresponding to the N atoms of the crystal (see Fig. 1.2). When the periodic potential is introduced, however, the energy levels are confined to certain allowed bands of energy, separated by regions in which no energy levels are allowed. For the inner electrons these allowed bands are extremely narrow and correspond to the atomic levels; for the valence electrons

[26] *Z. Phys.* (1928) **52**, 555.

the bands are quite broad. The arrangement of levels is shown in Fig. 1.3. Each band consists of many closely spaced levels and for many purposes may be regarded as a continuum. When we come to allocate electrons to these levels, however, we must remember that they are discrete and that there exists a definite number of them. Just as the inner levels in a heavy atom are all filled with electrons, so all the levels in the lower allowed bands are filled, and it is only the upper bands which may be wholly or partially unoccupied by electrons.

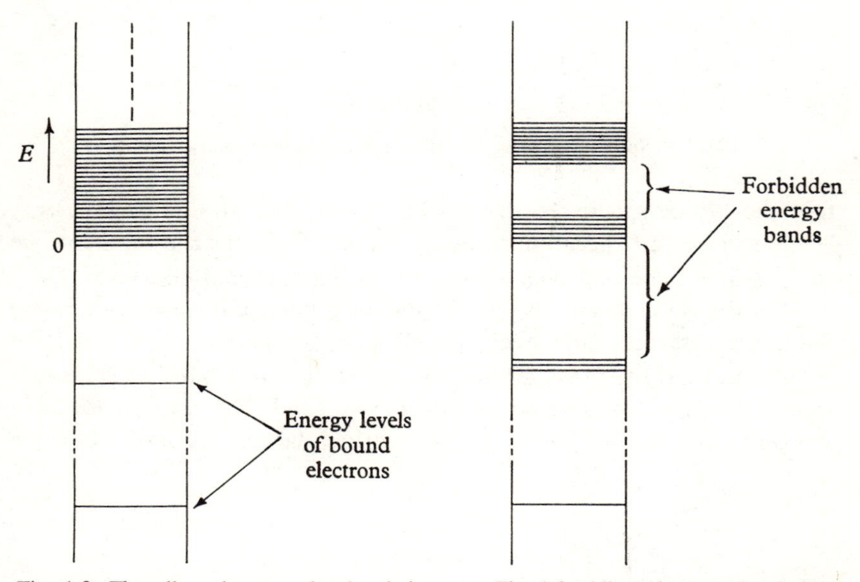

Fig. 1.2. The allowed energy levels of the Sommerfeld model of a metal. (The zero of energy is taken as that of a 'free' valence electron with no kinetic energy.)

Fig. 1.3. Allowed energy levels for a periodic lattice.

We may arrive at the energy band picture of a solid in another way. Suppose we have two atoms very far apart and consider a single level of each atom. The allowed levels for the combined system consist of a single doubly degenerate level, i.e. each electron has precisely the same energy. Now if we let the atoms approach, this degenerate level will be split into two by the interaction between the atoms. This is a very general result, corresponding to the splitting of the frequencies of two interacting identical oscillators in mechanics. As the atoms approach, the splitting between the levels will be increased (Fig. 1.4). If now we have N atoms in a crystal, and we assume the crystal to be expanded so

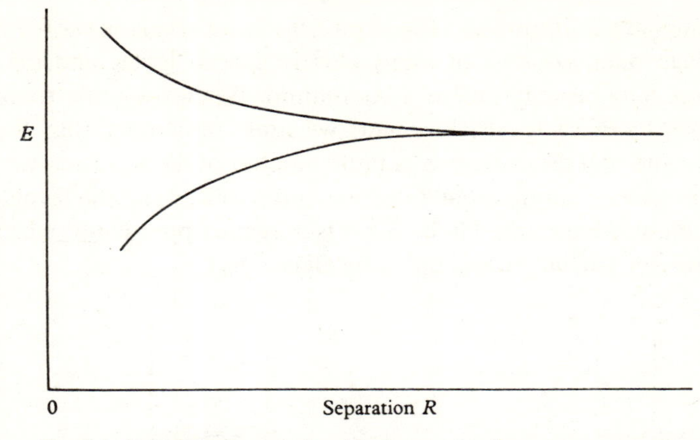

Fig. 1.4. Splitting of energy level by interaction between two atoms.

that the lattice spacing becomes very great, then the allowed energy levels will be just the atomic energy levels which, for the moment, we shall assume to be non-degenerate, i.e. each has a separate energy. Each level of the whole crystal is then N-fold degenerate. If we now decrease the lattice distance, each level will split into N separate levels under the perturbation of the other atoms, and we shall have a band of N closely spaced levels instead of a single degenerate level. The extreme levels in the band are shown in Fig. 1.5 for various lattice distances. For the

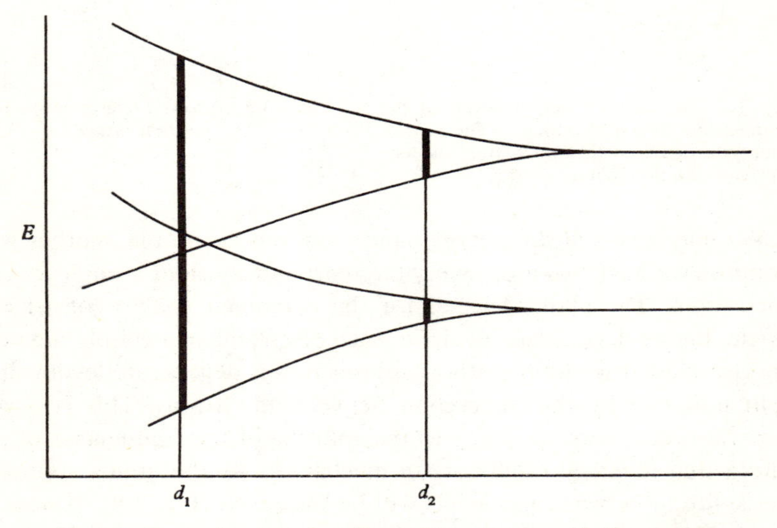

Fig. 1.5. Energy bands in crystalline solid as function of lattice spacing.

deep-lying levels the perturbation will be very small compared with the attractive force of the nucleus and the splitting will be very small – for the valence electrons the splitting may be quite large and in fact neighbouring bands may overlap (see Fig. 1.5).

Let us first consider the case when there is no overlapping. We must now decide how to allocate the various electrons to the bands. According to the Pauli principle we can allocate two electrons to each level – one for each of the two allowed values of the electron spin. Now consider a deep-lying level; each atom will have two electrons in this level, so that there are $2N$ electrons in the solid which at infinite separation would have the energy corresponding to this level. There are therefore $2N$ electrons to fit into the band. These exactly fill the band which has N levels, as we have seen, and two electrons can go in each. The bands corresponding to atomic levels in which there are paired electrons with opposite spin are therefore *completely filled.* Again, the various filled closed shells in an atom correspond to bands in which all the levels are occupied. It is only bands corresponding to the outer or valence electrons which may be partially occupied. This distinction between full and partially filled bands is quite fundamental; it is illustrated in Fig. 1.6. Corresponding to excited states of the atoms there will be higher energy bands which generally will be empty.

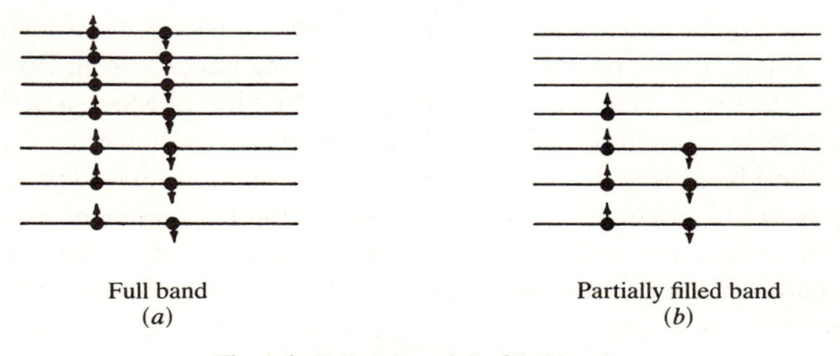

Full band
(a)

Partially filled band
(b)

Fig. 1.6. Full and partially filled bands.

We can now see why it is that the electrons in the lower bands corresponding to the 'core' electrons cannot take part in electrical conduction, although Bloch has shown that, in a sense, they are free to move unimpeded through a perfect crystal. To produce conduction, an electron must receive energy from an electric field, i.e. it must be accelerated. In terms of quantum mechanics it must be raised to a higher level. If, however, *all* the levels in a band are already occupied this cannot take place unless the electron is excited to a higher band in which there are free places. This will, however, require a lot of energy and will be most improbable. Thus we see that electrons can only contribute appreciably to conduction if there are available *neighbouring empty levels* into which they can go: completely filled bands do not contribute to conduction. This not only explains why 'inner' electrons do not contribute to conduction but it gave Wilson the clue to the essential difference between metals on the one hand and insulators and semiconductors on the other.

We must now consider the highest band in which we normally have occupied levels. In an alkali metal, such as sodium, there is only a single valence electron, the rest of the atom consisting of closed shells. The corresponding band in the solid to be occupied by these electrons, N in number, will have N levels. In the unexcited state these electrons will occupy the $\frac{1}{2}N$ lowest levels, two to each level. The band will be only half full, and there are plenty of neighbouring levels available for conduction. Thus we have a condition which approximates very closely to Sommerfeld's model when we consider *only* the valence electrons.

Let us now consider a crystal made up of atoms with two valence electrons in the same level, with opposite spins. These will now completely fill the band corresponding to this level and the highest occupied band should be completely full. Provided there is a forbidden energy gap between this band and the next highest, no conduction should take place, on the same basis as for the bands arising from the inner electrons. Thus we appear to have a situation in which a material might be an insulator. This would be a substance in which the highest occupied band is just filled and there is a substantial forbidden energy gap between this band and the next highest. On this simple picture one might expect the alkaline earth metals, such as magnesium, to be insulators. That they are not indicates that there is an overlap between two bands as indicated in Fig. 1.5 (at lattice spacing d_1). In this case, each band is only partially filled (see Fig. 1.7), and we have again the possibility of conduction.

The situation we have envisaged for an insulator does, however, arise in certain solids. Elements with atoms having four valence electrons, such as carbon (in the form of diamond), silicon and germanium, form bands, the highest one occupied being filled and the next highest separated from it by a forbidden energy gap. The situation in compounds is more complex, but if the number of valence electrons per molecule of the compound is even, we may have a completely filled

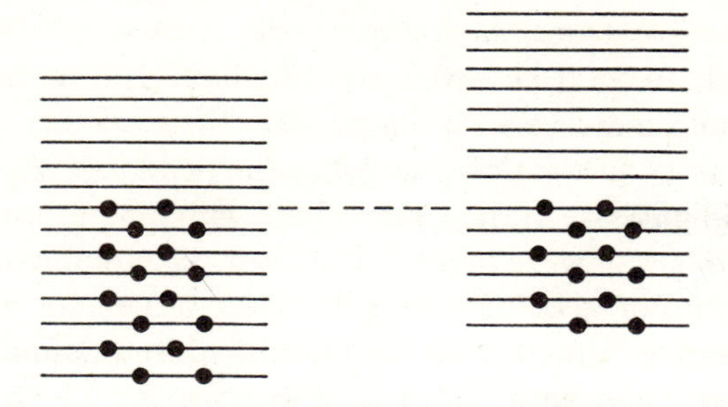

Fig. 1.7. Partial filling of two overlapping energy bands.

highest occupied band. We may note that this is *only* possible when the number of valence electrons is even.

The above considerations apply only to solids with a regular crystalline structure. Most of the substances with which we shall be concerned have such a structure, but may not always consist of large single crystals but of aggregates of very small crystals with random orientation. Exceptionally, they may be prepared in amorphous form (e.g. silicon layers) and interest in the properties of such forms has recently grown. We shall later discuss amorphous semiconductors (Chapter 15) briefly but shall mainly be concerned with materials in the form of good single crystals.

There is another aspect of the quantum theory which throws some light on the above ideas. Because of its wave properties an electron will not stay in a small potential hole, but will leak out into a neighbouring hole if the potential barrier between the two is small in thickness and not too high. It turns out that the packing of the atoms in actual solids is so close that an electron in any particular atom, particularly a valence electron, would easily escape from the potential hole confining it to that particular atom into that of a neighbouring atom. Thus it cannot be regarded as belonging to any particular atom in the crystal. This point was soon appreciated after the advent of wave-mechanics, and it was difficult to see how the electrons in an insulator may be regarded as 'tightly' bound. The theory of full bands shows that they need not be bound in this sense and may still make no contribution to conduction.

Wilson's theory of semiconductors is based on these principles. If the forbidden energy gap ΔE between the highest filled band and the next empty band is large, no electronic conduction can take place as no 'neighbouring' empty levels exist into which electrons may be accelerated. If, on the other hand, the value of ΔE is small there is the possibility that electrons may be thermally excited into the band above and that these excited electrons can conduct. The number of excited electrons would increase with temperature in a manner governed by a process having an 'activation energy' of the order of ΔE and we should expect a rapid increase of the conductivity with temperature. Substances behaving in this way were identified by Wilson as semiconductors. This definition turns out to be too restrictive since it depends on energy band theory, applicable only to crystalline solids, while a definition applicable to amorphous solids would also be desirable. As we shall see, (§ 3.2) such a definition can be given in terms of chemical bonds. Generally a semiconductor is a material in which, in its pure state, a small activation energy ΔE (say $0 < \Delta E <$ about 2 eV) is required to free electrons so that they can take part in conduction processes. As we shall also see later the effect of very small amounts of impurity can have a marked effect on this activation energy so that materials which have considerably greater values of ΔE may behave as semiconductors when they contain certain 'active' impurities.

1.3.1 Conduction processes

It was soon appreciated that, in practice, the processes of conduction were more complex and we must now consider these in more detail. A very interesting effect takes place when an electron is excited from the full band to an empty band above: the level occupied by the electron is left vacant so that other electrons may move into it. This position left unoccupied behaves rather like a bubble in a liquid under gravity. In an applied electric field it moves in the opposite direction to that in which an electron would move, and so appears to have a positive charge. It behaves, in fact, in many ways just like a positively charged particle and is generally called a 'positive hole' or simply a 'hole'. A more detailed discussion of this process will be given later (see § 2.4). Electrical conduction in an ideal semiconductor thus consists of motion of electrons in a nearly empty band, under the influence of an electric field, and of positive holes in a nearly full band.

This process went a long way towards explaining the properties of semiconductors at high temperatures and, in particular, the exponential increase in their conductivity with temperature. This is simply associated with an activation energy (in the chemical sense) of the order of ΔE required for the thermal creation of electron–hole pairs. We may regard the creation of a free electron (e), i.e. one in the empty or 'conduction' band, and a hole (h) in the full band as a reaction represented by the equation

$$(e)+(h)\rightleftarrows(eh) \tag{1}$$

in which the normal unexcited state is represented by (eh). If the densities of free electrons and holes n, p are both small compared with N, the density of atoms in the crystal, which is thus nearly equal to the number of unexcited states (eh), we have from the thermodynamical theory of chemical reactions

$$\frac{np}{N^2}=K(T), \tag{2}$$

where $K(T)$ is a function of the temperature only.[27]

Moreover, the function $K(T)$ may be written in the form

$$K(T)=[F(T)]^2\exp[-\Delta E/\boldsymbol{k}T] \tag{3}$$

where ΔE is the energy required for the process, $\boldsymbol{k}$ is Boltzmann's constant and T the absolute temperature. $F(T)$ is a relatively slowly varying function of T if $\Delta E \gg \boldsymbol{k}T$. Since in this case $n=p$ we have

$$n=p=A\exp[-\Delta E/2\boldsymbol{k}T], \tag{4}$$

where to a first approximation A may be regarded as constant. If we assume that the variation with T of the mobility of electrons and holes in an electric field is small compared with the variation in the exponential factor in (4) then we have for the conductivity σ, which is then simply proportional to the number of carriers, a variation of the form

$$\sigma=\sigma_0\exp[-\Delta E/2\boldsymbol{k}T]. \tag{5}$$

Such an exponential variation of conductivity had long been known for semiconductors at high temperature. At lower temperatures, however, a much less rapid increase in conductivity is generally observed and the value of the conductivity may vary enormously from sample to sample. To explain this Wilson proposed that the low-temperature behaviour was due to imperfections in the crystal due either to mechanical defects or to chemical impurities. The theory we have outlined indicates that there are no allowed levels in the forbidden energy gap, but this assumed a perfect crystal. Wilson showed that impurities may produce isolated levels in the forbidden energy gap, and it turns out that these

[27] See, for example, R. A. Smith, *The Physical Principles of Thermodynamics* (Chapman and Hall, 1952), p. 219.

levels may lie very near to the conduction band or very near to the full band.

If the impurity levels are separated by only a small energy gap from the empty band, electrons may be readily excited into this band and there they will conduct. Such impurities are called donor or n-type impurities. Again, if the impurity levels lie near the full band, and can accept electrons, these may be readily excited from this band and leave holes behind which will conduct. Such impurities are called acceptor or p-type impurities. Thus we must distinguish three processes of conduction.

(1) Intrinsic conduction in which equal numbers of electrons and holes are operative, these being created by excitation of electrons across the forbidden energy gap.

(2) Electron or n-type conduction in which electrons only are excited from donor levels near the conduction band.

(3) Hole or p-type conduction in which holes only are produced by excitation of electrons into acceptor levels near full band.

In general, all three processes will take place simultaneously but frequently one or other predominates. In this case the semiconductor is termed intrinsic, n-type, or p-type according to the predominant conduction process. We shall later have to consider this distinction more carefully as the situation is complicated by the different values of the mobility for holes and electrons. The n-type and p-type semiconductors correspond to the 'excess' and 'defect' semiconductors described by Wagner (see p. 5).

In many instances it is found that the energy required to excite an electron into the conduction band from a donor level is so small that the electrons from all the available donor levels are excited and are in the conduction band at room temperature. If the number of donors greatly exceeds the number of intrinsic electrons the number of carriers will hardly vary at all with the temperature and the variation of conductivity will be due only to the variation of mobility. Generally in this temperature range the mobility decreases as the temperature is raised: this accounts for the increase in resistance of metals with increased temperature, as in this case too the number of carriers is constant. Hence, for a semiconductor in this condition the resistance will increase with increasing temperature, till a temperature is reached at which the intrinsic electrons begin to predominate, when it will begin to fall exponentially in the manner once thought to be characteristic of semiconductors. Similar considerations also apply to p-type semiconductors.

This increase of resistance with temperature led to a great deal of confusion which was not really clarified till the extensive researches of K. Lark-Horowitz[28] and his colleagues at Purdue University were carried out on germanium from 1940 onward. This work showed that an increase of resistance with temperature cannot be taken as evidence that a substance is not a semiconductor. It was just this behaviour that led to doubts as to whether silicon is a semiconductor. The forbidden energy gap is so large for silicon (about 1.2 eV) that quite high temperatures (generally above 500 °C) have to be reached before intrinsic conduction sets in appreciably.

If the forbidden energy gap is greater still, say several electron volts, the temperature required to produce appreciable intrinsic conduction would be so high that the substance would melt before this condition is reached. We should therefore class such a substance as an insulator. In order to be a good insulator, however, the substance must also have no impurity energy levels which lie near the full or empty bands.

1.4 Control of carrier density

The most striking difference between metals and semiconductors is that, in the former, the number of carriers is large and constant, whereas in the latter the number is smaller and variable. This variable characteristic suggests that, in semiconductors, the number of carriers, and hence the conductivity, may be controlled. This control may be effected by control of impurity content, but the carrier density may also be varied for a material of fixed impurity content. One of the striking characteristics of semiconductors is that they show marked photo-conductive effects. This is readily understood in terms of the band model. When light of frequency high enough so that a quantum absorbed by a valence electron has sufficient energy to raise it from the top of the full band to the conduction band, extra carriers are created and these lead to increased conductivity (see § 10.9). Photo-conduction due to excitation of electrons from impurity levels has also been observed, but since these are much less numerous than the valence electrons, and are filled only at low temperatures, this effect is much less marked (see, however, § 10.7). Again, it turns out that electrons and holes may be injected into semiconductors from metallic contacts and by other means, and also that the carriers may be extracted under suitable conditions by an electric field.

[28] See review, ref. p. 2.

This ability to control the carrier density in semiconductors is the main reason for their great technological importance.

Generally, one type of impurity level, either donor or acceptor, predominates to a large extent in semiconductors used for practical purposes. When donors predominate, and greatly exceed the number of intrinsic carriers, then the electron concentration n greatly exceeds the hole concentration p. These concentrations are still connected by an equation like (2) under most practical conditions (see § 4.3), the product pn being independent of the impurity concentration. We therefore have

$$pn = n_i^2, \tag{6}$$

n_i being the concentration of intrinsic carriers at the given temperature, i.e. the carrier concentration in perfectly pure material. Equation (6) follows from equation (2) when we put $n = p = n_i$, the condition for an intrinsic semiconductor. Thus, if $n \gg n_i$ then $p \ll n_i$. When $n > p$ the electrons are referred to as the majority carriers and the holes as the minority carriers and vice versa. We thus see that except in near-intrinsic semiconductors the majority carriers greatly outnumber the minority carriers. It turns out, however, that the minority carriers are nevertheless, of great importance, as many electronic processes in semiconductor technology are controlled by the minority carriers, the density of which, being small, may be more readily varied.

We conclude this chapter with a numerical example, in order that the reader may appreciate the relative importance of the various conduction processes. In pure germanium the value of n_i, the number of intrinsic electrons or holes at room temperature, is about $2.5 \times 10^{13}\ \text{cm}^{-3}$. We may point out that intrinsic germanium at room temperature is now commonplace as the result of technological advances. The number of atoms of germanium per unit volume is $4.5 \times 10^{22}\ \text{cm}^{-3}$. Thus, if we have an impurity giving donor centres which are 'ionized' at room temperature, in a concentration of 1 p.p.m., we shall have $n = 4.5 \times 10^{16}\ \text{cm}^{-3}$ impurity electrons. Thus $p = n_i^2/n = 1.4 \times 10^{10}\ \text{cm}^{-3}$, and it will be seen that $n \gg p$. If the temperature is raised to 300 °C we have $n_i = 1.5 \times 10^{17}\ \text{cm}^{-3}$ and now intrinsic carriers predominate.

It will be clear from the above calculation that if germanium is to be intrinsic at room temperature it must be free from impurities giving levels near the conduction band to better than one part in 10^9. Thus it is not surprising to find that very few semiconductors have been purified to the extent required to give intrinsic conductivity at room temperature. Almost all the early work on semiconductors, except for measurements at high temperatures, was carried out in the condition in which the properties of the substance were dominated by impurities. At high temperatures other difficulties arise, particularly in compound semiconductors which tend to dissociate at the high temperatures. Thus, it is not surprising that confusing and misleading results have often been obtained. It will also be seen that the powerful techniques which have now been developed for the extreme purification of many substances have led to great advances in the study of the intrinsic properties of semiconductors.

2

Energy levels in crystalline solids

2.1 Wave mechanics of free electrons

In the previous chapter we have introduced the quantum theory of the electronic energy levels in crystalline solids. We must now consider in more detail the form of the energy bands arising from the crystalline lattice and also the energy levels due to impurities. We shall not discuss in detail the theoretical treatment of the allowed energy bands in solids. This subject is treated extensively in a number of standard text-books on the quantum theory of solids.[1] We shall simply give some of the more important fundamental results of this theory and indicate how they may be applied to semiconductors. We shall find that the theory enables us to describe the motion of electrons in solids by means of a small number of quantities which may be defined and that the detailed structure of the solid may be for many purposes neglected. The theory enables the values of these quantities to be calculated, but they may also be obtained (generally more accurately) experimentally and may be then regarded as parameters in a phenomenological treatment.

We shall first consider the motion of electrons in a perfect periodic crystal without impurities or imperfections. The electrons are supposed to move independently, and each obeys the Schrödinger equation for motion in a field of force in which it has a potential energy V

$$\nabla^2\Psi + \frac{8\pi^2 \boldsymbol{m}}{\boldsymbol{h}^2}[E - V]\Psi = 0, \tag{1}$$

[1] See, for example, A. H. Wilson, *The Theory of Metals* (Cambridge University Press, 1953); N. F. Mott and H. Jones, *The Theory of the Properties of Metals and Alloys* (Oxford University Press, 1936); F. Seitz, *Modern Theory of Solids* (McGraw-Hill, 1940). A treatment particularly applicable to semiconductors has been given by the author, *The Wave-Mechanics of Crystalline Solids* (Chapman and Hall, 2nd Edition, 1969); we shall subsequently refer to this as *W.M.C.S.*

Ψ is the wave-function for the electron and E is its energy. The probability of finding the electron in a volume $d\mathbf{V}$ is $|\Psi|^2\, d\mathbf{V}$ and the corresponding charge density $-\boldsymbol{e}|\Psi|^2$, $\boldsymbol{e}$ being the *magnitude* of the electronic charge. When V is a constant, the electron moves in a field-free space and the wave-function Ψ is given by

$$\Psi_{\mathbf{k}} = \exp\,[i\mathbf{k}\cdot\mathbf{r}] \tag{2}$$

which represents a plane wave. The energy E is then given by

$$E = \frac{\hbar k^2}{2\boldsymbol{m}} \tag{3}$$

where
$$k^2 = |\mathbf{k}|^2 = k_x^2 + k_y^2 + k_z^2 \tag{4}$$

and represents the kinetic energy of the electron. The wave-vector $\mathbf{k}$ is related to the electron's momentum $\mathbf{p}$ and velocity $\mathbf{v}$ by the equation

$$\mathbf{p} = \hbar\mathbf{k} = \boldsymbol{m}\mathbf{v}, \tag{5}$$

$\boldsymbol{m}$ being the mass of the electron in free space.[2]

When $V(\mathbf{r})$ is a periodic potential, with periodicity of the crystal lattice, a formally similar solution of the Schrödinger equation is found and we are able, as we shall see, to introduce quantities corresponding to the velocity and momentum of an electron moving in the crystal but averaged over a number of unit cells of the lattice.

2.2 Motion in a periodic potential

For a one-dimensional periodic potential, of period d, we have for the potential energy a function $V(x)$ such that

$$V(x) = V(x + rd), \tag{6}$$

where r is any integer. The solutions of the Schrödinger equation may then be shown to be of the form[3]

$$\psi_k(x) = u_k(x)\exp\,[ikx], \tag{7}$$

where $u(x)$ is a periodic function of x of period d and k is a constant, which we may use as a 'quantum number' to label the state described by the wave-function in equation (7). This type of wave-function corresponds to that given by equation (2) for a free electron. The physical interpretation of the quantity k is, however, not now so simple. It is

[2] *W.M.C.S.*, § 1.2.1. [3] F. Bloch, *Z. Phys.* (1928) **52**, 555.

certainly not proportional to the momentum, which is not a constant for motion in a periodic potential. The quantity P defined by means of the equation

$$P = k\hbar \tag{8}$$

is called the 'crystal momentum' to distinguish it from the real momentum, to which it is equal only when $V(x)$ is constant. P *is* a constant of the motion and may be used instead of k as a 'quantum number'. We shall see later that it has many of the properties of a momentum.

Wave-packets made up of functions like those given in equation (7) may be constructed to describe the motion of localized electrons. The velocity of an electron may then be shown to be equal to the group velocity $d\omega/dk$, ω being given in terms of the energy by the usual quantum equation $E = \hbar\omega$. The energy E is now, however, no longer a simple function of k but depends on the form of the potential function V. The velocity v of the electron is given by the equation[4]

$$v = \frac{d\omega}{dk} = \hbar^{-1}\frac{dE}{dk} \tag{9}$$

$$= \frac{dE}{dP}. \tag{10}$$

Only when $E = P^2/2m$ do we have $v = P/m$ as for a free electron.

We must now note one very important property of the wave-number k and also of the corresponding crystal momentum P. The quantity k is not uniquely defined by means of the wave-function in equation (7). We may write this equation in the form

$$\psi(x) = \{u_k(x) \exp[-2\pi irx/d]\} \exp[i(k + 2\pi r/d)x]$$
$$= u_{k'}(x) \exp[ik'x], \tag{11}$$

where
$$k' = k + 2\pi r/d. \tag{12}$$

The wave-function in equation (11) is a solution of the wave-equation for the same energy as the wave-function in (7), and $u_{k'}(x)$ is also a periodic function of period d, so the quantities k and k' are equivalent. The energy E is thus a periodic function of k period $2\pi/d$. This allows us to restrict the value of k to an interval $2\pi/d$, usually $-\pi/d \leqslant k \leqslant \pi/d$, called a zone.

Moreover, if $V(x)$ is an even function of x, as it may be taken, from considerations of symmetry, then changing the sign of x will not change

[4] *W.M.C.S.*, §§ 5.1, 5.2.

the energy. If we now also change the sign of k we get back to a wave-function corresponding to the same energy as before; $E(k)$ is therefore an *even* function of k. Thus in giving the energy $E(k)$ as a function of k we may further restrict k to the range $0 \leqslant k \leqslant \pi/d$. Such a representation is called a reduced representation and is illustrated in Fig. 2.1; in this representation the energy $E(k)$ turns out to be a multi-valued function of k as shown in Fig. 2.1, each branch of the curve corresponding to an allowed energy band. These bands, for one-dimensional problems, do not overlap, and are nearly always separated by a forbidden energy gap. It is this feature that enables this theory to account for the properties of semiconductors as already discussed in the previous chapter.

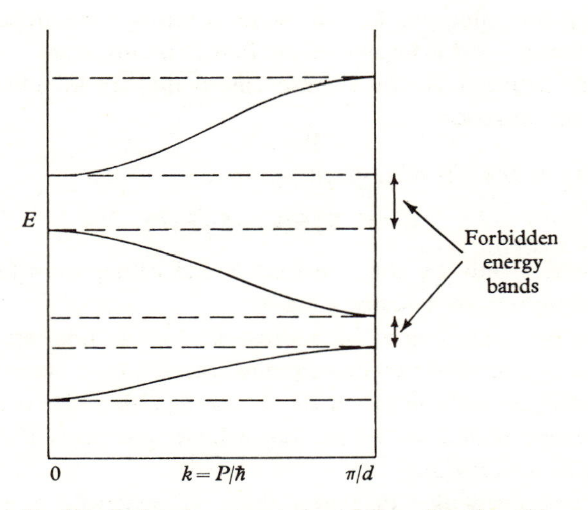

Fig. 2.1. Reduced representation of E as a function of k.

These considerations may readily be extended to three dimensions, to enable us to discuss the properties of real crystals. We shall assume, for simplicity, that three orthogonal axes (x, y, z) may be chosen in the crystal so that the lattice is periodic in (x, y, z). This may be done for most of the simple crystal forms. Let (d_1, d_2, d_3) represent the periodic spacings along the axes of the crystal, i.e. the unit cell is a parallelepiped whose sides are d_1, d_2, d_3. The potential V is then a periodic function such that

$$V(\mathbf{r}) = V(\mathbf{r} + \mathbf{d}), \tag{13}$$

where **d** is any vector of the form

$$\mathbf{d} = n_1 d_1 \mathbf{i} + n_2 d_2 \mathbf{j} + n_3 d_3 \mathbf{q},$$

n_1, n_2, n_3 being integers, and **i**, **j**, **q** unit vectors along the x-, y-, z-axes. The solution of the wave-equation may now be written in the form[5]

$$\psi(x, y, z) = u_k(x, y, z) \exp[i(\mathbf{k} \cdot \mathbf{r})], \tag{14}$$

where **k** is a constant vector whose components are (k_x, k_y, k_z) and $u_k(x, y, z)$ is a function with the periodicity of the lattice. The energy E is now a function of k_x, k_y, k_z; as before it is a periodic function of periods $2\pi/d_1$, $2\pi/d_2$, $2\pi/d_3$, in k_x, k_y, k_z and we may restrict the values of **k** in k-space to a certain zone. This zone is usually taken as the smallest volume of k-space centred on the origin which includes all non-equivalent values of **k**. The shape of this zone depends on the crystal structure, and is known as the first Brillouin zone.

We may define a vector **P**, also called the crystal momentum, by means of the equation[6]

$$\mathbf{P} = \mathbf{k}\hbar; \tag{15}$$

the velocity vector **v** is then given by

$$\mathbf{v} = \hbar^{-1} \nabla_k E(\mathbf{k}) = \nabla_P E(\mathbf{P}). \tag{16}$$

P and **k** differ only by the constant $\hbar$ and either may be used as a 'quantum number' to describe a state.

The energy cannot now be represented as a function of a single variable except in rather special circumstances. These occur when E is a function of k, the magnitude of the vector **k**. Otherwise it is customary to draw diagrams of E as a function of **k** along several of the principal directions in the crystal.

Along any particular direction there is generally a gap between adjacent bands, but the minimum in one direction may overlap the maximum of the next lowest band in another direction, and overlapping bands in three-dimensional problems are quite common. Only when we have a finite gap between the highest filled band and the lowest empty band do we have a semiconductor. We must now consider the number of states in a given energy band and see how they may be occupied by electrons.

For a cubic crystal, one of whose edges is of length L, it may be shown that, as for free electrons, the number of allowed values of k_x in the interval $-\pi/d_1 < k_x \leq \pi/d_1$ is L/d_1 and similarly for k_y and k_z. The total

[5] *W.M.C.S.*, §§ 4.1.1, 4.8, 4.9. [6] *W.M.C.S.*, § 5.1.

number of allowed values of **k** is therefore equal to $\mathbf{V}/d_1d_2d_3$ where **V** is the volume of the crystal. This is just equal to N, the number of unit cells in the crystal. This result is true for any crystal form.[7] If we have one atom per cell then N is also equal to the number of atoms in the crystal. Thus, if we have one electron per atom the electrons fill just half a band, i.e. $\frac{1}{2}N$ states, since, because of spin, two electrons may go in each level; if there are two electrons per atom they exactly fill a band. More generally when the number of electrons per atom is even we get completely filled bands, when all the electrons are in their lowest states. In general the form of the energy function $E(\mathbf{k})$ is quite complicated. However, for semiconductors (as distinct from metals), we shall almost always be concerned with nearly empty or nearly full bands so that the form of the function $E(\mathbf{k})$ near its maxima and minima only is of importance.

2.3 Form of the energy bands

The theoretical evaluation of the function $E(\mathbf{k})$ for even the simplest crystalline forms presents a very formidable problem in theoretical physics, and for semiconductors has been calculated in any detail for only a few. These we shall discuss further in later sections dealing with particular semiconductors (see Chapter 11). We shall now show that at the edges of a zone, in the one-dimensional case, and also at the centre of a zone, the group velocity is zero. Consider the points corresponding to $k = \pi/d$ and to $k = -\pi/d$. These are separated by $2\pi/d$ and so are equivalent, so that a wave-function with $k = \pi/d$ must represent a *standing* wave, otherwise wave-functions with $k = \pi/d$ and $k = -\pi/d$ would represent waves travelling in opposite directions and the two values of k could not be equivalent. There being no transport in a standing wave, it must correspond to an electron with zero velocity, i.e. the group velocity $\partial\omega/\partial k$ is zero, hence $\partial E/\partial k$ is zero. A similar argument shows that at symmetry points of the surface of a Brillouin zone, the normal derivative $\partial E/\partial k_n$ is zero. Thus, if we plot E as a function of k along any of the principal directions which are perpendicular to the surface of a zone we shall have E/k curves at the zone-edge like those shown in Fig. 2.1. From a discussion of Schrödinger's equation it may also be shown that $E(\mathbf{k})$ is a continuous function of **k** with continuous first and second derivatives, provided E is single valued, i.e. two branches do not meet. Since E is an even function

[7] *W.M.C.S.*, § 5.5.

it must therefore contain only quadratic terms near $k = 0$. Thus we also have $\partial E/\partial k = 0$ when $k = 0$ when we do not have degeneracy at $k = 0$, and again the E/k curves must be similar to that shown in Fig. 2.1 near $k = 0$. At $k = 0$ we may either have a maximum or a minimum. In practical cases other maxima or minima may occur; if they lie at intermediate values of E they will be relatively unimportant; if they are the lowest minimum or highest maximum they will be of paramount importance. Various situations may arise, as discussed below.

The simplest of all band structures occurs when the lowest unfilled band has a minimum at the centre of the zone and is single valued. We may then expand $E(k_x, k_y, k_z)$ in powers of k_x, k_y, k_z in the form

$$E(k_x, k_y, k_z) = E_0 + Ak_x^2 + Bk_y^2 + Ck_z^2 + \text{higher powers}, \qquad (17)$$

A, B, C being *positive* constants. In particular, if we have a crystal with cubic symmetry we may reasonably suppose that $A = B = C$ and, taking the zero of energy at the bottom of the band we may express E for small values of k in the form

$$E = \frac{\hbar^2}{2m_e}(k_x^2 + k_y^2 + k_z^2), \qquad (18)$$

where m_e is a constant, having the dimensions of a mass. Only when $V(x)$ is constant will m_e be equal to the free-electron mass. In terms of the crystal momentum we have

$$E = \frac{1}{2m_e}(P_x^2 + P_y^2 + P_z^2). \qquad (19)$$

The equations (18), (19) differ from those for free electrons only in having the quantity m_e instead of $\boldsymbol{m}$. Thus the electrons in this case behave like free electrons but with a different mass. The quantity m_e is generally referred to as the effective mass. In this case the energy bands are generally described as being 'spherical', the reason being that the constant energy surfaces are spheres in k-space. It was once thought that this condition would be quite common, but more recent research has shown that it is rather rare. An example occurs in the conduction band of the intermetallic semiconductor InSb. Before discussing alternative situations let us consider the form of the energy bands when the energy has a *maximum* $k = 0$. In this case we have, with appropriate choice of axes,

$$E = E_0' - A'k_x^2 - B'k_y^2 - C'k_z^2. \qquad (20)$$

When $A' = B' = C'$ this may be expressed in the form

$$E = E_0' - \frac{\hbar^2}{2m_h}(k_x^2 + k_y^2 + k_z^2). \tag{21}$$

In this case the electrons behave like free particles but with a negative effective mass. We shall later discuss the very important consequences of this.

The form of the energy bands corresponding to the above conditions is shown in Fig. 2.2. The energy is plotted as a function of k_x but will be similar for any other direction in the crystal. The situation in which the upper band is the conduction band and the lower the valence band represents the simplest possible band structure for a semiconductor. The forbidden energy gap, which we shall write as ΔE, is equal to $E_0 - E_0'$ (in general, $E_0' < E_0$). Unfortunately, no semiconductor is known which corresponds to this structure.

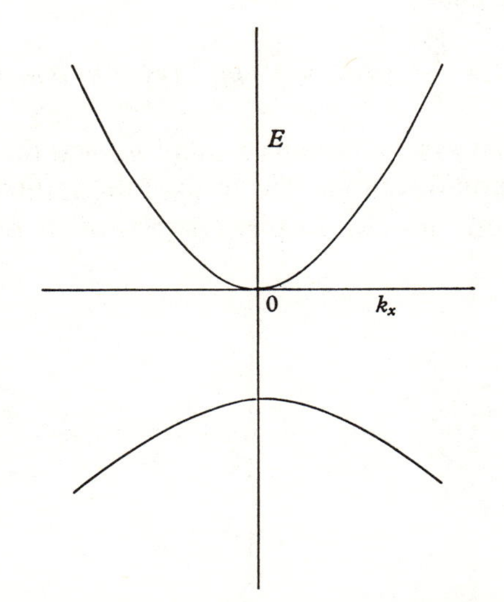

Fig. 2.2. 'Spherical' energy bands at the centre of the Brillouin zone.

A similar situation would arise if the maxima and minima were to occur at a point on the surface of the Brillouin zone. The condition when this occurs along the x-axis is shown in Fig. 2.3 for both conduction band and valence band. In this case there would not necessarily be spherical symmetry, but generally cylindrical symmetry about the x-axis for a cubic crystal. For a minimum we should have, taking the zero of

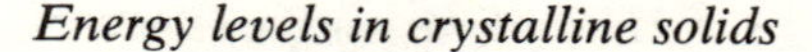

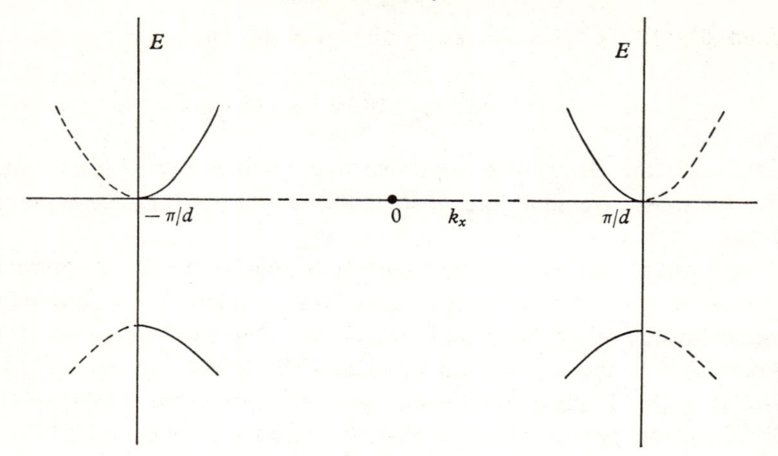

Fig. 2.3. Energy maximum or minimum at zone boundary.

energy at the minimum,

$$E = \frac{\hbar^2}{2}[(\pi/d - k_x)^2/m_1 + (k_y^2 + k_z^2)/m_2]. \tag{22}$$

In this case we have two constants m_1 and m_2 and the effective mass is different in different directions. The form of the section of the constant-energy surfaces by the (k_x, k_y)-plane is shown in Fig. 2.4; they are

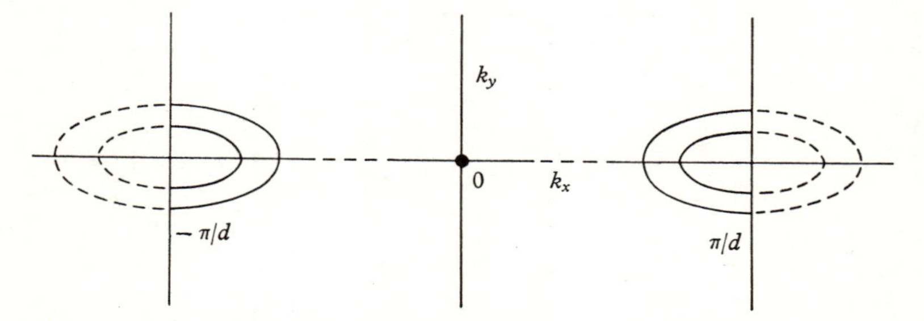

Fig. 2.4. Ellipsoidal constant-energy surfaces for minimum energy at zone boundary.

ellipsoidal in form. The dotted section represents the continuation of the ellipsoids for values of k outside the zone. This is equivalent to the part centred on $k_x = -\pi/d$ where d is the lattice spacing. If the crystal has cubic symmetry there will be other equivalent minima, for example, along the k_y-axis and the k_z-axis. There will thus be three equivalent minima in the conduction band at the points $(\pi/d, 0, 0)$, $(0, \pi/d, 0)$,

$(0, 0, \pi/d)$ in k-space. (Note that $(-\pi/d, 0, 0)$ corresponds to the same minimum as $(\pi/d, 0, 0)$.) In crystals with cubic symmetry, minima may occur along other lines of symmetry, for example along the $\langle 1, 1, 1\rangle$ directions for which

$$k_x = \pm k_y = \pm k_z.$$

If the minima are at the edges of the zone we have now four minima. This situation holds for the conduction band of Ge (see § 13.3). In such cases the constant-energy surfaces are multiple. For crystals having cubic symmetry they consist of a number of ellipsoids of revolution, when the energy is nearly equal to the minimum value for the band. The centres of these ellipsoids lie at the symmetry points corresponding to the minimum value of the energy.

Now let us suppose that the minima do not occur at the edge of the zone but that there is one (the deepest) at $(k_0, 0, 0)$ and others at equivalent positions, with $0 < k_0 < \pi/d$. There will then be minima also at $(-k_0, 0, 0)$, $(0, \pm k_0, 0)$, $(0, 0, \pm k_0)$ and these will all be distinct. There will thus be six minima. The energy near $(k_0, 0, 0)$ will then be given approximately by

$$E = \frac{\hbar^2}{2}[(k_0 - k_x)^2/m_1 + (k_y^2 + k_z^2)/m_2]. \tag{23}$$

For the state corresponding to $(k_0, 0, 0)$ the crystal momentum will not be zero (unless $k_0 = 0$) but equal to $k_0\hbar$. The velocity of an electron in this state will, however, be zero since we have from equation (16)

$$\left.\begin{aligned} v_x &= \frac{\hbar}{m_1}(k_x - k_0), \\ v_y &= \frac{\hbar}{m_2}k_y, \\ v_z &= \frac{\hbar}{m_2}k_z. \end{aligned}\right\} \tag{24}$$

The deepest minima may also occur inside the zone along other symmetry axes. For the conduction band of Si they occur along the $\langle 1, 0, 0\rangle$ axes. There will therefore be six such minima. Constant-energy surfaces corresponding to such conditions are shown in Fig. 2.5(c) for the (k_x, k_y)-plane. We shall discuss some other configurations when we deal with particular semiconductors. A number of examples of the form of the energy bands along particular directions in the crystal are shown

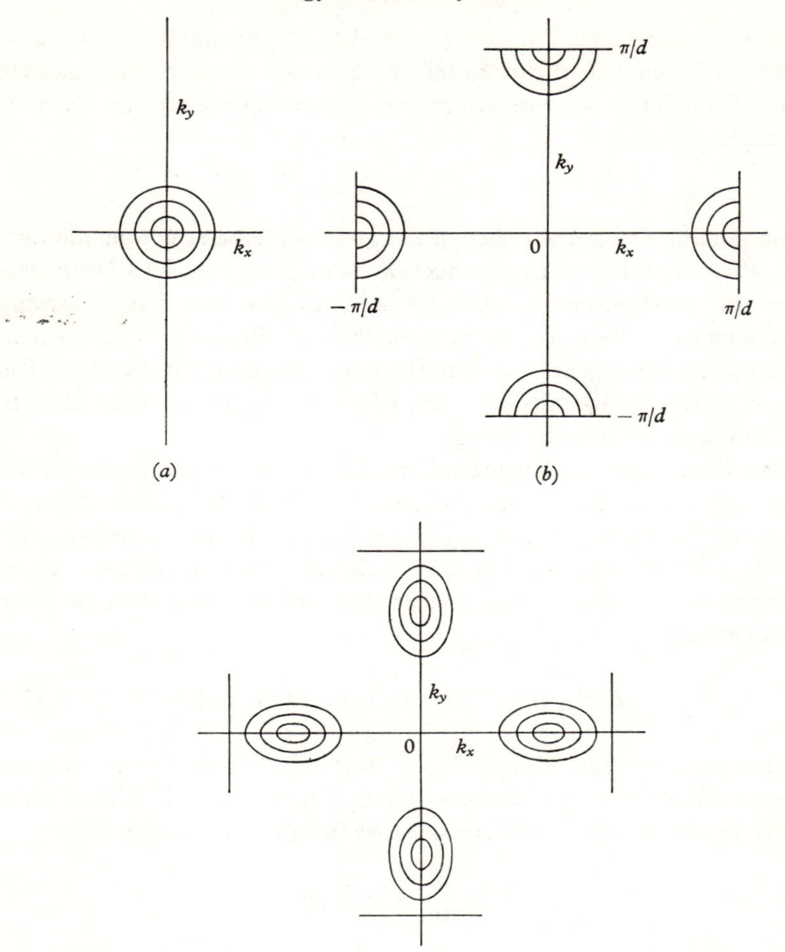

Fig. 2.5. Constant-energy surfaces shown as intersections with the (k_x, k_y) plane. (a) Spherical surfaces at $k=0$. (b) Ellipsoidal surfaces at $k=\pi/d$, etc. (c) Ellipsoidal surfaces at $k=k_0$ $(<\pi/d)$, etc.

in Fig. 2.5. It will be seen from equations (24) that the velocity v is zero at a minimum of the energy (or at a maximum). It may also be shown that under certain symmetry conditions it is also zero at the edges of a Brillouin zone.[8]

A complication arises when we have a condition of degeneracy, i.e. when two energy bands touch at a point. For example, in the valence

[8] *W.M.C.S.*, p. 174.

bands of Si and Ge we have two nearly parabolic bands touching at the centre of the zone as shown in Fig. 2.6, which shows the variation of the energy along the k_x direction. Either of the values for a given value of k_x are possible and at $k_x = 0$ they coincide. In this case a simple expansion of E in even powers of k_x, k_y, k_z is not possible and a more complex form of expression must be used to express E even for small values of k.

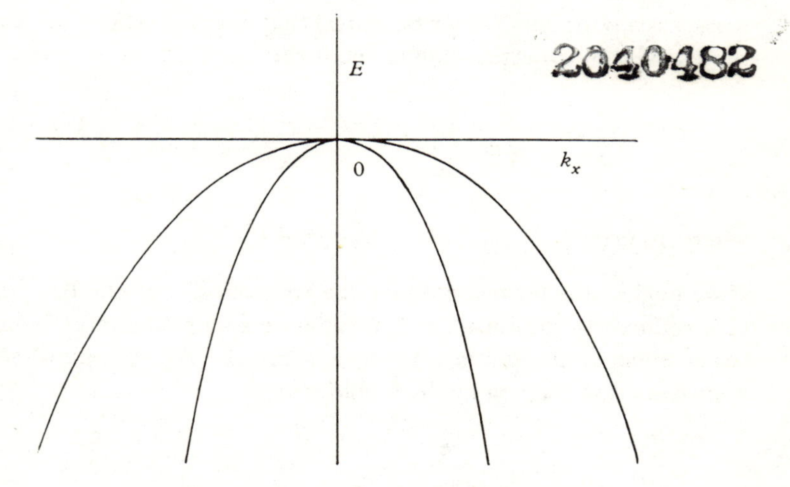

Fig. 2.6. Degenerate condition at $k = 0$ for k_x direction.

The result is that the constant-energy surfaces are distorted from the spherical shape. For many purposes, however, the degenerate bands can be represented as two 'spherical' bands with effective hole masses m_{h1} and m_{h2}. It usually turns out that one of the effective masses is much smaller than the other and the holes have come to be known as 'light holes' and 'heavy holes', much of the early work on their properties having been done on Si and Ge which have such a band structure (see § 13.3).

Crystals with cubic symmetry such as Si and Ge, with diamond or zincblende crystal structure have in fact bands which are *triply* degenerate at $k = 0$, if the coupling between the electron's translational motion and its spin are neglected. When this is taken into account a third band is split off from the former two and turns out to be almost 'spherical' in form with maximum lying ΔE_s below the common maximum of the other two bands.[9] This will have yet another effective hole

[9] *W.M.C.S.*, § 7.3.5.

mass m_{h3}. The energy ΔE_s is known as the spin-orbit splitting and is usually somewhat smaller than the forbidden energy gap ΔE. For some semiconductors, however, ΔE and ΔE_s are about the same and, as we shall see, this gives them some interesting properties.

Because of interaction between the degenerate bands the distortion is such that constant-energy surfaces are distorted from spherical shape and for some purposes an approximation in terms of two scalar effective masses is a poor one. It may be shown that near $\mathbf{k}=0$ the energy may be represented by an expression of the form[10]

$$E(k)=-\Delta E+\frac{\hbar^2}{2\boldsymbol{m}}[Ak^2\pm\{B^2k^4+C^2(k_x^2k_y^2+k_y^2k_z^2+k_z^2k_x^2)\}^{\frac{1}{2}}], \tag{25}$$

where, as usual $$k^2=k_x^2+k_y^2+k_z^2.$$

If we neglect the term containing the constant C we note that equation (25) reduces to the form for two separate spherical energy bands with scalar effective masses $\boldsymbol{m}/(A+B)$ and $\boldsymbol{m}/(A-B)$. A somewhat better approximation is given by the equations

$$\left.\begin{aligned}\frac{\boldsymbol{m}}{m_{h1}}&=A-(B^2+C^2/6)^{\frac{1}{2}},\\ \frac{\boldsymbol{m}}{m_{h2}}&=A+(B^2+C^2/6)^{\frac{1}{2}}.\end{aligned}\right\} \tag{26}$$

This is quite a good approximation for m_{h2}, the light-hole band being very nearly spherical; the warping of the heavy-hole band is much more marked.[11]

Were it not for spin–orbit splitting (see § 12.6) the valence band of Si and Ge would be triply degenerate at $\mathbf{k}=0$; the effect of the spin–orbit interaction is to lower the energy of one of the bands so that it is separated from the two degenerate bands already discussed. Since there is only a small interaction between this band and the others its constant-energy surfaces are nearly spherical. If we denote the spin–orbit splitting by E_{s0} we have for this band

$$E(\mathbf{k})=-\Delta E-E_{s0}-\frac{\hbar^2}{2\boldsymbol{m}}(Ak^2) \tag{27}$$

[10] *W.M.C.S.*, § 7.3.5.
[11] See, for example, F. Herman, *Proc. Inst. Radio Engrs.* (1955) **40**, 1703.

A being the same constant as in (25); this gives for the mass m_{h3} of the third hole the value of

$$\frac{m}{m_{h3}} = A \tag{26a}$$

Another form of band structure is that shown by the lead chalcogenides Pbs, PbSe, PbTe in which the valence and conduction bands have their maximum and minimum for the same value of **k** but this is not $\mathbf{k} = 0$. For these compounds it is at the zone edge in the $\langle 111 \rangle$ directions so that there are four maxima and four minima (see Fig. 13.7).

One final form of band structure we should mention, although it strictly does not belong to a semiconductor. It can happen that the top of the valence band just coincides with the bottom of the conduction band and the forbidden energy gap is precisely zero. This happens for example in the case of grey tin (α-Sn) and HgTe. The structure for α-Sn is shown in Fig. 13.5. In this case the material is known as a semimetal.

In general, when we do not have degeneracy, we may choose the axes to give expressions for E in one of the forms involving only squares of the components of **k**. If such a choice is not made we shall still be able to expand E near an extremum in the form

$$E = E_0 + \frac{\hbar^2}{2} \sum_{xyz} k_x k_y / m_{xy}, \tag{28}$$

the first-order terms not being present because the expansion is made at an extremum. The tensor $1/m_{xy}$ is then called the effective-mass tensor and may be written in the form

$$\frac{1}{m_{xy}} = \hbar^{-2} \frac{\partial^2 E}{\partial k_x \partial k_y}, \quad \text{etc.} \tag{28a}$$

2.4 Positive holes

In an ideal semiconductor, when all the electrons are in their lowest states, there are no electrons in the conduction band. This situation is, however, theoretically possible only at the absolute zero of temperature. At normal temperatures there will always be a few electrons in the conduction band due to thermal excitation from the full band. The instantaneous current density **I** due to the motion of a particular electron is proportional to its velocity **v** and parallel to it; we may calculate it as follows. Suppose we have N_v electrons (with velocity vector **v**) in a volume **V**. The number crossing a plane of unit area, whose normal is parallel to the direction of the vector $\mathbf{v}$, in unit time, is $N_v|\mathbf{v}|/\mathbf{V}$. The current density is thus equal to $-eN_v\mathbf{v}/\mathbf{V}$, and the current density $\mathbf{I}$ due to a single electron may therefore be expressed in the form

$$\mathbf{I} = -e\mathbf{v}/\mathbf{V}. \tag{29}$$

Now suppose we have an electron in a state whose wave-vector is equal to $\mathbf{k}$; the corresponding velocity $\mathbf{v}$ is given by equation (16). We may note that the velocity corresponding to a wave-vector $-\mathbf{k}$ is equal to $-\mathbf{v}$, since E is an even function of $\mathbf{k}$. If we have a number of electrons the total current density will be given by

$$\mathbf{I} = -(e/\mathbf{V}) \sum_s \mathbf{v}_s, \tag{30}$$

where the velocity vectors $\mathbf{v}_s$ correspond to the states occupied by the electrons. In particular, if the electrons just fill an allowed energy band we have $\mathbf{I} = 0$, since to each $\mathbf{k}_s$ there corresponds a $-\mathbf{k}_s$, giving an equal and opposite current.

Now suppose we have a band that is fully occupied by electrons except for a single state with wave-vector $\mathbf{k}_i$ and corresponding velocity $\mathbf{v}_i$. The current due to all the electrons in this band is then given by

$$\begin{aligned} \mathbf{I} &= -(e/\mathbf{V}) \sum_{s \neq i} \mathbf{v}_s \\ &= -(e/\mathbf{V}) \sum_{\text{all band}} \mathbf{v}_s + (e/\mathbf{V})\mathbf{v}_i. \end{aligned} \tag{31}$$

The sum on the left-hand side of equation (31) we have seen to be zero, so we have

$$\mathbf{I} = (e/\mathbf{V})\mathbf{v}_i. \tag{32}$$

This is just equal to the current which would arise from an electron occupying the empty state but with a *positive* charge $+e$. Such an empty state is called a positive hole. We shall see later than an effective mass may be ascribed to it equal in magnitude to that of an electron occupying the state. When we have a number of such positive holes the current density is clearly given by

$$\mathbf{I} = (e/\mathbf{V}) \sum_i \mathbf{v}_i. \tag{33}$$

If we have both electrons and holes present, as we shall have for a pure semiconductor, since each conduction band electron leaves behind a positive hole in the full band, the instantaneous current density will be given by

$$\mathbf{I} = -(e/\mathbf{V})\left\{\sum_j \mathbf{v}_j - \sum_i \mathbf{v}_i\right\}, \tag{34}$$

where the sum over j refers to electrons and the sum over i refers to holes.

2.5 Motion of electrons and holes in a crystal under the influence of an external field of force

An electron moving in an ideal crystal continues to do so with constant velocity when in a state represented by a wave-function of the form given in equation (7). Actually there will be fluctuations due to the crystalline field but the *observable* velocity will be constant. We now wish to determine the effect of an external field of force. Consider first an electric field $\mathscr{E}$. The rate of work done on the electron is given by

$$\frac{dE}{dt} = -e(\mathscr{E} \cdot \mathbf{v}). \tag{35}$$

Using equation (16) for $\mathbf{v}$ this may be written as

$$\left(\nabla_k E \cdot \frac{d\mathbf{k}}{dt}\right) = -e(\mathscr{E} \cdot \nabla_k E)\hbar^{-1}$$

giving

$$\hbar\frac{d\mathbf{k}}{dt} = \frac{d\mathbf{P}}{dt} = -e\mathscr{E}. \tag{36}$$

This is one of the most fundamental equations of the theory and shows that the crystal momentum increases at a rate equal to the applied force. The above derivation will not hold for forces which are at right angles to $\mathbf{v}$, such as those due to magnetic fields, and a more sophisticated treatment is necessary. Equation (36) is found to be quite general and may be written in the form[12]

$$\hbar\frac{d\mathbf{k}}{dt} = \frac{d\mathbf{P}}{dt} = \mathbf{F}, \tag{37}$$

where $\mathbf{F}$ is the force acting on the electron. This is the main justification for calling $\mathbf{P}$ the crystal momentum.

[12] *W.M.C.S.*, § 5.4.

When, in addition to the electric field, we have a magnetic field, whose induction is $\mathbf{B}$, then

$$\mathbf{F} = -e\mathscr{E} - e(\mathbf{v} \times \mathbf{B}). \tag{38}$$

If we differentiate the equation (16) for $\mathbf{v}$ with respect to the time we have

$$\frac{d\mathbf{v}}{dt} = \left(\nabla_k \mathbf{v} \cdot \frac{d\mathbf{k}}{dt}\right) = \hbar^{-2} \nabla_k (\nabla_k E \cdot \mathbf{F}). \tag{39}$$

The tensor $\hbar^{-2}\nabla_k \nabla_k E$ which we may write as $1/m_{\eta\zeta}$, where η and ζ can take the values x, y, z, is called the effective-mass tensor and reduces to the form given in equation (28a) for the particularly simple form of $E(\mathbf{k})$ to which the latter equation refers.

The vector components of equation (39) may be written in the form

$$\frac{dv_x}{dt} = \hbar^{-2}\left(\frac{\partial^2 E}{\partial k_x^2} F_x + \frac{\partial^2 E}{\partial k_x \partial k_y} F_y + \frac{\partial^2 E}{\partial k_x \partial k_z} F_z\right), \quad \text{etc.} \tag{40}$$

When the non-diagonal elements of the tensor are zero and we write the diagonal elements as $1/m_1$, $1/m_2$, $1/m_3$, equations (40) reduce to the simple form

$$\left.\begin{aligned} m_1 \frac{dv_x}{dt} &= F_x, \\ m_2 \frac{dv_y}{dt} &= F_y, \\ m_3 \frac{dv_z}{dt} &= F_z. \end{aligned}\right\} \tag{41}$$

If $m_1 = m_2 = m_3 = m_e$ we have a further simplification and obtain

$$m_e \frac{d\mathbf{v}}{dt} = \mathbf{F}. \tag{42}$$

In this case we have a single effective mass and the electron moves in the field of force as though it were a simple particle of mass m_e. When this is not so we must use the appropriate value of the effective mass for the particular direction of motion.

We must now discuss the motion of a positive hole. Suppose we first of all consider a particular electron in a full band. If at time t it has a velocity $\mathbf{v}_i$ and at time $t + \Delta t$ a velocity $\mathbf{v}_i + \Delta\mathbf{v}_i$ then

$$\Delta\mathbf{v}_i = \frac{d\mathbf{v}_i}{dt} \Delta t, \tag{43}$$

where $d\mathbf{v}_i/dt$ is given by equation (39). Thus if this electron were missing we should have the condition that at time t an electron would be missing with velocity $\mathbf{v}_i$ and at time $t+\Delta t$ an electron would be missing with velocity $\mathbf{v}_i+\Delta\mathbf{v}_i$, i.e. there would be a hole with a velocity $\mathbf{v}_i+\Delta\mathbf{v}_i$. Thus the hole moves with the *same* velocity as the electron would have done if it had been present. Near the top of the band the diagonal elements of the electron effective-mass tensor are negative, so let us write

$$-m'^{-1}_{\eta\zeta} = \hbar^{-2}\nabla_k\nabla_k E \tag{44}$$

and call $1/m'_{\eta\zeta}$ the hole-mass tensor, the diagonal elements now being positive. Thus we have for the hole in the purely diagonal case

$$v_x = -\partial E/\partial P_x = m_h^{-1} P_x$$

and

$$m_h \frac{dv_x}{dt} = -F_x, \quad \text{etc.,} \tag{45}$$

where F_x is the x-component of the force on an *electron*. For electrical forces therefore the hole behaves as though it had a positive charge $+e$ and an effective mass equal to minus the effective mass of the missing electron, the latter being negative for the top of a band. Thus the hole behaves exactly like a particle with positive mass and positive charge.

Summarizing the above results we may say that in ideal crystals electrons and holes move respectively under the influence of applied fields like free particles with negative and positive charges respectively with the appropriate effective masses. In general, the effective mass of the hole will not be the same as that of an electron, since they arise from different bands. For the spherically symmetrical case, writing m_e for the effective mass of an electron and m_h for the effective mass of a hole

$$\frac{1}{m_e} = \hbar^{-2}\frac{\partial^2 E_c}{\partial k^2}, \quad \frac{1}{m_h} = -\hbar^{-2}\frac{\partial^2 E_v}{\partial k^2}, \tag{46}$$

where E_c and E_v are respectively the energies of electrons in the conduction and valence bands. m_e and m_h are inversely proportional to the curvature of the constant-energy surfaces in k-space. For many purposes we may therefore ignore the detailed crystalline structure and use the appropriate effective mass, which, as we shall see, may be determined experimentally.

The various methods for determining the effective-mass tensor will be discussed in some detail later, since this is one of the most important quantities for any semiconductor. When we know the form of the effective-mass tensor we may deduce a great deal about the general form of the energy bands.

As a simple example of the motion of an electron in a crystal under the influence of an external field of force we shall discuss briefly the most direct method of determining the effective mass m_e for the case of 'spherical' symmetry. Suppose we have a constant magnetic field whose induction B is along the z-axis. The equations of motion of an electron are from (42)

$$\left.\begin{aligned} m_e\dot{v}_x &= -eBv_y, \\ m_e\dot{v}_y &= eBv_x, \\ \dot{v}_z &= 0. \end{aligned}\right\} \tag{47}$$

The solution of these equations is well known. If $v_z = 0$ initially the particle describes a circle in the xy-plane. If r is the radius of the circle and v the velocity we have

$$m_e v^2/r = Bev,$$

or

$$v = Ber/m_e. \tag{48}$$

The electron describes the circle with frequency ν_c given by

$$\nu_c = \frac{Be}{2\pi m_e}. \tag{49}$$

This frequency is generally known as the cyclotron frequency; it is independent of the radius of the circle. If now, a small radio frequency field is also applied, the electron will absorb energy provided the frequency is equal to ν_c, and the radius of the circle will grow. This resonant condition may be observed and ν_c determined experimentally. This method for determining m_e was first discussed theoretically by J. G. Dorfman,[13] by R. B. Dingle[14] and by W. Shockley,[15] and successfully applied experimentally by G. Dresselhaus, A. F. Kip and C. Kittel.[16] Germanium was used in these experiments and it was shown that in this case the effective mass of neither holes nor electrons could be represented by a single scalar quantity m_e. The precise determination of the effective-mass tensors for germanium will be discussed in § 13.3.

[13] *C. R. Acad. Sci. U.R.S.S.* (1951) **81**, 765.
[14] *Proc. Roy. Soc.* A (1952) **212**, 38.
[15] *Phys. Rev.* (1953) **90**, 491.
[16] *Phys. Rev.* (1953) **92**, 827.

2.6 Energy-level diagrams

It is frequently convenient to represent the energy levels in a semiconductor on an energy-level diagram as shown in Fig. 2.7. The energy is

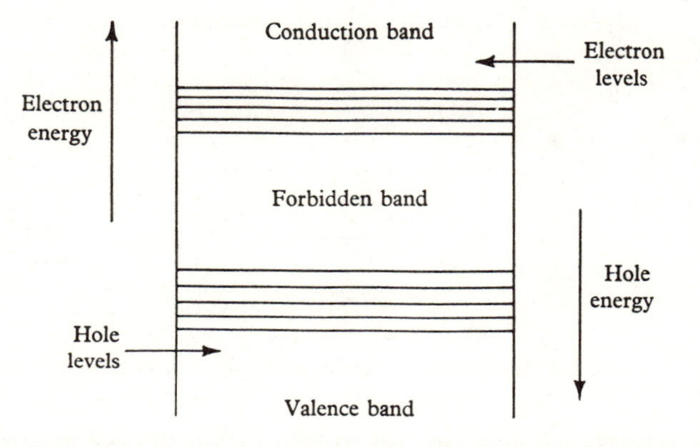

Fig. 2.7. Energy-level diagram for electrons and holes in the conduction and valence bands of a semiconductor.

shown as increasing vertically upward for electrons, the levels in the conduction band being indicated in the diagram. The same diagram may also be used for the energy of positive holes provided we now take the energy as increasing in a downward direction. The reason for this may be seen as follows. Consider as an example the band form shown in Fig. 2.2. Let E_0 represent the energy of all the electrons in a completely filled band, for which the E–k relationship is given by $E = E_v(\mathbf{k})$, the zero of the function $E_v(\mathbf{k})$ being taken at the top of the band. For small values of $\mathbf{k}$, $E_v(\mathbf{k})$ may be expressed in the form

$$E_v(\mathbf{k}) = -\frac{\hbar^2}{2}(k_x^2/m_1' + k_y^2/m_2' + k_z^2/m_3'). \tag{50}$$

For the band immediately above, i.e. the conduction band, let the E–k relationship be given by $E = \Delta E + E_c(\mathbf{k})$. Near the bottom of this band we shall have

$$E_c(\mathbf{k}) = \frac{\hbar^2}{2}(k_x^2/m_1 + k_y^2/m_2 + k_z^2/m_3). \tag{51}$$

ΔE is the forbidden energy gap between the bands (see Fig. 2.8).

Now suppose we take an electron from a state in the valence band characterized by a wave-vector $\mathbf{k}'$ and raise it to a state in the full band with wave-vector $\mathbf{k}$, we shall have created a hole–electron pair. This

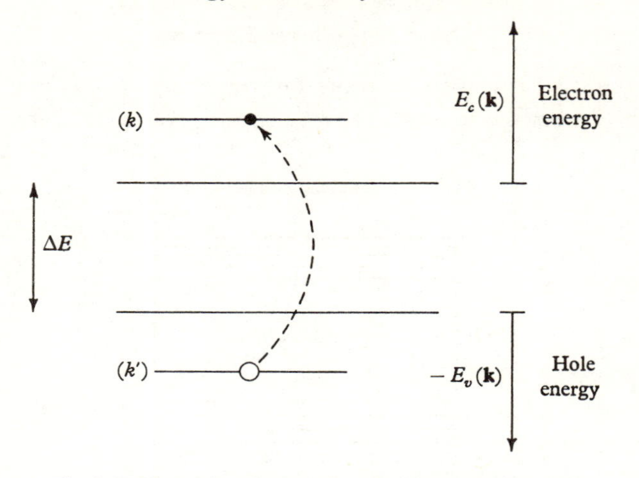

Fig. 2.8. Illustrating the creation of an electron–hole pair.

process may be achieved by adsorption of vibrational energy from the crystal lattice or by absorption of a quantum of radiation. The energy of the system will now be given by

$$E = E_0 + \Delta E + E_c(\mathbf{k}) - E_v(\mathbf{k}'), \tag{52}$$

where E_0 is the energy before the transition takes place. The second and third terms in the right-hand side of equation (52) clearly represent the energy of the electron and the last term the energy of the hole. We may note that $-E_v(\mathbf{k}')$ is positive, since we have taken the zero of $E_v(\mathbf{k})$ at the maximum. Thus we may use the same energy diagram for holes and electrons provided we measure the energy of the holes in the opposite direction to that for electrons. For the approximate forms of $E(\mathbf{k})$ we have for the *change* in energy

$$E - E_0 = \Delta E + \frac{\hbar^2}{2}(k_x^2/m_1 + k_y^2/m_2 + k_z^2/m_3)$$
$$+ \frac{\hbar^2}{2}(k_x'^2/m_1' + k_y'^2/m_2' + k_z'^2/m_3'). \tag{53}$$

ΔE is the minimum energy required to take an electron from the valence band to the conduction band. This, as we shall see, may be measured by various methods, the simplest, in principle, being to observe the long-wave limit of absorption of radiation causing a transition between the bands. ΔE is the most important of all the parameters associated with a semiconductor since it largely determines the conductivity and its variation with temperature. The second and third terms in equation (53) may be regarded respectively as the kinetic energies of the free electron and hole. The equation may also be written in terms of the crystal momentum **P** in the form

$$E-E_0=\Delta E+\tfrac{1}{2}(P_x^2/m_1+P_y^2/m_2+P_z^2/m_3)$$
$$+\tfrac{1}{2}(P_x'^2/m_1'+P_y'^2/m_2'+P_z'^2/m_3'). \tag{54}$$

When $m_1=m_2=m_3=m_e$ and $m_1'=m_2'=m_3'=m_h$ we have

$$E-E_0=\Delta E+\tfrac{1}{2}P^2/m_e+\tfrac{1}{2}P'^2/m_h, \tag{55}$$

where P and P' are respectively the magnitudes of the crystal momenta for the electron and hole. The similarity of this equation with the classical form will be obvious.

2.7 Resistance to motion of electrons and holes in a crystal

So far, we have neglected the effect of impurities and imperfections in the crystal and also the thermal vibrations of the atoms which displace them from their ideal positions in the crystalline lattice. All these prevent the electrons and holes from continuing their steady motion as quasi-free particles, and introduce scattering. Their effect may be described in terms of a change of the wave-vector **k** or the velocity **v**, and may also be described in terms of particle-collision theory. This we shall consider after we have discussed in more detail the nature of impurities and imperfections in a crystal.

As we shall see, the considerations of this chapter will enable us to discuss the motion of electrons and holes in nearly perfect crystals. For most of the applications of semiconductors in modern electronics the materials used are in fact high quality crystals. Such materials give electrons and holes high mobility and long free lifetime so that under the influence of electric and magnetic fields, as well as light flux and thermal gradients, they can move appreciable distances. Such movement is not, however, necessary for *all* applications: amorphous or polycrystalline materials have some important uses and are usually much cheaper. We shall find that although some of these concepts will still be of use for such materials some rather different ideas like hopping of electrons between lattice sites will have to be introduced to account for transport in such materials.

3

Impurities and imperfections in crystals

3.1 Types of imperfection

In the previous chapter the crystals discussed were assumed to have all their atoms at sites which lie on a perfect periodic lattice. This condition does not hold for real crystals. We shall leave aside for the moment the thermal vibrations of the atoms about their positions of equilibrium which clearly destroy the *perfect* periodicity and shall consider various other types of imperfection.

3.1.1 Impurities

The most obvious type of imperfection is due to the presence of foreign atoms in the crystal. No substance can be made perfectly pure, and although great strides have been made in this direction, mainly through research on semiconductors, even the purest crystals contain many foreign atoms. An impurity content of 1 part in 10^9 still leaves about

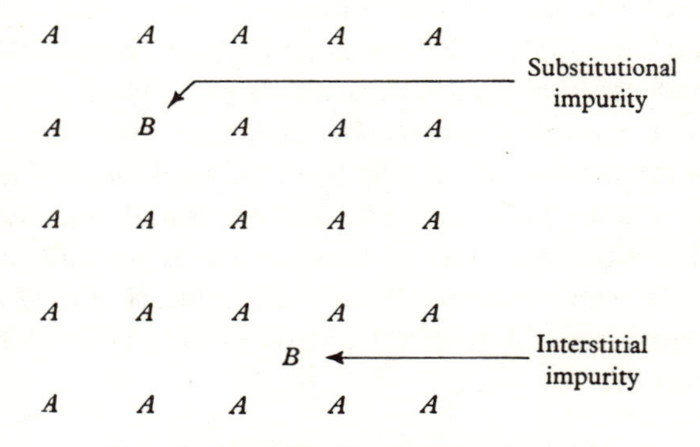

Fig. 3.1. Impurities in a monatomic crystal.

10^{14} impurity atoms per cm^3 in a crystal, since most crystals have between 10^{22} and 10^{23} atoms per cm^3. The bulk of the atoms making up the crystal we call 'atoms of the host crystal'. The foreign atoms are generally called impurities, and they play, as we shall see, a vital part in determining the properties of semiconductors. We must distinguish two types of impurity, substitutional impurities, which replace atoms of the host crystal on their lattice sites, and interstitial impurities which occupy positions in between the lattice sites. These two types of imperfection are illustrated in Fig. 3.1 for a monatomic crystal made up of atoms of type A with a few impurity atoms of type B.

3.1.2 Interstitial atoms and vacancies

Another type of imperfection occurs when an atom of the host crystal is displaced from a lattice site into an interstitial position. Such an imperfection is called an interstitial atom (to distinguish it from an interstitial impurity) and is illustrated in Fig. 3.2. When this occurs, a vacant lattice site may be left, and this forms another type of imperfection, generally called a vacancy, and is also illustrated in Fig. 3.2. More complex configurations may clearly exist in which two or more vacancies occur at neighbouring lattice sites.

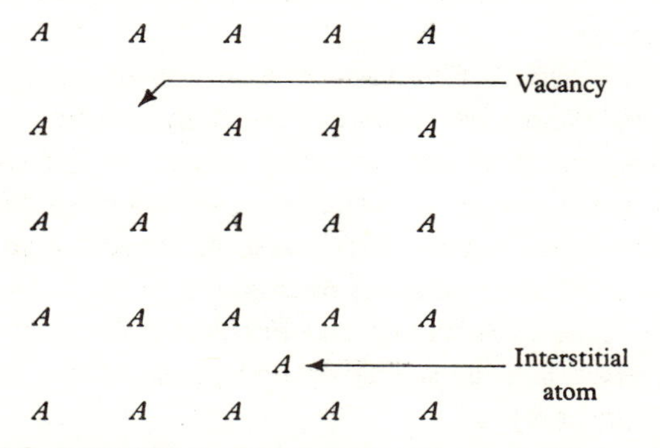

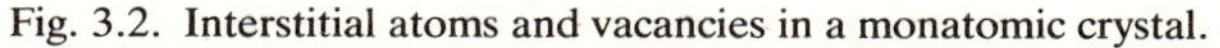
Fig. 3.2. Interstitial atoms and vacancies in a monatomic crystal.

If the atoms removed from lattice sites migrate to the surface of the crystal, they may extend the crystal by occupying regular lattice sites on the surface. Thus we may have vacancies without corresponding interstitial atoms; such defects are known as Schottky defects. On the other hand, vacancies and interstitial atoms may be formed in pairs; these defects are known as Frenkel defects. Thus we see that the number of vacancies is not necessarily equal to the number of interstitial atoms. All the above imperfections may also have variations according to their electrical charge, i.e. they may in some cases trap one or more electrons or positive holes and may exist in different states.

At high temperatures these imperfections will be mobile, and a thermodynamic equilibrium will be set up between the various types of imperfection. The smaller the energy of formation the larger the number of defects of a particular type. For example, if the energy of formation of a Schottky type of defect is much less than that of a Frenkel defect, the former will predominate and there will be relatively few interstitial atoms. In a close-packed lattice it is generally difficult to fit an atom of the crystal into an interstitial position. Impurity atoms which are small go readily into interstitial positions, but large atoms do this only with difficulty and tend to occupy lattice sites. Generally, one of the alternative types of imperfection will greatly outnumber the other.

Although thermodynamic equilibrium may exist at temperatures not far below the melting point of a crystal, the time required for the establishment of such equilibrium at lower temperatures may be very long, and the imperfections may be 'frozen in'. Thus the number of imperfections will depend on the thermal history of the crystal. By holding the crystal for some time at an intermediate temperature the number of imperfections may be considerably reduced. This process is known as annealing.

In thermal equilibrium, the number of defects of any given type may be calculated by means of statistical mechanics when the energy of formation is known. Alternatively, the energy of formation may be deduced if by some means or other the variation of the number of defects with temperature can be observed. A discussion of this in terms of the free energy of formation has been given by N. F. Mott and R. W. Gurney,[1] in the case of Schottky and Frenkel defects. It is shown, for example, that when Schottky-type defects predominate the number of vacancies N_v is given by

$$N_v = NF \exp[-W_v/\mathbf{k}T], \quad (1)$$

where N is the number of atoms in the crystal, and W_v the energy of formation of a vacancy; F is a slowly varying function of T which may have a numerical value of the order of 10^3 to 10^4. Values of W_v are usually of the order of 1 to 2 eV so that, *in equilibrium*, N_v would be

[1] *Electronic Processes in Ionic Crystals* (Oxford University Press, 1940), ch. 2.

quite small at ordinary temperatures but would increase rapidly with temperature.

3.1.3 Dislocations

The imperfections discussed above are all mainly point imperfections. A small number of neighbouring imperfections may group to form a cluster or short line of imperfections but these are generally sharply localized. We must now consider imperfections which run as lines (not necessarily straight, but frequently so) through the crystal. Such imperfections are known as dislocations. When a crystal is subject to stress it generally yields by crystal planes slipping over one another. The whole plane does not slip at once but does so along a curve which spreads gradually over the plane. The line separating the slipped from the unslipped portion is called a dislocation. When the direction of slip is at right angles to the line, the imperfection is called an edge dislocation and may be shown to be equivalent to the insertion of an extra plane of atoms in the crystal, as shown in Fig. 3.3. When the direction of slip is along the line the imperfection is called a screw dislocation. It is beyond the scope of this book to deal with the interesting geometry of dislocations; they play a central part in all modern work on strength of materials and are dealt with in a number of text-books.[2] They have also

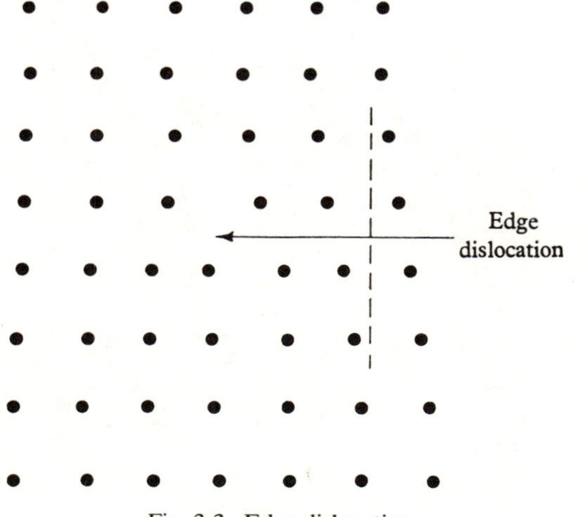

Fig. 3.3. Edge dislocation.

[2] See, for example, A. H. Cottrell, *Dislocations and Plastic Flow in Crystals* (Oxford University Press, 1953); *Theory of Crystal Dislocations* (Gordon and Breach, 1964); W. T. Read, *Dislocations in Crystals* (McGraw-Hill, 1953).

become of importance in the study of semiconductors and will be further discussed in § 8.11.

Where a dislocation line meets the surface of a crystal the regular order of the atoms is disturbed and the surface may be differentially attacked by an etching solution. Under proper conditions small pits, known as etch pits, are formed at the intersection of the dislocation with the surface and may readily be seen by means of a low-power microscope. Fig. 3.4 shows such etch pits for edge dislocations in InSb. The number of dislocations per unit area gives a measure of the perfection of the crystal (see § 8.11).

A full account of the properties of dislocations in crystals having the diamond type of lattice, and in particular for Si and Ge, has been given by H. Alexander and P. Haasen.[3] Semiconductors such as InSb which have the zincblende structure are rather more complex and the types of dislocation of importance to their electrical properties have been discussed by H. C. Gatos, M. C. Finn and M. C. Lavine.[4]

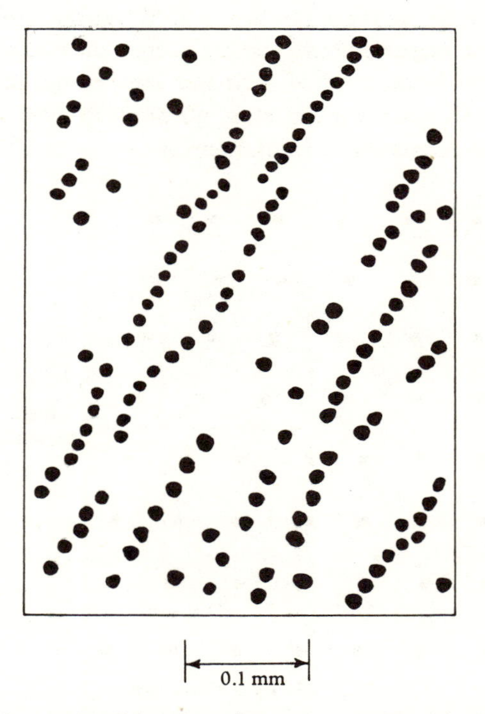

Fig. 3.4. Etch pits due to edge dislocations. (Drawn from a photomicrograph obtained by W. Bardsley and R. L. Bell.[5])

[3] *Solid State Physics* (Academic Press, 1968) **22**, 27.
[4] *J. Appl. Phys.* (1961) **32**, 1174.
[5] *J. Electron. Cont.* (1957) **3**, 103.

3.1.4 Polygonization and dislocation walls

Dislocations may be spread in a fairly random array throughout a crystal, but it is found that, on annealing the crystal, they frequently tend to line up to form dislocation walls. These walls sometimes form a network throughout the crystal. When intersected by a plane they show up a network of polygons, as has been beautifully demonstrated by J. W. Mitchell,[6] using photolytic silver deposited on dislocations in silver bromide. Some dislocation walls are also shown in Fig. 3.4.

A particularly striking form of a wall of parallel edge dislocations is formed at a small-angle grain boundary. This is a common form of rather more gross imperfection which occurs in crystals. Two sections of the crystal instead of having their atomic planes parallel, as in a perfect crystal, may have them making a small angle θ with each other, as is shown in Fig. 3.5. This is equivalent to inserting a series of extra atomic

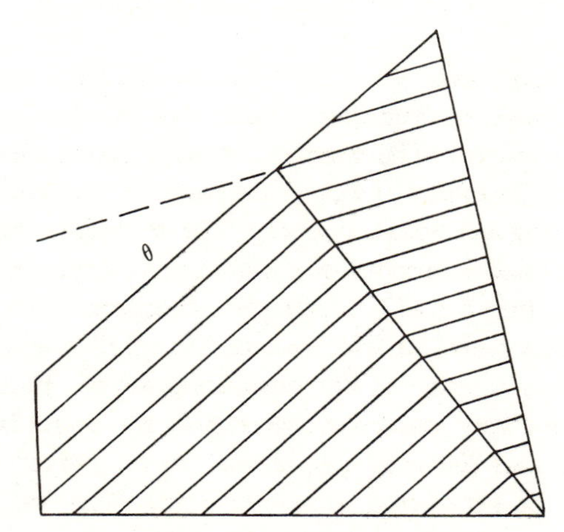

Fig. 3.5. Small-angle grain boundary.

planes and so is also equivalent to a parallel set of edge dislocations (see Fig. 3.6). The distance d between the dislocations is equal to a/θ, when θ is small, where a is the lattice spacing in the direction of slip. The number of dislocations per unit length of the grain boundary is thus equal to θ/a.

When a straight crystalline rod of rectangular cross-section is bent into an arc of a circle of large radius, the bent crystal may be regarded as being made up of a large number of small crystals each making a small

[6] *Phil. Mag.* Ser. 7 (1953) **44**, 223.

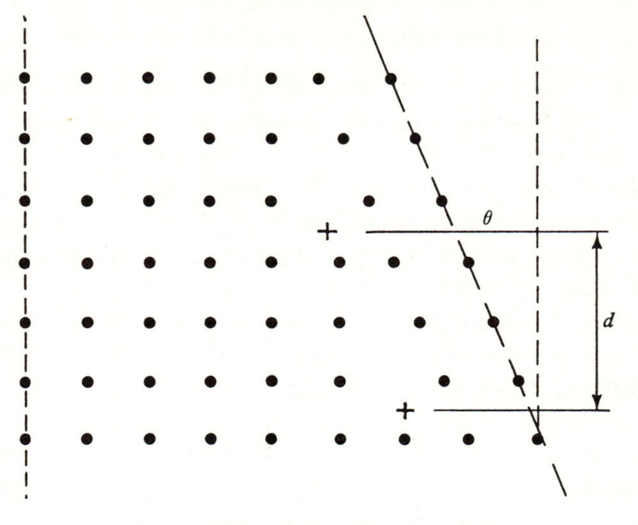

Fig. 3.6. Dislocations at grain boundary.

angle with its neighbour. Thus the bending produces edge dislocations parallel to the axis of bending. These will be distributed uniformly throughout the crystal. The number of edge dislocations may be calculated and this enables a check to be made of the etching techniques used for obtaining a dislocation count. The technique may also be used for inserting a known number of dislocations parallel to a given direction in a crystal (see § 8.11). Crystals are frequently made up of a large number of grains between which there are small-angle boundaries. If, however, the angle between corresponding atomic planes at the grain boundaries is not very small the crystal may no longer be regarded as a 'single' crystal and the material is said to be polycrystalline. In the extreme case, neighbouring grains or crystallites may be randomly orientated and there may be interstices between them. Much of the early work on semiconductors was carried out on such material, but the modern tendency is to use good single crystals which have no marked grain boundaries.

3.2 Chemical binding in semiconductors

Before proceeding to a more detailed study of impurities and imperfections in semiconductors it is instructive to consider the form of chemical binding which holds the atoms of the crystal together. Indeed, an alternative theoretical approach to the electronic structure of solids may be made by this means. We shall, however, restrict ourselves to some elementary considerations which throw light on the rather special properties of semiconductors.

3.2.1 Ionic bonds

One of the most fundamental ideas connected both with atomic structure and chemical binding is the formation of closed shells of electrons. By far the most important closed shell is that made up of eight electrons, two in s-states and six in p-states, as exemplified by the outer electrons of the rare gas atoms Ne, A, Kr, etc. In these shells all the available s-states and p-states are filled, each allowed state having two electrons, one of each spin. Such closed shells of eight electrons form very stable configurations of low energy and it is not surprising that they tend to form even when atoms are brought together. This concept enables us to understand the simplest form of chemical binding, ionic binding, as exemplified by the alkali halides. Let us consider, for example, a substance like NaCl. Each Na atom has only one electron in its outer shell, while each Cl atom has seven. There is a strong tendency for the single electron of the Na atom to migrate to the Cl atom and form a closed group of eight electrons, so that Na^+ positive ions and Cl^- negative ions are formed. These attract one another by simple electrostatic forces which may be shown to account for most of the binding energy of the crystal. Most solids of this type crystallize, like NaCl, in a face-centred cubic form or, like CsCl, in a body-centred cubic form. When the crystal structure is known, the binding energy may be calculated as has been shown by M. Born[7] and his colleagues. The arrangement of the atoms in one of the principal planes of such a crystal is shown in Fig. 3.7.

Clearly, in addition to the purely electrostatic forces which tend to pull atoms of the crystal together, there must be repulsive forces which balance them and establish equilibrium. The closed shell idea enables us to account for the repulsive forces as well. If the ions approach sufficiently close there will be a tendency for electrons from the closed shell of one of the ions to penetrate into the region of space occupied by the closed shell of the other. This is strongly resisted, since because of the Pauli principle, there is no 'room' for such extra electrons. The repulsive force thus set up varies much more rapidly with distance than

[7] M. Born and K. Huang, *The Dynamical Theory of Crystal Lattices* (Oxford University Press, 1954).

the purely electrostatic forces and it is not at all a bad approximation to regard the ions as impenetrable spheres of a certain radius. Thus we have the idea of ionic radius. The distance between two nearest neighbouring ions is thus equal to the sum of their ionic radii; this is illustrated in Fig. 3.8.

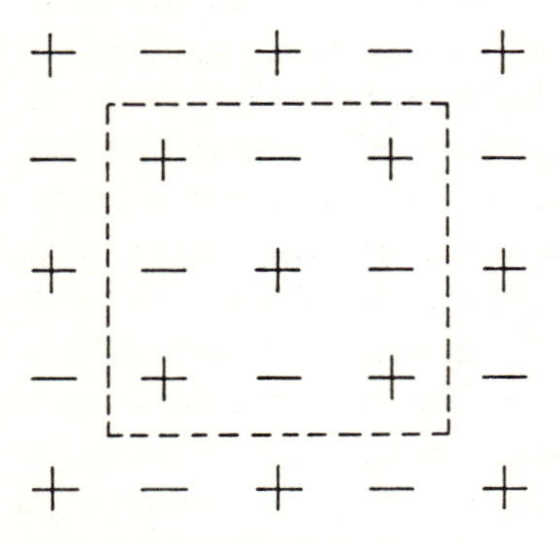

Fig. 3.7. Arrangements of ions in a {100} plane of an ionic crystal.

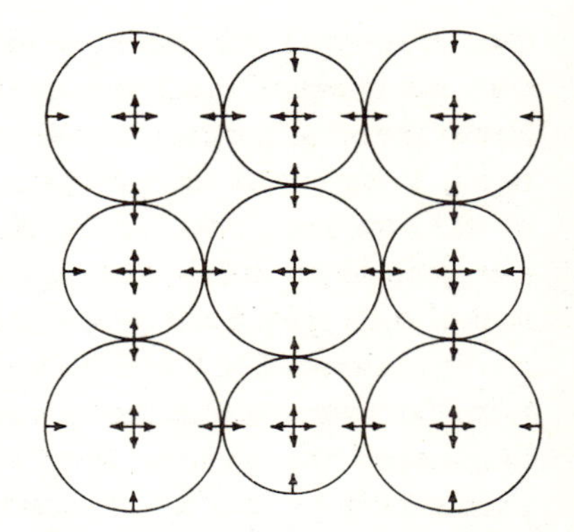

Fig. 3.8. Illustrating attractive and repulsive forces in an ionic crystal.

The principal feature of such ionic crystals is that they are generally transparent in the visible region of the spectrum. They have a small electrical conductivity at elevated temperatures due to ionic transport, but from an electronic point of view may be regarded as insulators. This arises from the wide gap between their highest filled band and the next empty band (there are eight valence electrons to fit in the bands so we expect a completely filled band). The energy required to produce a strong absorption takes place at wavelengths well in the ultra-violet. This radiation may also produce photo-conductivity. On the ionic bond theory the process of producing a free electron consists of taking one of the electrons from the negative ion and placing it on a distant positive ion. The neutral alkali metal atom thus created is to be regarded as the seat of the free electron. Since no energy transfer is involved, the extra electron may jump from one alkali metal ion to another and thus move through the crystal. It is this process that is described by the wave-functions in the form of travelling waves discussed in the previous chapter.

The neutral halogen atom left behind may also be regarded as a positive hole and this also, in principle, should be mobile. The ability of holes to move in such crystals is, however, generally found to be somewhat less than for free electrons. The creation of a hole–electron pair is illustrated in Fig. 3.9 for NaCl, where we have shown two representations of the same state of the crystal.

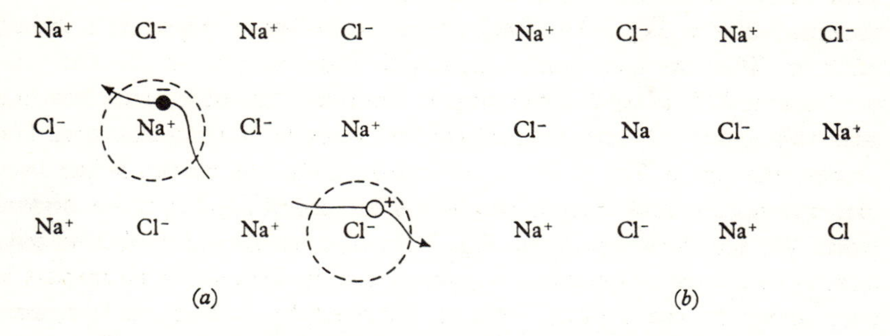

Fig. 3.9. Creation of hole–electron pair in an ionic crystal (NaCl).

3.2.2 Homopolar bonds

The formation of polar molecules such as HCl can, to some extent, be explained by similar means, but the existence of non-polar molecules such as H_2 requires an entirely different explanation. This was one of the difficult problems of structural chemistry before the advent of quantum mechanics and its solution was one of the outstanding successes of the theory. In their famous treatment of the hydrogen molecule W. Heitler and F. London[8] showed that when two electrons are shared between two identical atoms they provide strong bonding provided their spins are anti-parallel; a repulsive force is produced when the spins are parallel. The forces involved, while basically electrostatic in nature, are of purely quantum mechanical origin, and arise from the fact that identical particles cannot be labelled individually. The idea of the two-electron or homopolar bond now plays a quite fundamental part in structural organic chemistry. Bonds due to the sharing of single electrons, as in the H_2^+ ion, also exist but are generally weaker than the double bond.

The concept of the double electron bond, taken together with the closed shell of eight electrons already discussed, enables us to describe the chemical binding of a very important group of semiconductors, namely, the element semiconductors silicon, germanium and grey tin, and also of the insulator diamond. These elements belong to the fourth column of the periodic table and each has four valence electrons. In the atomic state, two of these electrons are in s-states and two in p-states.

[8] *Z. Phys.* (1927) **44**, 455.

When a solid is formed, these electrons are shared with nearest neighbours. If we have *four* nearest neighbours we shall then have, taking account of the shared electrons, a group of eight electrons associated with each atom, and hence a very stable structure. This is illustrated schematically in Fig. 3.10. Each of the bond lines represents a shared electron. They are arranged in pairs, with opposite spin, so that the spins are balanced. This type of bonding is sometimes called valence bonding and this group of semiconductors the valence semiconductors. The representation in Fig. 3.10 is, of course, only schematic, being two-dimensional; in real crystals the bonds are distributed in three dimensions. We see, however, from Fig. 3.10, that we should expect to have exactly four nearest neighbours, symmetrically disposed with respect to each atom of the crystal. This is achieved by having each nearest neighbour at the corner of a regular tetrahedron with the atom at the centre, as shown in Fig. 3.11. A regular lattice, known as the diamond lattice, may be built up in this way. The elements Ge and Si form crystals having this structure as well as diamond.[9]

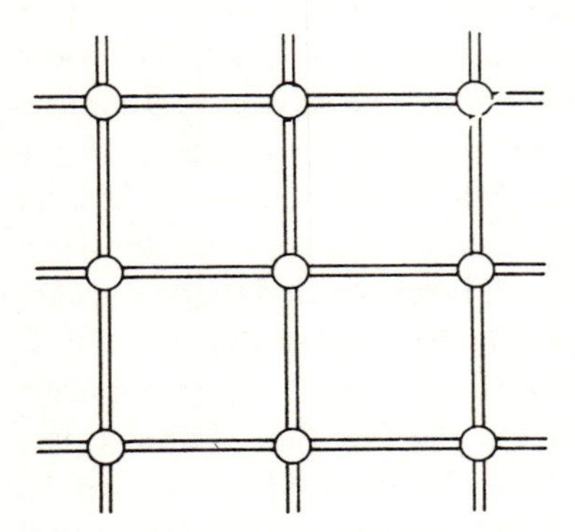

Fig. 3.10. Schematic representation of bonding of atoms with four valence electrons.

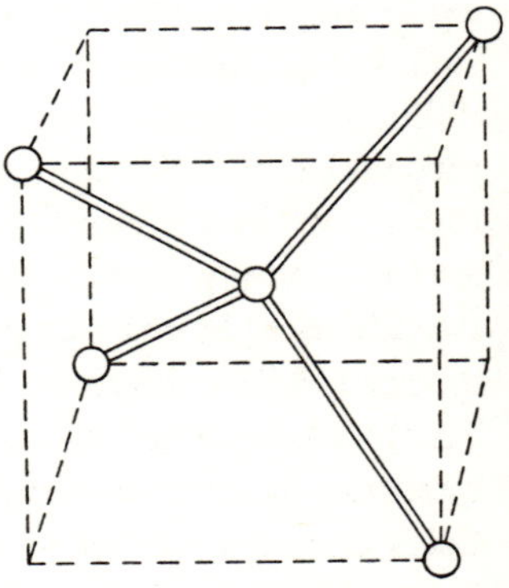

Fig. 3.11. Tetrahedral bonding.

We must now consider how to relate this bond picture to the conduction and valence bands. It is fairly clear that the valence band corresponds to the valence electrons forming the bonds. In the case of Ge, if we remove one of the electrons, we shall have a deficit which may jump from one atom to the next without energy change, and so move through the crystal. This deficit will correspond to the movement of a

[9] *W.M.C.S.*, § 3.8.1.

positive charge, i.e. movement of a Ge^+ ion, and corresponds to the movement of a positive hole in the valence band. An electron detached from a bond and placed on a distant Ge atom will correspond to an electron in the conduction band and its movement will correspond to that of a Ge^- ion. The forbidden energy gap between the valence and conduction bands will correspond to the minimum energy required to remove an electron from a bond and to place it in a *distant* Ge atom (where it will be supernumerary to the closed shell of eight). This is illustrated in Fig. 3.12 where two representations of the formation of a hole–electron pair are shown for Ge.

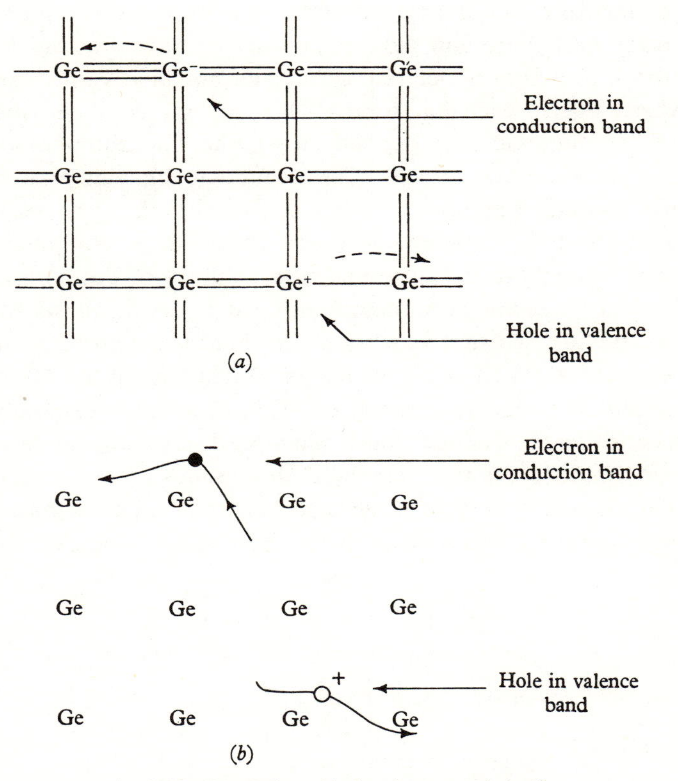

Fig. 3.12. Formation of hole–electron pair in Ge.

The energy required to form a hole–electron pair decreases as we move down the fourth column of the periodic table. For diamond it is about 5.6 eV, so that pure diamond is an insulator at ordinary temperatures, the gap being too great for there to be an appreciable number of electrons in the conduction band due to thermal excitation. For Si the energy gap has a value of approximately 1.12 eV and for Ge a value of about 0.67 eV. The elements are generally referred to as the group IV semiconductors.

3.2.3 Mixed bonds

The only semiconductors which have bonds which are purely homopolar in character are the group IV semiconductors. All the others have bonds which are a mixture of homopolar type and ionic type. This type of mixture is well known for molecules. For example, in the HF molecule all the electrons do not stay on the fluorine atom *all* the time, leaving the proton quite bare, but they spend some of their time moving round the proton in a kind of 'shared' orbit. The bonding is thus partly ionic and partly due to exchange forces. The principal characteristics of the type of bonding will then depend on whether the homopolar or ionic bonding predominates. A class of semiconductors having predominantly valence bonding but also having clearly an ionic component is the so-called group III–V semiconductors, consisting of compounds made up from elements in the third and fifth columns of the periodic table. That valence bonding is predominant in most cases follows from their crystal structure, which is very similar to that for the group IV semiconductors, each atom having four nearest neighbours of the other type at the corners of a regular tetrahedron. This structure is generally known as the zincblende structure and is illustrated schematically in Fig. 3.13 for InSb. In order to make up the closed groups of eight electrons for each lattice site, it is clear that each column V element must give an electron to each column III element. For InSb, therefore, the lattice

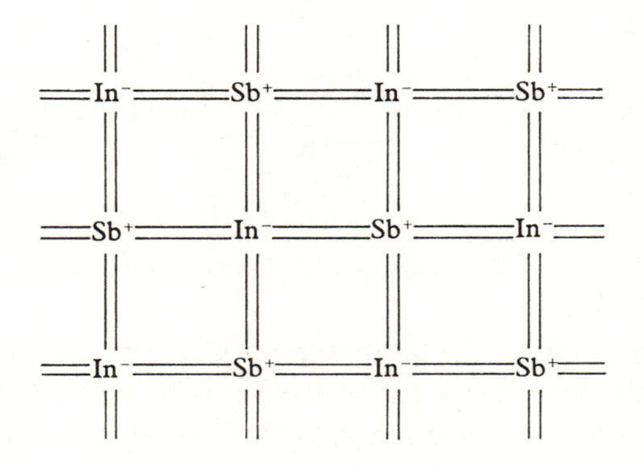

Fig. 3.13. Two-dimensional schematic representation of the structure of InSb.

sites may be regarded as occupied alternately by In^- and Sb^+ ions. In addition to the homopolar binding there is thus also ionic binding due to the electrostatic attraction between the ions.

An example of a class of semiconductor whose bonding is mainly ionic in character is provided by the lead salts PbS, PbSe, PbTe, which have been extensively studied. The crystal of PbS may, to a first approximation, be regarded as being made up of Pb^{2+} and S^{2-} ions. There is, however, a good deal of evidence that there is also a certain amount of homopolar binding in such crystals. Substances in which the bonding is predominantly homopolar generally form crystals with either the zincblende structure or with one of the related structures, while those with predominantly ionic bonding form crystals with one of the cubic structures, rather like the alkali halides. This is true for many group II–VI compounds but not for all. Semiconductors of this type are generally known as polar semiconductors.

There are two ways in which closed groups may be formed in a compound XY, if the atom X has two valence electrons and the atom Y has six. A polar compound of the form $X^{2+}Y^{2-}$ may be formed as described above. This we might call a 0–8 structure. Alternatively, a compound of the form $X^{2-}Y^{2+}$ may be formed giving a 4–4 structure as for the group III–V semiconductors, the bonding being now principally of the valence type but with an additional ionic component. Whereas the lead salts PbS, PbTe, PbSe are compounds of the former type, the corresponding beryllium salts BeS, BeSe, BeTe are of the latter type. As might be expected, they have crystal structures like the III–V compounds, namely, the zincblende structure. It is not easy to decide *a priori* in which way such a compound is likely to form. If the metallic atom is strongly electropositive the ionic structure is more likely.

There are several other types of bonding such as the metallic bond, the Van der Waals bond, etc., which play a part in the binding of solids. An account of these and of the methods of distinguishing between the various types of bond, in particular between ionic and homopolar bonds, has been given by R. C. Evans.[10]

3.3 Alternative approach to semiconductor designation

The bond picture of a semiconductor which we have introduced provides us with an alternative means of designating materials as semiconductors. That given in § 1.1, depending on the value of the forbidden

[10] *Crystal Chemistry* (Cambridge University Press, 2nd edition, 1964).

energy gap, applies only to materials which have a precise energy band structure as described in the previous chapter. For amorphous materials, on the other hand, although a form of energy band structure may be found, it is not nearly so precise as for crystalline materials (see § 15.2). The concept of breaking a chemical bond, however, is not dependent on long-range order in the atomic arrangements. We may therefore designate a material a semiconductor if a small amount of energy ΔE (say $0<\Delta E<2$ eV) is required to break one of the chemical bonds and release an electron so that it can take part in conduction. It is, of course, postulated that without such bond breaking no electrons are available, the material being such that closed shells operate in the bonding system. This says nothing about how effectively an electron released thermally or optically from a bond will take part in electrical conduction. If it is very immobile the material will behave like an insulator. Nevertheless the value of ΔE would, in principle, be obtained by measuring the variation of absorption of infra-red radiation or light with wavelength, and the semiconducting property established in this way. This therefore seems to give a rather more fundamental definition of a semiconductor.

A full discussion of chemical bonds in semiconductors has been given by E. Mooser and W. B. Pearson.[11] The relationship between the bond approach and the energy band approach has been treated in a very illuminating way by J. C. Phillips.[12]

3.4 Substitutional impurities in semiconductors

One of the simplest examples of a substitutional impurity, and one of great technological importance, occurs for the group IV semiconductors when some of the group IV atoms are replaced by atoms of elements taken from the third or fifth columns of the periodic table. Let us consider the latter first of all, for example an As impurity in Ge. The As atom has five valence electrons. Four of these will combine with the four unpaired electrons from the four nearest Ge atoms to form double electron bonds as before. The fifth electron being *de trop*, so to speak, is only very lightly bound to the As atom. As we shall see, we may estimate the strength of this binding approximately. The binding energy is so small that at room temperature practically all As atoms in Ge lose the extra electron, which is then free to wander through the crystal lattice. The As atom is then said to be ionized. At very low tempera-

[11] *Progress in Semiconductors* (1960) **5**, 103.
[12] *Bonds and Bands in Semiconductors* (Academic Press, 1973).

tures, on the other hand, the electron is bound to the As atom. This is illustrated in Fig. 3.14. In terms of the band picture we may say that each As atom introduces a single energy level just below the bottom of

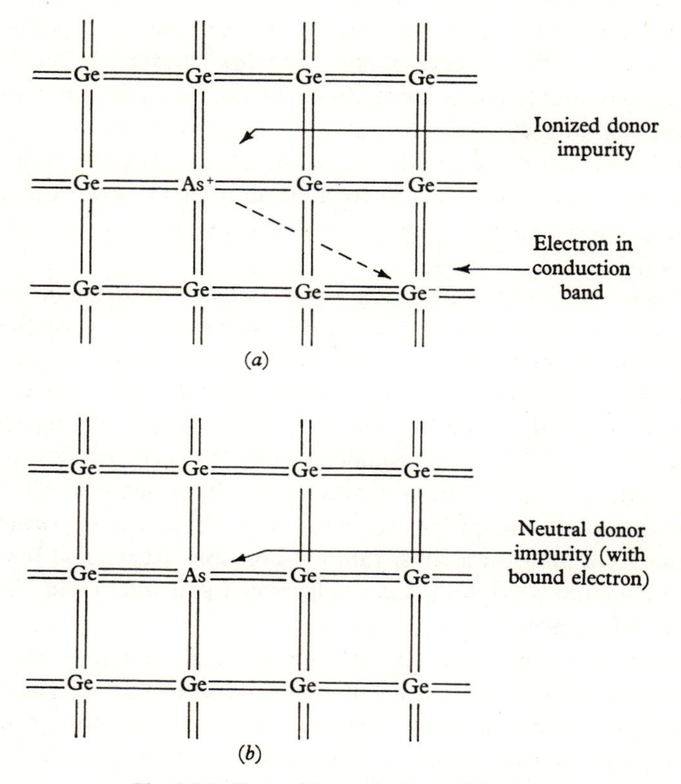

Fig. 3.14. Group V atom in Ge or Si lattice.

the conduction band. At low temperatures this level is occupied. At normal temperatures it is empty, there being for each empty level one free electron in the conduction band. Such an impurity is therefore called a *donor* impurity. This is illustrated in Fig. 3.15. The same applies

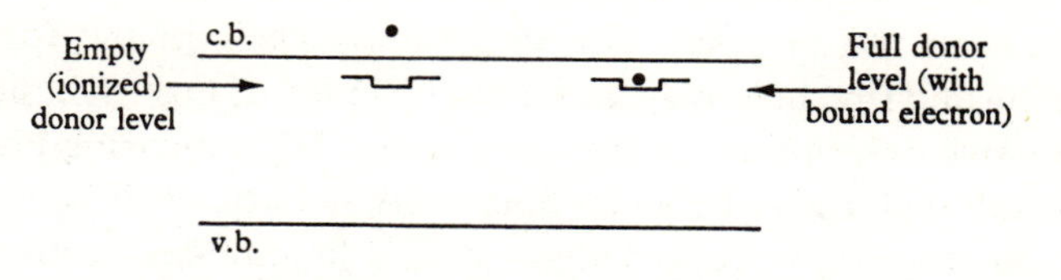

Fig. 3.15. Band picture of group V substitutional impurity.

to other group V atoms such as P and Sb. The evidence that the group V impurities go into the lattice substitutionally comes from various observations, such as their rate of diffusion, but the fact that they all behave as outlined above is also very strong evidence.

It is interesting to note that we may regard the positive ion As^+, left behind when the extra electron goes into the conduction band, as a positive hole bound to an As atom. Since the As^+ ion does not move this is an *immobile* hole. It is sometimes convenient to use this picture, since it fits in with the idea of the creation of an electron–hole pair. Frequently it is more convenient to think only of the electrons in the conduction band and any free holes which might be in the valence band.

Let us now consider what happens when an element such as Ga, taken from the third column of the periodic table, is substituted in the Si or Ge lattice. Group III atoms have only three valence electrons so, in order to make up the four double electron bonds, an extra electron must be taken from one of the group IV atoms. This creates a positive hole as is illustrated in Fig. 3.16. A very small amount of energy (which we shall later estimate) is required to take an electron from one of the Ge=Ge bonds to put it into one of the Ga=Ge bonds. At low temperatures we have thus a Ga *atom* in a substitutional position. At normal temperatures the Ga atom will have an extra electron, i.e. it will be Ga^-, and we shall have a free positive hole.

Two alternative descriptions may be given in terms of the band picture. At very low temperatures the impurity level may be regarded as an empty *electron* level, just above the valence band, capable of accepting an electron. As the temperature is raised, electrons from the valence band will be excited so as to occupy these levels, which will normally be filled at room temperature; such levels are therefore called acceptor levels. This is illustrated in Fig. 3.17. Alternatively, we might say that at low temperatures a hole is bound to the Ga^- ion and is released into the valence band at higher temperatures. This is illustrated in Fig. 3.18. Impurity levels, represented as hole levels, are indicated thus $\_\sqcap\_$ to remind us that hole energies increase *downward* and thus that holes tend to *rise* in the diagram. An acceptor which is in the form of a negative ion and free hole is again said to be ionized; Figs. 3.17 and 3.18 are alternative descriptions of the same states. By comparing Figs. 3.15 and 3.18 the reciprocity between holes and electrons will be seen. An ionized acceptor may be regarded as a bound (immobile) electron plus a free hole.

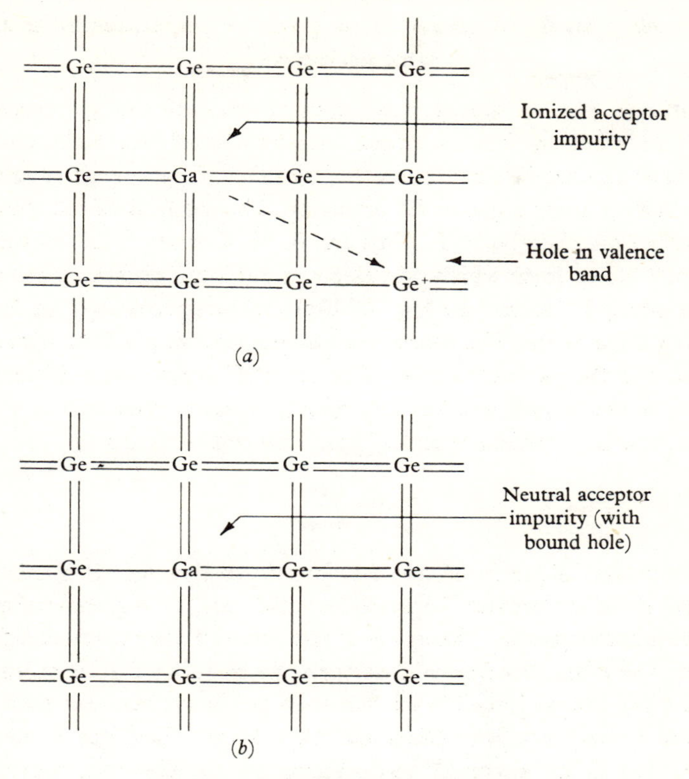

Fig. 3.16. Group III atom in Ge lattice.

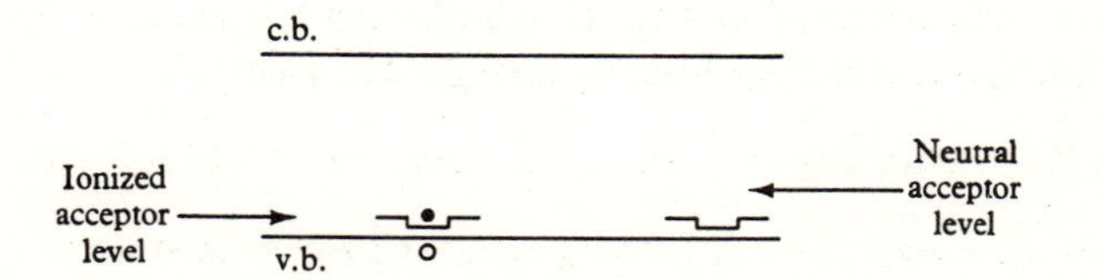

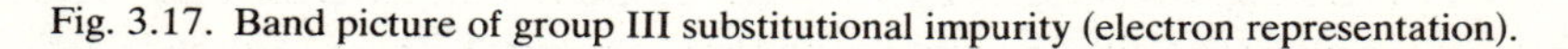

Fig. 3.17. Band picture of group III substitutional impurity (electron representation).

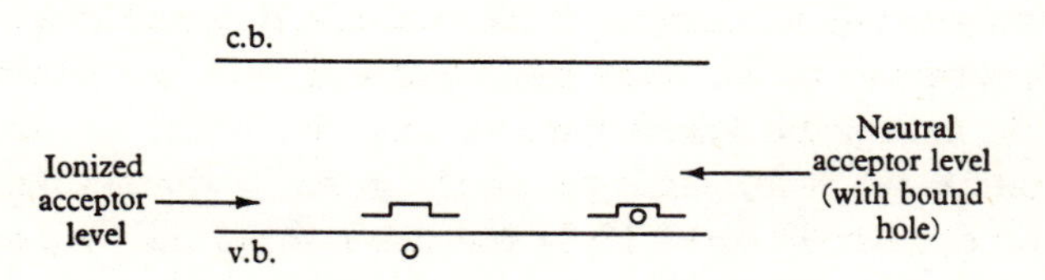

Fig. 3.18. Alternative form of Fig. 3.17 using *hole* levels.

3.4.1 Energy levels of group III or group V impurities in group IV semiconductors

We shall now give an approximate calculation of the energy required to release an electron from a donor impurity level due to a group V impurity in a group IV semiconductor and also of the energy required to release a hole from a group III impurity. The method of calculation is essentially that given by N. F. Mott and R. W. Gurney.[13] Let us consider a group V atom, from which one electron has been removed, as replacing one group IV atom (see Fig. 3.14(*a*)), to be specific, say an As^+ ion replacing a Ge atom. The electronic valence bonding will be almost the same as for the perfect crystal. The force F acting on a distant free electron in the crystal will be very nearly equal to that due to a single positive charge located at the As^+ ion. This will be given by

$$F = -\frac{e^2}{4\pi\epsilon r^2}, \tag{2}$$

where r is the distance of the electron from the As^+ ion and ϵ the permittivity of the crystal. This will be valid only if r is greater than two or three lattice spacings. When r is of the order of the nearest-neighbour distance the extra electron will perturb the bonding electrons appreciably and (2) will no longer hold. We shall see, however, that such small values of r have very little effect when ϵ is large. The force F may now be regarded as an 'external' force acting on the electron, regarded as moving in a perfect crystal, and the theory of § 2.5 applied. Thus we consider an electron of effective mass m_e moving in the field F. The problem is then reduced to that of a hydrogen atom, and we obtain a series of energy levels with binding energy W_n given by

$$W_n = m_e\epsilon_0^2 W_H/n^2\epsilon^2 \boldsymbol{m} = [(m_e/\boldsymbol{m})\epsilon_0^2/n^2\epsilon^2] \times 13.5 \text{ eV}, \tag{3}$$

where n is an integer and ϵ_0 is the permittivity of free space.

For Ge and Si this simple model requires modification to take account of the departure of the energy bands from spherical symmetry; we shall discuss this refinement later. If, however, we take for Ge an average scalar effective mass m_e equal to $0.2\boldsymbol{m}$ (which is required to account for other phenomena, as we shall see), and $\epsilon/\epsilon_0 = 16$ we obtain $W_n = 0.01/n^2$ eV, i.e. for the lowest state we have $W_1 = 0.01$ eV, and to this approximation W_1 is the same for all the group V donors. In practice there is some variation (see § 13.3), but this is small and the agreement

[13] *Electronic Processes in Ionic Crystals* (Oxford University Press, 1940), p. 83.

with the simple model is surprisingly good. For group V donors in Si, taking $m_e = 0.4\boldsymbol{m}$, we have $W_n = 0.04/n^2$ eV, the experimental values lying between 0.045 eV for *B* and 0.065 eV for Ga (with the exception of In for which $W_1 = 0.16$ eV (see § 13.3)). The quantity W_1 gives the depth of the impurity level below the conduction band, and this is seen to be small compared with the width of the forbidden energy band both for Si and Ge.

An exactly similar argument applies to a free hole moving in the field of a group III acceptor, e.g. Ga in Ge. The Ga^- ion produces at a distance a Coulomb field in which the hole moves. The formula for the impurity energy levels is the same as that given by equation (3) with m_e replaced by m_h the effective mass of the hole. The situation for Ge and Si, for which the best data regarding such impurity levels are available, is complicated further in this case by the degeneracy of the valence band. We defer till later a discussion of the effect of band structure on impurity levels and simply note now that, for Ge, W_1 is again about the same in magnitude as for donors. For Si we have W_1 approximately equal to 0.05 eV. Here W_1 represents the height of the acceptor level above the valence band, and again we see that for group III acceptor impurities these generally lie quite close to the valence band.

It is interesting to estimate the extent of the spread of the wave-functions of electrons or holes bound to the impurity centres. A measure of this is the 'radius of the first Bohr orbit' a_n, corresponding to the hydrogen-like wave-functions. This is given by

$$a_n = (\boldsymbol{m}/m_e)\epsilon a_0 n^2/\epsilon_0, \tag{4}$$

where a_0 is the 'radius of the first Bohr orbit' for hydrogen (equal to 0.53×10^{-8} cm). For group V donors in Ge, $a_1 = 80a_0$, so that we see that the wave-function is spread over many lattice spacings. In this case the contribution from the field at values of r of the order of a_0 may be shown to be small. This is the justification of the use of a Coulomb field in the approximate calculation of W_n. For group V donors in Si, $a_1 = 30a_0$ and although the wave-function covers many lattice sites the agreement with the simple theory may be expected to be not quite so good as for Ge. For excited states the agreement should be better because of the factor n^2. This turns out to be so (see § 13.3).

Some modification to the simple treatment based on the hydrogen model is required for a semiconductor with several equivalent minima in the conduction band. Moreover, as we have seen, the approximations we have used break down if a_n is not large compared with the lattice spacing. This will be so for impurities whose ground states lie further from the conduction or valence bands. In this case a rather more sophisticated type of calculation is required. We defer discussion of this to § 11.5.1.

3.4.2 Energy levels of other impurities in group IV semiconductors

For impurities with more than one excess or defect electron no *simple* means of estimating impurity levels is available. Experimentally it is found that such donor impurities generally have levels which lie farther from the conduction band, and that acceptor impurities have levels which are farther from the valence band. Several impurities of this type give rise to more than one level according to whether they are occupied by more than a single electron. For example, Au in Ge may give rise to an acceptor level 0.15 eV above the valence band and also to a donor level 0.20 eV below the conduction band. Such impurities are usually called 'amphoteric'. A few impurities apparently give rise to no easily observable impurity levels lying between the bands. These may occur as interstitial atoms which have an ionization energy in the crystal comparable with or even greater than the forbidden energy gap and do not accept an extra electron to form a stable negative ion in the crystal. Hydrogen in Ge and Si appears to be such an impurity. Another example of such behaviour is Si in Ge or Ge in Si in very small quantities. This is not surprising in view of the very similar electronic structure of the outer shells of the two elements.

Vacant lattice sites appear to play very little part in giving rise to free electrons or holes in Si and Ge. They tend to form traps with energy levels well removed from the valence or conduction bands. There is a tendency for vacancies to collect at dislocations and they undoubtedly play an important part in determining the diffusion rate of impurities in these semiconductors.

3.4.3 Impurities in polar semiconductors

One of the main differences between compound semiconductors and the element semiconductors is due to the fact that they contain, in their chemically pure state, more than one kind of atom. Deviations from perfect stoichiometric composition may take place and give rise to imperfections in the crystal lattice. For example, in a perfect crystal of PbS there is exactly the same number of Pb atoms as S atoms. It is well known, however, that PbS may exist as a homogeneous crystalline phase with composition $Pb_{1+\delta}S$, δ being a fraction having a value up to about 10^{-3}. It turns out that each excess Pb atom acts as a donor impurity giving an energy level very close to the conduction band. A value of δ of 10^{-3} would correspond to over 10^{19} donors per cm^3 – a very large number. Thus the problem of purification of compound semiconductors is far more difficult than for the element semiconductors, in that not only must foreign atoms be removed but the compound must be very nearly stoichiometric in composition.

In a diatomic polar semiconductor of stoichiometric composition vacant anion and cation lattice sites must exist in equal numbers. This is simply a consequence of the condition that the crystal should be electrically neutral. The vacant anion and cation sites we shall denote by V_A and V_C. When, however, there is an excess of one constituent taken up interstitially, this equality no longer holds. If there is an excess of cations in substitutional positions there clearly must be an equal excess of anion vacancies (see Fig. 3.19). Moreover, a large excess of cations would fill

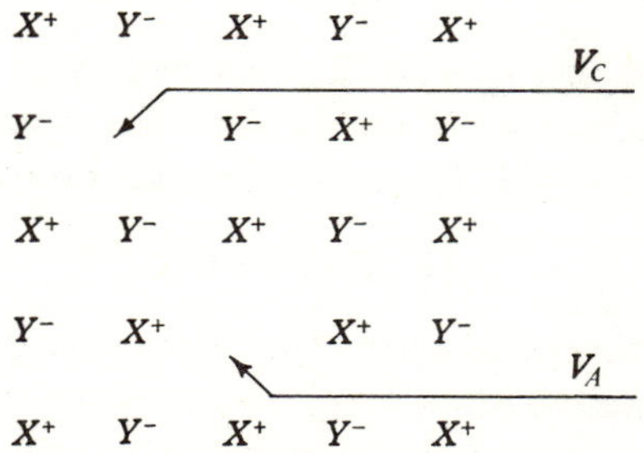

X^+ Y^- X^+ Y^- X^+
V_C
Y^- Y^- X^+ Y^-
X^+ Y^- X^+ Y^- X^+
Y^- X^+ X^+ Y^-
V_A
X^+ Y^- X^+ Y^- X^+

(*a*) Pair of vacancies V_A, V_C

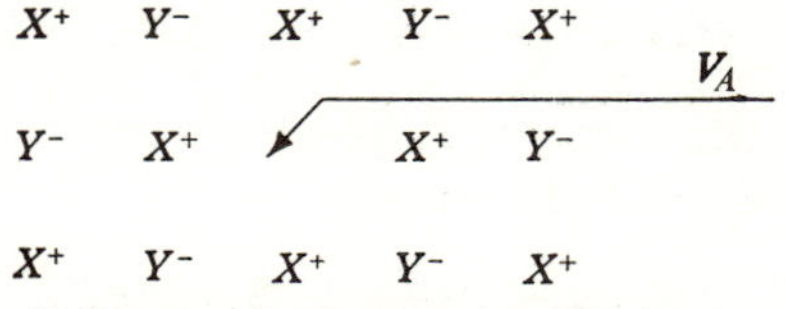

X^+ Y^- X^+ Y^- X^+
V_A
Y^- X^+ X^+ Y^-
X^+ Y^- X^+ Y^- X^+

(*b*) Vacancy V_A due to excess metal atom

Fig. 3.19. Anion and cation vacancies in polar crystals.

up most of the cation vacancies, so that if we have many anion vacancies we expect to have few cation vacancies. In thermodynamic equilibrium there is a 'mass-action' type of law relating the number of anion vacancies per unit volume $[V_A]$ to the number of cation vacancies $[V_C]$. It is of the form

$$[V_A][V_C]=K(T), \tag{5}$$

where $K(T)$ is a function of the temperature. This relationship follows from the fact that a cation vacancy may combine with an anion vacancy at the surface of the crystal and so annihilate each other. This, however, neglects the formation of double vacancies; these will indeed be formed, but the energy of formation is generally assumed to be large, so that the number formed will be small.[14] Equation (5) follows from the observation that the annihilation of an anion and cation vacancy may be written as a chemical reaction of the form

$$V_A+V_C=S+\Delta W, \tag{6}$$

where ΔW refers to the energy released and S represents a pair of empty surface sites. The chemical mass-action equation[15] for this reaction will then be

$$\frac{[V_A][V_C]}{[S]}=f(T). \tag{7}$$

Since the variation of $[S]$ is small, provided $[V_A]$ and $[V_C]$ are small compared with the number of atoms per unit volume, equation (7) may be written in the form (5). This may be also expressed in the form

$$[V_A][V_C]=V_s^2, \tag{7a}$$

where V_s is the number of each type of vacancy per unit volume in a stoichiometric crystal. Thus if $[V_A]\gg V_s$ then we must have $[V_C]\ll V_s$, and vice versa. The form of the function $K(T)$ may generally be written as[15]

$$K(T)=[F(T)]^2\exp[-\Delta W/\boldsymbol{k}T], \tag{8}$$

where $F(T)$ is a function of T which varies slowly compared with $\exp[-\Delta W/2\boldsymbol{k}T]$. We therefore have

$$V_s=(F(T)\exp[-\Delta W/2\boldsymbol{k}T]. \tag{9}$$

[14] A full discussion of the method of treating vacancies by means of the formulae for chemical equilibrium is given by A. L. G. Rees, *Chemistry of the Defect Solid State* (Methuen Monographs, 1954). See also F. A. Kröger and H. J. Vink, *Solid State Physics* (Academic Press, 1956) **3**, 307; H. J. Vink, *Rendiconti della Scuola Enrico Fermi, XXII Corso* (Academic Press, 1963), p. 68.

[15] This type of equation is derived in most text-books on Thermodynamics. See, for example, R. A. Smith, *Physical Principles of Thermodynamics* (Chapman and Hall, 1952), p. 214.

In crystals for which ΔW is large we thus have a small number of vacancies; for example, this appears to be so for InSb (see § 13.5). For the lead salts PbS, PbSe, PbTe, however, vacancies appear to be formed readily (see § 13.6).

We must now consider in more detail the effect of having an excess of one of the constituents. Consider first a diatomic ionic crystal having ions of unit valency. Such crystals are usually ionic conductors such as NaCl, which we may take as typical. The excess Na atoms are thought to occupy normal lattice sites. This they must do as Na^+ ions, and so for each excess Na atom we have one electron to dispose of. For each excess atom we shall have created a vacant site which would normally be occupied by a Cl^- ion. Such a site has an effective *positive* charge of one electron unit with respect to a perfect crystal and will therefore attract an electron. The electrons released by the Na atoms are then trapped by the vacant anion sites forming a unit of charge $-\boldsymbol{e}$, i.e. a unit neutral with respect to the perfect lattice; such a site is denoted by V_A^- and is illustrated for NaCl in Fig. 3.20. In the alkali halides, the binding energy

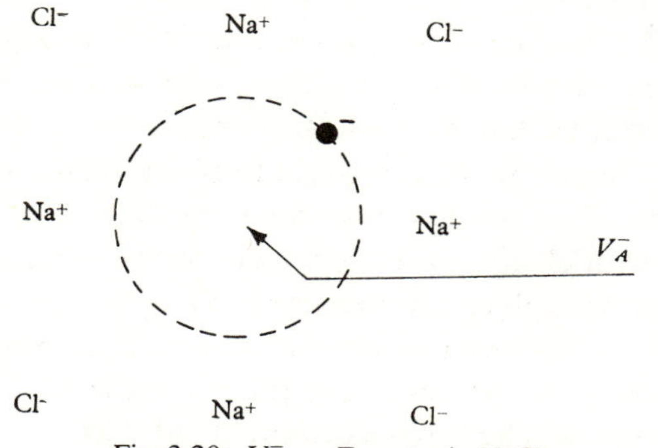

Fig. 3.20. V_A^- or F-centre in NaCl.

of the electron is of the order of half an electron volt and these charged vacancies, known as F-centres, give rise to the coloration of alkali halide crystals by excess metal. In this case the F-centres are capable of capturing a second electron to form a centre denoted by V_A^{-2} and called an F'-centre. If we have an excess of anions, e.g. Cl^-, then each halogen atom must release a positive hole which will be trapped at a vacant cation site forming a V_C^+ centre.

For semiconductors, we are frequently concerned with divalent ions, and the binding energy of the electron trapped at an anion vacancy is so small that it is released into the conduction band, except at very low temperatures. For example, in PbS each excess lead atom is known, at ordinary temperatures, to give rise to one electron in the conduction band. However, if we assume that the crystal of PbS is made up of Pb^{2+} and S^{2-} ions then *two* electrons are freed from each excess Pb^{2+} ion site. The vacancy V_A created is now doubly charged with respect to a perfect crystal. It may trap an electron into a deep level to form V_A^- which is singly charged with respect to a perfect crystal. This will then trap a second electron to form V_A^{2-}. At ordinary temperatures the centre V_A^{2-} breaks into V_A^- and one free electron. In the same way we may have the centres V_C^{2+} and V_C^+ which are of importance when we have an excess of anions, for example with an excess of S in PbS. In this case each excess S atom gives rise to a free positive hole. This is not always so. In CdS, excess Cd gives free electrons at room temperature, but excess S does not give free holes. This means that the acceptor level corresponding to V_C^{2+} is too far from the valence band to be ionized at room temperature.

Ions of different valency also give rise to donor and acceptor levels in polar semiconductors; for example, Bi is thought to enter PbS substitutionally as Bi^{3+}. At room temperature each Pb^{2+} ion replaced will give rise to one free electron which is easily detached from the Bi^{2+} ion. A centre (Bi^{3+}) will be left with effective charge $+\boldsymbol{e}$ with respect to the perfect crystal. Similarly, Ag is thought to go into PbS substitutionally as Ag^+, giving one free positive hole for each Pb^{2+} ion replaced, and a centre (Ag^+) with charge $-\boldsymbol{e}$ with respect to the perfect lattice. Again the cation may be replaced; for example, Cl is thought to enter PbS as Cl^-, each S^{2-} ion replaced giving rise to a free electron and a centre (Cl^-) with charge $+\boldsymbol{e}$ with respect to the perfect crystal.[16] The effect of such substitutions will be further discussed in § 13.6.

3.4.4 Impurities in the group III–V semiconductors

Similar considerations apply to compounds such as the group III–V semiconductors which have mainly valence binding but also ionic character (see § 13.5). Replacement of a group V atom by one of higher valency, e.g. substitution of Sb by Te in InSb, gives rise to donor levels near the conduction band, the Sb^+ ion presumably being replaced substitutionally by a Te^{2+} ion (Te has six valence electrons) and a free

[16] See p. 68, Kröger and Vink, *loc. cit.*

electron. Again Zn in InSb gives positive holes. Here the In^- ion is replaced by Zn^{2-} (Zn has two valence electrons) and a positive hole. If we substitute a group IV element in a group III–V semiconductor it is not clear which of the ions should be replaced. Ge in InSb seems to form shallow acceptor levels so presumably replaces one Sb^+ ion plus one electron. In many of the group III–V compounds the energy of formation of lattice vacancies appears to be high, as there is a strong tendency to form very nearly stoichiometric compounds. This is particularly so for InSb (see § 13.5) and is fortunate since it makes the preparation of pure single crystals very much easier.

3.4.5 Other types of imperfection

There are many other types of imperfection which can occur in crystalline solids. We have mentioned only a few of the more common ones which are of particular importance in determining the properties of semiconductors. All of the imperfections discussed so far may be inserted by chemical means. For example, vacancies can be created by introducing atoms of different valency in polar crystals as we have seen in § 3.4.3. The introduction of large numbers of vacancies and interstitials can also be caused by particular radiation – either by heavy particles or electrons or even by γ-rays through the energetic electrons which they produce. When the numbers are undesirably large this is known as radiation damage and can wreck a semiconductor device.

In smaller doses, however, radiation has proved a most useful tool in the study of lattice imperfections. Not only may single vacancies be produced, but di-vacancies and even tiny cavities consisting of a large number of vacancies. Under such conditions the vacancies tend to group and form dislocations, the crystal being locally under considerable strain (see § 3.1.3). This also can happen as a result of the introduction of large numbers (say $>10^{19}$ cm^{-3}) of impurities into the crystal.

Heavy particle radiation may also be used to introduce into a crystal chemical impurities in the form of ions which have been accelerated to high energies and so penetrate some distance below the surface of the crystal on which they are incident. Under these conditions the crystal usually receives considerable radiation damage but a good deal of this may be removed by careful annealing, leaving the desired impurities in substitutional or interstitial sites. The depth of penetration is usually fairly small but sufficient for device development and this technique is coming into extensive use for 'doping' of semiconductors. The subject has been treated in detail in a number of articles and books.[17]

We shall later discuss how the various kinds of imperfections affect the optical and transport properties of semiconductors. We shall also show how the energy levels of deeper and more complex impurities and imperfections may be measured and also how they may be treated theoretically (see § 11.5).

A very extensive literature[18] has grown up dealing with imperfections caused by radiation damage. This is not surprising since these markedly affect the performance of semiconductor devices.

3.5 Excitons

In our previous discussion of the motion of electrons in a perfect crystal we neglected the mutual repulsion of the electrons. Advanced theoretical treatments have been given to justify this apparently rather drastic neglect.[19] Similarly, we have neglected the mutual repulsion of the positive holes and apparently also the attraction between electrons and positive holes. For the latter this is not entirely true, as a considerable part of the energy of formation of an electron–hole pair consists of separating the pair against their Coulomb attraction. The mutual interaction of electrons gives rise to scattering and may be of importance when we have a high density of current carriers.[20] The attraction between a hole and an electron leads, however, to very interesting effects when the free carrier concentration is not too high.

If we apply the idea that a free hole and electron move in the crystal under an electric field as though they were particles, each with its appropriate effective mass, we see that a hole and electron could form a system with energy levels, and could move as a unit through the crystal lattice. Such a system is called an exciton.[21] It corresponds, in a sense, to

[17] See, for example, J. W. Mayer, L. Eriksson and J. A. Davies, *Ion Implantation in Semiconductors* (Academic Press, 1970); G. Dearnaley, J. H. Freeman, R. S. Nelson and J. Stephen, *Ion Implantation* (North Holland, 1973); F. Bassani, G. Iadonisi and B. Prezcosi, *Rept. Prog. Phys.* (1974) **37**, 1099; *Lattice Defects in Semiconductors*, ed. F. A. Huntley (Inst. Phys. Conf. Series, 1974); *Point Defects in Solids*, ed. J. H. Crawford and L. M. Slifkin (Plenum Press, 1975), vol. 2; A. M. Stoneham, *Theory of Defects in Solids* (Oxford University Press, 1975).

[18] See, for example, *Radiation Effects in Semiconductors*, ed. J. W. Corbett and G. D. Watkins (Gordon and Breach, 1971); *Radiation Damage and Defects in Semiconductors*, ed. J. E. Whitehouse (Inst. Phys. Conf. Series, 1973).

[19] See, for example, article by D. Pines, in *Solid State Physics* (Academic Press, 1955) **1**.

[20] H. Fröhlich and S. Doniach, *Proc. Phys. Soc.* B (1956) **69**, 961.

[21] J. Frenkel, *Phys. Rev.* (1931) **37**, 17, 1276.

an excited state of an atom of the crystal being passed on to neighbouring atoms by quantum mechanical resonance.[22]

If we assume, for simplicity, that we may use scalar effective masses m_e, m_h for the electron and hole, we again have the simple 'hydrogen' problem to solve for two particles moving under their Coulomb attraction. If we neglect for the moment the motion of the centre of mass of the particles we see that the energy of the system is W_{ex}^n, referred to the condition when both particles are at infinite separation, where

$$W_{ex}^n = -(m_r/\boldsymbol{m})\epsilon_0^2 W_H/\epsilon^2 n^2, \tag{10}$$

and m_r is the 'reduced' mass of the two particles $m_e m_h/(m_e + m_h)$. W_H, n and ϵ have the same meaning as in equation (3) with which equation (10) should be compared. If we take $m_e = m_h$ then $m_r = \frac{1}{2}m_e$ and $W_{ex}^n = -\frac{1}{2}W_n$. The binding energy of an exciton in its lowest state in a group IV semiconductor is thus of the same order of magnitude as the depth of a group V donor below the conduction band. For Ge we should expect the exciton binding energy to be about 0.005 eV. This is very nearly equal to the observed value 0.0036 eV (see § 10.6).

Let us now consider rather more carefully the meaning of the energy W_{ex}. If we consider the energy level diagram for a *single* electron, as in Fig. 3.15, the lowest exciton level would be at depth W_{ex}^1 *below* the conduction band, and the excited states at depths $W_{ex}^1/4$, $W_{ex}^1/9$, etc. We have in fact an infinite number of levels leading to a continuum as $n \to \infty$; the continuum corresponds to an unbound hole–electron pair and so to the conduction band. The above considerations appear to contradict the predictions already made on the basis of Bloch's theorem (see § 2.2) that there are *no* allowed states between the valence and conduction bands for a *perfect* crystal. The reason is, of course, that in the treatment based on Bloch's theorem we have dealt only with *single* electrons (or holes) and the introduction of the exciton represents a higher degree of approximation.

Let us consider what is the first excited state of a perfect crystal, which has a finite gap between a full valence band and conduction band. The lowest energy state of the crystal, the 'ground' state, so far as the electronic energy is concerned, occurs when all the electrons occupy the valence band, and completely fill it, and there are no electrons in the conduction band. If we neglect excitons, the first excited state would appear to be when we have one electron in the lowest allowed state of

[22] G. Wannier, *Phys. Rev.* (1937) **52**, 191.

the conduction band and one electron missing from the highest allowed state of the valence band, i.e. one free electron and one free hole with effectively zero kinetic energy. Clearly, however, this condition is unstable, so long as the electron remains in the conduction band, and lower energy states exist in which the electron and hole come together; these are just the exciton states. The lowest state of all, apart from the 'ground' state, corresponds to the exciton being in its lowest energy state. The first excited state of the crystal therefore corresponds to having a single exciton in its lowest state and lies an amount $\Delta E - |W_{ex}^1|$ above the 'ground' state.

It is not strictly correct to show the exciton energy on a single-electron diagram, since both an electron and hole are involved, but this is sometimes done, as shown in Fig. 3.21. Moreover, we have, so far,

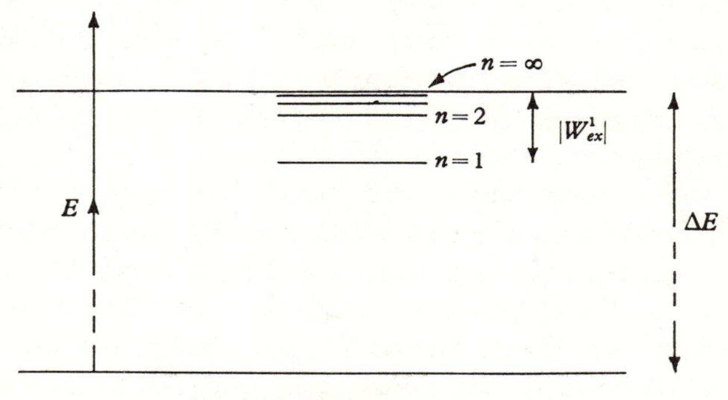

Fig. 3.21. Exciton energy levels.

neglected the motion of the centre of mass of the exciton. This may be represented by a wave-function of the form

$$\psi(\mathbf{R}) = \exp\,[i(\mathbf{K}\cdot\mathbf{R})], \tag{11}$$

where $\mathbf{R}$ is the position vector of the centre of mass, and $\mathbf{K}$ is a vector related to the momentum associated with the motion of the centre of mass in the same way as $\mathbf{k}$ is to the crystal momentum of an electron. For small values of $\mathbf{K}$ the kinetic energy of the exciton W_{ex}^0 is given by

$$W_{ex}^0 = \hbar^2|\mathbf{K}|^2/2M, \tag{12}$$

where $M = m_e + m_h$. The exciton energy levels are therefore each spread out into energy *bands* when we take into account the motion of the centre of mass, and the exciton energy should be plotted as a function of $\mathbf{K}$ on a diagram similar to that for electrons. $\mathbf{K}$ will be limited as is $\mathbf{k}$ by the periodicity of the lattice. We should therefore strictly speak of exciton bands rather than of exciton levels. When excitons are formed, however, by the absorption of a single quantum of radiation, the conservation of momentum requires the exciton to be formed with very little momentum, since the momentum of the quantum is small. The exciton may therefore sometimes be regarded as having energy equal to one of the values of W_{ex}. This condition, as we shall see, can hold only when the top of the valence band and bottom of the conduction band correspond to the same crystal momentum, e.g. are both at $\mathbf{k}=0$. In the simple case the absorption spectrum of the solid will consist of a series of lines, the first corresponding to a frequency ν given by $h\nu = \Delta E - |W^1_{ex}|$ and tending to a limit at a frequency ν given $h\nu = \Delta E$ (see Fig. 3.22).

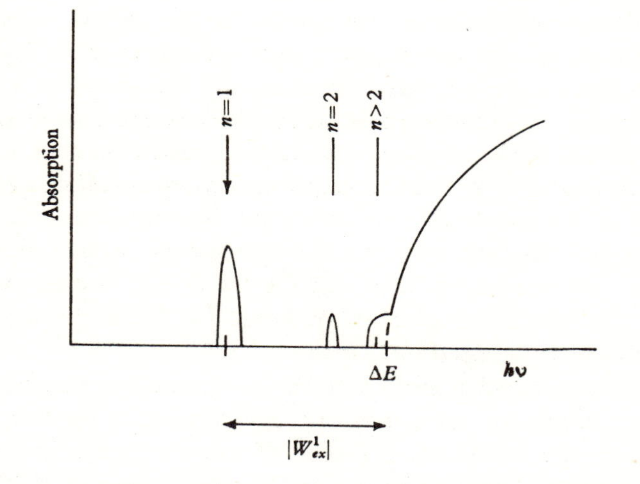

Fig. 3.22. Theoretical form of absorption spectrum of crystalline solid.

Above this limit there will be continuous absorption due to transitions involving the transfer of electrons from the valence to the conduction band. Due to vibration of the crystalline lattice and to structural imperfections, the exciton absorption lines will be somewhat broadened, and may merge into the continuum. This form of absorption spectrum was also predicted by R. Peierls[23] using a more general method to discuss the excited states of a solid. The main experimental evidence for exciton formation comes from the observation that absorption near the long-wave edge of the main absorption band (see Fig. 3.22) does not lead to

[23] *Ann. Phys.* (1932) **13**, 905.

the creation of free carriers at very low temperatures. At high temperatures the excitons may be thermally dissociated into free electrons and holes. Evidence also comes from the form of absorption spectra near the fundamental edge and from series of line spectra which have been interpreted as due to exciton formation. These we shall discuss in § 10.6. A review of experimental work up to 1955 (in English) has been given by E. F. Gross.[24]

Absorption line spectra in the form of 'hydrogenic' series have been observed in a number of compounds with fairly large energy gaps,[25] for example, CdS, HgI_2, PbI_2, CdI_2. Line spectra in Cu_2O have been observed by J. H. Apfel and L. N. Hadley,[26] by E. F. Gross and B. P. Zakharchenya,[27] and by S. Nikitine,[28] and have been interpreted as due to excitons. For some time it was thought that there would be a good chance of observing exciton line spectra in Ge and Si, as they are available in the form of very perfect and pure single crystals. The reason why such spectra are not easily observed only became clear when the band structure of these substances became known. Here we do not have the valence band maximum and conduction band minimum occurring for the same value of **k**, and the simple conditions postulated above do not hold. In this case it may be shown that absorption due to exciton formation may take the form of absorption bands. Such exciton bands have been observed both in Si and Ge by G. G. Macfarlane, T. P. McLean, J. E. Quarrington and V. Roberts.[29] A full discussion of their properties will be deferred to § 10.6.

It is perhaps somewhat illogical to have discussed excitons in a section devoted to impurities, since they are characteristic of the host crystal, and do not depend on the presence of imperfections. Some of the properties of the exciton levels are, however, similar to those of impurity levels and it is interesting to make a comparison of the two at this stage. In particular, both lead to energy levels in the 'forbidden' energy gap.

[24] *Nuovo Cim.* (Supplement) (1956) **3**, 672.
[25] See, for example, E. F. Gross and A. Kaplianski, *Zh. Tekh. Fiz.* (1955) **25**, 2061.
[26] *Phys. Rev.* (1955) **100**, 1689.
[27] *C. R. Acad. Sci. U.R.S.S.* (1953) **90**, 745.
[28] *J. Phys. Radium* (1956) **17**, 817.
[29] *Phys. Rev.* (1957) **108**, 1377; *ibid.* (1958) **111**, 1245

4

Carrier concentrations in thermal equilibrium

4.1 Distribution of electrons between the various energy levels

In the two previous chapters we have discussed the various energy levels which the electrons in a semiconducting crystal may occupy. For most purposes we may ignore the levels lying deeper than those of the valence band. For some problems, for example that of absorption of X-rays, they have to be taken into account, but for all the problems we shall discuss these lower levels may be regarded as filled by electrons, and may be ignored. Indeed, it is only the upper levels in the valence band, the lower levels in the conduction band, and any intermediate levels due to impurities, which are of importance. Our problem is to consider all the electrons that would normally occupy the valence band and any impurity levels in the condition of lowest energy, which corresponds to the absolute zero of temperature, and to see how they are distributed between the various levels for any value of the absolute temperature T. To do this it is necessary to use a fundamental equation of statistical mechanics which gives an expression for the probability that any particular level will be occupied by an electron. Let the energy associated with a particular level, measured from an arbitrary but fixed datum, be E. Then the probability $P_e(E)$ that the level will be occupied by an electron is given by

$$P_e(E) = \frac{1}{\exp[(E-E_F)/\boldsymbol{k}T]+1}. \tag{1}$$

E_F is a constant which is determined so that the total expectation number of electrons is equal to the actual number of electrons involved. The function in equation (1) gives the so-called Fermi–Dirac distribution, and is valid only for particles such as electrons, which obey the

Pauli Principle.[1] Equation (1) may be written in the form

$$P_e(E) = f[(E - E_F)/\boldsymbol{k}T], \tag{1a}$$

where
$$f(x) = 1/(e^x + 1). \tag{2}$$

The function $f(x)$ is called the Fermi–Dirac function and is shown in Fig. 4.1 as a function of x. In the derivation of equation (1) it is assumed that each level is non-degenerate. If a level has degeneracy g, and the occupation of one of the degenerate levels by an electron does not affect the occupancy of the others, then we have

$$P_e(E) = gf[(E - E_F)/\boldsymbol{k}T]. \tag{3}$$

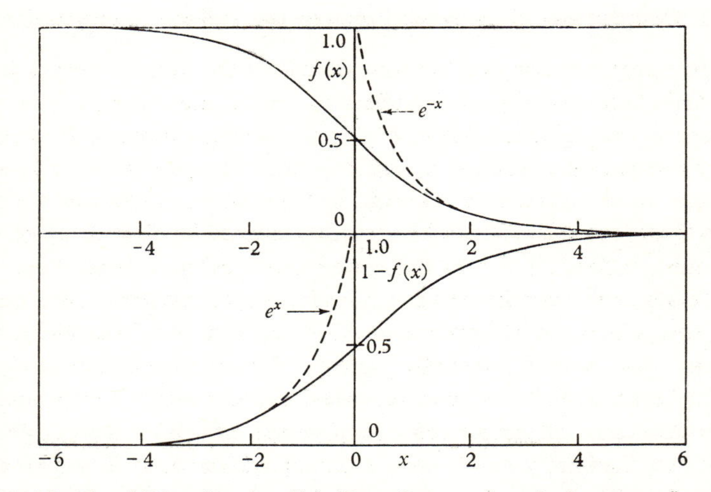

Fig. 4.1. The Fermi–Dirac function $f(x)$. (Dotted curves show the functions e^x and e^{-x}.)

This follows at once if we consider the group of levels as g separate levels very close together. The allowed energy levels for free electrons in a perfect crystal are of this type and each may be regarded as doubly degenerate, corresponding to the two allowed values of the spin. Occupancy of such a level by one electron of one spin does not prevent its occupation by an electron of the opposite spin, but does prevent its occupancy by an electron of the same spin. When we come to consider impurity levels, however, we shall have to consider degenerate energy

[1] The Fermi–Dirac function is derived in all standard text-books on quantum statistical mechanics. For a short and elegant account see E. Schrödinger, *Statistical Thermodynamics* (Cambridge University Press, 1946), ch. 7.

levels for which occupancy by one electron rules out the occupancy by a second electron altogether.

The function $f(x)$ will be seen to have the following properties: if x is large and positive, $f(x)$ is small, and the probability of occupation of a level is therefore small if $(E-E_F) \gg kT$. Again, if x is large and negative, $f(x) \simeq 1$ and so a level is almost certainly occupied if $(E-E_F) \ll kT$. We also note that $f(0)=\frac{1}{2}$ so that, if $E=E_F$, $P_e(E)=\frac{1}{2}$. If there happens to be an allowed level with $E=E_F$ it is equally likely to be occupied or empty. In any case, all levels with $E>E_F$ are more likely to be empty than occupied and all levels with $E<E_F$ are more likely to be occupied than empty. The energy corresponding to E_F is called the Fermi level, and this property may be used as a definition of E_F. Actually E_F is related to the thermodynamic potential and so is constant for a system made up of different 'phases', e.g. two different semiconductors in contact (see ref. on p. 78, note 1). For large positive values of x the function $f(x)$ is approximately equal to e^{-x}, so we have, when $(E-E_F) \gg kT$,

$$P_e(E)=\exp\,[-(E-E_F)/kT]=A\exp\,[-E/kT], \tag{4}$$

where A is a normalizing constant. Equation (4) gives just the value of $P(E)$ given by classical statistical mechanics in which the Pauli exclusion principle is not used. We thus see that the Fermi–Dirac distribution tends to the classical one for values of E such that $E-E_F$ is somewhat greater than kT. Under this condition the system is said to be non-degenerate.

The probability $P_h(E)$ that a level is *not* occupied by an electron is clearly given by

$$P_h(E)=1-P_e(E)=\frac{1}{\exp\,[(E_F-E)/kT]+1}. \tag{5}$$

The function $P_h(E)$ gives the probability of occupation of a level by a positive hole, and once again we see the similarity between positive holes and electrons if we measure energy in opposite directions for electrons and holes. When $(E-E_F) \ll kT$ we have $P_h(E) \simeq \exp\,[(E-E_F)/kT]$.

4.2 Intrinsic semiconductors

Let us now apply the above equations to determine the electron distribution in an intrinsic semiconductor having no impurity levels in the forbidden energy gap, whose width is ΔE. Let us take the zero of energy as the energy of the lowest allowed level in the conduction band. Let $N_c(E)\,dE$ be the number of allowed levels, for unit volume, in the conduction band, corresponding to values of the energy lying between E and $E+dE$. Similarly, $N_v(E)\,dE$ is defined for the valence band. Then the number of electrons $n(E)\,dE$ in the conduction band with energies between E and $E+dE$ is given by

$$n(E)\,dE = 2N_c(E)P_e(E)\,dE. \tag{6}$$

The factor 2 is included to take account of the spin degeneracy of each level, i.e. two electrons (of opposite spin) may occupy each. The total number n_i of electrons per unit volume in the conduction band is given by

$$n_i = 2\int_0^{E_t} N_c(E)P_e(E)\,dE, \tag{7}$$

where E_t is the energy corresponding to the top of the conduction band. Similarly, the number of holes $p(E)\,dE$ in the valence band occupying *electronic energy* levels with energy between E and $E+dE$ is given by

$$p(E)\,dE = 2N_v(E)P_h(E)\,dE. \tag{8}$$

The total number of holes per unit volume in the valence band, p_i, is then given by

$$p_i = 2\int_{E_b}^{-\Delta E} N_v(E)P_h(E)\,dE, \tag{9}$$

where E_b is the energy corresponding to the bottom of the valence band. For an intrinsic semiconductor, the total number of electrons in the conduction band must be equal to the total number of holes in the valence band, since they are created and destroyed in pairs, so that we have $n_i = p_i$. This equation determines the Fermi energy E_F and hence the probability functions $P_e(E)$, $P_h(E)$.

To proceed further, we require to know the form of the functions $N_c(E)$, $N_v(E)$. Let us first consider the simplest form of semiconductor in which we have a single minimum with spherical symmetry in the conduction band, and hence a scalar effective electron mass m_e, and also a single maximum with spherical symmetry in the valence band with effective hole mass m_h, as discussed in § 2.3. It may be shown quite generally by a consideration of the boundary conditions imposed on the wave-vector $\mathbf{k}$, and irrespective of the form of $E(\mathbf{k})$, that the number of allowed states for a crystal of volume $\mathbf{V}$ in the region of $\mathbf{k}$-space $d\mathbf{k}$ is given by

$$N(\mathbf{k})\,d\mathbf{k} = \mathbf{V}\,d\mathbf{k}/8\pi^3. \tag{10}$$

For electrons in free space having momentum $\mathbf{p} = \hbar\mathbf{k}$ this corresponds to one level per volume $\boldsymbol{h}^3$ of momentum space and unit volume of co-ordinate space, a rather interesting result.[2] The number of levels having the *magnitude* k of the wave-vector between k and $k + dk$ is then given by

$$N(k)\,dk = \mathbf{V}k^2\,dk/2\pi^2. \tag{11}$$

For a semiconductor with scalar effective mass m_e in the conduction band, since $E = \hbar^2 k^2/2m_e$ we may derive at once the number of levels for unit volume between E and $(E + dE)$ (having *one* value of spin)

$$N_c(E)\,dE = 2\pi(2m_e)^{\frac{3}{2}}\boldsymbol{h}^{-3}E^{\frac{1}{2}}\,dE. \tag{12a}$$

Similarly for a valence band with scalar effective mass m_h we have

$$N_v(E)\,dE = 2\pi(2m_h)^{\frac{3}{2}}\boldsymbol{h}^{-3}(-\Delta E - E)^{\frac{1}{2}}\,dE. \tag{12b}$$

It is clear from physical considerations that, provided $\Delta E/\boldsymbol{k}T$ is large, there will be only a small probability of occupation of levels in the conduction and valence bands by electrons and holes, respectively. We shall therefore assume a condition of non-degeneracy in both the valence band and conduction band in order to determine E_F, and shall then examine the conditions under which this assumption is valid. Inserting the value of $N_c(E)$ from equation (12a) in equation (7) we have

$$n_i = 4\pi(2m_e)^{\frac{3}{2}}\boldsymbol{h}^{-3}\int_0^{\infty}\frac{E^{\frac{1}{2}}\,dE}{\exp\left[(E - E_F)/\boldsymbol{k}T\right] + 1}. \tag{13}$$

In this equation we have made some approximations. We have assumed that equation (12a) is valid for *all* values of E. This will only be justified if $(E - E_F) \ll \boldsymbol{k}T$ so that only small values of E contribute appreciably to the integral. For the same reason we have extended the upper limit to $+\infty$. Similarly we have

$$p_i = 4\pi(2m_h)^{\frac{3}{2}}\boldsymbol{h}^{-3}\int_{-\infty}^{-\Delta E}\frac{(-\Delta E - E)^{\frac{1}{2}}\,dE}{\exp\left[(E_F - E)/\boldsymbol{k}T\right] + 1}. \tag{13a}$$

[2] *W.M.C.S.*, § 5.5.

We may now approximate by replacing the Fermi–Dirac function in (13) and (13a) by an exponential, since we suppose that $(E-E_F)/\boldsymbol{k}T \gg 1$ in (13) and $(E_F-E)/\boldsymbol{k}T \gg 1$ in (13a). Thus we have on writing

$$x = E/\boldsymbol{k}T$$

in (13)

$$n_i = 4\pi(2m_e)^{\frac{3}{2}}\boldsymbol{h}^{-3}(\boldsymbol{k}T)^{\frac{3}{2}} \exp[E_F/\boldsymbol{k}T]\int_0^\infty x^{\frac{1}{2}}\,e^{-x}\,dx, \tag{14}$$

$$= 2(2\pi m_e\boldsymbol{k}T/\boldsymbol{h}^2)^{\frac{3}{2}} \exp[E_F/\boldsymbol{k}T], \tag{15}$$

$$= N_c \exp[E_F/\boldsymbol{k}T]. \tag{15a}$$

Similarly, writing x for $(-\Delta E - E)/\boldsymbol{k}T$ in (13a) we have

$$p_i = 4\pi(2m_h)^{\frac{3}{2}}\boldsymbol{h}^{-3}(\boldsymbol{k}T)^{\frac{3}{2}} \exp[-(E_F+\Delta E)/\boldsymbol{k}T]\int_0^\infty x^{\frac{1}{2}}\,e^{-x}\,dx$$

$$= 2(2\pi m_h\boldsymbol{k}T/\boldsymbol{h}^2)^{\frac{3}{2}} \exp[-(E_F+\Delta E)/\boldsymbol{k}T], \tag{16}$$

$$= N_v \exp[-(E_F+\Delta E)/\boldsymbol{k}T]. \tag{16a}$$

Equating p_i and n_i we obtain

$$m_e^{\frac{3}{2}} \exp[E_F/\boldsymbol{k}T] = m_h^{\frac{3}{2}} \exp[-(E_F+\Delta E)/\boldsymbol{k}T],$$

or

$$E_F = -\tfrac{1}{2}\Delta E + \tfrac{3}{4}\boldsymbol{k}T \ln(m_h/m_e). \tag{17}$$

If $m_h = m_e$ the Fermi level lies exactly midway between the valence and conduction bands, and for most intrinsic semiconductors the deviation from this position is small at ordinary temperatures. For a semiconductor such as InSb, however, for which $m_h/m_e \simeq 20$ and $\Delta E \simeq 0.2$ eV, the Fermi level will be shifted well towards the conduction band at room temperature ($\boldsymbol{k}T \simeq 0.025$ eV).

On inserting the value of E_F from equation (17) in equation (15) we obtain the value of n_i or p_i, given by the equation

$$n_i = p_i = 2(2\pi\boldsymbol{k}T/\boldsymbol{h}^2)^{\frac{3}{2}}(m_e m_h)^{\frac{3}{4}} \exp[-\Delta E/2\boldsymbol{k}T]. \tag{18}$$

If we insert the numerical values of $\boldsymbol{h}$ and $\boldsymbol{k}$ we may write equation (18) in the form

$$n_i = p_i = 4.82\times10^{15}T^{\frac{3}{2}}(m_e m_h/\boldsymbol{m}^2)^{\frac{3}{4}} \exp[-\Delta E/2\boldsymbol{k}T]\ \text{cm}^{-3}. \tag{18a}$$

In Fig. 4.4 n_i is shown as a function of $1/T$ (curve (1)) for Ge, for which $\Delta E \simeq 0.665$ eV (see § 13.3). It will be seen that on a logarithmic plot the variation of n_i with $1/T$ appears to be nearly linear. A small curvature is in fact present due to the factor $T^{\frac{3}{2}}$ but is hardly noticeable on the scale of Fig. 4.4. In this case, for the conduction band, $(E-E_F)>15\boldsymbol{k}T$, and also for the valence band $(E_F-E)>15\boldsymbol{k}T$, so that the non-degenerate approximation holds quite well.

When the exponential approximations we have made are not permissible then n_i may be written in the form

$$n_i=4\pi(2m_e\boldsymbol{k}T/\boldsymbol{h}^2)^{\frac{3}{2}}F_{\frac{1}{2}}(E_F/\boldsymbol{k}T), \tag{19}$$

where the function $F_n(\zeta)$ is defined by the integral

$$F_n(\zeta)=\int_0^\infty \frac{x^n\,dx}{\exp[(x-\zeta)]+1}.$$

These functions have been tabulated by J. McDougall and E. C. Stoner[3] for small half-integral values of n. Similarly we have

$$p_i=4\pi(2m_h\boldsymbol{k}T/\boldsymbol{h}^2)^{\frac{3}{2}}F_{\frac{1}{2}}[-(E_F+\Delta E)/\boldsymbol{k}T]. \tag{20}$$

The Fermi energy E_F is then determined by the equation

$$m_e^{\frac{3}{2}}F_{\frac{1}{2}}[E_F/\boldsymbol{k}T]=m_h^{\frac{3}{2}}F_{\frac{1}{2}}[-(E_F+\Delta E)/\boldsymbol{k}T]. \tag{21}$$

Generally this equation has to be solved by numerical methods. When $\zeta>5$ the function $F_{\frac{1}{2}}(\zeta)$ is approximately equal to $\frac{2}{3}\zeta^{\frac{3}{2}}$ and the system is said to be fully degenerate, but this condition applies to metals and not to normal intrinsic semiconductors. When ζ is large and negative $F_{\frac{1}{2}}(\zeta)$ is approximately equal to $\frac{1}{2}\pi^{\frac{1}{2}}e^{\zeta}$ and we have the non-degenerate condition. W. Ehrenberg[4] has proposed an approximation given by

$$F_{\frac{1}{2}}(\zeta)=2\pi^{\frac{1}{2}}e^{\zeta}/(4+e^{\zeta}). \tag{22}$$

The calculated values of n_i/N_c using the exponential approximation, Ehrenberg's approximation, and the exact form given by equation (19) are shown as functions of $E_F/\boldsymbol{k}T$ in Fig. 4.2. It will be seen that the classical approximation is really quite good for values of $E_F/\boldsymbol{k}T<-1$, i.e. so long as the Fermi level is at least $\boldsymbol{k}T$ below the conduction band, corresponding to $n<0.4N_c$. For larger values of n, degeneracy makes a significant difference. Ehrenberg's approximation will be seen to be valid up to about $E_F/\boldsymbol{k}T=2$.

A knowledge of the quantities m_e and m_h is necessary before accurate values of N_c and N_v (see equations (15*a*), (16*a*)) can be calculated, and this is available for only a few semiconductors. The order of magnitude of n_i and p_i may, however, be determined when ΔE is known.

[3] *Phil. Trans.* A (1938) **237**, 67.   [4] *Proc. Phys. Soc.* A (1950) **63**, 75.

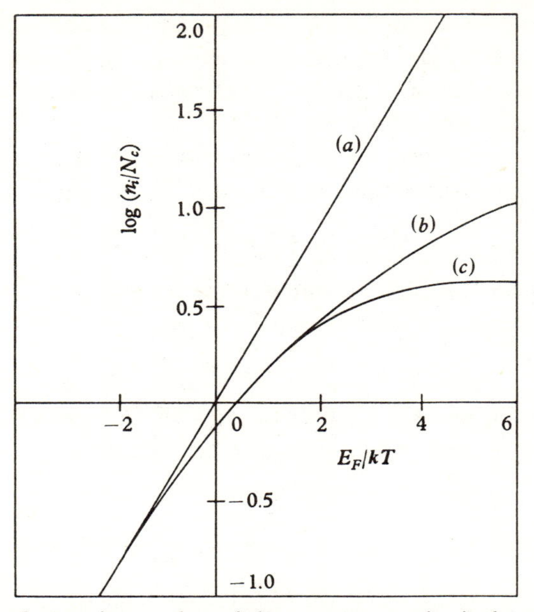

Fig. 4.2. Exact and approximate values of electron concentration in the conduction band n as a function of $E_F/\boldsymbol{k}T$. (*a*) Classical approximation. (*b*) Exact value. (*c*) Ehrenberg's approximation.

Alternatively, we may use the variation of n_i to determine ΔE. Certainly the simplest method to do this is to observe the variation of intrinsic conductivity with the temperature. If ΔE is fairly large, most of the variation of conductivity will come from the variation of the number of carriers. If therefore we assume the conductivity σ to be proportional to n_i (since $n_i = p_i$) and plot ln σ against $1/T$, we see from equation (16) that we should get a nearly straight line (neglecting the $T^{\frac{3}{2}}$ term) of slope $\Delta E/2\boldsymbol{k}$. This method has been widely used, but is full of pitfalls when used indiscriminately; not the least difficulty is that one must be quite certain that intrinsic conductivity is being observed. We shall discuss later the various methods for determining ΔE.

When we have several equivalent minima in the conduction band or equivalent maxima in the valence band, as discussed in § 2.3, we must integrate over each of these. Each will give the same contribution so that we should multiply equation (15) for n_i by M_c the number of equivalent minima in the conduction band. Similarly, equation (16) for p_i should be multiplied by M_v the number of equivalent maxima in the valence band. In such conditions we generally do not have spherical symmetry in $\boldsymbol{k}$-space. If the effective-mass tensor for electrons in the conduction band can be expressed in diagonal form, so that the constant-energy surfaces are ellipsoids of revolution, the effective electron masses being m_1 and m_2 we may use equation (23) of § 2.3 for E. By means of the simple transformation

$$k'_x = (m/m_1)k_x, \quad k'_y = (m/m_2)k_y, \quad k'_z = (m/m_2)k_z$$

we may again reduce the expression for E to 'spherical' form and proceed as before, obtaining the expression (12a) for $N(E)$ but with m_e replaced by the 'density of states' effective mass m_{de}, where $m^3_{de} = m_1m_2^2$. We therefore obtain for n_i

$$n_i = 2m_1^{\frac{1}{2}}m_2M_c\boldsymbol{h}^{-3}(2\pi\boldsymbol{k}T)^{\frac{3}{2}}\exp[E_F/\boldsymbol{k}T]. \tag{23}$$

If we have a degenerate valence band at $\boldsymbol{k} = 0$ (as for Ge and Si) the constant-energy surfaces are rather complex and an exact calculation is rather cumbersome.[5] If, however, we approximate to the actual surfaces by two spherical constant-energy surfaces with effective masses m_{h1} and m_{h2}, the calculation is again easily made. We simply integrate over the states corresponding to each surface, treated independently, and obtain

$$p_i = 2(m_{h1}^{\frac{3}{2}} + m_{h_2}^{\frac{3}{2}})\boldsymbol{h}^{-3}(2\pi\boldsymbol{k}T)^{\frac{3}{2}}\exp[-(\Delta E + E_F)/\boldsymbol{k}T]. \tag{24}$$

The 'density-of-states' effective mass is then given by $m_{dh}^{\frac{3}{2}} = m_{h_1}^{\frac{3}{2}} + m_{h_2}^{\frac{3}{2}}$. In each case the Fermi energy E_F is obtained by equating n_i to p_i as before.

Since the equations (15) and (23) may be written in the form (cf. equation (15a))

$$n_i = N_c \exp[E_F/\boldsymbol{k}T] \tag{25}$$

the conduction band may be regarded, to this approximation, as a single level with degeneracy N_c, placed at the bottom of the band. This follows from the approximate form of equation (3) when $E = 0$ and $-E_F \gg \boldsymbol{k}T$. In this case the average number of electrons occupying a degenerate level with $E = 0$ is $g \exp[E_F/\boldsymbol{k}T]$ and we see from equation (25) that $g = N_c$. Similarly, equations (16) or (24) may be written in the form

$$p_i = N_v \exp[-(\Delta E + E_F)/\boldsymbol{k}T]. \tag{26}$$

The valence band may thus be regarded as a single degenerate level at $E = -\Delta E$ and having degeneracy N_v. Since for an intrinsic semiconductor $n_i = p_i$ we have under conditions of non-degeneracy

$$n_i^2 = N_cN_v \exp[-\Delta E/\boldsymbol{k}T]. \tag{27}$$

[5] This has been carried out by B. Lax and J. G. Mavroides for Ge and Si, *Phys. Rev.* (1955) **100**, 1650.

In terms of N_c and N_v the equation for the Fermi level is

$$E_F = -\tfrac{1}{2}\Delta E + \tfrac{1}{2}\boldsymbol{k}T \ln (N_v/N_c). \tag{28}$$

4.3 Semiconductors with impurity levels

We must now take account of any impurity levels that lie between the valence and conduction bands and here we are faced at once with a difficulty. In writing down the expression for the number of electrons in an impurity level we must take account of the fact that, although two values of the spin may be allowed, the level cannot be regarded as doubly degenerate since the presence of one electron prevents another from occupying the level. A number of situations may arise and we shall consider some of them in detail. We may note that the analysis used in § 4.2 to determine the number of electrons in the conduction band still applies if the Fermi level is more than, say, $2\boldsymbol{k}T$ below the conduction band, so that the electrons in this band are non-degenerate. Thus n, the concentration of electrons in the conduction band, is still as given by equation (15a) so that we have

$$n = N_c \exp [E_F/\boldsymbol{k}T], \tag{29}$$

but E_F is no longer given by equation (28), or its equivalent, but now depends on the impurity concentration. Similarly, if the holes in the valence band are non-degenerate, we have for p, the concentration of holes in the valence band,

$$p = N_v \exp [-(E_F + \Delta E)/\boldsymbol{k}T]. \tag{30}$$

Thus under non-degenerate conditions we still have the relationship

$$np = N_c N_v \exp [-\Delta E/\boldsymbol{k}T] = n_i^2 \tag{31}$$

(cf. equation (27)). This is a most important equation and shows that the product np is independent of the impurity concentration or distribution *provided the resulting concentrations of electrons and holes in the conduction and valence bands are not degenerate.*

There is another instructive way of arriving at the form of equation (31) though in less detail. If we regard the combination of an electron (e) and a hole (h) as a chemical reaction to give a normal atom or ion of the crystal (N) we may write the reaction in the form

$$(e) + (h) = (N) + \Delta E. \tag{32}$$

The law of mass action then gives (cf. § 1.3.1, equations (2), (3))

$$np = K(T), \tag{33}$$

where $K(T)$ is of the form of $[F(T)]^2 \exp[-\Delta E/\boldsymbol{k}T]$ and $F(T)$ is a slowly varying function of T. Equation (33) is equivalent to (31) which gives the form of the function $F(T)$. If we know the number of electrons in the conduction band we may very readily see whether the semiconductor is degenerate. When $T = 300$ °K, for example, N_c is approximately equal to 2.5×10^{19} cm^{-3} if $m_e \simeq m_h \simeq \boldsymbol{m}$. Thus if n is less than 10^{19} cm^{-3} we shall have $E_F < -2\boldsymbol{k}T$. If, on the other hand, n is about 2×10^{19} cm^{-3} $E_F > -2\boldsymbol{k}T$ and the approximation is invalid, the electrons in the conduction band becoming degenerate. If, as for InSb, $m_e \ll \boldsymbol{m}$, the conduction band electrons become degenerate for considerably smaller concentrations.

Before dealing with more complex situations let us consider a semiconductor with N_d donor levels per unit volume, which lie just below the conduction band, the impurity ionization energy ϵ_d being very small compared with the forbidden energy gap. We have seen that such levels are quite common, e.g. in Ge $\epsilon_d \simeq 0.01$ eV with $\Delta E \simeq 0.7$ eV. We shall assume that the Fermi level lies well below the conduction band, i.e. $E_F \ll -\boldsymbol{k}T$. In this case practically all the impurities will be ionized, and will give rise to electrons in the conduction band. Let us first of all see how many impurities may be present before this condition is seriously violated. First of all let us suppose $N_d \gg n_i$ so that we may neglect the 'intrinsic' electrons, i.e. those excited from the valence band. We then simply have $n = N_d$. In this condition the number of electrons in the conduction band does not vary appreciably with temperature and we have what is known as the saturated extrinsic condition. Such a semiconductor is called an n-type extrinsic semiconductor.

The position of the Fermi level in this condition is given by

$$n = N_d = N_c \exp[E_F/\boldsymbol{k}T],$$

or

$$E_F = \boldsymbol{k}T \ln(N_d/N_c). \tag{34}$$

Equation (34) is a good approximation only if $E_F < -\boldsymbol{k}T$, i.e. if N_d is less than about $\frac{1}{2}N_c$. For Ge at 300 °K we should have $N_d < 10^{19}$ cm^{-3}, but if $N_d < 10^{14}$ cm^{-3} the intrinsic electrons are no longer negligible. At room temperature, therefore, we see that there is quite a large range of values of N_d for which the saturated extrinsic condition holds.

It will be clear from equation (34) that, as N_d increases, the Fermi level will rise from near the mid-point of the energy gap to within a distance of the order of $\boldsymbol{k}T$ from the conduction band as $N_d \to N_c$. For $N_d \geqslant N_c$ the electrons in the conduction band will become degenerate and an equation like (21) will have to be used to determine the position of the Fermi level. The semiconductor will then behave rather like a metal.

It is of interest to examine what happens in the more general case when N_d may be of the same order of magnitude as n_i, but $N_d \ll N_c$. Clearly the Fermi level will be well away from the conduction band if $\Delta E \gg \boldsymbol{k}T$, so that we may assume all the impurities to be ionized. We then have a relationship between n and p which is simply

$$n - p = N_d \tag{35}$$

and expresses the condition for charge neutrality. Taken together with equation (31) it enables us to express n and p in terms of n_i and N_d, and we have therefore

$$\left.\begin{aligned} n &= \tfrac{1}{2}N_d[1+(1+4n_i^2/N_d^2)^{\frac{1}{2}}], \\ p &= \frac{2n_i^2}{N_d}[1+(1+4n_i^2/N_d^2)^{\frac{1}{2}}]^{-1}. \end{aligned}\right\} \tag{36}$$

If we have $N_d \gg n_i$ (but also $N_d < N_c$) we have approximately

$$\left.\begin{aligned} n &= N_d + n_i^2/N_d, \\ p &= n_i^2/N_d. \end{aligned}\right\} \tag{37}$$

We note that the number of holes in this case is much less than for intrinsic material (in the ratio n_i/N_d). For example, if $n_i = 2.37 \times 10^{13}\ \text{cm}^{-3}$ (as for Ge) and $N_d \simeq 10^{16}\ \text{cm}^{-3}$, we have $n \simeq 10^{16}\ \text{cm}^{-3}$ and $p \simeq 5.6 \times 10^{10}\ \text{cm}^{-3}$. The electrons in this case are called the majority carriers and the holes the minority carriers. Although the minority carrier density may be small compared with the majority carrier density the minority carriers may still be very important. If, on the other hand, $N_d \ll n_i$ then we have approximately

$$\left.\begin{aligned} n &= n_i + \tfrac{1}{2}N_d, \\ p &= n_i - \tfrac{1}{2}N_d. \end{aligned}\right\} \tag{38}$$

Now suppose we have N_a acceptor impurities per unit volume instead of the donor impurities, their levels lying just above the valence band. We shall again suppose that the Fermi level is at a height of at least several times $\boldsymbol{k}T$ above the valence band. The acceptors will then all be occupied by electrons or 'ionized', leaving approximately N_a free holes in the valence band and we shall now have

$$p = N_a = N_v \exp[-(\Delta E + E_F)/\boldsymbol{k}T]. \tag{39}$$

Equation (39) is approximately valid only if N_a/N_v is less than about $\frac{1}{2}$. However, we see that as N_a increases from zero to about N_v, the Fermi level drops from near the middle of the forbidden band and approaches the valence band.

Clearly we have now in place of equation (35) the equation

$$p - n = N_a. \tag{40}$$

This, together with equation (31), gives

$$\left.\begin{aligned} n &= \frac{2n_i^2}{N_a}[1+(1+4n_i^2/N_a^2)^{\frac{1}{2}}]^{-1}, \\ p &= \tfrac{1}{2}N_a[1+(1+4n_i^2/N_a^2)^{\frac{1}{2}}]. \end{aligned}\right\} \tag{41}$$

Again, when $N_a \geqslant n_i$ we have approximately

$$\left.\begin{aligned} n &= n_i^2/N_a, \\ p &= N_a + n_i^2/N_a. \end{aligned}\right\} \tag{42}$$

If $n_i \gg N_a$ we have approximately

$$\left.\begin{aligned} n &= n_i - \tfrac{1}{2}N_a, \\ p &= n_i + \tfrac{1}{2}N_a. \end{aligned}\right\} \tag{43}$$

Now let us suppose we have both donors and acceptors present. First, suppose $(N_d - N_a) \gg n_i$. When $N_a = 0$ the Fermi level will be appreciably above the middle of the forbidden energy gap. As acceptors are added they will have a very high probability of being occupied by electrons, i.e. 'ionized', lying well below the Fermi level. The electrons filling the acceptor levels will have dropped down from donor levels (or from the conduction band) each acceptor taking one electron. The number of *free* holes will be very small since the Fermi level is still high. The number of effective donors will be equal to $N_d - N_a$ and the position of the Fermi level will be given by

$$N_d - N_a = N_c \exp[E_F/\boldsymbol{k}T]. \tag{44}$$

As N_a approaches N_d, however, the acceptors will tend to neutralize the donors and the Fermi level will move towards the intrinsic position.

When $N_a > N_d$ we shall likewise have $N_a - N_d$ effective acceptors, all the electrons from the donors filling acceptors, and the Fermi level will move down towards the valence band. This is illustrated in Fig. 4.3. (The actual form of the curve depends on the temperature.) A series of curves showing n and p for various values of $N_d - N_a$ are shown in Fig. 4.4 for Ge.

Up till now we have assumed that the donor and acceptor levels lie respectively very close to the conduction and valence bands; they will, however, in general, be separated from these by energies ϵ_d, ϵ_a. Except at very low temperatures the impurities will be fully ionized when ϵ_d and ϵ_a are small, unless $|N_d - N_a| > N_c$ or N_v. We must now consider the very important case when the temperature is so low that full ionization will not take place, and also deal with donors and acceptors whose energy levels are separated from the conduction and valence bands by amounts which are not small compared with ΔE. In order to treat this more complex situation we must consider the degeneracy of the impurity levels.

If each level can accept two electrons, one of either spin, then the probability of occupation for an electron of either spin is given by equation (1) with E replaced by $-\epsilon_d$ or $-\Delta E + \epsilon_a$. The average number n_d of electrons in the donor level is then given by

$$n_d = \frac{2N_d}{\exp\left[(E - E_F)/\boldsymbol{k}T\right] + 1}. \tag{45}$$

This condition hardly ever occurs for the donors and acceptors in semiconductors.

Two conditions are more common:

(i) An impurity level can accept *one* electron of either spin, or can have no electron.

(ii) An impurity level can have two paired electrons or one of either spin. This condition may be stated in terms of holes in the same form as condition (i).

For condition (i) it may be shown that the probability $P(E)$ of an electron of *either* spin occupying the level is given by

$$P(E) = \frac{1}{1 + \frac{1}{2}\exp\left[(E - E_F)/\boldsymbol{k}T\right]}. \tag{46}$$

Condition (i) would apply in a group IV semiconductor to group V donors with five valence electrons. Four of these electrons will have their spins paired in the valence binding states and the fifth will be able

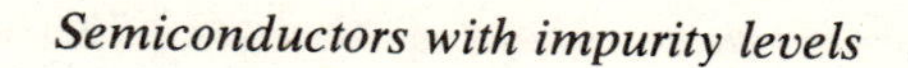

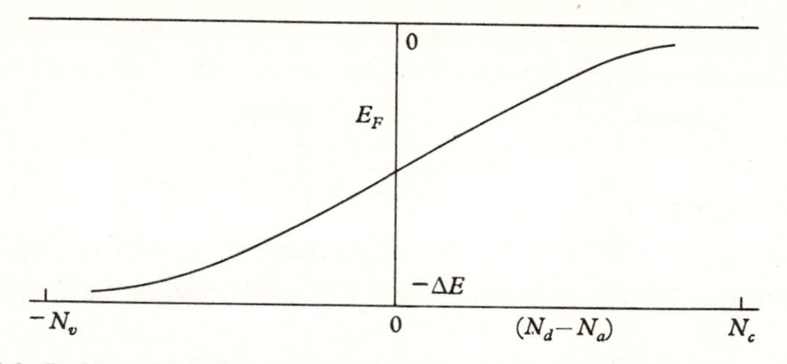

Fig. 4.3. Position of Fermi level as a function of donor and acceptor concentration.

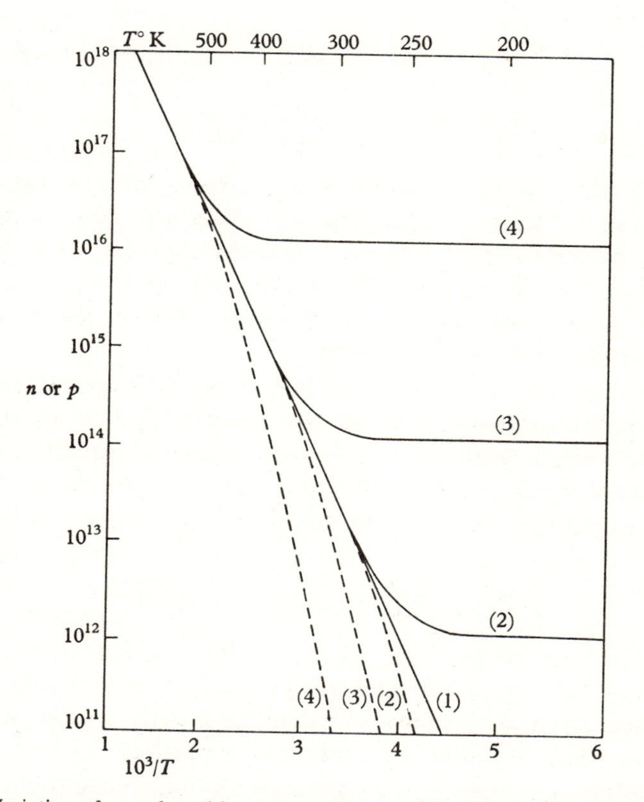

Fig. 4.4. Variation of n and p with temperature for Ge. (Full line curves give n, dotted curves give p.) (1) $N_d-N_a=0$ (intrinsic line), (2) $N_d-N_a=10^{12}$ cm^{-3}, (3) $N_d-N_a=10^{14}$ cm^{-3}, (4) $N_d-N_a=10^{16}$ cm^{-3}.

to take up either value of the spin. If we then simply have N_d donor impurities the number of un-ionized donors n_d will be given by

$$n_d = \frac{N_d}{1+\frac{1}{2}\exp[(E-E_F)/\boldsymbol{k}T]} \tag{47}$$

with $E = -\epsilon_d$.

For condition (ii) it may also be shown that the probability $P(E)$ of the level being occupied by two electrons with paired spins is given by

$$P(E) = \frac{1}{1+2\exp[(E-E_F)/\boldsymbol{k}T]}. \tag{48}$$

Condition (ii) would apply in a group IV semiconductor to group III acceptors having four electrons with paired spins when an extra electron is attached (ionized condition). The number of un-ionized acceptors n_a is given by

$$\begin{aligned} n_a &= N_a[1-P(E)] \\ &= \frac{N_a}{1+\frac{1}{2}\exp[(E_F-E)/\boldsymbol{k}T]} \end{aligned} \tag{49}$$

with $E = -\Delta E + \epsilon_a$. The similarity of equations (47) and (49) interpreted respectively in terms of electrons and holes will now be apparent. Various methods of deriving equations (46) and (48) or their equivalent have been given. These equations are, however, just as fundamental as equation (1) and may be derived in the same way from the fundamental treatment of Fermi–Dirac statistics.[6]

In deriving equations (47) and (49) from the expressions for the probability of occupation $P(E)$ we have assumed that we are concerned only with a single level. As we have seen, however, impurity levels may have excited states, and these should be included in equation (47) with the appropriate probabilities. We then have

$$n_d = \frac{N_d}{1+[\sum g_r \exp\{(E_F-E_r)/\boldsymbol{k}T\}]^{-1}}, \tag{50}$$

where E_r is the energy of the rth excited state; g_r is a number to take account of degeneracy and spin and $g_0 = 2$, $E_0 = -\epsilon_d$, for donors which accept one electron. A similar equation holds for acceptors. In the two situations which are most important from a practical point of view, the excited states are not of great significance in this connection. The first

[6] See, for example, A. H. Wilson, *The Theory of Metals* (Cambridge University Press, 2nd edition, 1953), pp. 326–8.

situation occurs when $(E_0 - E_F)/\boldsymbol{k}T \gg 1$. All the terms of the sum in equation (50) are then small and n_d/N_d is small, nearly all the impurity levels being ionized. The second situation occurs at very low temperatures. If $(E_1 - E_0)/\boldsymbol{k}T \gg 1$ the first term in the sum will predominate and equation (47) will be a good approximation. We shall, therefore, in the following neglect the excited states. Their effect has been treated in some detail by K. S. Shifrin[7] and also by E. Burstein, E. E. Bell, J. W. Davisson and M. Lax.[8]

In order to calculate the electron and hole concentrations n and p let us consider first a semiconductor having N_d donor impurities all with energy levels at depth ϵ_d below the conduction band, and N_a acceptor impurities with energy levels ϵ_a above the valence band. The equation expressing the condition for electrical neutrality, and which determines the position of the Fermi level, is now given by

$$n + n_d + N_a = p + N_d + n_a. \tag{51}$$

Using equations (29), (30), (47), (49) this may be expressed in terms of E_F, giving a quartic equation for $\exp[(E_F/\boldsymbol{k}T)]$. In order to obtain simple analytical solutions we must make approximations valid under certain conditions. We shall first of all suppose that $|N_a - N_d| \ll N_c$ or N_v. Let us assume that the Fermi level lies several times $\boldsymbol{k}T$ below both the conduction band and the impurity level, but above the intrinsic Fermi level, and examine the consequences. We see that n_a will be quite negligible compared with N_a and may be neglected in equation (51). So long as the Fermi level does not approach $-\epsilon_d$, n_d will also be small compared with N_d. This is the condition when practically all the acceptors have captured electrons and all the donors are ionized. We then have

$$N_c \exp[(E_F/\boldsymbol{k}T)] - N_v \exp[-(E_F + \Delta E)/\boldsymbol{k}T] = N_d - N_a. \tag{52}$$

This condition can only hold if $N_d > N_a$, since our assumption that the Fermi level lies above the intrinsic Fermi level makes the right-hand side of equation (52) positive. Clearly, we may have a similar equation if the Fermi level lies below the intrinsic Fermi level with $N_a > N_d$. Equation (52) is equivalent to equation (35) with N_d replaced by $(N_d - N_a)$ and leads to the results previously discussed. At high temperatures the terms on the left of equation (52) predominate and we tend to an intrinsic condition. As the temperature is lowered, however, the second

[7] *J. Tech. Phys., Moscow* (1944) **14**, 43. [8] *J. Phys. Chem.* (1953) **57**, 849.

term on the left, representing the free holes, becomes small compared with the first and we have

$$E_F = \boldsymbol{k}T \ln\left(\frac{N_d - N_a}{N_c}\right). \tag{53}$$

For sufficiently small values of T the assumption that the Fermi level lies well below the donor impurity level breaks down and this approximation no longer holds. As the temperature is lowered the Fermi level rises towards the donor level and the donors begin to be partially filled with electrons.

Fortunately, in this condition, we may neglect the contribution from the free holes as well as having n_a small. We no longer assume n_d small compared with N_d and obtain for the equation which determines the Fermi level

$$n = N_d - N_a - \frac{N_d}{1 + \frac{1}{2}\exp[-(\epsilon_d + E_F)/\boldsymbol{k}T]}. \tag{54}$$

Writing $n = N_c \exp[(E_F/\boldsymbol{k}T)]$ we obtain a quadratic equation for $\exp[(E_F/\boldsymbol{k}T)]$ or for n, namely

$$n^2 + n(N_a + N_c') - N_c'(N_d - N_a) = 0, \tag{55}$$

where we have written N_c' for $\frac{1}{2}N_c \exp[-\epsilon_d/\boldsymbol{k}T]$. The appropriate solution of equation (55) is

$$n = -\tfrac{1}{2}(N_a + N_c') + \tfrac{1}{2}\{(N_a + N_c')^2 + 4N_c'(N_d - N_a)\}^{\frac{1}{2}}. \tag{56}$$

When $\epsilon_d/\boldsymbol{k}T \gg 1$, N_c' will be small and we may approximate to the solution (56). Care must be taken, however, in doing this since an expansion of the square-root on the right-hand side in powers of N_c' is permissible only if $N_c' \ll N_a$. This form of approximation is therefore only valid for partially compensated semiconductors. We have then approximately

$$n = \frac{N_c'(N_d - N_a)}{N_a} = \frac{N_d - N_a}{2N_a} N_c \exp[-\epsilon_d/\boldsymbol{k}T]. \tag{57}$$

When N_a is very small so that $N_a \ll N_c' \ll N_d$ we have approximately

$$n = (N_c' N_d)^{\frac{1}{2}} = \frac{1}{\sqrt{2}}(N_d N_c)^{\frac{1}{2}} \exp[-\epsilon_d/2\boldsymbol{k}T]. \tag{58}$$

These equations were obtained by essentially the above method by J. H. de Boer and W. C. van Geel.[9] They may also be obtained by minimizing

[9] *Physica* (1935) **2**, 186.

the free energy change due to excitation of electrons from the donor centres, as shown by N. F. Mott and R. W. Gurney.[10] We note that, under the above approximate conditions, straight lines are obtained of slope $\epsilon_d/\boldsymbol{k}T$ or $\frac{1}{2}\epsilon_d/\boldsymbol{k}T$ when $\ln n$ is plotted against $1/T$. This has been used to obtain values for ϵ_d for donors in Ge and Si, but clearly the method must be used with care to ensure that the appropriate conditions hold. Exactly similar equations are readily obtained for a semiconductor in which acceptor impurities predominate.

Equations (57) and (58) indicate that as the temperature is lowered indefinitely all the free carriers should disappear. A great reduction in carrier concentration is generally found in the liquid-hydrogen and liquid-helium range of temperatures and curves whose slopes give values for ϵ_d or ϵ_a may frequently be obtained. At the lowest temperatures, however, it is found that all the free carriers do not seem to disappear as expected from simple theory. This phenomenon, generally called impurity band conduction, will be discussed later (see § 5.4). Neglecting such effects, we see that the general shape of the curves giving the majority carrier concentration as a function of T^{-1} will be as shown in Fig. 4.5.

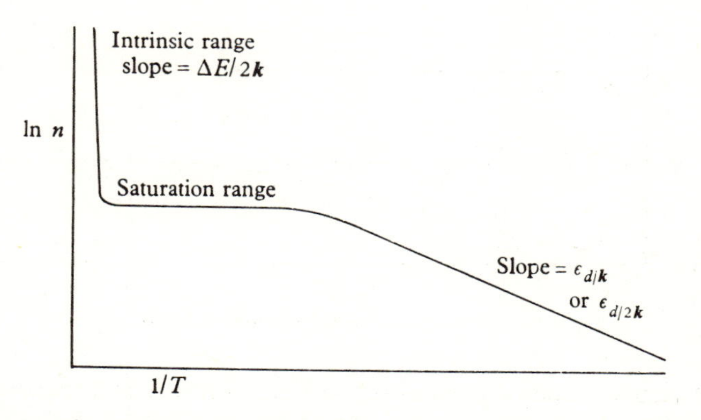

Fig. 4.5. Variation of carrier concentration for an n-type semiconductor at low temperatures.

The position of the Fermi level for the conditions leading to equation (57) is given by the equation

$$E_F = -\epsilon_d \boldsymbol{k}T \ln\left(\frac{N_d - N_a}{2N_a}\right). \tag{59}$$

[10] *Electronic Processes in Ionic Crystals* (Oxford University Press, 1940), pp. 156–60.

It therefore lies slightly above the level $E=-\epsilon_d$. For conditions leading to equation (58), e.g. $N_a=0$, we have

$$E_F=-\tfrac{1}{2}\epsilon_d+\tfrac{1}{2}kT\ln(N_d/N_c). \tag{60}$$

For a semiconductor containing only donor impurities the Fermi level at low temperatures, therefore, lies about half way between the donor level and the conduction band. Similar equations hold for the condition $N_a>N_d$. In this case the Fermi level either lies just below the acceptor level or half way between the acceptor level and the valence band. The general form of the variation of Fermi level with temperature is shown in Fig. 4.6. The above calculations have been based on the assumption

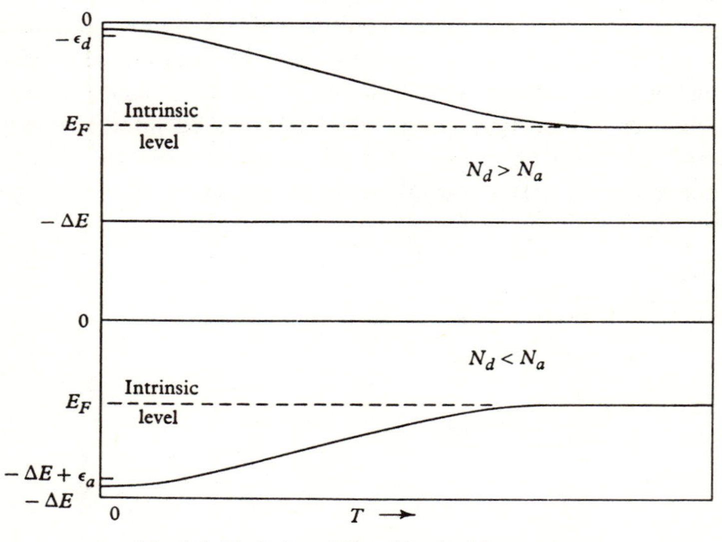

Fig. 4.6. Variation of Fermi level with temperature.

that the semiconductor is non-degenerate. When it is degenerate over some range of temperature the calculations are more complex.[11] At sufficiently low temperatures the degeneracy will be removed due to the removal of electrons from the conduction band.

[11] See, for example, J. S. Blakemore, *Semiconductor Statistics* (Pergamon Press, 1962).

5

Electron transport phenomena

5.1 Collisions with crystalline imperfections – relaxation time τ

In this chapter we shall consider various transport phenomena which arise from the motions of electrons and positive holes in semiconductors under the influence of electric and magnetic fields. The discussion will be based on the effective-mass concept introduced in § 2.5, making use of the fundamental equation

$$\dot{\mathbf{P}} = \hbar\dot{\mathbf{k}} = \mathbf{F}, \tag{1}$$

where **F** is the force acting on an electron; a similar equation holds for positive holes (see § 2.5). Let us consider for simplicity motion in the x-direction and suppose we have a conduction band with a minimum at $\mathbf{k}=0$ and a single maximum at the edge of the zone at $k_x = k_d$. Near $k_x = 0$ the effective mass m_e will be positive and we shall have

$$m_e \frac{dv_x}{dt} = F_x. \tag{1a}$$

If $F_x > 0$ the electron will be accelerated in the x-direction. Under the action of the force F_x the value of k_x will increase steadily till it reaches k_d. It will then pass out of the first zone and may be regarded as re-entering it at $k_x = -k_d$ and again passing through it. As k_x approaches k_d the effective mass becomes negative (see § 2.5) and the electron will be slowed down and will come to rest as $k_x \to k_d$ (since for $k_x = k_d$, $v_x = 0$). It will then begin to be accelerated in a backward direction coming to rest again when k_x is next equal to zero. This is just Bragg reflection in the crystal planes.[1] The electron would therefore appear to oscillate backwards and forwards in the x-direction. That this

[1] *W.M.C.S.*, § 5.8.

does not take place in practice is simply due to the fact that real crystals contain so many imperfections (including lattice vibrations) that the electron would make many collisions with these long before the value of k_x had increased appreciably. Only in very high electric fields does the average value of k for all the electrons change appreciably. The value of $\mathbf{k}$ for an individual electron may, however, change very markedly at a collision. The amount of energy imparted by an electron to a heavy ion is quite small, and it can be shown that the energy imparted to lattice vibrations by an electron is also quite small (see § 12.2). In an elastic collision an electron may be regarded as changing its $\mathbf{k}$-vector from one point on a constant-energy surface in k-space to another point on the surface. The motion of an electron in a crystal is thus rather similar to the motion of a molecule in a gas. It executes a free path between collisions moving in a perfectly random fashion when no field is applied. Under the influence of a field a drift motion is set up rather like a wind in a gas. The length of actual path described by the electron is very much greater than the distance it has drifted under the field, and its actual velocity of motion much greater than the drift velocity. This is illustrated in Fig. 5.1.

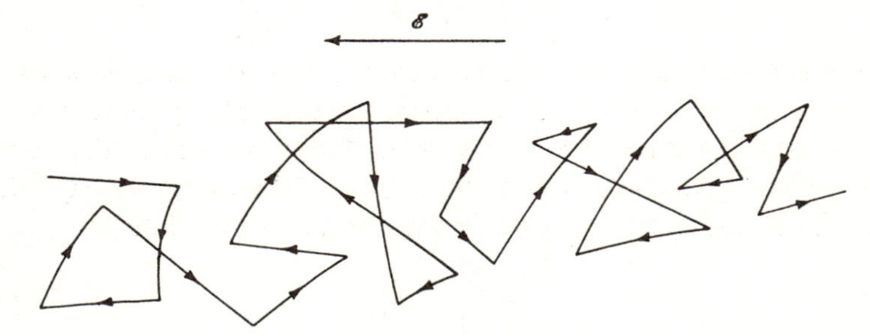

Fig. 5.1. Path of an electron in a crystal under the influence of an electric field $\mathscr{E}$.

It is convenient to define various quantities in order to describe the collision process. In the quantum theory of collisions the usual quantity is the effective cross-section which gives a measure of the probability of a particular type of encounter taking place. For example, we may define a quantity $\sigma(\theta, \phi)\, d\omega$ giving the probability of a collision which takes an electron from motion in the direction $\theta = 0$ to a direction (θ, ϕ) in the small solid angle $d\omega$, θ and ϕ being spherical-polar co-ordinates. From this we may define a total cross-section σ_t, given by the equation

$$\sigma_t = \int \sigma(\theta, \phi)\, d\omega. \tag{2}$$

Other useful quantities are the mean free time τ and the mean free path length $l = v\tau$, where v is the velocity. If we have n electrons travelling with velocity v in a given direction the number of collisions made by the electrons in a small interval of time dt will be proportional to n and to dt. The number of electrons $n(t)$ which have *not* made a collision at time t therefore satisfies an equation of the form

$$\frac{dn}{dt} = -\frac{n}{\tau(v)}, \tag{3}$$

where $\tau(v)$ is a quantity independent of t, and as we shall see, is the mean time between collisions. We have then

$$n = n_0 \exp[-t/\tau], \tag{4}$$

where $n = n_0$ at $t = 0$. The probability that an electron has *not* made a collision is thus given by n/n_0 and is therefore equal to $\exp[-t/\tau]$. The average value of the time $\bar{t}$ between collisions is thus given by

$$\bar{t} = \tau^{-1} \int_0^\infty t \exp[-t/\tau]\, dt = \tau. \tag{5}$$

Thus we see that τ is just the mean free time defined above. It follows that the mean free path l is given by $l(v) = v\tau(v)$. It can also easily be seen that $1/l = N\sigma_t$, where N is the number of scattering centres per unit volume, so that $1/\tau = N\sigma_t v$.

The simplest case to consider is that of isotropic scattering, in which the direction of an electron immediately after a collision may be in any direction with equal probability. In this case the drift velocity acquired in an electric field is completely lost in scattering. As we shall see (§ 8.5), this is so for scattering by the thermal vibrations. For many other types of scattering, and in particular for scattering by the Coulomb field of ionized impurity centres, small-angle deviations predominate. The total cross-section σ_t defined in equation (2) is not so useful in this connection as a cross-section σ_c obtained by weighting $\sigma(\theta, \phi)$ by an appropriate function to take account of the change in drift velocity at the collision.

Let us consider an electron moving initially with velocity v_{z0} in the z-direction ($\theta = 0$), and, on making a collision, let it be scattered in the direction (θ, ϕ). The angle between the initial and final directions being θ, the change in velocity δv_z in the z-direction is given by

$$\delta v_z = v_{z0}(1 - \cos\theta). \tag{6}$$

If $\sigma(\theta, \phi)$ depends only on θ, the average change in v_z per collision is given by

$$\overline{\delta v_z} = v_{z0} \int_0^\pi (1-\cos\theta)\sigma(\theta)\sin\theta\, d\theta \Big/ \int_0^\pi \sigma(\theta)\sin\theta\, d\theta. \tag{7}$$

Let us define the cross-section σ_c by means of the equation

$$\sigma_c(\theta) = 2\pi \int_0^\pi (1-\cos\theta)\sigma(\theta)\sin\theta\, d\theta, \tag{8}$$

then we have

$$\overline{\delta v_z} = v_{z0}\sigma_c/\sigma_t. \tag{9}$$

It will therefore require on the average σ_t/σ_c collisions completely to destroy the original drift velocity v_{z0}. The probability that an electron makes a collision in a time dt is $Nv\sigma_t\, dt$ and the probability of a velocity-destroying collision is $Nv\sigma_c\, dt$. The cross-section σ_c is therefore the appropriate one to use in transport problems.[2] Thus we may also define an effective relaxation time τ_c given by

$$\frac{1}{\tau_c} = \frac{1}{\tau}\overline{(1-\cos\theta)}, \tag{10}$$

where

$$\overline{(1-\cos\theta)} = \sigma_c/\sigma_t.$$

We shall now consider only collisions which completely destroy the drift velocity, taking account of the other condition by use of σ_c and τ_c. Suppose we have an electric field $\mathscr{E}$ in the x-direction. Consider an electron which has just made a collision at time $t=0$; its x-component of velocity at subsequent time t, provided it has not made a collision, is given by

$$v_x = v_{x0} - \mathscr{E}et/m_e, \tag{11}$$

where v_{x0} is the value of v_x at $x=0$, and m_e is the appropriate effective mass. We must now average equation (11) over all values of t, using the expression $\exp[-t/\tau]\, dt/\tau$ for the probability that it will make a collision in the time interval dt after a time t. We thus have

$$\begin{aligned}\overline{v_x} &= \overline{v_{x0}} - \frac{\mathscr{E}e}{\tau m_e}\int_0^\infty t\exp[-t/\tau]\, dt \\ &= \overline{v_{x0}} - \mathscr{E}e\tau/m_e. \end{aligned} \tag{12}$$

In view of our assumption that, on the average, collisions destroy any previous drift velocity we have $\overline{v_{x0}} = 0$. Hence, we have

$$\overline{v_x} = -e\mathscr{E}\tau/m_e. \tag{13}$$

[2] *W.M.C.S.*, § 10.3.1.

To this approximation we see that the average drift velocity $\overline{v_x}$ is proportional to the electric field $\mathscr{E}$. We may write equation (13) in the form

$$\overline{v_x} = -\mu_e \mathscr{E}, \tag{14}$$

where μ_e (a positive quantity) is the magnitude of the drift velocity for electrons produced by unit field, and is known as the electron mobility. We have for μ_e the equation

$$\mu_e = e\tau/m_e. \tag{15}$$

Similarly for positive holes we have

$$\overline{v'_x} = e\tau\mathscr{E}/m_h. \tag{16}$$

The hole mobility μ_h is then given by the equation

$$\mu_h = e\tau/m_h. \tag{16a}$$

Let us now consider the magnitude of some of these quantities. For pure Ge at room temperature we have $\mu_e = 3900\ \text{cm}^2\ \text{V}^{-1}\ \text{s}^{-1}$. (For purposes of calculation the M.K.S. value $0.39\ \text{m}^2\ \text{V}^{-1}\ \text{s}^{-1}$ is more convenient.) If we take $m_e = 0.3\boldsymbol{m}$, $e/m_e \simeq 6 \times 10^{11}\ \text{C kg}^{-1}$, it follows that $\tau = 6 \times 10^{-13}$ s. Also we have for the mean-square velocity $\overline{v^2}$

$$\tfrac{1}{2} m_e \overline{v^2} = \tfrac{3}{2} \boldsymbol{k} T.$$

Since the average velocity $\bar{v}$ does not differ greatly from $[\overline{v^2}]^{\frac{1}{2}}$ we have

$$\bar{v} \simeq \left(\frac{3\boldsymbol{k}T}{e}\right)^{\frac{1}{2}} \left(\frac{e}{m_e}\right)^{\frac{1}{2}}. \tag{17}$$

Now for $T = 300\ °\text{K}$, $\boldsymbol{k}T/e \simeq 1/40$ V. Thus $\bar{v} \simeq 2.5 \times 10^5\ \text{m s}^{-1}$, so that $l = 1.5 \times 10^{-7}\ \text{m} = 1.5 \times 10^{-5}$ cm, so the mean free path extends over several hundred lattice spacings. On the average, an electron describes a path of length $\bar{v}$ in a second, i.e. 2.5×10^7 cm, while in a field of $1\ \text{V cm}^{-1}$ it drifts a distance μ_e in the field, i.e. 3.9×10^3 cm, which is smaller by a factor of about 10^4.

5.2 Constant relaxation time τ

The values of τ for holes and electrons are not necessarily the same, and we must remember that τ is, in general, a function of the velocity v, so that we should average over all relevant values of v. We shall, however, first of all consider the simple case where τ is taken to be a constant, so that equations (13) and (16) may be applied to all the free electrons in the conduction band and holes in the valence band; as we have seen in § 1.3.1 the other electrons do not take part in conduction.

5.2.1 Electrical conductivity

Let there be n electrons in the conduction band and p holes in the valence band. The current density J_x in the x-direction is then given by the equation

$$J_x = -en\overline{v_x} + ep\overline{v'_x} \tag{18}$$

$$= e(n\mu_e + p\mu_h)\mathscr{E}_x. \tag{18a}$$

The electrical conductivity σ is given by the equation

$$\mathbf{J} = \sigma\boldsymbol{\mathscr{E}}. \tag{19}$$

So that we have

$$\sigma = e(n\mu_e + p\mu_h). \tag{20}$$

This important relationship connecting the conductivity with the electron and hole concentrations and mobilities will be used a great deal in our subsequent discussions.

For a semiconductor with spherical constant-energy surfaces and a single scalar effective mass, similar expressions hold for any direction in the crystal and σ is a scalar quantity. Let us, however, suppose that we have an n-type semiconductor ($n \gg p$) for which we have non-spherical constant-energy surfaces with a single minimum in the conduction band at $\mathbf{k} = 0$. For small values of k the energy will be given by an equation of the form

$$E = \frac{h^2}{8\pi^2}\left[\frac{k_x^2}{m_1} + \frac{k_y^2}{m_2} + \frac{k_z^2}{m_3}\right] \tag{21}$$

$$= \tfrac{1}{2}[m_1 v_x^2 + m_2 v_y^2 + m_3 v_z^2], \tag{21a}$$

cf. equation (17) of § 2.3. The equations of motion in an electric field with components ($\mathscr{E}_x$, $\mathscr{E}_y$, $\mathscr{E}_z$) along the axes, which we have chosen to make the effective-mass tensor diagonal, are

$$\left.\begin{aligned} m_1\dot{v}_x &= -e\mathscr{E}_x, \\ m_2\dot{v}_y &= -e\mathscr{E}_y, \\ m_3\dot{v}_z &= -e\mathscr{E}_z. \end{aligned}\right\} \tag{22}$$

By an obvious extension of the analysis carried out above we have

$$\left.\begin{aligned} J_x &= ne\mu_1\mathscr{E}_x, \\ J_y &= ne\mu_2\mathscr{E}_y, \\ J_z &= ne\mu_3\mathscr{E}_z, \end{aligned}\right\} \tag{23}$$

where $\mu_1 = e\tau/m_1$, $\mu_2 = e\tau/m_2$, $\mu_3 = e\tau/m_3$. The electrical conductivity is thus a tensor, in this case a diagonal one through special choice of axes. If we write

$$J_r = \sum_s \sigma_{rs}\mathscr{E}_s \tag{24}$$

with $r, s = x, y, z$ we have

$$\left.\begin{aligned} \sigma_{xx} &= ne\mu_1, \\ \sigma_{yy} &= ne\mu_2, \\ \sigma_{zz} &= ne\mu_3, \end{aligned}\right\} \quad \sigma_{xy} = \sigma_{yz} = \sigma_{zx} = 0. \tag{25}$$

We thus have an anisotropic conductor and, except along the axes, the current **J** will not be parallel to the field. Frequently we have a crystal with an axis of symmetry, which we may choose as one of the special axes. The other two may be any two mutually perpendicular lines at right angles to the symmetry axis. If we take z along the axis of symmetry we shall have $m_1 = m_2$, and consequently $\mu_1 = \mu_2$ and $\sigma_{xx} = \sigma_{yy} = \sigma_T$ and $\sigma_{zz} = \sigma_L$. Let the current vector **J** make an angle θ with the z-axis. The components of electric field will then be given by $\mathscr{E}_x = J \sin\theta/\sigma_T$, $\mathscr{E}_y = 0$, $\mathscr{E}_z = J\cos\theta/\sigma_L$, if we take the y-axis perpendicular to the z-axis and current direction. The components of field along **J** will be given by $J\sin^2\theta/\sigma_T + J\cos^2\theta/\sigma_L$. We have therefore

$$\frac{1}{\sigma} = \frac{\sin^2\theta}{\sigma_T} + \frac{\cos^2\theta}{\sigma_L}. \tag{26}$$

Tellurium is such a semiconductor with $\sigma_T = 1.95\sigma_L$.

If we have a number of equivalent minima in the conduction band, as for a semiconductor of the type illustrated in Fig. 2.5(*b*) or (*c*), we must sum over the electrons in all the minima. If there are M such minima symmetrically disposed we shall have n/M electrons per unit volume in each. The current density is then given by a symmetrical expression of the form

$$\left.\begin{matrix} J_x \\ J_y \\ J_z \end{matrix}\right\} = \tfrac{1}{3}ne(\mu_1 + \mu_2 + \mu_3)\left\{\begin{matrix} \mathscr{E}_x, \\ \mathscr{E}_y, \\ \mathscr{E}_z, \end{matrix}\right. \tag{27}$$

since by symmetry $J_y = J_z = 0$ when the field is along the x-axis. We thus again have a scalar conductivity and we may write $\mathbf{J} = \sigma\mathscr{E}$ with

$$\sigma = \tfrac{1}{3}ne(\mu_1 + \mu_2 + \mu_3). \tag{28}$$

If we write $\sigma = ne\mu_c$ we may call μ_c the conductivity mobility, and we have

$$\mu_c = \tfrac{1}{3}(\mu_1 + \mu_2 + \mu_3). \tag{29}$$

Again, if we write

$$\mu_c = \frac{e\tau}{m_c} \tag{30}$$

we may call m_c the conductivity effective mass, and we have

$$\frac{1}{m_c} = \frac{1}{3}\left(\frac{1}{m_1} + \frac{1}{m_2} + \frac{1}{m_3}\right). \tag{31}$$

When we have a crystal with cubic symmetry the constant-energy surfaces near each minimum will be spheroids (see § 2.3). Each spheroid may be represented by a diagonal effective-mass tensor having two equal components m_2 and a third component m_1. The magnitudes of m_1 amd m_2 will be the same for each spheroid but will correspond to different directions in the crystal. When we perform the sum over all the spheroids it will readily be seen that we again obtain equation (30) with

$$\frac{1}{m_c} = \frac{1}{3}\left(\frac{1}{m_1} + \frac{2}{m_2}\right). \tag{32}$$

The effective mass m_c differs somewhat from the 'density of states' effective mass m_d defined in obtaining equation (23) of § 4.2.

When the constant-energy ellipsoids do not lie along the ⟨100⟩ axes (which we may take as axes of co-ordinates), but along some other symmetrical directions, such as ⟨111⟩, we have for each ellipsoid a conductivity tensor with non-zero diagonal elements. When an average is taken over all the equivalent ellipsoids it will be found that all the non-diagonal elements are zero and the diagonal elements are again equal, leading to an isotropic scalar conductivity. We shall see that this is no longer true in the presence of a magnetic field, the magneto-resistance being anisotropic.

5.2.2 The Hall effect

When a magnetic field is applied to a conductor carrying a current, in a direction at right angles to the current, an e.m.f. is produced across the conductor in a direction perpendicular to the current and to the magnetic field. This effect, known as the Hall effect, after E. H. Hall[3] who discovered it in thin metallic foils in 1879, has become one of the most powerful tools for studying the electronic properties of semiconductors. Whereas electrical conductivity may, at the higher temperatures, have a component of ionic conductivity in addition to that due to electrons, the Hall effect due to the ionic motion is so small as to be quite negligible.

For simplicity let us consider an infinite semiconductor having an electric current density J in the x-direction and a magnetic field in the z-direction. We shall first of all deal with an n-type semiconductor with spherical constant-energy surfaces.

If we consider the effect of the magnetic field on the drift velocity of the electrons in an electric field in the xy-plane, we see that there must be a component of the field at right angles to the current flow (x-direction) to balance the transverse force due to the magnetic field. The drift velocity v_x is equal to $-J/ne$. The average transverse force on an electron in the y-direction is then eBv_x and the balancing field $\mathscr{E}_y$ is given by

$$e\mathscr{E}_y = eBv_x = -BJ/n, \tag{33}$$

where B is the magnetic induction. The component of electric field $\mathscr{E}_x$, parallel to the current, since $\overline{v_x} = -\mathscr{E}\mu_e$, is given by the equation

$$J = ne\mu_e\mathscr{E}_x. \tag{34}$$

The angle θ between the current and resultant electric field is then given by[4]

$$\tan\theta = \mathscr{E}_y/\mathscr{E}_x = -B\mu_e; \tag{35}$$

the angle θ (see Fig. 5.2) is called the Hall angle.

In measuring the voltage due to the Hall effect, a rectangular strip, whose thickness and width are small compared with the length, is generally used as shown in Fig. 5.3. We shall take the x-axis along the strip and the y-axis across its width. The Hall effect is observed by measuring the transverse voltage set up across the strip; this is the external voltage required to make the current flow entirely in the x-direction. The Hall effect is described by means of the Hall constant R

[3] *Amer. J. Maths.* (1879) **2**, 287; *Amer. J. Sci.* (1880) **19**, 200.

[4] Note that the product $B\mu_e$ is a dimensionless quantity. This is not obvious at first sight. In M.K.S. units μ_e has dimensions $m^2 V^{-1} s^{-1}$, while B, measured in teslas (T), has dimensions $V\,s\,m^{-2}$.

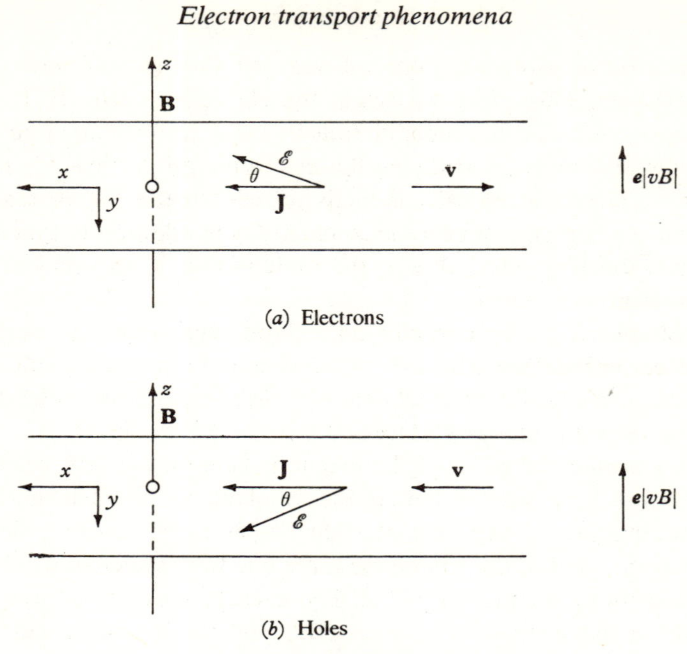

Fig. 5.2. Small field Hall effect.

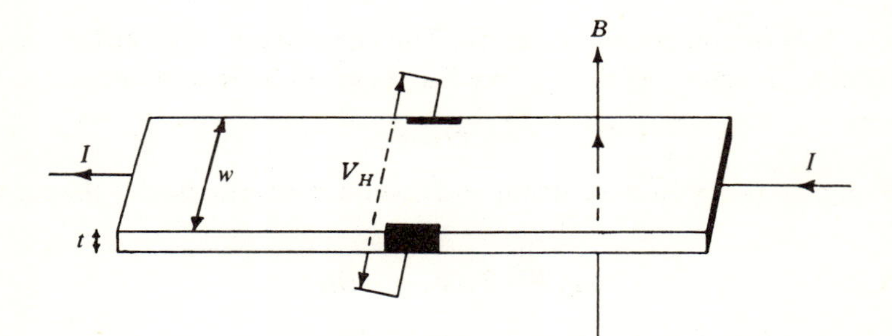

Fig. 5.3. Arrangement for Hall effect measurement.

defined in terms of the current density J by the equation

$$\mathscr{E}_H = RJB, \tag{36}$$

where $\mathscr{E}_H$ is the Hall field. If t is the thickness of the strip and w its width (see Fig. 5.3) we have, if V_H is the Hall voltage and I the current,

$$V_H = RIB/t, \tag{37}$$

or

$$R = tV_H/IB. \tag{38}$$

We have, therefore, since $\mathbf{J} = ne\mu_e$,

$$\mathscr{E}_H = Rne\mu_e = -B\mu_e\mathscr{E}_x,$$

so that

$$R = -1/ne. \tag{39}$$

As we shall see later, this equation holds exactly only when τ is not a function of the velocity (or energy). We generally have to include a numerical factor r which varies between about 1 and 2 according to the type of scattering which predominates, and also with the degree of degeneracy in the conduction band, and write

$$R = -r/ne. \tag{40}$$

Apart from this uncertainty, which we shall discuss later, we may obtain from a measurement of the Hall constant the important quantity n giving the electron concentration in the conduction band.

Similarly when holes predominate we have

$$\tan\theta = B\mu_h \tag{35a}$$

$$R = 1/pe \tag{39a}$$

or more accurately

$$R = r'/pe, \tag{40a}$$

the value of r' for holes not being necessarily the same as r for electrons (but both being positive numbers).

The sign of the Hall constant tells whether we have an n-type or p-type semiconductor. The dimensions of the Hall constant R will be seen from equation (39) to be L^3/Q, where Q represents electric charge. The appropriate M.K.S. units are therefore $\mathrm{m^3\,C^{-1}}$. Since n is frequently expressed in $\mathrm{cm^{-3}}$, R is also frequently expressed in $\mathrm{cm^3\,C^{-1}}$. Care must be taken to convert to $\mathrm{m^3\,C^{-1}}$ if equation (37) is used with B expressed in T. If centimetres are used for length and I and V expressed in amperes and volts with B in gauss, a factor 10^9 must be included in the right-hand side of equation (39).

From equations (39), (40), we note that, since for conduction principally by electrons or by holes $\sigma = ne\mu_e$ or $\sigma = pe\mu_h$,

$$|R|\sigma = \mu. \tag{41}$$

This relationship is again only exact when $r = 1$; more generally we have

$$|R|\sigma/r = \mu. \tag{42}$$

We may, however, *define* a quantity μ_H with the dimensions of mobility by means of the equation

$$|R|\sigma = \mu_H, \tag{43}$$

and μ_H is called the Hall mobility. The mobility defined in terms of the conductivity we write as μ_c. Thus when τ is constant we have $\mu_c = \mu_H$, otherwise we have $\mu_H = r\mu_c$. We shall later calculate some values of r.

Let us now make an estimate of the value of the Hall voltage for a typical semiconductor. Suppose we have a strip of n-type Ge 1 cm wide and 1 mm thick with $n = 5\times 10^{14}\ \mathrm{cm}^{-3}$. The Hall constant $-R$ will be approximately equal to 10^4 (equation (39)) $\mathrm{cm}^3\ \mathrm{C}^{-1}$. These are the units conventionally used for R. To calculate the voltage it is better to transform to M.K.S. units. Thus $R = 10^{-2}\ \mathrm{m}^3\ \mathrm{C}^{-1}$. If $B = 10^3\ \mathrm{G} = 10^{-1}\ \mathrm{T}$ and the current is equal to 1 mA we have from equation (37) $V_H =$ 1 mV.

We may note from Fig. 5.2 that the effect of the magnetic field is to make holes and electrons tend to drift to the *same* side of the current direction, namely to the right, whereas in an electric field they drift in *opposite* directions. When we have both electrons and holes present in appreciable quantities with *no magnetic field* the drift velocity v_x of both holes and electrons must be zero when J_x is zero – i.e. we cannot have equal and opposite hole and electron currents which balance. When we have a magnetic field present, however, it is possible for the electron and hole currents to balance so that we can have $J_y = 0$ but v_y for electrons and for holes not individually equal to zero. This follows from the observation that they tend to drift in the *same* direction transverse to the current flow in a magnetic field. We must therefore proceed rather more carefully when calculating the magnitude of the Hall effect for mixed conduction. This illustrates very well the essential difference between semiconductors and metals having one type of current carrier.

In the calculation of the Hall constant R for mainly holes or mainly electrons it has not been necessary to assume that the Hall angle θ is small. To obtain a *simple* formula for R corresponding to equations (39) and (39a) we must, as we shall see, assume that the Hall angle θ is small, i.e. both $B\mu_e \ll 1$ and $B\mu_h \ll 1$. We shall later remove this restriction in calculating R for mixed conduction but it is instructive to see why the simple treatment of single-carrier conduction cannot be applied when we have *both* holes and electrons present in comparable numbers.

We cannot now assume that v_y for holes or for electrons is zero under the condition $J_y = 0$ so we must use the equations of motion. For electrons we have

$$\left.\begin{aligned} m_e\dot{v}_x &= -e\mathscr{E}_x - ev_yB, \\ m_e\dot{v}_y &= -e\mathscr{E}_y + ev_xB. \end{aligned}\right\} \tag{44}$$

If $B\mu_e$ is small, the term Bev_y will be of the order of $Be\mu_e\mathscr{E}_y = (B\mu_e)^2e\mathscr{E}_x$ so will be very small and may be neglected. The first of equations (44) may then be replaced by the simpler expression

$$m_e\dot{v}_x = -e\mathscr{E}_x \tag{45}$$

and proceeding as before we get $\overline{v_x} = -\mu_e\mathscr{E}_x$. Using this result in the second equation of (44) we have

$$m_e\dot{v}_y = -e\mathscr{E}_y - eB\mu_e\mathscr{E}_x. \tag{46}$$

The right-hand side of equation (46) now being constant we may proceed as before and obtain

$$\overline{v_{ye}} = -\mu_e\mathscr{E}_y - \mu_e^2B\mathscr{E}_x. \tag{47}$$

Similarly for holes we have

$$\overline{v_{yh}} = \mu_h\mathscr{E}_y - \mu_h^2B\mathscr{E}_x. \tag{48}$$

The component of current density J_y in the y-direction is now obtained by adding the contributions from holes and electrons to give

$$\begin{aligned} J_y &= -ne\overline{v_y}e + pe\overline{v_y}h \\ &= ne\mu_e\mathscr{E}_y + ne\mu_e^2B\mathscr{E}_x + pe\mu_e\mathscr{E}_y - pe\mu_e^2B\mathscr{E}_x. \end{aligned} \tag{49}$$

We may *now* set $J_y = 0$ to obtain the ratio of the fields $\mathscr{E}_y/\mathscr{E}_x$, and so

$$\tan\theta = \mathscr{E}_y/\mathscr{E}_x = (p\mu_h^2 - n\mu_e^2)/(n\mu_e + p\mu_h). \tag{50}$$

The Hall constant R is then given by $R = \mathscr{E}_y/JB = \mathscr{E}_y/\mathscr{E}_x\sigma B$, and using the value for σ given by equation (20) we obtain

$$R = \frac{p\mu_h^2 - n\mu_e^2}{e(n\mu_e + p\mu_h)^2}. \tag{51}$$

The equations (50) and (51) reduce to (35) and (39) when $p \doteq 0$ and to (35a) and (39a) when $n = 0$.

Of particular interest is the value of the Hall coefficient for intrinsic material; we have then $n = p = n_i$ and

$$R_i = \frac{1}{en_i}\frac{\mu_h - \mu_e}{(\mu_e + \mu_h)}. \tag{52}$$

If we introduce the mobility ratio $b = \mu_e/\mu_h$, equation (51) may be written in the form

$$R = \frac{1}{e}\frac{p - b^2 n}{(nb + p)^2}. \tag{53}$$

Thus we have $R = 0$ when $p = b^2 n$ and we may note that the product $R\sigma$ is a function of the ratio p/n. The intrinsic Hall constant is given by

$$R_i = -\frac{1}{en_i}\frac{b-1}{b+1}. \tag{54}$$

If $b > 1$, as is usual, R_i is negative, i.e. the electron conduction predominates. The variation of Hall constant with temperature is shown for an n-type semiconductor (with $b > 1$) in Fig. 5.4 and for a p-type (with $b > 1$) in Fig. 5.5. We note the zero and change of sign of R in this case.

It is instructive now to calculate the drift velocities $\overline{v_{ye}}$ and $\overline{v_{yh}}$. Substituting the value of $\mathscr{E}_y$ in equations (47) and (48) we obtain

$$\overline{v_{ye}} = -p\mu_e\mu_h B\mathscr{E}_x(\mu_e + \mu_h)/(n\mu_e + p\mu_h), \tag{55}$$

$$\overline{v_{yh}} = -n\mu_e\mu_h B\mathscr{E}_x(\mu_e + \mu_h)/(n\mu_e + p\mu_h). \tag{56}$$

Again we see that both electrons and holes tend to drift to the right, i.e. in negative y-direction. We also see that $\overline{v_{ye}} = 0$ when $J_y = 0$ only when $p = 0$, and $\overline{v_{yh}} = 0$ when $J_y = 0$ only when $n = 0$, as assumed in our treatment of the conditions when either electrons or holes predominate.

In the derivation of these formulae we have assumed that θ is small both for holes and electrons, i.e. that $B\mu_e$ and $B\mu_h$ are both small. We may, however, remove this restricton by solving the equation of motion. This we now proceed to do. The equations of motion for electrons may be written in the form

$$\left.\begin{aligned} \dot{v}_x &= -(e/m_e)\mathscr{E}_x - \omega v_y, \\ \dot{v}_y &= -(e/m_e)\mathscr{E}_y + \omega v_x, \end{aligned}\right\} \tag{57}$$

where $\omega = eB/m_e$. $\omega/2\pi$ is just the cyclotron frequency introduced in § 2.5. These are the well-known equations describing the motion of electrons in crossed-electric and magnetic fields. If we use complex variables $Z = v_x + iv_y$ and $z = x + iy$, equations (57) may be written in the form

$$\dot{Z} - i\omega Z = -(e/m_e)(\mathscr{E}_x + i\mathscr{E}_y). \tag{58}$$

The general solution of this equation is

$$Z = Z_0\, e^{i\omega t} + (e/m_e)(\mathscr{E}_x + i\mathscr{E}_y)(1 - e^{i\omega t})/i\omega, \tag{59}$$

where $Z_0 = v_{x0} + iv_{y0}$, and reduces, as it should, to $Z = Z_0$ at $t = 0$. We now proceed to average Z over all collision times taking as before $\bar{Z}_0 = 0$, and obtain

$$\bar{Z} = \overline{v_x} + i\overline{v_y} = \frac{1}{\tau}\int_0^\infty Z\, e^{-t/\tau}\, dt$$

$$= (-e/m_e)(\mathscr{E}_x + i\mathscr{E}_y)\tau/(1 - i\omega\tau). \tag{60}$$

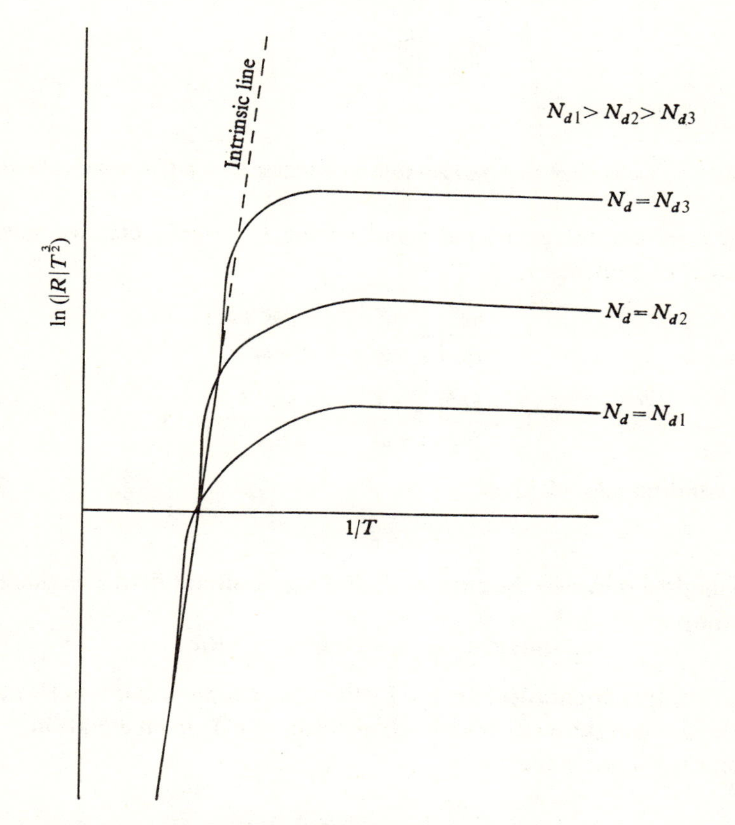

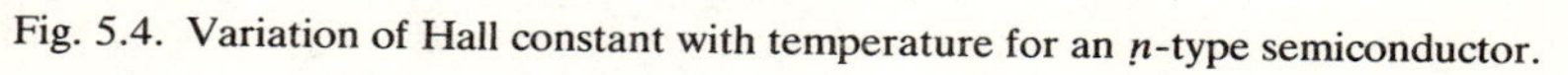
Fig. 5.4. Variation of Hall constant with temperature for an n-type semiconductor.

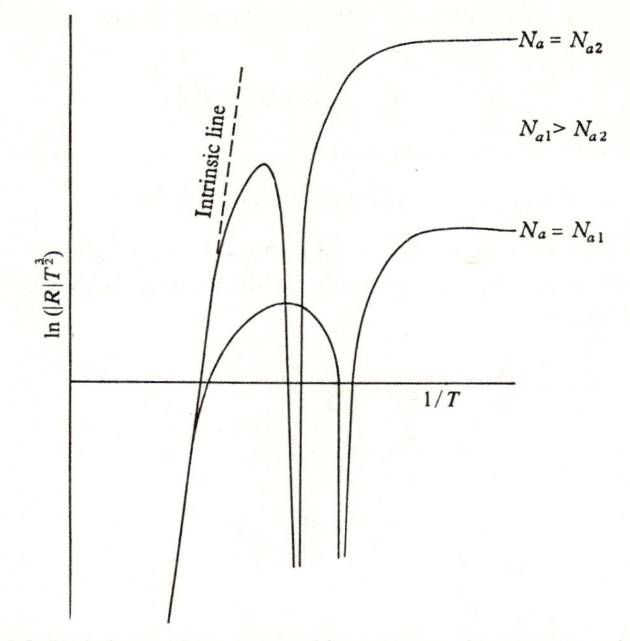

Fig. 5.5. Variation of Hall constant with temperature for a *p*-type semiconductor.

Taking real and imaginary parts and writing $J_x = -en\overline{v_x}$, etc., we have for the current densities J_x, J_y

$$J_x = \frac{ne^2}{m_e}\left\{\frac{\tau\mathscr{E}_x}{1+\omega^2\tau^2} - \frac{\omega\tau^2\mathscr{E}_y}{1+\omega^2\tau^2}\right\}, \tag{61}$$

$$J_y = \frac{ne^2}{m_e}\left\{\frac{\tau\mathscr{E}_y}{1+\omega^2\tau^2} + \frac{\omega\tau^2\mathscr{E}_x}{1+\omega^2\tau^2}\right\}. \tag{62}$$

The condition $J_y = 0$ gives

$$\mathscr{E}_y/\mathscr{E}_x = -\omega\tau. \tag{63}$$

The angle θ between the current J_x and the resultant field $\mathscr{E}$ is therefore given by

$$\tan\theta = -\omega\tau = -Be\tau/m_e = -B\mu_e. \tag{64}$$

This is simply equation (35), and justifies the simple treatment by means of which it was derived. When we substitute for $\mathscr{E}_y$ from equation (63) in equation (61) we have

$$J_x = \frac{ne^2\tau\mathscr{E}_x}{m_e} = ne\mu_e\mathscr{E}_x = \sigma\mathscr{E}_x, \tag{65}$$

where $\mu_e = e\tau/m_e$. We therefore have exactly the same results as before. We note that under the above conditions, namely, spherical energy surfaces and constant τ, there is no change of resistance due to a transverse magnetic field. We have also

$$R = -\frac{1}{ne}. \tag{66}$$

This independence of the magnetic field no longer holds when we have two types of carrier present. For holes we simply change the sign of ω in equations (61) and (62) and we obtain as, before, $R = 1/pe$. When both holes and electrons are present we must add their contributions to J_x and J_y, and we shall assume different values of τ, namely, τ_e and τ_h, for electrons and holes. We may write equations (61) and (62) in the form

$$\left.\begin{aligned} J_x &= A\mathscr{E}_x - D\mathscr{E}_y, \\ J_y &= A\mathscr{E}_y + D\mathscr{E}_x, \end{aligned}\right\} \tag{67}$$

so that, taking $J_y = 0$ and proceeding as above, we obtain

$$\left.\begin{aligned} -BR &= D/(A^2 + D^2), \\ \sigma &= (A^2 + D^2)/A. \end{aligned}\right\} \tag{68}$$

The equations, when we have two types of carrier, may be written in the form

$$\left.\begin{aligned} J_x &= (A_1 + A_2)\mathscr{E}_x - (D_1 + D_2)\mathscr{E}_y, \\ J_y &= (A_1 + A_2)\mathscr{E}_y + (D_1 + D_2)\mathscr{E}_x. \end{aligned}\right\} \tag{69}$$

When $J_y = 0$ we have

$$\mathscr{E}_x = -(A_1 + A_2)\mathscr{E}_y/(D_1 + D_2). \tag{70}$$

Substituting for $\mathscr{E}_x$ we have

$$BR = \frac{-(D_1 + D_2)}{(A_1 + A_2)^2 + (D_1 + D_2)^2}. \tag{71}$$

In terms of the conductivities σ_1, σ_2 and Hall coefficients R_1, R_2, which we should have with one kind of carrier only present, we may write A_1, A_2, and D_1, D_2 in the form

$$\begin{aligned} A_1 &= \sigma_1/(1 + \sigma_1^2 B^2 R_1^2), \\ D_1 &= -\sigma_1^2 B R_1/(1 + \sigma_1^2 B^2 R_1^2), \quad \text{etc.} \end{aligned}$$

After some algebraic manipulation equation (71) may be reduced to the form

$$R=\frac{R_1\sigma_1^2(1+\sigma_2^2B^2R_2^2)+R_2\sigma_2^2(1+\sigma_1^2B^2R_1^2)}{(\sigma_1+\sigma_2)^2+\sigma_1^2\sigma_2^2B^2(R_1+R_2)^2}. \tag{72}$$

Equation (72) was obtained by R. G. Chambers[5] and shown to be generally true even when τ is a function of v and for any two kinds of carrier. On substituting from equations (65) and (66) and similar ones for holes we obtain

$$R=\frac{(p-nb^2)+b^2\mu_h^2B^2(p-n)}{(bn+p)^2+b^2\mu_h^2B^2(p-n)^2}\cdot\frac{1}{e}. \tag{73}$$

Equation (73) is, however, strictly accurate only when τ is constant (see § 5.3.3). We have a zero Hall coefficient when

$$p=nb^2(1+\mu_h^2B^2)/(1+b^2\mu_h^2B^2). \tag{74}$$

Thus the value of p/n for which $R=0$ varies with the magnetic field unless μ_hB is much smaller than unity. For Ge, $b\mu_h=0.39\ \mathrm{m^2\,V^{-1}\,s^{-1}}$, and for $B=10^3\ \mathrm{G}=0.1\ \mathrm{T}$, $bB\mu_h=0.039$; the variation of R with B is therefore small. For $B=10^4$ G, however, the variation from the low-field value becomes appreciable. For very large values of B it is interesting to note that again we have

$$R\to\frac{1}{e(p-n)}. \tag{75}$$

In the above calculation we have assumed that the transverse component of current is zero. In practice, holes and electrons may recombine at the surface of a semiconductor and it is possible to have a condition in which $J_y\neq 0$. This effect has been studied by P. C. Banbury, H. K. Henisch and A. Many[6] and also by R. Landauer and J. Swanson,[7] and it has been shown that, provided certain precautions are taken, it is unlikely to make a significant difference in practical measurements.

5.2.3 Transverse magneto-resistance

We have seen that a semiconductor with spherical constant-energy surfaces shows no transverse magneto-resistance when τ is constant and we have only one type of carrier. This is no longer true when we have

[5] *Proc. Phys. Soc.* A (1952) **65**, 903.
[6] *Proc. Phys. Soc.* A (1953) **66**, 753.
[7] *Phys. Rev.* (1953) **91**, 555.

more than one type of carrier. We have from equations (69), (70)

$$\sigma = J_x/\mathscr{E}_x = (A_1+A_2)+(D_1+D_2)^2/(A_1+A_2). \tag{76}$$

From equation (71) we have then

$$\begin{aligned} R\sigma &= -(D_1+D_2)/B(A_1+A_2) \\ &= \frac{R_1\sigma_1^2(1+\sigma_2^2B^2R_2^2)+R_2\sigma_2^2(1+\sigma_1^2B^2R_1^2)}{(\sigma_1+\sigma_2)+B^2\sigma_1\sigma_2(\sigma_1R_1^2+\sigma_2R_2^2)} \end{aligned} \tag{77}$$

and hence we have

$$\sigma = \frac{(\sigma_1+\sigma_2)^2+\sigma_1^2\sigma_2^2B^2(R_1+R_2)^2}{(\sigma_1+\sigma_2)+B^2\sigma_1\sigma_2(\sigma_1R_1^2+\sigma_2R_2^2)}. \tag{78}$$

To the first order in B^2 we have, since for τ constant σ_1 and σ_2 are independent of B,

$$-\frac{\Delta\sigma}{\sigma_0B^2} = \frac{\Delta\rho}{\rho_0B^2} = \frac{-\sigma_1^2\sigma_2^2(R_1+R_2)^2}{\sigma_0^2} + \frac{\sigma_1\sigma_2(\sigma_1R_1^2+\sigma_2R_2^2)}{\sigma_0}, \tag{79}$$

where $\Delta\sigma$ is the change in conductivity from the zero-field value σ_0 ($\sigma_0=\sigma_1+\sigma_2$) and similarly $\Delta\rho$ is the change in resistivity from the zero-field value ρ_0. Inserting the values of σ_1, R_1 and σ_2, R_2 appropriate to holes and electrons we have

$$-\frac{\Delta\sigma}{\sigma_0B^2} = \frac{\Delta\rho}{\rho_0B^2} = \frac{np\mu_e\mu_h(\mu_e+\mu_h)^2}{(n\mu_e+p\mu_h)^2} \tag{80}$$

$$= \frac{npb\mu_h^2(1+b)^2}{(nb+p)^2}. \tag{81}$$

By using the expression for the small-field Hall coefficient R_0 given by equation (53), equation (81) may be written in the form

$$-\frac{\Delta\sigma}{\sigma_0} = \frac{\Delta\rho}{\rho_0} = \xi R_0^2\sigma_0^2B^2, \tag{82}$$

where

$$\xi = \frac{npb(1+b)^2}{(p-nb^2)^2}. \tag{83}$$

The quantity ξ is generally called the transverse magneto-resistance coefficient. It is equal to zero with either $n=0$ or $p=0$. Thus we see that this kind of semiconductor, which shows no transverse magneto-resistance when either n-type or p-type may do so in the mixed conduction range. It will show no longitudinal magneto-resistance since we have J_z independent of B.

5.3 Relaxation time a function of E

We must now consider the effect of the variation of the relaxation time τ with velocity. We shall deal first with an extrinsic semiconductor with spherical constant-energy surfaces. There are two methods of approach. The more fundamental is to use Boltzmann's equation which shows how the distribution function is disturbed from its equilibrium value (Fermi–Dirac distribution) by the application of external fields. This method is treated in detail in advanced text-books on the theory of solids.[8] Alternatively, we may use the solutions of the equations of motion (equations (61), (62)) and average them with a suitable weighting function.[9] It might be thought at first sight that for electrons we should simply average the equations as they stand using the weighting function

$$N(E)f_0[(E-E_F)/kT]$$

giving the number of occupied states per unit energy interval, f_0 being the Fermi function defined in § 4.1. This, however, is incorrect. It does not take account of the fact that *occupation* of neighbouring states will prevent transitions to them. Even when the system is non-degenerate it is incorrect, giving too little weight to the more energetic electrons. For the non-degenerate case the correct weighting function turns out to be $EN(E)f_0[(E-E_F)/kT]$, the function f_0 being replaced by an exponential.[9]

5.3.1 Electrical conduction

Let us consider first of all only those electrons with energy between E and $E+dE$ and suppose that there are $\Delta n(E)$ of them. For these electrons the contribution to the current density J will be

$$\Delta J = \Delta n(E)e^2\tau(E)/m_e. \tag{84}$$

We assume τ to be a function of the scalar velocity v and hence of the energy E. To obtain the total current density we sum over all these groups of electrons, but using a weighting factor $W(E)$. In this way we obtain

$$J = ne^2\langle\tau\rangle/m_e \tag{85}$$

[8] See, for example, A. H. Wilson, *The Theory of Metals* (Cambridge University Press, 2nd edition, 1953); J. M. Ziman, *Electrons and Phonons* (Oxford University Press, 1960); *W.M.C.S.*, §§ 10.5, 10.6.

[9] This method has been adopted by Harvey Brooks, *Advances in Electronics and Electron Physics* (Academic Press, 1955) **7**, 85; see also *W.M.C.S.*, § 10.6.

where, for a non-degenerate n-type semiconductor, we replace $\sum \Delta n(E)\tau(E)W(E)$ by[10] $n\langle\tau\rangle$ where

$$\langle\tau\rangle = \frac{\int_0^\infty EN(E)\exp[-E/\boldsymbol{k}T]\tau(E)\,dE}{\int_0^\infty EN(E)\exp[-E/\boldsymbol{k}T]\,dE}. \tag{86}$$

If we have 'spherical' energy bands $N(E)\propto E^{\frac{1}{2}}$ (see § 4.2) and

$$\langle\tau\rangle = \frac{\int_0^\infty E^{\frac{3}{2}}\exp[-E/\boldsymbol{k}T]\tau(E)\,dE}{\int_0^\infty E^{\frac{3}{2}}\exp[-E/\boldsymbol{k}T]\,dE}. \tag{87}$$

We may therefore use our previous analysis of conductivity, replacing the constant relaxation time τ by the average $\langle\tau\rangle$.

When we do not have a completely non-degenerate situation we must use a rather more complex form of weighting function $W(E)$ given by[11]

$$W(E) = Ef_0(1-f_0) \tag{88}$$

which, apart from a factor which may be ignored, may be written as

$$W(E) = E\frac{\partial f_0}{\partial E}, \tag{89}$$

so that for 'spherical' bands

$$\langle\tau\rangle = \frac{\int_0^\infty E^{\frac{3}{2}}\frac{\partial f_0}{\partial E}\tau(E)\,dE}{\int_0^\infty E^{\frac{3}{2}}\frac{\partial f_0}{\partial E}\,dE}. \tag{90}$$

When we have a condition of strong degeneracy we have, very approximately, $f_0 = 1$, $E < E_F$, and $f_0 = 0$, $E \geqslant E_F$. The integrals involving $\partial f_0/\partial E$ therefore give a contribution only near $E = E_F$, and the other terms may consequently be taken outside. Thus we have

$$\int_0^\infty \tau(E)E^{\frac{3}{2}}\frac{\partial f_0}{\partial E}\,dE \simeq -\tau(E_F)E_F^{\frac{3}{2}},$$

$$\int_0^\infty E^{\frac{3}{2}}\frac{\partial f_0}{\partial E}\,dE \simeq \int_0^{E_F} E^{\frac{1}{2}}\,dE = -E_F^{\frac{3}{2}},$$

[10] *W.M.C.S.*, § 10.6. [11] *W.M.C.S.*, § 10.6.

and hence, we have approximately

$$J_x = e^2 \mathscr{E} n\tau(E_F)/m_e. \tag{91}$$

Thus we see that all the electrons appear to take part in the conductivity, but the relaxation time is that appropriate for an energy corresponding to the Fermi level.

When the constant-energy surfaces are not spherical the weighting function for averaging must be modified. We shall now only consider the non-degenerate case in which f_0 may be replaced by an exponential approximation. Let us suppose that the constant-energy surfaces are ellipsoids as in § 2.1. We now include field components along the (x, y, z) directions. It may readily be shown that [12]

$$\left.\begin{aligned} \sigma_{xx} &= n(e^2/m_1)\langle\tau\rangle, \\ \sigma_{yy} &= n(e^2/m_2)\langle\tau\rangle, \\ \sigma_{zz} &= n(e^2/m_3)\langle\tau\rangle, \end{aligned}\right\} \tag{92}$$

where $\langle\tau\rangle$ is the same average quantity as defined in equation (87). Thus we see that the weighting function Ef_0 may also be used for the non-spherical case. Equations (92) are the same as equations (25) derived for a constant value of τ provided we use the average value $\langle\tau\rangle$. The appropriate mobilities are given by

$$\left.\begin{aligned} \mu_{c1} &= (e/m_1)\langle\tau\rangle, \\ \mu_{c2} &= (e/m_2)\langle\tau\rangle, \\ \mu_{c3} &= (e/m_3)\langle\tau\rangle. \end{aligned}\right\} \tag{93}$$

The formula (31) for the conductivity effective mass still applies, as do the deductions concerning crystals with cubic symmetry made in § 5.2.1.

5.3.2 Variation of τ with energy E

For a semiconductor with spherical constant-energy surfaces it may be shown that the mean free path l due to scattering by the lattice vibrations is independent of the electron's velocity (see § 8.5). We therefore have $\tau_l \propto v^{-1}$ or $\tau_l \propto E^{-\frac{1}{2}}$. For scattering due to ionized impurity centres it may be shown that $\tau_i \propto E^{\frac{3}{2}}$ (see § 8.9). Thus we see that we get a quite different average for the two types of scattering. For the impurity scattering the effect of the very slow electrons or holes predominates,

[12] *W.M.C.S.*, § 10.6.

but for lattice scattering the faster electrons or holes are given more weight. When both forms of scattering are present we should combine them to give a value of τ given by[13]

$$\frac{1}{\tau}=\frac{1}{\tau_l}+\frac{1}{\tau_i}, \tag{94}$$

since the total cross-section for scattering will be the sum of the two (see § 5.1). Over a considerable range of energies it is found that for many semiconductors we may write $\tau = aE^{-s}$, where s is a constant, and a may vary with the temperature. The value of s is generally derived from the observed variation of mobility with temperature. We have for the non-degenerate case with $\tau = aE^{-s}$

$$\langle\tau\rangle = a\int_0^{\infty} E^{\frac{3}{2}-s}\exp\left[-E/\boldsymbol{k}T\right]dE \Big/ \int_0^{\infty} E^{\frac{3}{2}}\exp\left[-E/\boldsymbol{k}T\right]dE. \tag{95}$$

When s is a fraction, the integral in the numerator of equation (95) may be expressed in terms of the Γ-function defined by means of the integral

$$\Gamma(x)=\int_0^{\infty} t^{x-1}\,e^{-t}\,dt$$

from which it is readily shown that $\Gamma(x+1)=x\Gamma(x)$; when n is an integer $\Gamma(n+1)=n!$. When $x=\frac{1}{2}$ we have

$$\Gamma(\tfrac{1}{2})=\int_0^{\infty} t^{-\frac{1}{2}}e^{-t}\,dt = 2\int_0^{\infty} e^{-y^2}\,dy = \pi^{\frac{1}{2}}.$$

We have, therefore, putting $t = E/\boldsymbol{k}T$,

$$\langle\tau\rangle = a(\boldsymbol{k}T)^{-s}\Gamma(\tfrac{5}{2}-s)/\Gamma(\tfrac{5}{2}). \tag{96}$$

For lattice scattering with $s=\frac{1}{2}$ we may write $\tau = l/v$, where l is the mean free path and is independent of the velocity v, being of the form $l = A/T$, where A is a constant (see § 8.5, equation (24)). We therefore have

$$\langle\tau\rangle = \frac{4m_e^{\frac{1}{2}}l}{3(2\pi\boldsymbol{k}T)^{\frac{1}{2}}} \tag{97}$$

and

$$\mu_l = \frac{e\langle\tau\rangle}{m_e} = \frac{4el}{3(2\pi m_e\boldsymbol{k}T)^{\frac{1}{2}}}. \tag{98}$$

[13] This is not strictly correct because of the different variations of τ_l and τ_i with energy. The averages have been carried out with $\tau_l \propto E^{-\frac{1}{2}}$ for various amounts of impurity scattering by H. Jones, *Phys. Rev.* (1951) **81**, 149, V. A. Johnson and K. Lark-Horowitz, *ibid.* (1951) **82**, 977. The problem of combining the two types of scattering has been further discussed by Esther M. Conwell, *Proc. I.R.E.* (1952) **11**, 1327.

Since $l = A/T$ we may write μ_l in the form

$$\mu_l = BT^{-\frac{3}{2}}, \tag{99}$$

where B is a constant.

For ionized impurity scattering we have (see § 8.9)

$$\langle\tau\rangle = 8a(\boldsymbol{k}T)^{\frac{3}{2}}/\pi^{\frac{1}{2}}. \tag{100}$$

In this case we have approximately

$$a = C/\ln(dT),$$

where C and d are constants (see § 8.9) so that

$$\mu_i = DT^{\frac{3}{2}}/\ln(dT), \tag{101}$$

where D and d are constants.

When we combine the two forms of scattering we may take for the mobility μ_c

$$\frac{1}{\mu_c} = \frac{1}{\mu_l} + \frac{1}{\mu_i}, \tag{102}$$

i.e. we combine the mobilities in the same way as we combine the relaxation times.[14] Thus we see that, at high temperatures, lattice scattering should predominate but at a sufficiently low temperature, depending on the density of ionized impurity centres, the impurity scattering should take over. This is generally indicated by a deviation from the simple power law which usually holds for the variation of mobility with temperature at the higher temperatures. It is generally found that equation (99) does not apply even at higher temperatures. The reasons for this are mainly due to the fact that its validity depends on the assumption of spherical constant-energy surfaces. We shall postpone discussion of this to § 8.7.

For pure lattice scattering when we have strong degeneracy we have

$$\langle\tau\rangle = \tau(E_F) \propto lE_F^{-\frac{1}{2}}, \tag{103}$$

so that $\langle\tau\rangle/l$ varies only slowly with T and may be regarded as a constant. We have then, since l is inversely proportional to T,

$$\mu_c = \mu_0(T_0/T). \tag{104}$$

Thus if we have an extrinsic semiconductor with large and nearly constant value of n, we have

$$\sigma = ne\mu_0(T_0/T) = \sigma_0(T_0/T). \tag{105}$$

[14] But see p. 119.

The resistivity is therefore proportional to T and the semiconductor behaves like a metal, for which a variation of this kind holds except at very low temperatures. This shows that the original criterion for placing an electrical conductor in the class of semiconductors, namely, that its conductivity should increase as the temperature is raised, is not a reliable criterion and holds only in the near-intrinsic range when n or p is increasing with the temperature more rapidly than the mobility is decreasing. For lattice scattering leading to a $T^{-\frac{3}{2}}$ law for the conductivity mobility, the variation of conductivity with temperature in the intrinsic range will be seen from equation (18) of § 4.2 to be given by

$$\sigma = \sigma_0 \exp[-\Delta E/2\boldsymbol{k}T], \tag{106}$$

where σ_0 is a constant.

Even when we have a semiconductor so pure that impurity scattering may be neglected (e.g. pure Ge and Si) it is found that the mobility does not obey the $T^{-\frac{3}{2}}$ law either for holes or for electrons. For electrons in Ge and Si this is now thought to be due to the fact that we have a number of minima or 'valleys' in the conduction band, and scattering of electrons may be taking place between the different 'valleys'. This has the effect of modifying the $T^{-\frac{3}{2}}$ law and although a simple power law for τ does not seem to be valid over a wide range of energies, a value of s can be found so that the law $\tau = \tau_0 E^{-s}$ holds over a considerable range and enables this form to be used to calculate various quantities such as conductivity, Hall constant, etc. We postpone a further discussion of this so-called 'inter-valley' scattering to § 8.7.

When we no longer have spherical constant-energy surfaces we use equations (92) for the conductivity tensor, the average value of τ being the same as above, as we have shown in § 5.3.1. The discussion of this case given in § 5.2.1 still applies provided we use the averaged value of τ. We assumed that τ depended only on the *magnitude* of the velocity. This analysis may be generalized by using a tensor form for τ to take account of variation of τ with direction in the crystal, as has been shown by C. Herring and E. Vogt.[15] For transport in *steady* fields the effect of the anisotropy in τ is to weight each of the components of the effective-mass tensor with the corresponding component of the relaxation-time tensor, i.e. it is the quantities m_r/τ_r which appear in the various transport equations.

In the treatment we have given of electrical conduction we have assumed that the electrical field is not large enough greatly to upset the

[15] *Phys. Rev.* (1956) **101**, 944.

energy distribution of electrons and holes. We shall later (see § 12.1) consider what happens when this is no longer so and we have the situation in which 'hot electrons' are created, i.e. electrons with energy considerably in excess of the average thermal equilibrium value. Moreover we have neglected quantum effects which occur in very high magnetic fields and which also modify the equilibrium distribution to an appreciable extent (see §§ 12.5.1, 12.8). Both of these topics we shall defer till later.

5.3.3 Hall effect for semiconductor with spherical constant-energy surfaces

When we come to consider the effect of magnetic fields on the motion of electrons and holes in semiconductors two methods of approach to the problem are again open to us. We may include terms due to the magnetic field in the expressions for the force components in Boltzmann's equation and may then seek solutions to as high a degree of approximation as is desired. This method, which is undoubtedly the most logical and rigorous, has been used extensively in advanced treatments of transport phenomena both in metals and in semiconductors.[16] The second approach is to return to the equations of motion of the electron as averaged over times between collisions, and to average these with respect to τ regarded as a function of v or E as in § 5.3.

We return to equations (61) and (62) giving the values of the current densities J_x, J_y in the presence of a constant magnetic field, of induction B, directed along the z-axis. To begin with, we shall assume that $\omega = eB/m_e$ is so small that $\omega\tau \ll 1$ for all values of τ of importance, and neglect the terms in $\omega^2\tau^2$ in the denominators of equations (61), (62), which apply when we have one type of carrier and spherical constant-energy surfaces. We then have, on averaging over all values of E, using the weighting function given in equation (87) or (88), as appropriate,

$$J_x = \frac{ne^2}{m_e}[\langle\tau\rangle\mathscr{E}_x - \omega\langle\tau^2\rangle\mathscr{E}_y],$$

$$J_y = \frac{ne^2}{m_e}[\langle\tau\rangle\mathscr{E}_y + \omega\langle\tau^2\rangle\mathscr{E}_x]. \qquad (107)$$

[16] See *W.M.C.S.*, § 10.8.

The condition that $J_y = 0$ now gives for the Hall angle θ the equation

$$\tan\theta = \frac{\mathscr{E}_y}{\mathscr{E}_x} = -\omega\frac{\langle\tau^2\rangle}{\langle\tau\rangle}. \tag{108}$$

Also we have, again neglecting terms in ω^2,

$$\mathscr{E}_y = -\frac{B}{ne}\frac{\langle\tau^2\rangle J_x}{\langle\tau\rangle^2} = RBJ_x. \tag{109}$$

The Hall constant R is therefore now given by

$$R = -\frac{r}{ne}, \tag{110}$$

where

$$r = \langle\tau^2\rangle/\langle\tau\rangle^2. \tag{111}$$

This is the generalization of equation (39) when τ is a function of E. For positive holes we have

$$R = \frac{r}{pe}, \tag{112}$$

where r again is given by equation (111) with the appropriate values of τ for holes.

If we have lattice scattering for which $\tau = aE^{-\frac{1}{2}}$ (see § 5.3.2) we have

$$\begin{aligned}\langle\tau^2\rangle &= a^2\int_0^\infty E^{\frac{1}{2}}\exp[-E/\boldsymbol{k}T]\,dE\bigg/\int_0^\infty E^{\frac{3}{2}}\exp[-E/\boldsymbol{k}T]\,dE\\ &= 2a^2/3\boldsymbol{k}T.\end{aligned} \tag{113}$$

From equation (96) for $\langle\tau\rangle$ we have $r = 3\pi/8 = 1.18$. Thus we have

$$R = -\frac{3\pi}{8ne} \quad \text{or} \quad \frac{3\pi}{8pe}. \tag{114}$$

More generally, when we have $\tau = aE^{-s}$, we have

$$\begin{aligned}\langle\tau^2\rangle &= a^2\int_0^\infty E^{\frac{3}{2}-2s}\exp[-E/\boldsymbol{k}T]\,dE\bigg/\int_0^\infty E^{\frac{3}{2}}\exp[-E/\boldsymbol{k}T]\,dE\\ &= a^2(\boldsymbol{k}T]^{-2s}\Gamma(\tfrac{5}{2}-2s)/\Gamma(\tfrac{5}{2}).\end{aligned} \tag{115}$$

From equation (96) for $\langle\tau\rangle$ we obtain

$$r = \Gamma(\tfrac{5}{2}-2s)\Gamma(\tfrac{5}{2})/[\Gamma(\tfrac{5}{2}-s)]^2. \tag{116}$$

For ionized impurity scattering with $s = -\frac{3}{2}$ we have $r = 315\pi/512 = 1.93$. When we have a degenerate semiconductor $\langle\tau^2\rangle = [\tau(E_F)]^2$ and we have $r=1$ as for a constant value of τ. Also for very large values of B we have, as before, $R \to -1/ne$ or $1/pe$, as may be seen by taking ω very large in equations (107).

If we have several different kinds of carriers we must sum equations (107) over the various types. In particular, if we have both electrons and holes we must add their contribution to the total current. Provided the relaxation times for holes and for electrons vary in the same way with the energy E it will readily be seen by proceeding as in § 5.2.2 that

$$R = \frac{r}{e} \frac{p - b^2 n}{(p+bn)^2}, \tag{117}$$

r again being given by equation (116).

When it is not permissible to neglect the terms $(\omega\tau)^2$ in the denominators of equations (61), (62) we may write the average current densities in the form

$$\left. \begin{aligned} J_x &= A\mathscr{E}_x - D\mathscr{E}_y, \\ J_y &= D\mathscr{E}_x + A\mathscr{E}_y, \end{aligned} \right\} \tag{118}$$

where

$$A = \frac{ne^2}{m_e}\left\langle \frac{\tau}{1+\omega^2\tau^2} \right\rangle, \tag{119}$$

$$D = \frac{ne^2\omega}{m_e}\left\langle \frac{\tau^2}{1+\omega^2\tau^2} \right\rangle. \tag{120}$$

If $\tau = aE^{-s}$ we have

$$A = \frac{4ane^2}{3\pi^{\frac{1}{2}} m_e (kT)^s} \int_0^\infty \frac{x^{\frac{3}{2}+s} e^{-x}\, dx}{x^{2s} + a^2\omega^2 (kT)^{-2s}}, \tag{121}$$

$$D = \frac{4a^2ne^2}{3\pi^{\frac{1}{2}} m_e (kT)^{2s}} \int_0^\infty \frac{x^{\frac{3}{2}} e^{-x}\, dx}{x^{2s} + a^2\omega^2 (kT)^{-2s}}. \tag{122}$$

In terms of A and D the Hall constant is given by (cf. equation (68))

$$R = -D/(A^2 + D^2)B. \tag{123}$$

For lattice scattering with $s = \frac{1}{2}$ we have, writing $\mu_{c0} = e\langle\tau\rangle/m_e$

$$A = ne\mu_{c0} I_2(\alpha), \tag{124}$$

$$D = \frac{3\pi^{\frac{1}{2}}}{4} ne\mu_{c0}^2 B I_{\frac{3}{2}}(\alpha), \tag{125}$$

where

$$\alpha = \frac{9\pi}{16} B^2 \mu_{c0}^2 \tag{126}$$

and

$$I_r(\alpha)=\int_0^\infty \frac{x^r e^{-x}\,dx}{x+\alpha}. \tag{127}$$

We have then

$$R=-\frac{3\pi^{\frac{1}{2}}}{4ne}I_{\frac{3}{2}}(\alpha)\Big/\left\{I_2^2(\alpha)+\frac{9\pi}{16}\mu_{c0}^2B^2I_{\frac{3}{2}}^2(\alpha)\right\}. \tag{128}$$

As $B\to 0$, $I_2\to 1$ and $I_{\frac{3}{2}}\to \pi^{\frac{1}{2}}/2$ so that, as we should expect,

$$R\to -3\pi/8ne.$$

When we have mixed conduction we use equation (71) with the values of A and D appropriate to each type of carrier. For holes the sign of D is changed and this gives

$$R=-\frac{3\pi^{\frac{1}{2}}}{4e}\frac{nb^2I_{\frac{3}{2}}(\alpha_n)-pI_{\frac{3}{2}}(\alpha_p)}{\{nbI_2(\alpha_n)+pI_2(\alpha_p)\}^2+\alpha_p\{nb^2I_{\frac{3}{2}}(\alpha_n)-pI_{\frac{3}{2}}(\alpha_p)\}^2}, \tag{129}$$

where α_n and α_p are the values of α appropriate to electrons and holes and b is the ratio of μ_{c0} for electrons to μ_{c0} for holes. For very large values of B we have $R\to 1/e(p-n)$.

5.3.4 Hall effect for semiconductor with multiple energy maxima or minima

We must now consider the generalization of the above treatment which must be made when we have a number of equivalent minima in the conduction band or a number of equivalent maxima in the valence band. For simplicity we shall treat the case for which the extrema lie along three axes at right angles as for the conduction band of Si. We shall consider first a simple ellipsoidal constant-energy surface, the magnetic field being along the z-axis. The effective-mass components for the x, y, z-axes we shall take as m_1, m_2, m_3. We shall assume τ to be isotropic and to depend only on E. We have seen from our study of conductivity under these conditions that the same weighting function may be used as before. We write the equations of motion now in the form

$$\left.\begin{aligned}\dot{v}_x&=-(e/m_1)\mathscr{E}_x-\omega_1 v_y,\\ \dot{v}_y&=-(e/m_2)\mathscr{E}_y+\omega_2 v_y.\end{aligned}\right\} \tag{130}$$

To solve these we multiply the first by $\omega_2^{\frac{1}{2}}$ and the second by $\omega_1^{\frac{1}{2}}$ and write $Z=\omega_1^{\frac{1}{2}}v_x+i\omega_2^{\frac{1}{2}}v_y$. The equations (130) may then be written as a single equation of the form

$$\dot{Z} = -e\left(\frac{\omega_2^{\frac{1}{2}}}{m_1}\mathscr{E}_x + i\frac{\omega_1^{\frac{1}{2}}}{m_2}\mathscr{E}_y\right) + i\omega_1^{\frac{1}{2}}\omega_2^{\frac{1}{2}}Z. \tag{131}$$

The appropriate solution is

$$Z = Z_0\, e^{i\omega t} - e\left\{\frac{\omega_2^{\frac{1}{2}}\mathscr{E}_x}{m_1} + \frac{i\omega_1^{\frac{1}{2}}\mathscr{E}_y}{m_2}\right\}(1 - e^{i\omega t})/i\omega, \tag{132}$$

where we have written ω^2 for $\omega_1\omega_2$. We may proceed to average over various values of the collision time τ as in § 5.2.2 obtaining

$$\bar{Z} = -\tau e\left\{\frac{\omega_2^{\frac{1}{2}}\mathscr{E}_x}{m_1} + \frac{i\omega_1^{\frac{1}{2}}\mathscr{E}_y}{m_2}\right\}\Big/(1 - i\omega\tau). \tag{133}$$

Taking real and imaginary parts we have

$$-e\overline{v_x} = e^2\left\{\frac{\tau\mathscr{E}_x}{m_1(1+\omega^2\tau^2)} - \frac{\omega_1\mathscr{E}_y\tau^2}{m_2(1+\omega^2\tau^2)}\right\}, \tag{134}$$

$$-e\overline{v_y} = e^2\left\{\frac{\tau\mathscr{E}_y}{m_2(1+\omega^2\tau^2)} + \frac{\omega_2\mathscr{E}_x\tau^2}{m_1(1+\omega^2\tau^2)}\right\}. \tag{135}$$

We shall neglect terms in B^2 to obtain the low-field Hall constant, and carry out the usual averaging process for τ. If we have equivalent ellipsoids along the x, y, z-axes we must sum the contributions from all of these, permuting the masses m_1, m_2, m_3. We obtain

$$\begin{aligned} -e\Sigma\overline{v_x} = \frac{ne^2}{3}\bigg\{&\langle\tau\rangle\mathscr{E}_x\left(\frac{1}{m_1} + \frac{1}{m_2} + \frac{1}{m_3}\right) \\ &- eB\langle\tau^2\rangle\mathscr{E}_y(m_1+m_2+m_3)/m_1m_2m_3\bigg\}, \end{aligned} \tag{136}$$

$$\begin{aligned} -e\Sigma\overline{v_y} = \frac{ne^2}{3}\bigg\{&\langle\tau\rangle\mathscr{E}_y\left(\frac{1}{m_1} + \frac{1}{m_2} + \frac{1}{m_3}\right) \\ &+ eB\langle\tau^2\rangle\mathscr{E}_x(m_1+m_2+m_3)/m_1m_2m_3\bigg\}. \end{aligned} \tag{137}$$

Putting $J_y = -\Sigma e\overline{v_y} = 0$ and neglecting terms in B^2 we have

$$J_x = -nem_1m_2m_3\langle\tau\rangle^2\left(\frac{1}{m_1} + \frac{1}{m_2} + \frac{1}{m_3}\right)^2\mathscr{E}_y\Big/3B\langle\tau^2\rangle(m_1+m_2+m_3). \tag{138}$$

The Hall constant R is thus given by

$$R = -\frac{1}{ne}\frac{\langle\tau^2\rangle}{\langle\tau\rangle^2}\frac{3(m_1+m_2+m_3)}{\left(\dfrac{1}{m_1} + \dfrac{1}{m_2} + \dfrac{1}{m_3}\right)^2 m_1m_2m_3}. \tag{139}$$

If $m_2 = m_3 = m_T$ and $m_1 = m_L$ and we write $K = m_L/m_T$ we have

$$R = -\frac{r}{ne}\frac{3K(K+2)}{(2K+1)^2}, \tag{140}$$

where r is written as before for $\langle\tau^2\rangle/\langle\tau\rangle^2$. This equation also holds for any crystal with cubic symmetry, such as Ge (see § 5.3.6, equation (219)).[17] If, as before, we define the conductivity mobility μ_c by means of the equation $\sigma = ne\mu_c$ and the Hall mobility μ_H by means of the equation $R\sigma = \mu_H$ we have

$$\mu_H/\mu_c = 3rK(2+K)/(2K+1)^2. \tag{141}$$

The small-field Hall coefficients for each pair of principal axes are thus all equal. It follows that the small-field Hall coefficient measured with the current along any direction with the magnetic field at right angles gives the same value. This may be seen as follows: neglecting terms of the second order in the magnetic field we may write the equations for the current densities when the field is along the z-axis in the form

$$\left.\begin{aligned} J_x &= \sigma\mathscr{E}_x + R\sigma^2\mathscr{E}_y B_z, \\ J_y &= \sigma\mathscr{E}_y - R\sigma^2\mathscr{E}_x B_z, \\ J_z &= \sigma\mathscr{E}_z. \end{aligned}\right\} \tag{142}$$

Two similar sets of equations may be written for the current densities when the magnetic field is along the x-axis and along the y-axis; they have the same coefficients σ and $R\sigma^2$ as in equations (142) with suitable permutation. Now the general form of equations relating the current densities to the fields up to first order in $\mathscr{E}$ and $\mathbf{B}$ are, since σ is scalar to this approximation,

$$J_r = \sigma\mathscr{E}_r + \sum_{st} A_{rst}\mathscr{E}_s B_t. \tag{143}$$

From equations (142) and the two related sets of equations, putting in turn $B_x = B_y = 0$, $B_x = B_z = 0$, $B_y = B_z = 0$, we see that $A_{rst} = 0$ unless $r \neq s \neq t$; $|A_{rst}| = R\sigma^2$, $r \neq s \neq t$, and $A_{rst} = -A_{rts}$, etc. As a result we may write equation (143) in the vector form

$$\mathbf{J} = \sigma\mathscr{E} + R\sigma^2\mathscr{E} \times \mathbf{B}. \tag{144}$$

This is the same form as for an isotropic conductor and shows that the small-field Hall effect has also an isotropic behaviour. This is no longer

[17] For a more general treatment see B. Abeles and S. Meiboom, *Phys. Rev.* (1954) **95**, 31; C. Herring, *Bell Syst. Tech. J.* (1955) **34**, 237.

necessarily true when the magnetic field is so large that terms in B^2 have to be taken into account. For very large values of B it will readily be seen from equations (134), (135) that again we simply have $R \to -1/ne$ for electrons and $R \to 1/pe$ for holes.

For mixed conduction we may insert the values of R_1, R_2 for electrons and holes in equation (72). Neglecting the terms in B^2 we have

$$R = \frac{R_1\sigma_1^2 + R_2\sigma_2^2}{(\sigma_1+\sigma_2)^2}, \tag{145}$$

$$= -\frac{1}{e}\frac{\mu_{He}\mu_{ce}n - \mu_{Hh}\mu_{ch}p}{(n\mu_{ce} + p\mu_{ch})^2}, \tag{146}$$

$$= -\frac{1}{e}\frac{r_e b^2 n - r_h p}{(bn+p)^2}, \tag{147}$$

where $r_e = \mu_{He}/\mu_{ce}$, $r_h = \mu_{Hh}/\mu_{ch}$. μ_{He}, μ_{ce} are respectively the Hall and conductivity mobilities for electrons and μ_{Hh}, μ_{ch} the Hall and conductivity mobilities for holes. When the variation of τ with E is of the form $\tau = aE^{-s}$, and s has the same value for holes and electrons, equation (147) reduces to the form of equation (117) with $b = \mu_{ce}/\mu_{ch} = \mu_{He}/\mu_{Hh}$. In practice it is found that for many semiconductors the value of s is not quite the same for holes as for electrons and the more general form given by equation (147) must be used. For very large values of B we have $R \to R_1R_2/(R_1+R_2)$ giving again $R \to 1/e(p-n)$.

In the above treatment we have neglected magnetic quantum effects (magnetic quantization) already mentioned in connection with electrical conduction in very high magnetic fields. The effect of modification of the equilibrium distribution appears to have less effect on the Hall constant than on the conductivity and the result $R \to 1/e(p-n)$ appears to be true to a good degree of approximation when $B\mu_e \gg 1$ or $B\mu_h \gg 1$ (see § 12.8).

5.3.5 Magneto-resistance of semiconductor with spherical constant-energy surfaces

To obtain an expression for the change of resistance due to a magnetic field when τ is a function of the energy E we return to equations (61), (62). We shall consider only effects quadratic in B so we retain terms only up to this order of magnitude in the expansion of the expression for

J_x. We have then for a semiconductor with spherical constant-energy surfaces

$$J_x = \frac{ne^2}{m_e}[\tau\mathscr{E}_x - \omega\tau^2\mathscr{E}_y - \omega^2\tau^3\mathscr{E}_x], \tag{148}$$

$$J_y = \frac{ne^2}{m_e}[\tau\mathscr{E}_y + \omega\tau^2\mathscr{E}_x]. \tag{149}$$

Averaging over the allowed values of E with the usual weighting function, and putting $J_y = 0$, we have

$$J_x = \frac{ne^2}{m_e}\mathscr{E}_x[\langle\tau\rangle + \omega^2\langle\tau^2\rangle^2/\langle\tau\rangle - \omega^2\langle\tau^3\rangle]. \tag{150}$$

If we write

$$\sigma_0 = 1/\rho_0 = (ne^2/m_e)\langle\tau\rangle, \tag{151}$$

$$\sigma = \sigma_0(1 - \Delta\sigma/\sigma_0), \tag{152}$$

$$\rho = \rho_0(1 + \Delta\rho/\rho_0) \tag{153}$$

we have when $\Delta\sigma/\sigma_0 \ll 1$

$$-\frac{\Delta\sigma}{\sigma_0} = \frac{\Delta\rho}{\rho_0} = \frac{e^2B^2}{m_e^2}\frac{\langle\tau^3\rangle\langle\tau\rangle - \langle\tau^2\rangle^2}{\langle\tau\rangle^2}. \tag{154}$$

If R_0 is the small-field Hall coefficient (see equations (110), (111)) equation (154) may be expressed in the form

$$\frac{\Delta\rho}{\rho_0} = -\frac{\Delta\sigma}{\sigma_0} = \xi R_0^2\sigma_0^2B^2, \tag{155}$$

where the quantity ξ is the magneto-resistance coefficient and is given by the equation

$$\xi + 1 = \langle\tau^3\rangle\langle\tau\rangle/\langle\tau^2\rangle^2. \tag{156}$$

For a non-degenerate semiconductor with only lattice scattering and $\tau \propto E^{-\frac{1}{2}}$ we have $\xi + 1 = 4/\pi$ so that $\xi = 0.275$, as will readily be verified by expressing the averages in terms of integrals such as (87) which reduce to Γ-functions. For ionized impurity scattering with $\tau \propto E^{\frac{3}{2}}$, $1 + \xi = 32\,768/6615\pi$ so that $\xi = 0.57$. It may be shown that $\langle\tau^3\rangle\langle\tau\rangle$ is always greater than or equal to $\langle\tau^2\rangle^2$ so that $\xi \geqslant 0$. When $\tau \propto E^{-s}$ we have

$$\xi + 1 = \Gamma(\tfrac{5}{2} - 3s)\Gamma(\tfrac{5}{2} - s)/[\Gamma(\tfrac{5}{2} - 2s)]^2. \tag{157}$$

When τ is constant we have $\xi = 0$, as we found in § 5.2.3. This is effectively the case for a degenerate semiconductor for which, as we have seen in § 5.3.2, we may take for all the averages their value at the Fermi level. For a non-degenerate semiconductor with spherical constant-energy surfaces, we have a transverse magneto-resistance given by equation (155) but no longitudinal magneto-resistance, since J_z is not affected by the magnetic field, along the z-axis.

Returning to equations (61), (62), we see that for very large values of B we have on averaging

$$J_x = \frac{ne^2}{m_e}[-\mathscr{E}_y/\omega], \tag{158}$$

$$J_y = \frac{ne^2}{\omega^2 m_e}\left[\mathscr{E}_y\left\langle\frac{1}{\tau}\right\rangle + \omega\mathscr{E}_x\right]. \tag{159}$$

For $J_y = 0$ we have $\mathscr{E}_x = -\mathscr{E}_y\langle 1/\tau\rangle/\omega$, so that

$$J_x = \frac{ne^2}{m_e}\left\langle\frac{1}{\tau}\right\rangle^{-1}\mathscr{E}_x. \tag{160}$$

If we write σ_∞ for the large-field conductivity and ρ_∞ for the large-field resistivity, we have

$$\sigma_0/\sigma_\infty = \langle 1/\tau\rangle\langle\tau\rangle = \rho_\infty/\rho_0. \tag{161}$$

We thus see that as the magnetic field is increased the resistivity increases to a saturation value ρ_∞. For scattering with $\tau \propto E^{-s}$ we have

$$\rho_\infty/\rho_0 = \Gamma(\tfrac{5}{2}+s)\Gamma(\tfrac{5}{2}-s)/[\Gamma(\tfrac{5}{2})]^2. \tag{162}$$

For lattice scattering with $s = \frac{1}{2}$ we have $\rho_\infty/\rho_0 = (32/9)\pi$, while for ionized impurity scattering with $s = -\frac{3}{2}$ we have $\rho_\infty/\rho_0 = (32/3)\pi$.

For mixed conduction we may use the values of σ_1 and σ_2, the conductivities for each type of carrier, together with equation (78) to obtain σ, for any value of B. For very large values of B we clearly have

$$\sigma_\infty = \frac{\sigma_1\sigma_2(R_1+R_2)^2}{\sigma_1 R_1^2 + \sigma_2 R_2^2}, \tag{163}$$

σ_1, σ_2, R_1, R_2 having their values for large value of B. If σ_1, R_1 refer to electrons and σ_2, R_2 to holes we have, using equation (161) and the values $-1/ne$ and $1/pe$ for R_1 and R_2,

$$\frac{\sigma_\infty}{\sigma_0} = \frac{\rho_0}{\rho_\infty} = \frac{(n-p)^2}{(\sigma_{0e}+\sigma_{0h})(\sigma_1^{-1}p^2+\sigma_2^{-1}n^2)} \tag{164}$$

$$= \frac{(n-p)^2}{\left(n\frac{\langle\tau_e\rangle}{m_e}+p\frac{\langle\tau_h\rangle}{m_h}\right)\left(nm_e\left\langle\frac{1}{\tau_e}\right\rangle+pm_h\left\langle\frac{1}{\tau_h}\right\rangle\right)}. \tag{165}$$

If we write

$$\sigma_{\infty e}=\frac{ne^2}{m_e}\left\langle\frac{1}{\tau_e}\right\rangle^{-1}=ne\mu_{\infty e} \tag{166}$$

and

$$\sigma_{\infty h}=\frac{pe^2}{m_h}\left\langle\frac{1}{\tau_h}\right\rangle^{-1}=pe\mu_{\infty h} \tag{167}$$

we may write equation (165) in the form

$$\frac{\sigma_\infty}{\sigma_0}=\frac{\rho_0}{\rho_\infty}=\frac{(n-p)^2}{(n\mu_e+p\mu_h)(n/\mu_{\infty e}+p/\mu_{\infty h})}; \tag{168}$$

$\mu_{\infty e}$ and $\mu_{\infty h}$ may be called the high-field mobilities for holes and electrons. Since they depend on the average values of $1/\tau$, if we have more than one scattering mechanism, the various contributions to the high-field resistance are simply added. For high fields we do not require to define a Hall mobility and conductivity mobility, since $R_\infty\,\sigma_\infty=\mu_\infty$ for each type of carrier. For an intrinsic semiconductor the high-field Hall constant is infinite and the limiting value of the high-field conductivity is zero according to equation (168). We should point out, however, that fields for which $\omega\tau$ is very much greater than 1 are not easily produced, and in any case the equations on which the large-field values are based are not strictly valid when $\omega\langle\tau\rangle\gg 1$ because they neglect effects due to magnetic quantization. We shall discuss this topic later (see § 12.8). If we write $\omega\langle\tau\rangle$ in terms of the magnetic induction and the conductivity mobility we have $\omega\langle\tau\rangle=B\mu_c$. For high values of μ_c, say about $10\ \mathrm{m^2\,V^{-1}\,s^{-1}}$ (such as are obtainable at low temperatures) and $B=10$ T (10^5 G) we have $\omega\langle\tau\rangle=100$, and high-field conditions may be applied.

To obtain the low-field magneto-resistance for mixed conduction we may either use equations (148), (149) with added terms for the other types of carriers present or we may use equation (78) which is valid for variable τ as well as for constant τ. When we expand to the first order in B^2 we must include as well as the terms in equation (79) the first-order terms in σ_1 and σ_2. Thus if we write Δ_1 for $\Delta\sigma_1/\sigma_1B^2$, etc., we have

$$\frac{\Delta\sigma}{\sigma_0B^2}=\frac{\Delta_1\sigma_{10}+\Delta_2\sigma_{20}}{\sigma_{10}+\sigma_{20}}-\frac{\sigma_{10}\sigma_{20}(\sigma_{10}R_{10}^2+\sigma_{20}R_{20}^2)}{\sigma_{10}+\sigma_{20}}+\frac{\sigma_{10}^2\sigma_{20}^2(R_{10}+R_{20})^2}{(\sigma_{10}+\sigma_{20})^2}. \tag{169}$$

If σ_1 and σ_2 refer to holes and electrons the last two terms now reduce to

$$-\frac{np\mu_{ce}\mu_{cp}(\mu_{He}+\mu_{Hp})^2}{(n\mu_{ce}+p\mu_{ch})^2}. \tag{170}$$

This is the same as the value in equation (80) when τ is constant, for we have then $\mu_{ce}=\mu_{He}=\mu_e$, $\mu_{ch}=\mu_{Hh}=\mu_h$. The first term may be written in the form

$$-\frac{n\mu_{ce}\mu_{He}^2\xi_e+p\mu_{ch}\mu_{Hh}^2\xi_h}{n\mu_{ce}+p\mu_{ch}}, \tag{171}$$

where
$$1+\xi_e=\langle\tau_e^3\rangle\langle\tau_e\rangle/\langle\tau_e^2\rangle^2,$$
$$1+\xi_h=\langle\tau_h^3\rangle\langle\tau_h\rangle/\langle\tau_h^2\rangle^2.$$

We may therefore write the full expression for $\Delta\sigma/\sigma_0$ in the form

$$\frac{\Delta\sigma}{\sigma_0}=-B^2\mu_{Hh}^2\left\{\frac{npb(1+b')^2}{(nb+p)^2}+\frac{nbb'^2\xi_e+p\xi_h}{(nb+p)}\right\}, \tag{172}$$

where $b=\mu_{ce}/\mu_{ch}$, $b'=\mu_{He}/\mu_{Hh}$. Both terms on the right-hand side of equation (172) represent negative contributions to $\Delta\sigma$. The first is due to mixed conduction and vanishes when either $n=0$ or $p=0$; the second is due to the individual contributions of holes and electrons to the magneto-resistance and vanishes if τ is constant for both types of carrier.

The quantity ξ for mixed conduction, defined by means of the equation

$$\xi=-\Delta\sigma/\sigma_0^3R_0^2B^2, \tag{173}$$

may now be obtained by using for R_0 the value given by equation (146) for R. We have, combining equations (172) and (173),

$$\xi+1=\frac{(nb+p)\{nbb'^2(\xi_e+1)+p(\xi_h+1)\}}{(bb'n-p)^2}. \tag{174}$$

In terms of the average values of various powers of τ equation (174) may be written in the form

$$\xi+1=\frac{(n\langle\tau_e\rangle/m_e+p\langle\tau_h\rangle/m_h)(n\langle\tau_e^3\rangle/m_e^3+p\langle\tau_h^3\rangle/m_h^3)}{(n\langle\tau_e^2\rangle/m_e^2-p\langle\tau_h^2\rangle/m_h^2)^2}. \tag{175}$$

5.3.6 Magneto-resistance of semiconductor with constant-energy surfaces in the form of ellipsoids

The full theory of the magneto-resistance of a semiconductor with constant-energy surfaces in the form of ellipsoids is very complicated. It has been given by S. Meiboom and B. Abeles,[18] and by M. Shibuya.[19] Here we shall deal with only a few simple cases. First, let us suppose we have a single minimum at $k=0$ and the constant-energy surfaces ellipsoids with principal axes along the x, y, z-directions. We shall take the magnetic field along the z-axis. Expanding the equations (134) and (135) as far as terms in B^2, we have on averaging over τ

$$J_x = ne^2\left[\frac{\langle\tau\rangle}{m_1}\mathscr{E}_x - \frac{eB\langle\tau^2\rangle}{m_1m_2}\mathscr{E}_y - \frac{e^2B^2\langle\tau^3\rangle}{m_1^2m_2}\mathscr{E}_x\right],$$
$$J_y = ne^2\left[\frac{\langle\tau\rangle}{m_2}\mathscr{E}_y + \frac{eB\langle\tau^2\rangle}{m_1m_2}\mathscr{E}_x\right]. \tag{176}$$

Putting $J_y = 0$, as before, we obtain for J_x the equation

$$J_x = ne^2\left[\frac{\langle\tau\rangle}{m_1} + \frac{e^2B^2\langle\tau^2\rangle^2}{m_1^2m_2\langle\tau\rangle} - \frac{e^2B^2\langle\tau^3\rangle}{m_1^2m_2}\right]\mathscr{E}_x, \tag{177}$$

and we find that

$$\frac{\Delta\sigma}{\sigma_0} = -\frac{e^2B^2}{m_1m_2}\xi\langle\tau^2\rangle^2/\langle\tau\rangle^2, \tag{178}$$

where ξ is given by equation (156).

If we have a number of constant-energy ellipsoids with their axes of revolution along the co-ordinate axes, we must add to equations (176) the contributions from each. We shall take $m_1 = m_L$, $m_2 = m_3 = m_T$ and write $K = m_L/m_T$. This arrangement applies for Si. The second term in equation (177) may always be written in the form $B^2R_0^2\sigma_0^3$. When the terms from the other ellipsoids are added this is still so with the appropriate value of R_0. We therefore have on writing

$$\sigma_0 = \frac{ne^2\langle\tau\rangle}{3}\left(\frac{2}{m_T} + \frac{1}{m_L}\right),$$

$$J_x = \mathscr{E}_x\left[\sigma_0 + R_0^2\sigma_0^3B^2 - ne^4B^2\langle\tau^3\rangle\left(\frac{1}{m_Lm_T^2} + \frac{1}{m_Tm_L^2} + \frac{1}{m_T^3}\right)\right]. \tag{179}$$

[18] *Phys. Rev.* (1954) **95**, 31. [19] *Phys. Rev.* (1954) **95**, 1388.

Using for R_0 the value given by equation (139) for R, equation (179) may be written in the form

$$J_x = \mathscr{E}_x\left[\sigma_0 + R_0^2\sigma_0^3B^2 - \frac{R_0^2\sigma_0^3(K^2+K+1)(2K+1)\langle\tau^3\rangle\langle\tau\rangle}{K(2+K)^2\langle\tau^2\rangle^2}\right] \tag{180}$$

and we find that

$$-\frac{\Delta\sigma}{\sigma_0} = R_0^2\sigma_0^2B^2\{(1+\xi)F(K)-1\}, \tag{181}$$

where ξ is given by equation (156) and

$$F(K) = (K^2+K+1)(2K+1)/K(K+2)^2. \tag{182}$$

When $m_L = m_T$, $K = 1$ and $F(K) = 1$. Equation (181) then simply reduces to equation (155).

The magneto-resistance coefficient, which in this case is usually written ζ_{100}^{001}, the lower indices giving the current direction and the upper indices the magnetic field direction, is given by

$$\zeta_{100}^{001} = (1+\xi)F(K) - 1. \tag{183}$$

Since we have also under these conditions

$$J_z = \sigma_0\mathscr{E}_z \tag{184}$$

it follows that

$$\zeta_{100}^{001} = 0, \tag{185}$$

i.e. there is no longitudinal magneto-resistance in a $\langle 100\rangle$ direction. As we shall see, there are three independent magneto-resistance coefficients for a crystal with cubic symmetry, so we require another value of ζ before we can calculate the magneto-resistance for any direction of current and magnetic field. The calculation of the magneto-resistance coefficient when both current and field are in the (110) direction, i.e. $H_x = H_y$, $H_z = 0$; $J_x = J_y$, $J_z = 0$, may be carried out by solving the appropriate equations of motion and averaging. The algebra involved is rather tedious and will not be given. The result is

$$\zeta_{110}^{110} = (1+\xi)\frac{(K-1)^2(2K+1)}{2K(K+2)^2}. \tag{186}$$

A general expression for the current vector $\mathbf{J}$ in terms of the electric field vector $\mathscr{E}$ and magnetic induction $\mathbf{B}$ may be given in the following form when we consider only linear terms in $\mathscr{E}$ and linear and quadratic terms in $\mathbf{B}$,

$$\mathbf{J} = \sigma_0\mathscr{E} + R_0\sigma_0^2\mathscr{E}\times\mathbf{B} + M\mathscr{E}\mathbf{BB}, \tag{187}$$

where M is a fourth-order tensor. The first two terms follow from equation (144) and the third is a quite general term linear in $\mathscr{E}$ and quadratic in $\mathbf{B}$. In terms of the components of the vectors along three mutually perpendicular axes (x, y, z), which we shall specify by subscripts $r, s, t = 1, 2, 3$, we have

$$J_r = \sigma_0 \mathscr{E}_r + R_0 \sigma_0^2 (\mathscr{E}_s B_t - \mathscr{E}_t B_s) + M_{ruvw} \mathscr{E}_u B_v B_w \tag{188}$$

using the summation convention that repeated subscripts are to be summed. For crystals with cubic symmetry, a considerable number of the elements of the tensor M are zero, as may be seen by considering the effect of changing the direction of one of the vectors in a number of symmetrical configurations. We have in general only the following non-zero types of elements

$$M_{rrrr}, \quad M_{rrss} \quad (r \neq s), \quad M_{rsrs} \quad (r \neq s).$$

These in turn are unchanged by permutation of the subscripts r, s, t. We have included M_{rssr} in M_{rsrs} as it gives the same term. Thus we have three independent constants in the quadratic term of equation (188). When the constant-energy surfaces are ellipsoids of revolution along the x, y, z-axes we have also $M_{rrrr} = 0$ as follows from the fact that we have a zero longitudinal magneto-resistance coefficient along any of the co-ordinate axes. If the axes of the ellipsoids are along the $\langle 111 \rangle$ axes, there is, however, as we shall see, no direction in which the longitudinal magneto-resistance coefficient is zero and so $M_{rrrr} \neq 0$.

If we write $M_{rrrr} = S$, $M_{rrss} = P$, $M_{rsrs} = Q$, equation (188) becomes

$$J_r = \sigma_0 \mathscr{E}_r + R_0 \sigma_0^2 (\mathscr{E}_s B_t - \mathscr{E}_t B_s) + S \mathscr{E}_r B_r^2 + P \mathscr{E}_r (B_s^2 + B_t^2) + Q(\mathscr{E}_s B_r B_s + \mathscr{E}_t B_r B_t). \tag{189}$$

This will be seen to be equivalent to an equation derived by F. Seitz[20] on general grounds for any crystal of cubic symmetry

$$\mathbf{J} = \sigma_0 \mathscr{E} + \alpha \mathscr{E} \times \mathbf{B} + \beta \mathscr{E} B^2 + \gamma \mathbf{B}(\mathscr{E} \cdot \mathbf{B}) + \delta T \mathscr{E}, \tag{190}$$

where α, β, γ, δ are constants and T is a diagonal tensor having elements B_r^2, B_s^2, B_t^2. By comparison with equation (189) it will be seen that

$$\alpha = R_0 \sigma_0^2, \quad \beta = P, \quad \gamma = Q, \quad \delta = S - P - Q.$$

When measurements of magneto-resistance are made it is usual to keep the value of the current constant and observe the changes in

[20] *Phys. Rev.* (1950) **79**, 372.

potential difference as the magnetic field is applied. By inverting equation (190) we may obtain $\mathscr{E}$ as a function of the current density **J**. To the first order in **B** we clearly have

$$\mathscr{E} = \mathbf{J}/\sigma_0 - R_0\mathbf{J}\times\mathbf{B}. \tag{191}$$

Substituting this expression for $\mathscr{E}$ in the term of equation (189) linear in B, and $\mathbf{J}/\sigma_0$ for $\mathscr{E}$ in the quadratic term, we obtain

$$\sigma_0\mathscr{E} = \mathbf{J} - R_0\sigma_0\mathbf{J}\times\mathbf{B} + b\mathbf{J}B^2 + c\mathbf{B}(\mathbf{J}\cdot\mathbf{B}) + dT\mathbf{J}$$

using the result that

$$\mathbf{B}\times(\mathbf{J}\times\mathbf{B}) = \mathbf{J}B^2 - \mathbf{B}(\mathbf{J}\cdot\mathbf{B}).$$

The resistivity ρ is given by $(\mathscr{E}\cdot\mathbf{J})/J^2$ so we have

$$\begin{aligned}\frac{\Delta\rho}{\rho_0 B^2} &= \frac{(\mathbf{E}\cdot\mathbf{J}) - \rho_0}{B^2(\mathbf{E}_0\cdot\mathbf{J})}\\ &= b + c(\mathbf{J}\cdot\mathbf{B})^2/J^2B^2 + d(J_x^2B_x^2 + J_y^2B_y^2 + J_z^2B_z^2)/J^2B^2,\end{aligned} \tag{192}$$

where

$$\begin{aligned} b &= -R_0^2\sigma_0^2 - \beta/\sigma_0 = -R_0^2\sigma_0^2 - P/\sigma_0,\\ c &= R_0^2\sigma_0^2 - \gamma/\sigma_0 = R_0^2\sigma_0^2 - Q/\sigma_0,\\ d &= -\delta/\sigma_0 = (P+Q-S)/\sigma_0,\end{aligned}$$

S, P, Q, being the values of M_{rrrr}, M_{rrss}, M_{rsrs}, etc. If i, j, k are the direction cosines giving the direction of current flow and l, m, n, the direction cosines giving the direction of the magnetic field we have

$$\frac{\Delta\rho}{\rho_0 B^2} = b + c(il + jm + kn)^2 + d(i^2l^2 + j^2m^2 + k^2n^2). \tag{193}$$

This equation, proposed by J. Bardeen, was given by G. L. Pearson and H. Suhl,[21] who have used it in their discussion of magneto-resistance measurements on Si and Ge. It is most useful in determining the variation of magneto-resistance with crystal orientation.

If the constant-energy surfaces are ellipsoids of revolution with axes along the x, y, z-axes we have $S = 0$, and P and Q may be expressed in terms of the magneto-resistance coefficients ζ. For example, if **B** is along the z-direction and **J** along the x-direction we have simply $\Delta\sigma/\sigma_0B^2 = -b$ so that

$$R_0^2\sigma_0^3 + P = \Delta\sigma/B^2, \tag{194}$$

[21] *Phys. Rev.* (1951) **83**, 768.

or

$$\frac{\Delta\sigma}{R_0^2\sigma_0^3B^2} = 1 + P/R_0^2\sigma_0^3 = -\zeta_{100}^{001}. \tag{195}$$

Hence we have

$$M_{rrss} = P = -R_0^2\sigma_0^3(\zeta_{100}^{001} + 1), \tag{196}$$

$$= -R_0\sigma_0^3(1+\xi)(K^2+K+1)(2K+1)/K(K+2)^2. \tag{197}$$

Again, taking $B_x = B_y$, $B_z = 0$ and $J_x = J_y$, $J_z = 0$ we have

$$\frac{\Delta\sigma}{\sigma_0 B^2} = -(b+c+d/2)$$

$$= \tfrac{1}{2}(P+Q)/\sigma_0. \tag{198}$$

In terms of ζ_{110}^{110} we have

$$\tfrac{1}{2}(P+Q) = -R_0^2\sigma_0^3\zeta_{110}^{110}, \tag{199}$$

so that

$$M_{rsrs} = Q = R_0^2\sigma_0^3(\zeta_{100}^{001} - 2\zeta_{110}^{110} + 1), \tag{200}$$

$$= 3R_0\sigma_0^3(1+\xi)(2K+1)/K(K+2)^2. \tag{201}$$

Thus we have

$$M_{rsrs}/M_{rrss} = Q/P = -3K/(K^2+K+1). \tag{202}$$

The coefficients b, c, d are then given in terms of the quantities ζ by the following equations

$$\left.\begin{aligned} b &= R_0^2\sigma_0^2\zeta_{100}^{001}, \\ c &= R_0^2\sigma_0^2(2\zeta_{110}^{110} - \zeta_{100}^{001}), \\ d &= -2R_0^2\sigma_0^2\zeta_{110}^{110}. \end{aligned}\right\} \tag{203}$$

Another symmetrical arrangement of the constant-energy surfaces is in the form of equivalent ellipsoids along the $\langle 111\rangle$ axes. An example of this arrangement is the conduction band of Ge. To treat this problem we must extend the analysis of § 5.3.4 to deal with the situation in which the effective-mass tensor is no longer diagonal. Let us take the co-ordinate axes (x, y, z) along the $\langle 100\rangle$ directions in the crystal. Consider first a constant energy surface in the form of an ellipsoid of revolution whose axis lies along the (111) direction, and has transverse effective mass m_T and longitudinal effective mass m_L. Let us take a new co-ordinate system (x', y', z') with the x'-axis having direction cosines $(1/\sqrt{3}, 1/\sqrt{3}, 1/\sqrt{3})$, i.e. in the (111) direction. The y'-axis we shall take at right angles to this and having direction cosines $(1/\sqrt{2}, 0-1/\sqrt{2})$. The

z'-axis is the mutual perpendicular with direction cosines $(-1/\sqrt{6}, 2/\sqrt{6}, -1/\sqrt{6})$. With these axes the equation of an ellipsoid corresponding to energy E is

$$E = \frac{\hbar^2}{2}[k_x'^2/m_L + (k_y'^2 + k_z'^2)/m_T]. \tag{204}$$

We now transform to the x, y, z-co-ordinates using the transformation matrix

	x	y	z
x'	$1/\sqrt{3}$	$1/\sqrt{3}$	$1/\sqrt{3}$
y'	$1/\sqrt{2}$	0	$-1/\sqrt{2}$
z'	$-1/\sqrt{6}$	$2/\sqrt{6}$	$-1/\sqrt{6}$

By this means we obtain for the equation of the constant-energy ellipsoid

$$E = \frac{\hbar^2}{2}\left[\frac{1}{3}\left(\frac{2}{m_T} + \frac{1}{m_L}\right)(k_x^2 + k_y^2 + k_z^2) - \frac{2}{3}\left(\frac{1}{m_T} - \frac{1}{m_L}\right)(k_x k_y + k_y k_z + k_z k_x)\right]. \tag{205}$$

Let us write

$$\frac{1}{m_c} = \frac{1}{3}\left(\frac{2}{m_T} + \frac{1}{m_L}\right), \tag{206}$$

$$\frac{1}{m_d} = \frac{1}{3}\left(\frac{1}{m_T} - \frac{1}{m_L}\right), \tag{207}$$

m_c is the conductivity effective mass which we have previously used. For the effective-mass tensor we have

$$m_{xx} = m_{yy} = m_{zz} = m_c, \quad m_{xy} = m_{xz} = m_{yz} = -m_d.$$

We have therefore for the equations of motion (§2.5, equation (39)) when the magnetic field is along the z-axis

$$\left.\begin{aligned} \dot{v}_x &= -\frac{e}{m_c}(\mathscr{E}_x + Bv_y) + \frac{e}{m_d}(\mathscr{E}_y - Bv_x) + \frac{e\mathscr{E}_z}{m_d}, \\ \dot{v}_y &= -\frac{e}{m_c}(\mathscr{E}_y - Bv_x) + \frac{e}{m_d}(\mathscr{E}_x + Bv_y) + \frac{e}{m_d}\mathscr{E}_z, \\ \dot{v}_z &= -\frac{e}{m_c}\mathscr{E}_z + \frac{e}{m_d}(\mathscr{E}_x + Bv_y) + \frac{e}{m_d}(\mathscr{E}_y - Bv_x). \end{aligned}\right\} \tag{208}$$

If we write $\omega_c = eB/m_c$, $\omega_d = eB/m_d$ these equations may be rewritten in the form

$$\left.\begin{aligned}\dot{v}_x + \dot{v}_y &= -e\left(\frac{1}{m_c} - \frac{1}{m_d}\right)\mathscr{E}_x - e\left(\frac{1}{m_c} - \frac{1}{m_d}\right)\mathscr{E}_y + \frac{2e}{m_d}\mathscr{E}_z + (\omega_c - \omega_d)(v_x - v_y),\\ \dot{v}_x - \dot{v}_y &= -e\left(\frac{1}{m_c} + \frac{1}{m_d}\right)\mathscr{E}_y + e\left(\frac{1}{m_c} + \frac{1}{m_d}\right)\mathscr{E}_y + (\omega_c + \omega_d)(v_x + v_y).\end{aligned}\right\} \tag{209}$$

We may now solve these equations as in § 5.3.3 for variables $(v_x + v_y)$ and $(v_x - v_y)$ and take time averages as before, since the equations are of the form

$$\left.\begin{aligned}\dot{v}_x + \dot{v}_y &= F - \omega_1(v_x - v_y),\\ \dot{v}_x - \dot{v}_y &= G + \omega_2(v_x + v_y),\end{aligned}\right\} \tag{210}$$

where $\omega_1 = \omega_d - \omega_c$ and $\omega_2 = \omega_d + \omega_c$. Writing $\Omega^2 = \omega_1\omega_2 = \omega_d^2 - \omega_c^2$ we see by comparison with equations (134), (135) that the result is

$$\left.\begin{aligned}\overline{v_x} + \overline{v_y} &= F\left\langle\frac{\tau}{1+\Omega^2\tau^2}\right\rangle - \omega_1 G\left\langle\frac{\tau^2}{1+\Omega^2\tau^2}\right\rangle,\\ \overline{v_x} - \overline{v_y} &= G\left\langle\frac{\tau}{1+\Omega^2\tau^2}\right\rangle + \omega_2 F\left\langle\frac{\tau^2}{1+\Omega^2\tau^2}\right\rangle,\end{aligned}\right\} \tag{211}$$

giving

$$\left.\begin{aligned}\overline{v_x} &= \tfrac{1}{2}(F+G)\left\langle\frac{\tau}{1+\Omega^2\tau^2}\right\rangle + \tfrac{1}{2}(\omega_2 F - \omega_1 G)\left\langle\frac{\tau^2}{1+\Omega^2\tau^2}\right\rangle,\\ \overline{v_y} &= \tfrac{1}{2}(F-G)\left\langle\frac{\tau}{1+\Omega^2\tau^2}\right\rangle - \tfrac{1}{2}(\omega_2 F + \omega_1 G)\left\langle\frac{\tau^2}{1+\Omega^2\tau^2}\right\rangle.\end{aligned}\right\} \tag{212}$$

Since ω_1, ω_2, Ω are proportional to the magnetic induction B we may now expand in powers of B obtaining

$$\overline{v_x} = \tfrac{1}{2}(F+G)\langle\tau\rangle - \tfrac{1}{2}(\omega_2 F - \omega_1 G)\langle\tau^2\rangle - \tfrac{1}{2}(F+G)\Omega^2\langle\tau^3\rangle, \tag{213}$$

$$\overline{v_y} = \tfrac{1}{2}(F-G)\langle\tau\rangle - \tfrac{1}{2}(\omega_2 F + \omega_1 G)\langle\tau^2\rangle. \tag{214}$$

We must now average over the equivalent ellipsoids. This is done by replacing the effective masses m_{xy}, m_{yz}, m_{zx} by $\pm m_d$ in permutation. We find that

$$\bar{F} = -\frac{e}{m_c}(\mathscr{E}_x + \mathscr{E}_y),$$

$$\bar{G} = -\frac{e}{m_c}(\mathscr{E}_x - \mathscr{E}_y),$$

$$\overline{\omega_2 F} = -e^2 B\left(\frac{1}{m_c^2} - \frac{1}{m_d^2}\right)(\mathscr{E}_x + \mathscr{E}_y),$$

$$\overline{\omega_1 G} = -e^2 B\left(\frac{1}{m_c^2} - \frac{1}{m_d^2}\right)(\mathscr{E}_x - \mathscr{E}_y),$$

$$\overline{\Omega^2 F} = \frac{e^3 B^2}{m_c}\left(\frac{1}{m_c^2} - \frac{1}{m_d^2}\right)(\mathscr{E}_x + \mathscr{E}_y),$$

$$\overline{\Omega^2 G} = \frac{e^3 B^2}{m_c}\left(\frac{1}{m_c^2} - \frac{1}{m_d^2}\right)(\mathscr{E}_x - \mathscr{E}_y).$$

Thus, writing $J_x = -ne\bar{v}_x$, etc, we have

$$J_x = ne^2\left\{\frac{\langle\tau\rangle}{m_c}\mathscr{E}_x + eB\left(\frac{1}{m_c^2} - \frac{1}{m_d^2}\right)\langle\tau^2\rangle\mathscr{E}_y - \frac{e^2B^2}{m_c}\left(\frac{1}{m_c^2} - \frac{1}{m_d^2}\right)\langle\tau^3\rangle\mathscr{E}_x\right\}, \tag{215}$$

$$J_y = ne^2\left\{\frac{\langle\tau\rangle}{m_c}\mathscr{E}_y - eB\left(\frac{1}{m_c^2} - \frac{1}{m_d^2}\right)\langle\tau^2\rangle\mathscr{E}_x\right\}. \tag{216}$$

From these we obtain at once

$$\sigma_0 = ne^2\langle\tau\rangle/m_c, \tag{217}$$

$$R_0 = -\frac{1}{ne}\frac{\langle\tau^2\rangle}{\langle\tau\rangle^2}m_c^2\left(\frac{1}{m_c^2} - \frac{1}{m_d^2}\right). \tag{218}$$

Writing $K = M_L/M_T$ we have

$$m_c = 3M_L/(2K+1) \quad m_d = 3M_L/(K-1),$$

and on inserting these we obtain

$$R_0 = -\frac{1}{ne}\frac{\langle\tau^2\rangle}{\langle\tau\rangle^2}\frac{3K(K+2)}{(2K+1)^2}. \tag{219}$$

These equations for R_0 and σ_0 are the same for a semiconductor whose constant-energy ellipsoids have their axes along the $\langle 100\rangle$ directions.

The magneto-resistance will be seen from equation (215) to be given by the equation

$$\frac{\Delta\sigma}{\sigma_0} = R_0^2\sigma_0^2 B^2 - \frac{ne^4 B^2}{m_c\sigma_0}\langle\tau^3\rangle\left(\frac{1}{m_c^2} - \frac{1}{m_d^2}\right), \tag{220}$$

so that

$$-\frac{\Delta\sigma}{R_0^2\sigma_0^3 B^2} = \frac{\langle\tau^3\rangle\langle\tau\rangle}{\langle\tau^2\rangle^2}\frac{(2K+1)^2}{3K(K+2)} - 1. \tag{221}$$

Thus we have $\quad 1 + \zeta_{100}^{001} = (2K+1)^2(1+\xi)/3K(K+2). \qquad (222)$

From equations (212) we may also obtain the limiting value of σ as $B \to \infty$. We have

$$\frac{\sigma_0}{\sigma_\infty} = \left\langle\frac{1}{\tau}\right\rangle\langle\tau\rangle m_c^2\left(\frac{1}{m_c^2} - \frac{1}{m_d^2}\right), \tag{223}$$

$$= \left\langle\frac{1}{\tau}\right\rangle\langle\tau\rangle\frac{(2K+1)^2}{3K(K+2)}, \tag{224}$$

and also $R_\infty = -1/ne$ as before.

The values of ζ and σ_∞ for other directions are more tedious to calculate. They are given by[22]

$$\zeta_{100}^{100} = 2(\xi+1)(K-1)^2(2K+1)/3K(K+2)^2, \tag{225}$$

$$\zeta_{110}^{110} = \tfrac{1}{2}\zeta_{100}^{100}. \tag{226}$$

Also we have

$$\left.\frac{\sigma_0}{\sigma_\infty}\right)_{100}^{100} = (2K+1)(K+2)/9K, \tag{227}$$

$$\left.\frac{\sigma_0}{\sigma_\infty}\right)_{110}^{110} = (2K+1)^2/3K(K+2). \tag{228}$$

We note that in this case ζ_{100}^{100} is not equal to zero so that we have a change of resistance with magnetic field when both field and current lie along one of the $\langle 100\rangle$ directions.

The coefficients M_{ruvw} in equation (188) may be calculated from the values of ζ_{100}^{001}, ζ_{100}^{100}, ζ_{110}^{110}. We have as before (equation (196))

$$M_{rrss} = P = -R_0^2\sigma_0^3(\zeta_{100}^{001} + 1), \tag{229}$$

$$= -R_0^2\sigma_0^3(1+\xi)(2K+1)^2/3K(K+2). \tag{230}$$

[22] Harvey Brooks, *Advances in Electronics and Electron Physics* (1955) **7**, 85; M. Glicksman, *Progress in Semiconductors* (Heywood, 1958) **3**, 1.

For the conditions determining ζ_{110}^{110} we have now

$$\frac{\Delta\sigma}{\sigma_0 B^2} = -(b+c+\tfrac{1}{2}d) = \tfrac{1}{2}(P+Q+S), \tag{231}$$

so that
$$P+Q+S = -2R_0^2\sigma_0^3\zeta_{110}^{110} = -R_0^2\sigma_0^3\zeta_{100}^{100}. \tag{232}$$

For the conditions determining ζ_{100}^{100} we have

$$\frac{\Delta\sigma}{\sigma_0 B^2} = -(b+c+d) = S/\sigma_0, \tag{233}$$

so that
$$M_{rrrr} = S = -R_0^2\sigma_0^3\zeta_{100}^{100}, \tag{234}$$

$$= -2R_0^2\sigma_0^3(1+\xi)(K-1)^2(2K+1)/3K(K+2)^2. \tag{235}$$

It also follows that $P = -Q$ so that

$$M_{rrss} = -M_{rsrs}. \tag{236}$$

The constants b, c, d may now be expressed in terms of the quantities ζ. We have

$$\left.\begin{aligned} b &= R_0^2\sigma_0^2\zeta_{100}^{001}, \\ c &= -R_0^2\sigma_0^2\zeta_{100}^{001}, \\ d &= R_0^2\sigma_0^2\zeta_{100}^{100}. \end{aligned}\right\} \tag{237}$$

An extensive account of the magneto-resistance of semiconductors having constant-energy surfaces in the form of ellipsoids and also having multiple minima (such as conduction bands of Si and Ge) has been given by M. Glicksman.[23] A treatment of the even more complex problem of semiconductors with warped constant-energy surfaces (such as the valence bands of Si and Ge) has been given by B. Donovan and R. Herbert.[24] The subject has been reviewed by A. C. Beer.[25]

The effects of magnetic quantization already mentioned in connection with conductivity and Hall constant measurements with very high magnetic fields will also affect the values of the magneto-resistance coefficients. Measurements are usually made, however, at sufficiently low fields for the effect of the magnetic quantization to be unimportant (see § 12.8).

[23] *Progress in Semiconductors* (Heywood, 1958) **3**, 1.
[24] *Electronic Structure in Solids*, ed. E. D. Haidemenakis (Plenum Press, 1969), pp. 79–82.
[25] *Solid State Physics* (Academic Press, 1963), Suppl. 4.

5.4 Electrical conduction at very low temperatures

At first sight we should expect the electrical conductivity of a semiconductor to become very small at temperatures so low that all the free carriers are frozen into donor or acceptor centres. The condition for this is that kT should be considerably less than the impurity ionization energy ϵ_d. For shadow impurity centres (such as group V elements in Ge) with $\epsilon_d \simeq 0.01$ eV this would imply that T should be somewhat less than 100 °K. Thus at 10 °K or lower we should expect such donors to be largely un-ionized and so would not provide electrons for conduction. When the donor concentration is large, say of the order of 10^{19} cm^{-3}, the wave-functions of the electrons overlap and form what is known as an impurity band, i.e. a quasi-continuous narrow band of levels within the forbidden energy gap (see Fig. 5.6). If we have one electron per

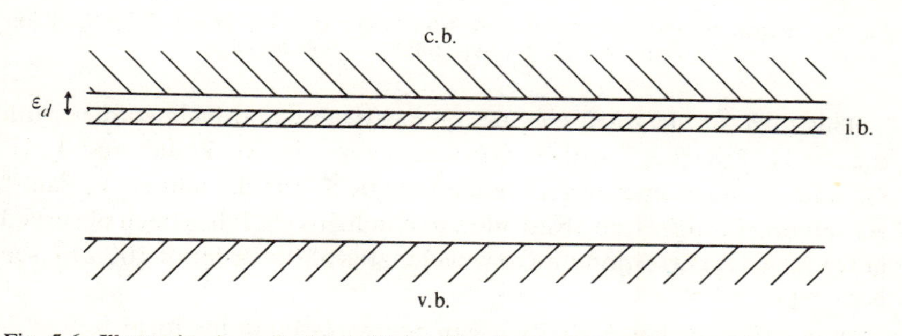

Fig. 5.6. Illustrating impurity band (i.b.) arising from a high concentration of donors at depth ϵ_d (in low concentration) below the conduction band.

donor this band will be only half filled and conduction will take place as in a metal. It turns out that because of the narrowness of the band the mobility is low but is sufficient to provide appreciable conductivity which is readily observable.

It is found, however, that conduction occurs for much smaller concentrations of donors or acceptors than 10^{19} cm^{-1}, down to 10^{15}–10^{16} cm^{-3}, and in this case impurity band conduction cannot explain the value of the conductivity. It is found that the conductivity at low temperatures is greatest when *both* donors or acceptors are present. The semiconductor is then said to be 'compensated'. For a p-type semiconductor with $N_a > N_d$ the compensation ratio K is equal to N_d/N_a. Let us consider what happens in a p-type semiconductor for which $K \neq 0$. At low temperatures such that $kT \ll \epsilon_d$ we shall have all the N_d donors empty since their electrons will drop to N_d of the acceptors to neutralize the holes that otherwise would occupy them at low temperature. Thus we have N_d ionized acceptors and $N_a - N_d$ un-ionized acceptors (see Fig. 5.7). The ionized acceptors will be spread throughout the material and holes from un-ionized acceptors will be able to hop by a tunnelling process from one vacant site to another.

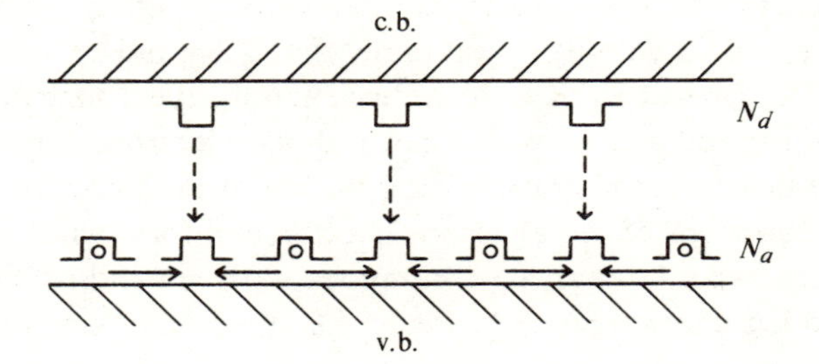

Fig. 5.7. Illustrating partial compensation of N_a acceptors by N_d donors ($N_a > N_d$) at low temperature and with possibility of hole hopping.

This form of electrical conduction has been extensively studied with d.c. by H. Fritzche,[26] with low-frequency a.c. by M. Pollak and T. H. Geballe,[27] and at microwave frequencies by S. Tanaka and H. Y. Fan.[28] By relating far infra-red absorption to conductivity, it has been observed at frequencies corresponding to wavelengths of the order of 100 μm (see § 10.2.1).

In general, the conductivity σ can be expressed in the form

$$\sigma = A_1 \exp[-\epsilon_1/kT] + A_2 \exp[-\epsilon_2/kT] + A_3 \exp[-\epsilon_3/kT]. \tag{238}$$

The first term arises from thermal impurity ionization and is insignificant at the lowest temperatures. The second comes from impurity band conduction and is unimportant if N_d or N_a are not greater than about 10^{16} cm^{-3}. The third term corresponds to hopping conduction and generally is the most important. Fritzche (*loc cit.*) found that for d.c. A_3 depends largely on the majority carrier concentration while ϵ_3 depends on the compensation ratio $K = N_a/N_d$ (for n-type) or $K = N_d/N_a$ (for p-type). As $K \to 0$, ϵ_3 becomes large. ϵ_3 also varies with frequency, becoming smaller as the frequency increases. Indeed, for frequencies corresponding to wavelengths between 200 μm and 800 μm σ varies very little with temperature indicating that for such frequencies ϵ_3 is

[26] *Phys. Rev.* (1955) **99**, 406.
[27] *Phys. Rev.* (1961) **122**, 1472.
[28] *Phys. Rev.* (1964) **132**, A564.

small (see § 10.2.1). Similar results were obtained by J. S. Blakemore[29] who has discussed the problems involved in the simultaneous determination of N_a and N_d from measurements of the Hall constant over an extensive temperature range. These were carried out with Ge. D. A. Woodbury and J. S. Blakemore[30] have made measurements with GaAs doped with Mn and found evidence of all three types of conduction.

Theoretical studies have been made both of impurity band and hopping conduction by N. F. Mott and W. D. Twose[31] and agree surprisingly well with the observations in view of a considerable uncertainty regarding compensation ratio, which is difficult to measure.

The activation energy ϵ_3 is thought to be due to the electric field attracting an electron or hole back towards a centre from which it is hopping. The effect of the field of homopolar pairs has been studied by K. Uchinokura and S. Tamaka[32] using frequencies of the order of 9 GHz with $T = 10$ °K and again good agreement with theory is found.

The theoretical aspects of these conduction processes are somewhat similar to those concerned with conduction in non-crystalline solids and in this connection have been discussed by N. F. Mott and A. E. Davis.[33]

[29] *Phil. Mag.* (1959) **4**, 560.
[30] *Phys. Rev.* (1973) B **8**, 3803.
[31] *Adv. in Phys.* (1961) **10**, 107.
[32] *Phys. Rev.* (1967) **153**, 828.
[33] *Electronic Processes in Non-crystalline Solids* (Oxford University Press, 1971).

6

Thermal effects in semiconductors

6.1 Thermal conductivity

In the previous chapter we have considered the transport of electric charge in semiconductors due to electric and magnetic fields; in the present chapter we shall consider some of the effects due to temperature gradients. If we have a temperature gradient, the equilibrium distribution function $f(\mathbf{k})$, giving the distribution of the electrons between allowed energy levels with various values of the wave-vector $\mathbf{k}$, will be disturbed from its equilibrium value. This will give rise to transport of energy in the form of heat as well as electric charge, and clearly electrical and thermal effects will interact. In the simplest case, when we have no external fields, we shall get a transport of kinetic energy due to the variation in the numbers of electrons of given energy along the temperature gradient; this is the phenomenon of thermal conduction.

When we have temperature gradients we cannot use the simple equations of motion averaged over the equilibrium energy distribution and to obtain equations for the electric current density and heat flow we must make use of the variation of $f(\mathbf{k})$ given by Boltzmann's equation. This is treated in standard works on transport in solids and gases and we shall not give the detailed analysis leading to the modified equation for the current density $\mathbf{J}$. (See *W.M.C.S.* §§ 10.5, 10.6, 10.7.) Equation (85) of § 5.3.1. must be replaced by

$$J_x = \frac{ne}{m_e}\left[e\mathscr{E}_x + T\frac{\partial}{\partial T}\left(\frac{E_F}{T}\right)\frac{dT}{dx}\langle\tau\rangle\right] + \frac{ne}{Tm_e}\frac{dT}{dx}\langle\tau E\rangle \qquad (1)$$

where, as before, E_F is the Fermi energy. To obtain the heat current W_x

we simply replace $-e$ by E, before averaging, and obtain

$$W_x=-\frac{n}{m_e}\left[e\mathscr{E}_x+T\frac{\partial}{\partial T}\left(\frac{E_F}{T}\right)\frac{dT}{dx}\langle\tau E\rangle\right]-\frac{n}{Tm_e}\frac{dT}{dx}\langle\tau E^2\rangle. \tag{2}$$

If no electric current flows we may eliminate $\mathscr{E}_x$ and obtain an expression for W_x (we note that $\mathscr{E}_x\neq 0$ if $dT/dx\neq 0$); in this way we obtain

$$W_x=-\frac{n}{m_eT}\frac{dT}{dx}\frac{\langle\tau\rangle\langle E^2\tau\rangle-\langle E\tau\rangle^2}{\langle\tau\rangle}. \tag{3}$$

It will be seen that as well as the average value $\langle\tau\rangle$ we now have also the averages $\langle\tau E\rangle$ and $\langle\tau E^2\rangle$, the weighting functions being the same as before.

For a semiconductor there are two separate contributions to the thermal conduction, that due to the vibrations of the atoms about their positions of thermal equilibrium (lattice vibrations, for a crystal) and that due to the mobile charge carriers, electrons or holes. The thermal conductivity due to the former we shall denote by κ_L and the latter by κ_e for a strongly n-type semiconductor, κ_h for a strongly p-type semiconductor and κ_m when we have mixed conduction. The thermal conductivity due to electrons κ_e is then given by

$$\kappa_e=-W_x\bigg/\frac{dT}{dx}=\frac{n[\langle\tau\rangle\langle E^2\tau\rangle-\langle E\tau\rangle^2]}{Tm_e\langle\tau\rangle}. \tag{4}$$

The electrical conductivity σ being equal to $ne^2\langle\tau\rangle/m_e$, we have

$$\kappa_e=L\sigma T=\mathscr{L}(k^2/e^2)\sigma T, \tag{5}$$

where

$$\mathscr{L}=\frac{\langle\tau\rangle\langle E^2\tau\rangle-\langle E\tau\rangle^2}{k^2T^2\langle\tau\rangle^2}. \tag{6}$$

$\mathscr{L}$ is a pure number and $L=\mathscr{L}(k^2/e^2)$ is known as the Lorentz ratio. For a non-degenerate semiconductor with $\tau=aE^{-s}$ we have

$$\left.\begin{aligned}\langle\tau\rangle&=a\Gamma(\tfrac{5}{2}-s)/\Gamma(\tfrac{5}{2}),\\ \langle\tau E\rangle&=akT\Gamma(\tfrac{7}{2}-s)/\Gamma(\tfrac{5}{2}),\\ \langle\tau E^2\rangle&=ak^2T^2\Gamma(\tfrac{9}{2}-s)/\Gamma(\tfrac{5}{2}).\end{aligned}\right\} \tag{7}$$

Since $\Gamma(1+x)=x\Gamma(x)$ we have $\mathscr{L}=\frac{5}{2}-s$. For scattering by the acoustical modes with $s=\frac{1}{2}$ we have $\mathscr{L}=2$. (The value derived from classical electron theory is $\mathscr{L}=3$.)

For a fully degenerate semiconductor a better approximation than that previously used to obtain $\langle\tau\rangle$ must be obtained; to this approximation equation (6) gives $L=0$, since

$$\langle E\tau\rangle = E_F\tau(E_F) \quad \text{and} \quad \langle E^2\tau\rangle = E_F^2\tau(E_F)$$

to the first order in $E_F/\boldsymbol{k}T$ (cf. § 5.3.1, equation (91)). To evaluate $\langle\tau\rangle$, $\langle E\tau\rangle$, $\langle E^2\tau\rangle$, which are now given by equations of the form

$$\langle\tau\rangle = -E_F^{-\frac{3}{2}}\int_0^\infty \tau E^{\frac{3}{2}}\frac{\partial f_0}{\partial E}\,dE \tag{8}$$

(cf. equation (90) of § 5.3.1), we use a better approximation to

$$-\int_0^\infty \phi(E)\frac{\partial f_0}{\partial E}\,dE$$

than $\phi(E_F)$, which we previously used. Expanding $\phi(E)$ in powers of $(E-E_F)$ we have

$$\phi(E) = \phi(E_F) + (E-E_F)\phi'(E_F) + \tfrac{1}{2}(E-E_F)^2\phi''(E_F) + \cdots \tag{9}$$

On integrating we also obtain the values

$$\int_0^\infty \frac{\partial f_0}{\partial E}\,dE = -1,$$

$$\int_0^\infty (E-E_F)\frac{\partial f_0}{\partial E}\,dE = 0,$$

$$\int_0^\infty (E-E_F)^2\frac{\partial f_0}{\partial E}\,dE = -\frac{\pi^2}{3}\boldsymbol{k}^2T^2.$$

Thus we have

$$-\int_0^\infty \phi(E)\frac{\partial f_0}{\partial E}\,dE = \phi(E_F) + \frac{\pi^2}{6}(\boldsymbol{k}T)^2\phi''(E_F) + \cdots \tag{10}$$

and on inserting the appropriate values of $\phi(E)$ in equation (9) corresponding to $\langle\tau\rangle$, $\langle E\tau\rangle$, $\langle E^2\tau\rangle$ we have after some algebraic manipulation

$$\langle\tau\rangle\langle E^2\tau\rangle - \langle E\tau\rangle^2 = \tfrac{1}{3}\pi^2\boldsymbol{k}^2T^2\langle\tau\rangle^2, \tag{11}$$

so that $\mathscr{L} = \frac{1}{3}\pi^2$. For the intermediate case, corresponding to weak degeneracy, the integrals have to be evaluated numerically; if $\tau = aE^{-s}$, provided $s<\frac{3}{2}$, we have

$$\frac{\langle E^n\tau\rangle}{\langle\tau\rangle}=\frac{(\frac{3}{2}+n-s)F_{\frac{1}{2}+n-s}}{(\frac{3}{2}-s)F_{\frac{1}{2}-s}}, \tag{12}$$

where
$$F_r(E_F)=\int_0^\infty E^r f_0\,dE. \tag{13}$$

The functions F_r have been tabulated by J. McDougall and E. C. Stoner[1] for certain half-integral values of r, and by P. Rhodes[2] for integral values of r. The quantity $\mathscr{L}$ has been calculated for $s=\frac{1}{2}$ for various values of E_F/kT and is shown in Fig. 6.1.

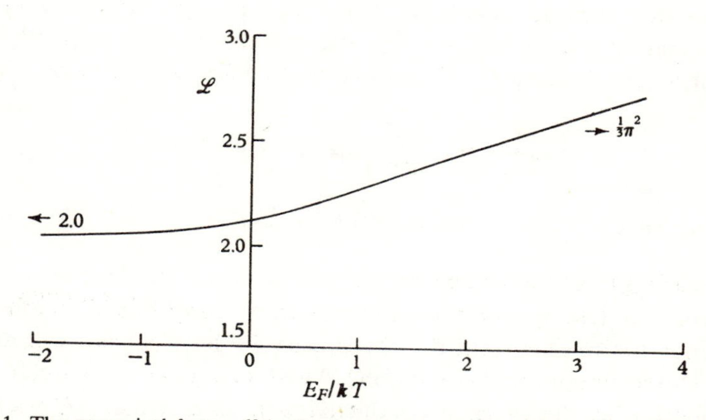

Fig. 6.1. The numerical factor $\mathscr{L}$ in the Lorentz number $L=\mathscr{L}(k/e)^2$ as a function of E_F/kT, with $\tau=aE^{-\frac{1}{2}}$.

For a p-type semiconductor it is more convenient to use the distribution function $f_0'(E)$ for holes given by

$$f_0'(E)=1-f_0(E). \tag{14}$$

If we write $E_F'=-\Delta E-E_F$, where ΔE is the forbidden energy gap, and E' for the kinetic energy of a hole, so that the energy E of the level for which the hole represents a vacancy is equal to $-\Delta E-E'$, we have

$$f_0'(E',E_F')=f_0(E,E_F). \tag{14a}$$

[1] *Phil. Trans.* A (1938), **237**, 67.

[2] *Proc. Roy. Soc.* A (1950) **204**, 396.

For the current density J_x we now obtain the expression

$$J_x = p\mu_h\left\{\boldsymbol{e}\mathscr{E}_x - \left[T\frac{\partial}{\partial T}\left(\frac{E'_F}{T}\right) + \boldsymbol{k}(\tfrac{5}{2} - s')\right]\frac{dT}{dx}\right\} \tag{15}$$

assuming we have $\tau_h = a'E^{-s'}$. To obtain the heat current we must now replace $\boldsymbol{e}$ by $E' + \Delta E$, and we thus obtain

$$W_x = \frac{p\mu_h(\tfrac{5}{2} - s')\boldsymbol{k}T}{\boldsymbol{e}}\left\{\boldsymbol{e}\mathscr{E}_x - \left[T\frac{\partial}{\partial T}\left(\frac{E'_F}{T}\right) + \boldsymbol{k}(\tfrac{7}{2} - s')\right]\frac{dT}{dx}\right\}$$
$$+\frac{p\mu_h\Delta E}{\boldsymbol{e}}\left\{\boldsymbol{e}\mathscr{E}_x - \left[T\frac{\partial}{\partial T}\left(\frac{E'_F}{T}\right) + \boldsymbol{k}(\tfrac{5}{2} - s')\right]\frac{dT}{dx}\right\}. \tag{16}$$

The second term in equation (16) represents the transport of the recombination energy of holes and electrons; it is zero when the hole current is zero. Putting $J_x = 0$ and substituting for $\boldsymbol{e}\mathscr{E}_x$ we have

$$\kappa_h = -W_x\Big/\frac{dT}{dx} = p\mu_h(\tfrac{5}{2} - s')\boldsymbol{k}^2T/\boldsymbol{e}. \tag{17}$$

Thus we have
$$\kappa_h = (\tfrac{5}{2} - s')(\boldsymbol{k}^2/\boldsymbol{e}^2)\sigma T \tag{18}$$

which differs from the value of κ_e only by having s' in place of s.

A very interesting situation arises when we have mixed conduction, for, when the total current density is zero the hole current will not be zero. The second term in equation (16) will then make a relatively large contribution to the thermal conductivity if, as is usual, $\Delta E \gg \boldsymbol{k}T$. To obtain the total electric and heat currents we must now add the contributions given by equations (1), (2), (15), (16). When we do this and eliminate $\mathscr{E}_x$ we obtain, if we neglect a small contribution due to the variation of ΔE with T, for the thermal conductivity κ_m due to mixed electron and hole conduction

$$\frac{k_m}{\sigma T} = \frac{\boldsymbol{k}^2(\tfrac{5}{2} - s)n\mu_e + \boldsymbol{k}^2(\tfrac{5}{2} - s')p\mu_h}{\boldsymbol{e}^2(n\mu_e + p\mu_h)} + \frac{\boldsymbol{k}^2[5 - s - s' + \Delta E/\boldsymbol{k}T]^2 np\mu_e\mu_h}{\boldsymbol{e}^2(n\mu_e + p\mu_h)^2}. \tag{19}$$

If we have $s = s' = \tfrac{1}{2}$ and include the lattice contribution κ_L the total thermal conductivity κ is given by the equation

$$\frac{\kappa}{\sigma T} = \frac{\kappa_L}{\sigma T} + \frac{2\boldsymbol{k}^2}{\boldsymbol{e}^2} + \left(4 + \frac{\Delta E}{\boldsymbol{k}T}\right)^2 \boldsymbol{k}^2 np\mu_e\mu_h/\boldsymbol{e}^2(n\mu_e + p\mu_h)^2, \tag{20}$$

a result quoted by E. H. Putley.[3] The first term gives the lattice contribution and the third term on the right-hand side of equation (20) represents the contribution due to mixed conduction. The latter may be quite large when $\Delta E \gg \boldsymbol{k}T$ and $n \simeq p$. It does not explain a considerable increase in thermal conductivity observed in some semiconductors at low temperatures. This is undoubtedly an effect due to transport of phonons (see § 8.4), and has been studied experimentally in InSb by G. Busch and M. Schneider;[4] it has also been studied theoretically by D. ter Haar and A. Neaves.[5]

The contribution to the thermal conductivity of semiconductors of the conduction electrons and holes is in general quite small; for example, if we take $\mathscr{L} = 2$ we have $\kappa_e = 2(\boldsymbol{k}^2/e^2)\sigma T = 4.4 \times 10^{-6}\sigma$ for $T = 300\,°\mathrm{K}$. (If σ is expressed in $\Omega^{-1}\,\mathrm{m}^{-1}$, κ_e is in $\mathrm{W\,m^{-1}\,deg^{-1}}$.) Let us consider two fairly extreme cases. For good single crystal Ge,

$$\kappa = 0.15\ \mathrm{cal\,s^{-1}\,cm^{-1}\,deg^{-1}} \simeq 60\ \mathrm{W\,m^{-1}\,deg^{-1}}.$$

Thus for Ge with a resistivity of $1\,\Omega\,\mathrm{cm}$ ($\sigma = 100\,\Omega^{-1}\,\mathrm{m}^{-1}$) we have $\kappa_e = 4.4 \times 10^{-4}\ \mathrm{W\,m^{-1}\,deg^{-1}}$, so that $\kappa_e/\kappa \sim 10^{-5}$. For Bi_2Te_3 with $\kappa = 6 \times 10^{-3}\ \mathrm{cal\,s^{-1}\,cm^{-1}\,deg^{-1}} = 2.4\ \mathrm{W\,m^{-1}\,deg^{-1}}$ and $\sigma = 10^5\,\Omega^{-1}\,\mathrm{m}^{-1}$ we have $\kappa_e = 0.45\ \mathrm{W\,m^{-1}\,deg^{-1}}$ so that $\kappa_e/\kappa = 0.2$.

Reviews of the work on thermal conductivity in semiconductors have been given by A. V. Ioffe and A. F. Ioffe[6] and by J. Appel.[7] (See also the book by J. R. Drabble and H. J. Goldsmid.[8])

6.2 Thermo-electric power

We have seen that when we have a temperature gradient along a conductor an e.m.f. is set up even when no electric current flows. For an n-type semiconductor, putting $J_x = 0$ in equation (1), we see that the electric field $\mathscr{E}_x$ is given by

$$\mathscr{E}_x = -\left[T\frac{d}{dT}\left(\frac{E_F}{T}\right) + \frac{1}{eT}\frac{\langle E\tau\rangle}{\langle\tau_e\rangle}\right]\frac{dT}{dx}. \tag{21}$$

The electric field is related to the Thomson coefficient $\mathscr{T}$ which gives the rate of reversible heating per unit volume and is defined by means of the

[3] *Proc. Phys. Soc.* B (1955) **58**, 35. [4] *Physica* (1954) **20**, 1084.
[5] *Advances in Physics* (*Phil. Mag.* Supplement) (1956) **5**, 241.
[6] *C.R. Akad. Sci. U.R.S.S.* (1954) **98**, 757. See also A. F. Ioffe, *Canad. J. Phys.* (1956) **34**, 1342.
[7] *Progress in Semiconductors* (Heywood, 1960) **5**, 141.
[8] *Thermal Conduction in Semiconductors* (Pergamon Press, 1961).

equation

$$\frac{dW_x}{dt} = -\mathscr{T} J_x \frac{dT}{dx}. \tag{22}$$

The rate of heating per unit volume may also be expressed as

$$\frac{dW_x}{dt} = J_x \mathscr{E}_x - \frac{\partial W_x}{\partial x} = J_x \mathscr{E}_x - \frac{\partial W_x}{\partial T}\frac{dT}{dx}. \tag{23}$$

Solving equation (1) for $\mathscr{E}_x$ and substituting in equation (23) we obtain, on retaining only the terms *linear* in J_x,

$$\frac{dW_x}{dT} = -\frac{J_x}{e}\left[T\frac{d}{dT}\left(\frac{E_F}{T}\right) - T\frac{d}{dT}\left(\frac{\langle E\tau\rangle}{T\langle\tau\rangle}\right)\right]\frac{dT}{dx}. \tag{24}$$

Comparing equation (24) with (22) we see that

$$\mathscr{T} = -T\frac{d}{dT}\left[\frac{\langle E\tau_e\rangle - E_F\langle\tau_e\rangle}{eT\langle\tau_e\rangle}\right]. \tag{25}$$

Similarly, if we have $dT/dx = 0$ then $W_x \neq 0$ but is proportional to J_x. This is the effect known as Thomson heating and is closely related to the thermo-electric effect. A quantity commonly used to describe these effects[9] is the thermo-electric power $\mathscr{P}_{ab}$. This refers to two materials denoted by a and b. If we have a circuit containing two junctions A, B as shown in Fig. 6.2, at temperatures T_2, and T_1, the material b being

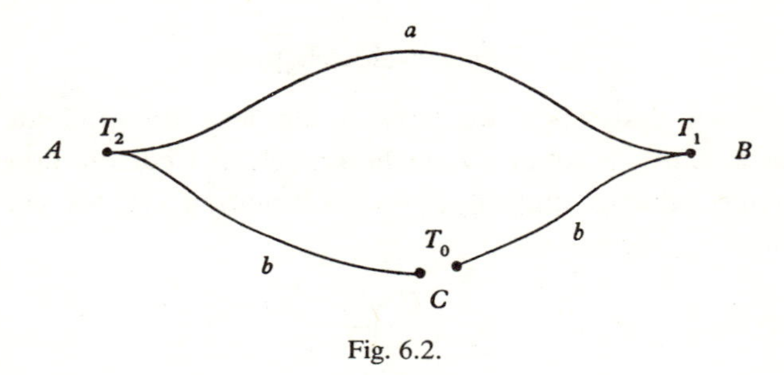

Fig. 6.2.

broken at C, which is at temperature T_0, then the open circuit voltage $\mathscr{V}_0$ at C is given by[9]

$$\mathscr{V}_0 = \int_{T_1}^{T_2} \mathscr{P}_{ab}(T)\,dT. \tag{26}$$

[9] See, for example, R. A. Smith, *The Physical Principles of Thermodynamics* (Chapman and Hall, 1952) pp. 84–6.

If $T_2 = T_1 + \Delta T$, when ΔT is small we have

$$\mathcal{V}_0 = \mathcal{P}_{ab}\Delta T. \tag{27}$$

If $\mathcal{T}_a$, $\mathcal{T}_b$ are the Thomson coefficients for the two materials we have[9]

$$\mathcal{T}_a - \mathcal{T}_b = T\frac{d\mathcal{P}_{ab}}{dT}. \tag{28}$$

If we choose for the material b one for which $\mathcal{T}_b = 0$ (usually lead), and for the thermo-electric power relative to this material write $\mathcal{P}_{ab} = \mathcal{P}_a$, $\mathcal{P}_a$ is sometimes called the absolute thermo-electric power. In this case we have[9]

$$\mathcal{T}_a = T\frac{d\mathcal{P}_a}{dT}. \tag{29}$$

Comparing this equation with equation (25) we see that

$$\mathcal{P} = \frac{E_F\langle\tau_e\rangle - \langle\tau_e E\rangle}{kT\langle\tau_e\rangle}\left(\frac{k}{e}\right). \tag{30}$$

For a semiconductor with scalar effective mass m_e and relaxation time τ of the form $\tau = aE^{-s}$ we have

$$\mathcal{P} = -\{(\tfrac{5}{2} - s) - E_F/kT\}(k/e). \tag{31}$$

Since under the assumed conditions $E_F/kT < 0$, $\mathcal{P}$ is *negative* for an n-type semiconductor, since generally $s < \frac{5}{2}$. (If $s > \frac{5}{2}$ the integral for $\langle\tau\rangle$ diverges.) For any two materials a, b we have then

$$\mathcal{P}_{ab} = \mathcal{P}_a - \mathcal{P}_b. \tag{32}$$

If the two materials a, b, consist of the same n-type semiconductor but with different values of n, we have (equation (29) of § 4.3)

$$E_F^{(a)} = -kT\ln(N_c/n_a), \quad E_F^{(b)} = -kT\ln(N_c/n_b),$$

so that

$$\mathcal{P}_{ab} = -(k/e)\ln(n_b/n_a). \tag{33}$$

Equation (31) may also be written in terms of the electron concentration n in the form

$$\mathcal{P} = -(k/e)[(\tfrac{5}{2} - s) + \ln(N_c/n)], \tag{34}$$

where $N_c = 2(2\pi m_e kT/h^2)^{\frac{3}{2}}$. Since $(k/e) \simeq 100\ \mu\text{V deg}^{-1}$, thermo-electric power of semiconductors is high, of the order of $1\ \text{mV deg}^{-1}$, whereas for metals $\mathcal{P}_{ab}$ is of the order of a few $\mu\text{V deg}^{-1}$. For a semiconductor–metal junction $\mathcal{P}_{ab}$ is thus approximately equal to the absolute thermo-electric power $\mathcal{P}_a$ of the semiconductor.

For a *p*–type non-degenerate semiconductor we have from equation (15)

$$\mathcal{P} = \left[\frac{\langle E\tau_p\rangle - E'_F\langle\tau_p\rangle}{eT\langle\tau_p\rangle}\right],$$

If $\tau = a'E^{-s'}$
$$\mathcal{P} = (\boldsymbol{k}/\boldsymbol{e})[(\tfrac{5}{2}-s') - E'_F/\boldsymbol{k}T], \tag{35}$$
$$= (\boldsymbol{k}/\boldsymbol{e})[(\tfrac{5}{2}-s') + E_F/\boldsymbol{k}T + \Delta E/\boldsymbol{k}T], \tag{36}$$
$$= (\boldsymbol{k}/\boldsymbol{e})[(\tfrac{5}{2}-s') + \ln(N_v/p)], \tag{37}$$

where
$$N_v = 2(2\pi m_h \boldsymbol{k}T/\boldsymbol{h}^2)^{\frac{3}{2}}.$$

Under the assumed conditions $E'_F/\boldsymbol{k}T < 0$ so that $\mathcal{P}$ is *positive*, since generally $s' < \frac{5}{2}$. The sign of the thermo-electric power therefore gives a simple direct test as to whether a semiconductor is *n*-type or *p*-type.

When we have mixed conduction, by combining equations (1) and (15), we obtain for $\mathscr{E}_x$ the equation

$$\mathscr{E}_x = T\frac{d}{dT}\left[\frac{n\langle E\tau_e\rangle/m_e - nE_F\langle\tau_e\rangle/m_e - p\langle E\tau_h\rangle/m_h + pE'_F\langle\tau_h\rangle/m_h}{eT(n\langle\tau_3\rangle/m_e + p\langle\tau_h\rangle/m_h)}\right] \tag{38}$$

so that

$$\mathcal{P} = \frac{(n/m_e)(E_F\langle\tau_e\rangle - \langle E\tau_e\rangle) - (p/m_h)(E'_F\langle\tau_h\rangle - \langle E\tau_h\rangle)}{eT(n\langle\tau_e\rangle/m_e + p\langle\tau_h\rangle/m_h)} \tag{39}$$

$$= -\frac{\boldsymbol{k}}{\boldsymbol{e}}\left[\frac{n\mu_e(\frac{5}{2} - s - E_F/\boldsymbol{k}T) - p\mu_h(\frac{5}{2} - s' - E'_F/\boldsymbol{k}T)}{n\mu_e + p\mu_h}\right] \tag{40}$$

if $\tau_e = aE^{-s}$ and $\tau_h = a'E^{-s'}$. If we write $\mathcal{P}_e$, σ_e for the electron contribution to conductivity and thermo-electric power, assuming the holes to be absent, and similarly $\mathcal{P}_h$, σ_h for the contributions from the holes, assuming the electrons to be absent, we have

$$\mathcal{P} = \frac{\mathcal{P}_e\sigma_e + \mathcal{P}_h\sigma_h}{\sigma_e + \sigma_h}. \tag{41}$$

Thus we see that the contributions combine as though the thermo-electric voltage were in parallel, each being produced by a generator whose admittance is proportional to the conductivity arising from a single band. Expressing E_F and E'_F in terms of n and p we have

$$\mathscr{P} = -\frac{k}{e}\left[\frac{n\mu_e\{(\frac{5}{2}-s)+\ln(N_c/n)\}-p\mu_h\{(\frac{5}{2}-s')-\ln(N_v/p)\}}{n\mu_e+p\mu_h}\right]. \quad (42)$$

When we have more than one kind of carrier in a single band (as in p-type Ge) the contributions to $\mathscr{P}$ from the various types of carriers are combined as above with appropriate change of sign.

When the semiconductor is completely degenerate we have from equation (10)

$$\langle\tau\rangle = \tau + (\tfrac{1}{6}\pi^2)(kT)^2(\tau E^{\frac{3}{2}})''_{E_F}E_F^{-\frac{3}{2}} + \ldots$$

$$= \tau[1 + A(\tfrac{3}{2}-s)(\tfrac{1}{2}-s)] + \ldots, \quad (43)$$

$$\langle E\tau\rangle = \tau E_F[1 + A(\tfrac{5}{2}-s)(\tfrac{3}{2}-s)] + \ldots, \quad (44)$$

$$\langle E^2\tau\rangle = \tau E_F^2[1 + A(\tfrac{7}{2}-s)(\tfrac{5}{2}-s)] + \ldots, \quad (45)$$

where

$$A = \frac{\pi^2}{6}\left(\frac{kT}{E_F}\right)^2.$$

The value of τ on the right-hand side of these equations is that for $E = E_F$. From these we obtain the equation

$$\frac{\langle\tau E\rangle - E_F\langle\tau\rangle}{\langle\tau\rangle} = \frac{\pi^2}{3}(\tfrac{3}{2}-s)\frac{k^2T^2}{E_F}. \quad (46)$$

For electrons we have therefore

$$\mathscr{P}_e = -\frac{\pi^2 k}{e}\left(\frac{kT}{E_F}\right)(\tfrac{1}{2}-\tfrac{1}{3}s). \quad (47)$$

Under the assumed conditions $E_F/kT > 0$, so that $\mathscr{P}_e$ is again negative. For holes we have

$$\mathscr{P}_h = \frac{\pi^2 k}{e}\left(\frac{kT}{E'_F}\right)(\tfrac{1}{2}-\tfrac{1}{3}s'), \quad (48)$$

and $\mathscr{P}_h$ is again positive since for the assumed conditions $E'_F/kT > 0$. For mixed conduction, we again apply the formula given in equation (41). We see that the values of $\mathscr{P}_e$ and $\mathscr{P}_h$ are greatly reduced as compared with the non-degenerate condition, since we have assumed $E_F/kT \gg 1$ and $E'_F/kT \gg 1$. For the intermediate case of partial degeneracy the quantities $\langle\tau\rangle$, $\langle\tau E\rangle$, etc, must be evaluated using the exact value of f_0. We have, using equation (12)

$$\left.\begin{aligned}\langle E\tau_e\rangle/\langle\tau_e\rangle &= (\tfrac{5}{2}-s)F_{\frac{3}{2}-s}/(\tfrac{3}{2}-s)F_{\frac{1}{2}-s},\\ \langle E\tau_h\rangle/\langle\tau_h\rangle &= (\tfrac{5}{2}-s')F_{\frac{3}{2}-s'}/(\tfrac{3}{2}-s')F_{\frac{1}{2}-s'}.\end{aligned}\right\} \tag{49}$$

For electron conduction we obtain the equation

$$\mathscr{P}_e = \frac{\boldsymbol{k}}{e}\left[\frac{E_F-(\tfrac{5}{2}-s)F_{\frac{3}{2}-s}/(\tfrac{3}{2}-s)F_{\frac{1}{2}-s}}{\boldsymbol{k}T}\right] \tag{50}$$

and for holes $$\mathscr{P}_h = \frac{\boldsymbol{k}}{e}\left[\frac{-E'_F+(\tfrac{5}{2}-s')F_{\frac{3}{2}-s'}/(\tfrac{3}{2}-s')F_{\frac{1}{2}-s'}}{\boldsymbol{k}T}\right], \tag{51}$$

where the functions $F_r(E_F)$ are given by equation (13). These are equivalent to expressions given by J. Tauc.[10] For mixed conduction equation (41) still holds with these values of $\mathscr{P}_e$ and $\mathscr{P}_h$.

When we have both ionized impurity scattering and lattice scattering the relaxation time τ may no longer be expressed in the form $\tau = aE^{-s}$ but has the form $\tau = aE^{-\frac{1}{2}}/(1+cE^{-2})$ (see § 5.3.2). In this case the integrals for $\langle\tau\rangle$ and $\langle\tau E\rangle$ have to be evaluated numerically. Values of $\mathscr{P}$ for various values of c, which determines the ratio of lattice to ionized impurity scattering, have been calculated by V. A. Johnson and K. Lark-Horowitz.[11]

We must now consider the effect of having non-scalar effective masses. The expression for the current density J_x (equation (1)) contains the effective electron mass in two places, first, in the factor ne/m_e, and secondly, in the Fermi energy E_F. The effect of the first factor is exactly the same as we have found in electric conduction and gives rise to expressions for the mobility involving the different components of the mass tensor. For single-band conduction, the mobility does not appear in the expressions for the thermo-electric power so these are unmodified. For mixed conduction the effect of the mass tensor is reflected in the mobilities and provided the appropriate values are used the formulae obtained still hold. We now must consider the effect on E_F. We have seen in § 4.2 that we must use the 'density of states' effective mass and must also take account of the fact that there may be more than one maximum or minimum in the band. For an n-type semiconductor with constant-energy surfaces in the form of ellipsoids of revolution, and having cubic symmetry, we have

$$N_c = 2m_L^{\frac{1}{2}}m_T M_c \boldsymbol{h}^{-3}(2\pi\boldsymbol{k}T)^{\frac{3}{2}}, \tag{52}$$

[10] *Phys. Rev.* (1954) **95**, 1394.

[11] *Phys. Rev.* (1953) **92**, 226.

where m_L and m_T are respectively the longitudinal and transverse effective masses and M_c is the number of equivalent minima in the band. With this value of N_c equation (34) still holds.

While the above formulae describe very well the measurements on semiconductors at ordinary temperatures they fail to account for a large increase in the absolute magnitude of the thermo-electric power observed in some pure materials at very low temperatures, for example, in germanium by H. P. R. Frederikse[12] and by T. H. Geballe and G. W. Hull.[13] This effect has been discussed theoretically by Frederikse (*loc. cit.*) and also by a number of other authors,[14,15,16] and has been shown to be due to electron–phonon interaction as already discussed in connection with thermal conductivity. The phenomenon is known as 'phonon drag'. When we have a temperature gradient more phonons carry energy from the hot to the cool end of the specimen than in the reverse direction. The electrons are therefore no longer scattered isotropically and more electrons are urged by the phonons towards the hot end, thus setting up an electric field in addition to that set up by the temperature gradient as already discussed. The effect is to give a marked increase in thermo-electric power at low temperatures. At higher temperatures, phonon–phonon scattering tends to restore the isotropic distribution and the effect is small. C. Herring (*loc. cit.*) has given a very full analysis of the existing experimental data for Ge and has shown that the measurements are well described by the theory. A number of reviews of theoretical work on thermo-electric power in semiconductors, and critical comparison with the available experimental data, have been given.[17] In particular, the thermo-electric properties of the III–V group semiconductors have been described in detail by R. W. Ure.[18]

6.3 Thermo-magnetic effects

We must now consider a number of phenomena which occur when a magnetic field is applied in the presence of both an electric field and a

[12] *Phys. Rev.* (1953) **91**, 491; *ibid.* (1953) **92**, 248.
[13] *Phys. Rev.* (1954) **94**, 1134.
[14] C. Herring, *Phys. Rev.* (1953) **92**, 857; *ibid.* (1954) **96**, 1163.
[15] E. H. Sondheimer, *Proc. Roy. Soc.* A, (1956) **234**, 391.
[16] D. ter Harr and A. Neaves, *Advances in Physics* (*Phil. Mag.* Supplement) (1956) **5**, 241.
[17] See, for example, V. A. Johnson, *Photo- and Thermoelectric Effects in Semiconductors* (Pergamon Press, 1962); *The Thermoelectric Properties of Semiconductors*, ed. V. A. Kutasov (Consultants Bureau, N.Y., 1964).
[18] *Semiconductors and Semimetals* (Academic Press, 1972) **8**, 67.

temperature gradient. We shall suppose the magnetic field to be directed along the z-axis and consider only fields and gradients in the (x, y)-plane. There are a number of effects which appear under this condition, the most important being the so-called transverse effects. When we have a current in the x-direction, a temperature gradient, as well as an electric field (the Hall field), is set up in the y-direction. This we have neglected in our previous treatment of the Hall effect. The Hall voltage which we have obtained is that appropriate to isothermal conditions in which $\partial T/\partial y$ is held equal to zero and corresponds to what is generally known as the *isothermal* Hall effect. Under adiabatic conditions a temperature gradient $\partial T/\partial y$ is set up and the corresponding thermo-electric voltage is added to the Hall voltage. The voltage obtained in this case then gives a measure of the *adiabatic* Hall effect. Spurious thermo-electric voltages are normally eliminated in Hall effect measurements by reversing the direction of both the current and the magnetic field. This does not, however, eliminate the above effect which is known as the Ettingshausen effect. As we shall see, the temperature gradient is inversely proportional to the total thermal conductivity. When we take account of the lattice contribution to the thermal conductivity the ratio of this to the electrical conductivity is high for semiconductors as compared with metals. On the other hand, the thermo-electric power for semiconductors is also high. Even so, in measurements on semiconductors, voltages due to the Ettingshausen effect usually amount to less than 1% of the Hall voltage, so that the difference between adiabatic and isothermal Hall effect is generally small. When measured with an alternating current of frequency greater than about 50 Hz the temperature gradient cannot be established and the isothermal Hall voltage is measured.

A simple physical explanation of the effect may be given as follows. The average force on an electron due to the magnetic field is proportional to its drift velocity, so there is a tendency for the faster electrons to be driven towards the side of the specimen at which they accumulate to produce the Hall voltage. The side which becomes negative for an n-type specimen becomes colder. Since the thermo-electric power for n-type material is negative the voltage produced by the Ettingshausen effect adds to the Hall voltage; clearly this also happens for p-type material.

When we have a thermal current due to a temperature gradient in the x-direction (but no electric current) we have an electric field developed in the y-direction. This is the thermal analogue of the Hall effect and is known as the Nernst effect. There is a net flow of the faster electrons from the hot to the colder end and these are urged to one side by the magnetic field, producing a transverse electric field. Under these conditions a temperature gradient is also produced in the y-direction. This effect, known as the Righi–Leduc effect, is the thermal analogue of the Ettingshausen effect.

To study these effects we require to modify Boltzmann's equation by including the force term due to the magnetic field, whose induction we assume to be $(0, 0, B)$. We must replace $\mathscr{E}$ by $\mathscr{E}+\mathbf{v}\times\mathbf{B}$ and we should then proceed to find the solution of Boltzmann's equation for any value of B, as shown by A. H. Wilson;[19] the formulae for the electric and heat currents are then derived as before. An alternative method of solution is to note that the equation is the same to the first order as when we have no temperature gradients provided we replace $\mathscr{E}$ by $\mathscr{E}'$ for electrons and $\mathscr{E}''$ for holes, where

$$\mathscr{E}' = \mathscr{E} + \left\{\frac{T}{e}\frac{\partial}{\partial T}\left(\frac{E_F}{T}\right) + \frac{E}{eT}\right\}\frac{\partial T}{\partial x}, \tag{53}$$

$$\mathscr{E}'' = \mathscr{E} - \left\{\frac{T}{e}\frac{\partial}{\partial T}\left(\frac{E'_F}{T}\right) + \frac{E'}{eT}\right\}\frac{\partial T}{\partial x}, \tag{54}$$

E'_F and E' being the quantities defined in § 6.1. We may now use our previous solutions for conduction in the presence of a magnetic field, obtained from the equations of motion, those for the electric current densities J_x, J_y being given by equation (107) of § 5.3.3. Here we shall consider only effects linear in the magnetic field and so shall neglect terms of the order of $\omega_e^2\tau^2$ where, as before, $\omega_e = eB/m_e$. We thus have for electrons

$$J_x = \frac{ne^2}{m_e}[\tau_e\mathscr{E}'_x - \omega_e\tau_e^2\mathscr{E}'_y], \tag{55}$$

$$J_y = \frac{ne^2}{m_e}[\tau_e\mathscr{E}'_y + \omega_e\tau_e^2\mathscr{E}'_x]. \tag{56}$$

In order to obtain the heat transfer equations for electrons we replace $-e$ by E; we have then

$$W_x = -\frac{ne}{m_e}[E\tau_e\mathscr{E}'_x - \omega_e E\tau_e^2\mathscr{E}'_y], \tag{57}$$

$$W_y = -\frac{ne}{m_e}[E\tau_e\mathscr{E}'_y + \omega_e E\tau_e^2\mathscr{E}'_x]. \tag{58}$$

[19] *The Theory of Metals* (Cambridge University Press, 1954), p. 210.

To calculate the heat transport for holes we have to replace a factor e by $E'+\Delta E$ in equations (55), (56) having previously replaced ω_e by $-\omega_h$, where $\omega_h = eB/m_h$. We thus obtain the equations

$$J_x = \frac{pe^2}{m_h}[\tau_h \mathscr{E}''_x + \omega_h \tau_h^2 \mathscr{E}''_y], \tag{59}$$

$$J_y = \frac{pe^2}{m_h}[\tau_h \mathscr{E}''_y - \omega_h \tau_h^2 \mathscr{E}''_x], \tag{60}$$

and

$$W_x = \frac{ne}{m_h}[E'\tau_h \mathscr{E}''_x + \omega_h E'\tau_h^2 \mathscr{E}''_y] + \frac{ne\,\Delta E}{m_h}[\tau_h \mathscr{E}''_x + \omega_h \tau_h^2 \mathscr{E}''_y], \tag{61}$$

$$W_y = \frac{ne}{m_h}[E'\tau_h \mathscr{E}''_y - \omega_h E'\tau_h^2 \mathscr{E}''_x] + \frac{ne\,\Delta E}{m_h}[\tau_h \mathscr{E}''_y - \omega_h \tau_h^2 \mathscr{E}''_x]. \tag{62}$$

We now proceed to perform the usual average with respect to E; the quantities which occur will be of the form $\langle E^n \tau \rangle$ which we have used in §§ 6.1 and 6.2 and also $\langle E^n \tau^2 \rangle$, and the equations (59)–(62) may be expressed in terms of these quantities as follows. Let us write them first in the form

$$J_x = \sigma_{11}\mathscr{E}_x + \sigma_{12}\mathscr{E}_y + \alpha\frac{\partial T}{\partial x} + \beta\frac{\partial T}{\partial y}, \tag{63}$$

$$J_y = -\sigma_{12}\mathscr{E}_x + \sigma_{11}\mathscr{E}_y - \beta\frac{\partial T}{\partial x} + \alpha\frac{\partial T}{\partial y}, \tag{64}$$

$$W_x = \gamma\mathscr{E}_x + \delta\mathscr{E}_y + \kappa_{11}\frac{\partial T}{\partial x} + \kappa_{12}\frac{\partial T}{\partial y}, \tag{65}$$

$$W_y = -\delta\mathscr{E}_x + \gamma\mathscr{E}_y - \kappa_{12}\frac{\partial T}{\partial x} + \kappa_{11}\frac{\partial T}{\partial y} \tag{66}$$

(the equality of some coefficients will be seen from the form of the equations). Comparing these with equations (59)–(62) we see that for electrons we have

$$\sigma_{11} = e^2 n\langle \tau_e \rangle / m_e = ne\mu_e = \sigma_e, \tag{67}$$

$$\sigma_{12} = -\omega_e e^2 n \langle \tau_e^2 \rangle / m_e = -B(ne\mu_e)^2 r/en = \sigma_e^2 R_e B, \tag{68}$$

$$\alpha = \frac{en}{Tm_e}\langle E\tau_e \rangle + \frac{Ten}{m_e}\frac{d}{dT}\left(\frac{E_F}{T}\right)\langle \tau_e \rangle, \tag{69}$$

$$\beta = -\frac{en\omega_e}{m_e T}\langle E\tau_e^2\rangle - \frac{en\omega_e T}{m_e}\frac{d}{dT}\left(\frac{E_F}{T}\right)\langle\tau_e^2\rangle, \tag{70}$$

$$\gamma = -\frac{en}{m_e}\langle E\tau_e\rangle, \tag{71}$$

$$\delta = \frac{en\omega_e}{m_e}\langle E\tau_e^2\rangle, \tag{72}$$

$$\kappa_{11} = -\frac{n}{m_e T}\langle E^2\tau_e\rangle - \frac{nT}{m_e}\frac{d}{dT}\left(\frac{E_F}{T}\right)\langle E\tau_e\rangle - \kappa_L, \tag{73}$$

$$\kappa_{12} = \frac{n\omega_e}{m_e T}\langle E^2\tau_e^2\rangle + \frac{n\omega_e T}{m_e}\frac{d}{dT}\left(\frac{E_F}{T}\right)\langle E\tau_e^2\rangle. \tag{74}$$

We have included the lattice contribution to the thermal conductivity κ_L in κ_{11}. The quantity r in the equation for σ_{12} is equal to $\langle\tau^2\rangle/\langle\tau\rangle^2$ and is the factor which occurs in the formula for the Hall constant R; it is given when τ is of the form aE^{-s} by equation (116) of § 5.3.3. For a relaxation time of this form we have also from equations (7)

$$\left.\begin{aligned}\langle E\tau_e\rangle &= (\tfrac{5}{2}-s)\boldsymbol{k}T\langle\tau_e\rangle,\\ \langle E^2\tau_e\rangle &= (\tfrac{7}{2}-s)(\tfrac{5}{2}-s)\boldsymbol{k}^2T^2\langle\tau_e\rangle.\end{aligned}\right\} \tag{75}$$

Similarly we have

$$\left.\begin{aligned}\langle E\tau_e^2\rangle &= (\tfrac{5}{2}-2s)\boldsymbol{k}T\langle\tau_e^2\rangle = (\tfrac{5}{2}-2s)\boldsymbol{k}Tr\langle\tau_e\rangle^2,\\ \langle E^2\tau_e^2\rangle &= (\tfrac{7}{2}-2s)(\tfrac{5}{2}-2s)\boldsymbol{k}^2T^2\langle\tau_e^2\rangle = (\tfrac{7}{2}-2s)(\tfrac{5}{2}-2s)\boldsymbol{k}^2T^2r\langle\tau_e\rangle^2.\end{aligned}\right\} \tag{76}$$

For holes we have

$$\sigma_{11} = e^2p\langle\tau_h\rangle/m_h = pe\mu_h = \sigma_h, \tag{77}$$

$$\sigma_{12} = \omega_h e^2 p\langle\tau_h^2\rangle/m_h = \sigma_h^2 R_h B, \tag{78}$$

$$\alpha = -\frac{ep}{Tm_h}\langle E\tau_h\rangle - \frac{ep\,\Delta E}{Tm_h}\langle\tau_h\rangle + \frac{Tep}{m_h}\frac{d}{dT}\left(\frac{E_F}{T}\right)\langle\tau_h\rangle, \tag{79}$$

$$\beta = -\frac{ep\omega_h}{m_h T}\langle E\tau_h^2\rangle - \frac{ep\omega_h\,\Delta E}{m_h T}\langle\tau_h^2\rangle + \frac{ep\omega_h T}{m_h}\frac{d}{dT}\left(\frac{E_F}{T}\right)\langle\tau_h^2\rangle, \tag{80}$$

$$\gamma = \frac{ep}{m_h}\langle E\tau_h\rangle + \frac{ep\,\Delta E}{m_h}\langle\tau_h\rangle, \tag{81}$$

$$\delta = \frac{ep\omega_h}{m_h}\langle E\tau_h^2\rangle + \frac{ep\omega_h\,\Delta E}{m_h}\langle\tau_h^2\rangle, \tag{82}$$

$$\kappa_{11} = -\frac{p}{m_h T}\langle E^2\tau_h\rangle - \frac{2p\,\Delta E}{m_h T}\langle E\tau_h\rangle - \frac{p\,\Delta E^2}{m_h T}\langle\tau_h\rangle$$

$$+\frac{pT}{m_h}\frac{d}{dT}\left(\frac{E_F}{T}\right)\langle E\tau_h\rangle + \frac{pT\,\Delta E}{m_h}\frac{d}{dT}\left(\frac{E_F}{T}\right)\langle\tau_h\rangle - \kappa_L, \tag{83}$$

$$\kappa_{12} = -\frac{p\omega_h}{m_h T}\langle E^2\tau_h^2\rangle - \frac{2p\omega_h\,\Delta E}{m_h T}\langle E\tau_h^2\rangle - \frac{p\omega_h\,\Delta E^2}{m_h T}\langle\tau_h^2\rangle$$

$$+\frac{p\omega_h T}{m_h}\frac{d}{dT}\left(\frac{E_F}{T}\right)\langle E\tau_h^2\rangle + \frac{p\omega_h T\,\Delta E}{m_h}\frac{d}{dT}\left(\frac{E_F}{T}\right)\langle\tau_h^2\rangle. \tag{84}$$

In writing down these equations we have changed from E'_F to E_F by means of the relationship of $E'_F = -E_F - \Delta E$, and neglected the variation of ΔE with T. We have dropped the dash on E in forming the average as it is clear what is meant. Also by $\langle E\tau_h\rangle$ we mean a positive quantity. For mixed conduction equations (77)–(84) are added to equations (67)–(74) except that κ_L *is only included once*. We note that the thermal conductivity κ is given by the equation

$$\kappa = -\kappa_{11} + \alpha\gamma/\sigma_{11}. \tag{85}$$

6.3.1 The Ettingshausen effect

The Ettingshausen effect is usually obtained with no temperature gradient in the x-direction. We therefore take $\partial T/\partial x = 0$ and also assume that $J_y = W_y = 0$. Equations (64) and (66) may then be used to eliminate E_y and to solve for $\partial T/\partial y$, so that we have

$$\frac{\partial T}{\partial y} = \frac{(\delta - \sigma_{12}\gamma/\sigma_{11})\mathscr{E}_x}{\kappa_{11} - \alpha\gamma/\sigma_{11}}. \tag{86}$$

In (86) we may use the approximate form $E_x = J_x/\sigma_{11}$ as correct to the first order in B, and on substituting for the other quantities from equations (67)–(74) we have

$$B^{-1}\frac{\partial T}{\partial y} = -\frac{\mu_e J_x}{\kappa e}\left[\frac{\langle E\tau_e^2\rangle}{\langle\tau_e\rangle^2} - \frac{\langle\tau_e^2\rangle\langle E\tau_e\rangle}{\langle\tau_e\rangle^3}\right]. \tag{87}$$

The Ettingshausen coefficient P is defined by means of the equation

$$\frac{\partial T}{\partial y} = -PBJ_x, \tag{88}$$

so that we have for n-type conduction

$$P_e = \frac{\mu_e}{e\kappa}\left[\frac{\langle E\tau_e^2\rangle}{\langle\tau_e\rangle^2} - \frac{\langle\tau_e^2\rangle\langle E\tau_e\rangle}{\langle\tau_e\rangle^3}\right]. \tag{89}$$

When τ has the form $\tau = aE^{-s}$ we obtain ($r = \langle\tau_e^2\rangle/\langle\tau_e\rangle^2$)

$$P_e = -\frac{rskT\mu_e}{e\kappa}. \tag{90}$$

For pure lattice scattering with $s = \frac{1}{2}$ we have $r = \frac{3}{8}\pi$ so that

$$P_e = -\frac{3\pi kT\mu_e}{16e\kappa}. \tag{91}$$

We note that except for any variation of μ_e with n, P is independent of n.

For p-type conduction it may be similarly shown that

$$P = \frac{\mu_h}{e\kappa}\left[\frac{\langle E\tau_h^2\rangle}{\langle\tau_h\rangle^2} - \frac{\langle\tau_h^2\rangle\langle E\tau_h\rangle}{\langle\tau_h\rangle^3}\right]. \tag{92}$$

We note that there is no change in sign as for the thermo-electric power. When τ_h has the form $a'E^{-s'}$ we obtain $(r' = \langle\tau_h^2\rangle/\langle\tau_h\rangle)^2$

$$P_h = -\frac{r's'kT\mu_h}{e\kappa}, \tag{93}$$

so that for lattice scattering with $s' = \frac{1}{2}$

$$P_h = -\frac{3\pi kT\mu_h}{16e\kappa}. \tag{94}$$

The voltages generated are illustrated in Fig. 6.3 for both types of conduction. It will be seen that the thermo-electric voltage generated by the Ettingshausen effect is in the same direction as the Hall voltage if $s > 0$. In some of the published work on these effects equations (91) and (94) are written in a form which assumed κ to be given by the *electronic* contribution to the thermal conductivity. This is incorrect, the quantity κ which appears in equations (89)–(94) being the *total* thermal conductivity as was pointed out by E. H. Putley[20] in connection with measurements on lead selenide.

The Ettingshausen coefficient is proportional to s (equation (90)) if $\tau = aE^{-s}$; this shows that for pure lattice scattering by the acoustical

[20] E. H. Putley, *Proc. Phys. Soc.* B (1955) **58**, 35.

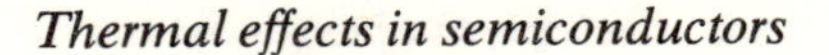

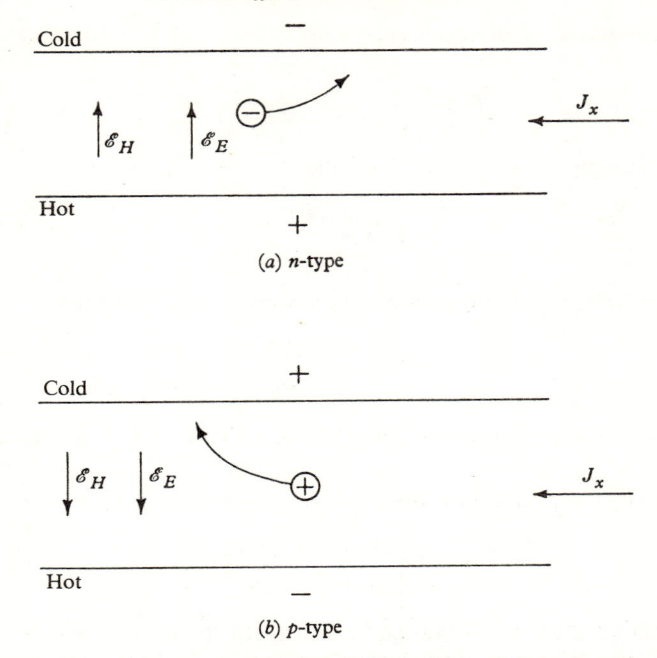

(*a*) *n*-type

(*b*) *p*-type

Fig. 6.3. Voltages generated by isothermal Hall effect and Ettingshausen effect.

modes ($s=\frac{1}{2}$) it is negative, whereas for ionized impurity scattering ($s=-\frac{3}{2}$) it is positive. For mixed scattering the sign will depend on the ratio of lattice to impurity scattering. If impurity scattering predominates in a semiconductor at low temperatures, giving a positive value of P, on raising the temperature a change in sign may be noted as lattice scattering becomes predominant. A change in sign due to mixed conduction will now be discussed.

The value of P for mixed conduction is obtained by adding respectively to the equations (64) and (66) the equations appropriate to hole conduction, the quantities σ_{11}, α, etc. being given by equations (77)–(84). $\partial T/\partial y$ is then found, as before, by putting $W_y=J_y=0$. The expression obtained for P in the general case is rather complex and will not be quoted; the value when $s=s'=\frac{1}{2}$ has been given by E. H. Putley (*loc. cit.*) in the form

$$P=-\frac{3\pi}{16}\frac{\boldsymbol{k}T}{e}\frac{(n^2\mu_e^3+p^2\mu_h^3)-np\mu_e\mu_h(\mu_e+\mu_h)(7+2\Delta E/\boldsymbol{k}T)}{\kappa(n\mu_e+p\mu_h)^2}. \tag{95}$$

It would appear that, in this case, because of the large value of the term involving $\Delta E/kT$ when $n \simeq p$, P would become positive for near-intrinsic material. This effect has been observed by Putley (*loc. cit.*).

6.3.2 The Nernst effect

The isothermal Nernst effect is obtained with $J_x = 0$ and $\partial T/\partial y = 0$; we also assume that $J_y = 0$. Using equations (63) and (64) we may eliminate $\mathscr{E}_x$ and solve for $\mathscr{E}_y$, and so obtain

$$\mathscr{E}_y = \frac{\beta\sigma_{11} - \alpha\sigma_{12}}{\sigma_{11}^2 + \sigma_{12}^2} \frac{\partial T}{\partial x}, \tag{96}$$

since the term $\sigma_{12}^2 \propto B^2$ may be omitted. The Nernst coefficient Q is defined by means of the equation

$$\mathscr{E}_y = -QB \frac{\partial T}{\partial x}. \tag{97}$$

Substituting for σ_{11}, σ_{12}, α, β we have

$$Q_e = \frac{\mu_e}{Te}\left[\frac{\langle E\tau_e^2\rangle}{\langle\tau_e\rangle^2} - \frac{\langle\tau_e^2\rangle\langle E\tau_e\rangle}{\langle\tau_e\rangle^3}\right]. \tag{98}$$

When $\tau = aE^{-s}$ we have

$$Q_e = -\frac{k\mu_e rs}{e} = \frac{ks\mu_{He}}{e} \tag{99}$$

and for the special case $s = \frac{1}{2}$

$$Q_e = -\frac{3\pi k\mu_e}{16e}. \tag{99a}$$

For p-type conduction we may show in a similar manner that

$$Q_h = \frac{\mu_e}{Te}\left[\frac{\langle E\tau_h^2\rangle}{\langle\tau_h\rangle^2} - \frac{\langle\tau_h^2\rangle\langle E\tau_h\rangle}{\langle\tau_h\rangle^3}\right]. \tag{100}$$

When $\tau_h = a'E^{-s'}$ we have

$$Q_h = -\frac{k\mu_h r's'}{e} = -\frac{ks'\mu_{Hh}}{e} \tag{101}$$

and for $s' = \frac{1}{2}$

$$Q_h = -\frac{3\pi k\mu_h}{16e}. \tag{102}$$

From equations (98) and (100) it will be seen that under these conditions we may write

$$Q = \kappa P/T. \tag{103}$$

This is a well-known relationship derived by P. W. Bridgman[21] using classical thermodynamics, and being a thermodynamic relationship holds quite generally. The value of Q for mixed conduction may be thus obtained from that of the Ettingshausen coefficient.

We see that the Nernst coefficient may also change sign when ionized impurity scattering predominates, and that it may also change sign due to mixed conduction. The condition for this will be seen from equation (95), when we have $n \simeq p$, to be

$$\frac{b^3+1}{b(b+1)} < 7 + 2\Delta E/\boldsymbol{k}T, \tag{104}$$

where $b = \mu_e/\mu_h$. Unless b is very large this condition will generally be satisfied.

The phenomenon of 'phonon drag' mentioned in connection with thermal conductivity and thermo-electric power also affects the value of the Nernst coefficient. This effect has been discussed theoretically by J. E. Parrott[22] who has shown that under certain circumstances it may also change the sign of the Nernst coefficient, particularly at low temperatures. This effect should be readily distinguishable from the change in sign due to mixed conduction which usually takes place at high temperatures.

For the Nernst effect and Ettingshausen effect, as for the Hall effect, a distinction may be made between isothermal and adiabatic coefficients, although for semiconductors the difference is small. The isothermal definition is normally used in experimental work even though strict isothermal conditions cannot be maintained.

6.3.3 The Righi–Leduc effect

The Righi–Leduc effect is obtained when $J_x = 0$, and we also take $J_y = W_y = 0$. The transverse temperature gradient $\partial T/\partial y$ and longitudinal gradient $\partial T/\partial x$ are related through the Righi–Leduc coefficient S by the equation

$$\frac{\partial T}{\partial y} = SB\frac{\partial T}{\partial x}. \tag{105}$$

[21] *Thermodynamics of Electrical Phenomena in Metals* (Macmillan, 1934).
[22] *Proc. Phys. Soc.* (1958) **71**, 82.

For n-type conduction, we use equations (64) and (66) to eliminate $\mathscr{E}_y$ and to solve for $\partial T/\partial y$ in terms of $\partial T/\partial x$. Since to the first order in B we may take $\mathscr{E}_x = -\alpha(\partial T/\partial x)/\sigma_{11}$, we obtain the equation

$$\kappa\frac{\partial T}{\partial y} = \frac{\partial T}{\partial x}\left\{-\kappa_{12} + \frac{\beta\gamma + \alpha\delta}{\sigma_{11}} - \frac{\alpha\gamma\sigma_{12}}{\sigma_{11}^2}\right\}. \tag{106}$$

Inserting the values for α, γ, β, δ, etc. in equation (106) we obtain for S_e the value

$$S_e = -\frac{n\mu_e^2}{Te\kappa}\left[\frac{\langle E^2\tau_e^2\rangle}{\langle\tau_e\rangle^2} + \frac{\langle E\tau_e\rangle^2\langle\tau_e^2\rangle}{\langle\tau_e\rangle^4} - \frac{2\langle E\tau_e^2\rangle\langle E\tau_e\rangle}{\langle\tau_e\rangle^3}\right]. \tag{107}$$

If $\tau_e = aE^{-s}$ equation (107) reduces to the form

$$S_e = -n\boldsymbol{k}^2\mu_e^2 Tr(5 - 4s + 2s^2)/2e\kappa, \tag{108}$$

and for lattice scattering with $s = \frac{1}{2}$ we have $r = \frac{3}{8}\pi$ and

$$S_e = -21\pi\boldsymbol{k}^2 Tn\mu_e^2/32e\kappa. \tag{109}$$

For p-type conduction we may show in the same way that

$$S_h = \frac{p\mu_h^2}{Te\kappa}\left[\frac{\langle E^2\tau_h^2\rangle}{\langle\tau_h\rangle^2} + \frac{\langle E\tau_h\rangle^2\langle\tau_h^2\rangle}{\langle\tau_h\rangle^4} - \frac{2\langle E\tau_h^2\rangle\langle E\tau_h\rangle}{\langle\tau_h\rangle^3}\right]. \tag{110}$$

When $\tau_h = a'E^{-s'}$ we have

$$S_h = p\boldsymbol{k}^2\mu_h^2 Tr'(5 - 4s' - 2s'^2)/2e\kappa \tag{111}$$

and for $s' = \frac{1}{2}$, we have $r' = \frac{3}{8}\pi$, so that

$$S_h = 21\pi\boldsymbol{k}^2 Tp\mu_h^2/32e\kappa. \tag{112}$$

We note that S is negative for conduction by electrons and positive for conduction by holes. The quantities R (Hall constant) and S are not zero when τ does not vary with E ($s = 0$), while P and Q are both zero under this condition.

6.4 Condition of degeneracy

When we have a strongly degenerate condition the values for P, Q, S may be obtained by expanding the quantities $\langle E^n\tau\rangle$, $\langle E^n\tau^2\rangle$ using equation (12). We require to take account of terms of the order of $\boldsymbol{k}^2T^2/E_F^2$, since P, Q, are zero in the zero-order approximation. To the first approximation we have

$$\langle\tau^2\rangle = \tau^2[1 + A(\tfrac{3}{2} - 2s)(\tfrac{1}{2} - 2s)],$$

$$\langle\tau^2 E\rangle = \tau^2 E_F[1 + A(\tfrac{5}{2} - 2s)(\tfrac{3}{2} - 2s)],$$

$$\langle\tau^2 E^2\rangle = \tau^2 E_F^2[1 + A(\tfrac{7}{2} - 2s)(\tfrac{5}{2} - 2s)],$$

where $A = \frac{1}{6}\pi^2(\boldsymbol{k}T/E_F)^2$. Using these relationships we obtain for the Ettingshausen coefficient P_e the value

$$P_e = -\frac{\pi^2 s \boldsymbol{k}^2 T^2 \mu_e}{3E_F e\kappa}. \tag{113}$$

Similarly, we have for P_h the value

$$P_h = -\frac{\pi^2 s' \boldsymbol{k}^2 T^2 \mu_h}{3E'_F e\kappa}. \tag{114}$$

Equation (103) still holds for the relationship between P and the Nernst coefficient Q, so that we have

$$Q_e = -\frac{\pi^2 s \boldsymbol{k}^2 T \mu_e}{3eE_F}, \tag{115}$$

$$Q_h = -\frac{\pi^2 s' \boldsymbol{k}^2 T \mu_h}{3eE'_F}. \tag{116}$$

We see that P and Q are reduced approximately by the ratio $\boldsymbol{k}T/E_F$ or $\boldsymbol{k}T/E'_F$ from their value for the non-degenerate condition. For the Righi–Leduc coefficient S we have similarly

$$S_e = -\frac{\pi^2 n \mu_e^2 \boldsymbol{k}^2 T}{3e\kappa}, \tag{117}$$

$$S_h = \frac{\pi^2 p \mu_h^2 \boldsymbol{k}^2 T}{3e\kappa}. \tag{118}$$

Formulae for these coefficients quoted in the literature have sometimes been reduced to a form which is equivalent to the above with κ replaced by κ_e or κ_h. These apply to metals when $\kappa \simeq \kappa_e$, but not to semiconductors. The value of κ to be used in this case is the total thermal conductivity including κ_L. The formulae given by equations (113)–(118) apply only for the condition of strong degeneracy, i.e. when either $E_F/\boldsymbol{k}T \gg 1$ or $E'_F/\boldsymbol{k}T \gg 1$; for the intermediate case the quantities $\langle E^n\tau\rangle$, $\langle E^n\tau^2\rangle$ must be evaluated using integrals of the form given by equations (49).[23]

[23] For a full account of these effects see E. H. Putley, *The Hall Effect and Related Phenomena* (Butterworths, 1960).

6.5 Strong magnetic fields

When the magnetic field is sufficiently large to make $\omega\tau$ no longer negligible compared with unity, equations (55)–(58) must be replaced by equations corresponding to (61) and (62) of § 5.2.2. Each term in the right-hand sides of (55)–(58) is then multiplied by a factor $1/(1+\omega^2\tau^2)$ and this factor must be included in the averages. The quantities such as $\langle E^n\tau\rangle$ are then replaced by $\langle E^n\tau/(1+\omega^2\tau^2)\rangle$, etc. The rather simple relationships between $\langle E\tau\rangle$, $\langle\tau\rangle$, etc., when $\tau = ae^{-s}$ now no longer hold, and each average must be calculated in the non-degenerate case using integrals of the form

$$K_{rt} = \int_0^\infty \frac{\tau^t E^r \exp(-E/kT)\,dE}{1+\omega^2\tau^2}. \tag{119}$$

In this case numerical methods must be used for their evaluation for particular values of s. Care must also be taken to include terms of the order of B^2 which we have neglected in deriving the above formulae for P, Q, S, and equations (63)–(66) should be used to obtain the coefficients with the correct values of α, β, γ, etc. The resulting formulae are somewhat complex[24] and will not be given.

6.6 Relative magnitudes of the magnetic effects

The ratio of the voltage V_E arising from the Ettingshausen effect to the voltage V_H from the isothermal Hall effect, may be expressed in the form

$$V_E/V_H = \mathscr{P}P/R, \tag{120}$$

when conduction is due mainly to one type of carrier. If we let $P = qk/e$ this may be expressed in the form

$$V_E/V_H = Aq(\kappa_e/\kappa), \tag{121}$$

where A is a numerical factor (for $s = \frac{1}{2}$, $A = 4$). Also we may express the Righi–Leduc coefficient S in the form

$$S = C\mu(\kappa_e/\kappa), \tag{122}$$

where C is a constant of order unity (for $s = \frac{1}{2}$, $C = 21\pi/64$). Let us calculate various quantities for a specimen of a semiconductor of the dimensions and characteristics given below.

[24] See, for example, A. C. Beer, J. A. Armstrong and I. N. Greenberg, *Phys. Rev.* (1957) **107**, 1506; A. C. Beer, *Solid State Phys.*, Suppl. 4 (Academic Press, 1963).

Length l	$1\ \text{cm} = 10^{-2}\ \text{m}$
Width w	$1\ \text{mm} = 10^{-3}\ \text{m}$
Thickness t	$0.1\ \text{mm} = 10^{-4}\ \text{m}$
Current I	$10^{-3}\ \text{A}$
Current density J_x	$10^{4}\ \text{A m}^{-2}$
Thermal conductivity κ	$50\ \text{W m}^{-1}\ \text{deg}^{-1}$
Magnetic induction B_z	$10^{3}\ \text{G} = 0.1\ \text{T}$
Electrical conductivity σ	$1\ \Omega^{-1}\ \text{cm}^{-1} = 100\ \Omega^{-1}\ \text{m}^{-1}$
Electron mobility $\mu_3 \simeq R\sigma$	$2000\ \text{cm}^2\ \text{V}^{-1}\ \text{s}^{-1} = 0.2\ \text{m}^2\ \text{V}^{-1}\ \text{s}^{-1}$
Thermo-electric power	$-500\ \mu\text{V deg}^{-1}$
$\partial T/\partial x$	$10^{3}\ \text{deg m}^{-1}$

The following quantities have been calculated approximately from the above data:

Hall constant R	$2 \times 10^{-3}\ \text{m}^3\ \text{C}^{-1}$
Hall voltage V_H	$2\ \text{mV}$
Nernst coefficient Q	$-10^{-5}\ \text{m}^2\ \text{s}^{-1}\ \text{deg}^{-1}$
Nernst voltage V_N	$1\ \mu\text{V}$
Electronic thermal conductivity κ_e	$5 \times 10^{-4}\ \text{W m}^{-1}\ \text{deg}^{-1}$
κ_e/κ	10^{-5}
Ettingshausen coefficient P	$-6 \times 10^{-5}\ \text{deg m}^3\ \text{J}^{-1}$
dT_E/dy	$6 \times 10^{-2}\ \text{deg m}^{-1}$
T_E (temperature change across specimen due to Ettingshausen effect)	$6 \times 10^{-5}\ \text{deg}$
Ettingshausen voltage V_E	$0.03\ \mu\text{V}$
V_E/V_H	10^{-5}
Righi–Leduc coefficient S	$2 \times 10^{-6}\ \text{m}^2\ \text{V}^{-1}\ \text{s}^{-1}$
$\dfrac{\partial T_{RL}}{\partial y} \Big/ \dfrac{\partial T}{\partial x} = SB$	2×10^{-7}
Temperature change ΔT_{RL} due to Righi–Leduc effect	$2 \times 10^{-7}\ \text{deg}$
Righi–Leduc voltage V_{RL}	$\sim 10^{-10}\ \text{V}$

It will be seen that under these conditions both the Ettingshausen effect and Righi–Leduc effect give very small transverse voltages. These are approximately proportional to the electrical conductivity σ, and for a semiconductor with $\sigma = 10^5\ \Omega^{-1}\ \text{m}^{-1}$ and $\kappa = 5\ \text{W m}^{-1}\ \text{deg}^{-1}$ the voltage due to the Ettingshausen effect would become about 10% of the Hall voltage. It will also be clear from equations (121) and (122) why these effects are so much more important in metals for which $\kappa_e \simeq \kappa$.

7

Diffusion of electrons and positive holes

7.1 Inhomogeneous semiconductors

In Chapter 4 we have discussed the equilibrium conditions in a homogeneous semiconductor; we must now extend this treatment to include inhomogeneous material. Such material, mainly in the form of junctions between parts of a crystal with different impurity concentrations, plays an important part in technology based on semiconductors. In an inhomogeneous semiconductor the electron concentration n and hole concentration p will vary with position, and so will be functions of the spatial co-ordinates (x, y, z). In equilibrium we may define a Fermi level which is constant throughout the material; this follows from the relation between the Fermi energy and thermodynamic potential (see § 4.1), the condition for equilibrium under constant pressure being that the latter is constant.

Since the electron and hole concentrations vary with the co-ordinates (x, y, z) we shall have diffusion currents of electrons and holes. The number of electrons n_x^d crossing unit area in the x-direction per unit time is given in terms of the concentration gradient $\partial n/\partial x$ by an equation of the form

$$n_x^d = -D_e \frac{\partial n}{\partial x}, \tag{1}$$

the quantity D_e being known as the diffusion coefficient for electrons. The electron current density J_{ex}^d due to diffusion is thus given by the equation

$$J_{ex}^d = eD_e \frac{\partial n}{\partial x}. \tag{2}$$

Similarly, we may define a diffusion coefficient D_h for holes; the number of holes p_x^d crossing unit area per unit time in the x-direction is given by the equation

$$p_x^d = -D_h \frac{\partial p}{\partial x}. \tag{3}$$

The hole current density J_{dx}^h due to diffusion is thus given by the equation

$$J_{hx}^d = -eD_h \frac{\partial p}{\partial x}. \tag{4}$$

In conditions of equilibrium there can clearly be no flow of current (either of holes or electrons) so that in inhomogeneous material there must exist a static electric field to counteract the flow due to diffusion. The calculation of the value of this field is quite complex in general, but solutions for a number of special conditions may readily be obtained and will be discussed later. Equations (1)–(4) may be expressed in vector form so as to give the diffusion current in any direction. To obtain the total current density $\mathbf{J}_e$ due to electrons we have to add the conduction current; this gives

$$\mathbf{J}_e = ne\mu_e \mathscr{E} + eD_e \nabla n, \tag{5}$$

and similarly for the hole current density we have

$$\mathbf{J}_h = pe\mu_h \mathscr{E} - eD_h \nabla p. \tag{6}$$

The total current density $\mathbf{J}$ is thus given by the equation

$$\mathbf{J} = e(n\mu_e + p\mu_h)\mathscr{E} + e(D_e \nabla n - D_h \nabla p). \tag{7}$$

It is interesting at this stage to consider why diffusion currents are important in semiconductors and not in metals. The reason is that in semiconductors, n and p may each vary markedly without the total charge density varying. As we shall see, any large variation of charge density must be accompanied by very large electric fields – a consequence of Poisson's equation. In a metal, on the other hand, the electron density is practically constant.

7.2 Einstein's relationship

In equilibrium, the electron current density must be zero. Let us take the x-direction as that of the electric field; then we must have

$$n\mu_e \mathscr{E}_x = -D_e \frac{\partial n}{\partial x}. \tag{8}$$

If ψ is the electrostatic potential we have

$$\mathscr{E}_x = -\frac{\partial \psi}{\partial x} \tag{9}$$

and the potential energy of an electron is equal to $-e\psi$. For conditions of non-degeneracy, the concentration of electrons n will obey classical statistics and we have

$$n = \text{constant} \times \exp\left[e\psi/kT\right] \tag{10}$$

so that

$$\frac{1}{n}\frac{\partial n}{\partial x} = \frac{e}{kT}\frac{\partial \psi}{\partial x}. \tag{11}$$

Substituting for $\mathscr{E}_x$ and $\partial n/\partial x$ in equation (8) in terms of ψ we obtain the equation

$$D_e = (kT/e)\mu_e. \tag{12}$$

This equation, giving the diffusion coefficient D_e in terms of the conductivity mobility μ_e, is known as Einstein's relationship; it holds quite generally for any particles diffusing in an electric field. For holes we have

$$D_h = (kT/e)\mu_h, \tag{13}$$

which follows similarly from the condition that, in equilibrium, the total hole current should also be equal to zero. D_h and D_e have dimensions $\text{m}^2\,\text{s}^{-1}$ if μ is in $\text{m}^2\,\text{V}^{-1}\,\text{s}^{-1}$, or $\text{cm}^2\,\text{s}^{-1}$ if μ is in $\text{cm}^2\,\text{V}^{-1}\,\text{s}^{-1}$; for example, if $\mu = 10^3\,\text{cm}^2\,\text{V}^{-1}\,\text{s}^{-1}$, $D = 25\,\text{cm}^2\,\text{s}^{-1}$. Equation (12) holds only for conditions of non-degeneracy; it may, however, readily be extended to deal with a degenerate condition.

The form of the energy bands for inhomogeneous material is shown in Fig. 7.1. It is no longer convenient to take our energy zero at the bottom of the conduction band as the latter varies in height above the constant Fermi level. Let E_c represent the lowest energy in the conduction band

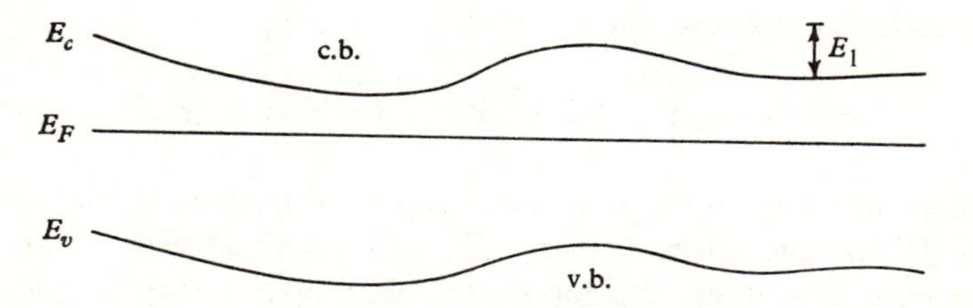

Fig. 7.1. Energy bands for inhomogeneous material.

and E_v the highest energy in the valence band. These will vary with position for inhomogeneous material. Then we have

$$E_c - E_v = \Delta E, \tag{14}$$

where ΔE is the forbidden energy gap, which we shall assume to be independent of impurity content, and so constant throughout the material. Let E be the energy of any electron level and E_1 its height above the bottom of the conduction band, so that $E = E_c + E_1$. The distribution function f_0 may then be written in the form

$$f_0 = \frac{1}{\exp[(E - E_F)/\boldsymbol{k}T] + 1} = \frac{1}{\exp[(E_1 + E_c - E_F)/\boldsymbol{k}T] + 1}. \tag{15}$$

We have then

$$n = \int_{E_c}^{\infty} N(E - E_c) f_0 \, dE = \int_0^{\infty} N(E_1) f_0 \, dE_1. \tag{16}$$

Clearly E_1 may be regarded as before, as the kinetic energy of an electron, whose potential energy may be taken, apart from an arbitrary constant, as E_c; thus E_c is related to the electrostatic potential ψ by means of the equation

$$E_c = \text{constant} - \boldsymbol{e}\psi. \tag{17}$$

The electric field $\mathscr{E}_x$ in the x-direction is therefore given by the equation

$$\mathscr{E}_x = -\frac{\partial \psi}{\partial x} = \frac{1}{\boldsymbol{e}} \frac{\partial E_c}{\partial x}. \tag{18}$$

We have now
$$\frac{\partial n}{\partial E_c} = \int_0^{\infty} N(E_1) \frac{\partial f_0}{\partial E_1} \, dE_1 \tag{19}$$

so that, using equation (18), we have

$$\frac{\partial n}{\partial x} = \boldsymbol{e}\mathscr{E}_x \int_0^{\infty} N(E_1) \frac{\partial f_0}{\partial E_1} \, dE_1. \tag{20}$$

Substituting in equation (8) we have

$$\boldsymbol{e}D_e = -\mu_e \int_0^{\infty} N(E_1) f_0 \, dE_1 \Big/ \int_0^{\infty} N(E_1) \frac{\partial f_0}{\partial E_1} \, dE_1, \tag{21}$$

which clearly reduces to the Einstein relationship given by equation (12) when f_0 has the form $C \exp[-E_1/\boldsymbol{k}T]$ which applies to the non-degenerate condition. We have seen that $N(E_1)$ may frequently be written in the form $N(E_1) = AE_1^{\frac{1}{2}}$. A partial integration of the denominator of equation (21) shows that this equation may be reduced to the form

$$eD_e = (2\mu_e kT)F_{\frac{1}{2}}[(E_F - E_c)/kT]/F_{-\frac{1}{2}}[(E_F - E_c)/kT], \tag{22}$$

where $F_r(x)$ are the functions defined in § 4.2 for various values of r.

7.3 Departures from thermal equilibrium

When an electric field is applied to a semiconductor, the thermal equilibrium is upset and a new function is required to describe the distribution of electrons between the various energy states. In this case we cannot use an energy-level diagram such as Fig. 7.1, with a constant Fermi level. When an external electric field $\mathscr{E}_x$ is applied to a homogeneous semiconductor the change in f_0 is quite small. In the non-degenerate state it may be shown that[1]

$$f = f_0\left(1 - \frac{ev_x\mathscr{E}_x\tau}{kT}\right) \tag{23}$$

and the voltage $v_x\tau\mathscr{E}_x$ is quite small, compared with kT/e for $T = 300\,°K$, unless $\mathscr{E}_x$ is of the order of $10^3\ \mathrm{V\,cm^{-1}}$. In any case, on integrating over all values of v_x we find that the value of n is unchanged from its equilibrium value. Quite large changes of n and p may, however, be obtained by shining light on the semiconductor, the wavelength being sufficiently short to cause band-to-band transitions (optical injection), or by applying a voltage of appropriate sign to injecting electrodes (see § 7.5.3). Let the disturbed electron concentration n be equal to $n_0 + \Delta n$, where n_0 is the equilibrium concentration of electrons; similarly, let $p = p_0 + \Delta p$, where p_0 is the equilibrium concentration of holes. Unless $\Delta n = \Delta p$, a space-charge ρ will be set up, given by the equation

$$\rho = e(\Delta p - \Delta n). \tag{24}$$

One of the very important results which we shall now prove is that Δn must be nearly equal to Δp unless a very strong electric field is present. This can only occur near the surface or in a region in which the impurity concentration varies very rapidly, giving rise to a high concentration gradient. We have, from Poisson's equation

$$\operatorname{div} \boldsymbol{\mathscr{E}} = \rho/\epsilon = e(\Delta p - \Delta n)/\epsilon, \tag{25}$$

where ϵ is the permittivity. For example, let $p = 10^{12}\ \mathrm{cm^{-3}} = 10^{18}\ \mathrm{m^{-3}}$,

[1] *W.M.C.S.*, § 10.6.

$(\Delta p - \Delta n) = 0.01p$, and $\epsilon = 10\epsilon_0 \simeq 10^{-10}$ F m^{-1}; we then have, if $\mathscr{E}_x = 0$ at $x = 0$, $\mathscr{E}_x = 10^7 x$ V m^{-1} and hence at $x = 1$ cm, $\mathscr{E}_x = 10^5$ V m^{-1}.

We may also consider the time variation of ρ in bulk material (neglecting diffusion for the moment). The equation of continuity for the electric current gives

$$\operatorname{div} \mathbf{J} = \operatorname{div} \sigma \mathscr{E} = -\frac{\partial \rho}{\partial t} = e \frac{\partial}{\partial t}(\Delta n - \Delta p) \tag{26}$$

so that, using equation (25), we have

$$\frac{\partial}{\partial t}(\Delta n - \Delta p) = -\frac{\sigma}{\epsilon}(\Delta n - \Delta p); \tag{27}$$

the solution of this equation is

$$(\Delta n - \Delta p) = (\Delta n - \Delta p)_{t=0} \exp[-t/\tau_0], \tag{28}$$

where $\tau_0 = \epsilon/\sigma$. If $\sigma = 100\ \Omega^{-1}$ m^{-1}, we have $\tau_0 \simeq 10^{-12}$ s, so that any space-charge set up in bulk material would very soon decay; this would appear to show that under *no* conditions can space-charge persist. A more accurate calculation, which we shall give later, and which takes diffusion into account, shows that a *very small* space-charge can persist and this may set up quite large internal electric fields. Our previous discussion of inhomogeneous semiconductors indicates that such space charges and fields exist whenever the donor and acceptor concentration varies. For the present we shall assume that $\Delta n = \Delta p$.

7.3.1 Quasi-Fermi levels

When the electron concentration n and hole concentration p depart from their equilibrium values n_0 and p_0, their values are no longer expressible in terms of a *single* quantity, namely the Fermi energy E_F. In equilibrium, at a given value of the temperature T, we have seen (§ 4.3) that in conditions of non-degeneracy n and p may be expressed in the form

$$n = N_c \exp[E_F/\boldsymbol{k}T] \tag{29}$$

$$p = N_v \exp[-(\Delta E + E_F)/\boldsymbol{k}T]. \tag{29a}$$

We may, however, *define* two new energy parameters E_{Fe} and E_{Fh} which enable us to express n and p in the same form as equations (29), (29a), namely

$$n = N_c \exp[E_{Fe}/\boldsymbol{k}T] \tag{30}$$

$$p = N_v \exp[-(\Delta E + E_{Fh})/\boldsymbol{k}T]. \tag{30a}$$

The quantities E_{Fe} and E_{Fh} are known respectively as the quasi Fermi levels for electrons in holes. In thermal equilibrium they are each equal to E_F but in general lie one above and one below the Fermi level. This is illustrated in Fig. 7.2 for the more usual situation, when $n > n_0$ and $p > p_0$ in which E_{Fe} lies above E_F and E_{Fh} lies below E_F. This will

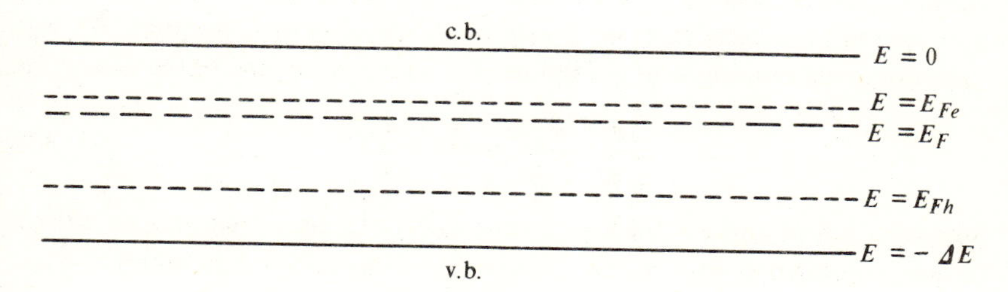

Fig. 7.2. Quasi-Fermi levels, $\Delta n = \Delta p > 0$.

readily be seen when we write E_{Fe} and E_{Fh} in the forms

$$E_{Fe} = \boldsymbol{k}T \ln (n/N_c) \tag{31}$$

$$E_{Fh} = -\Delta E - \boldsymbol{k}T \ln (p/N_v). \tag{31a}$$

If $n_0 \gg p_0$ and $\Delta n = \Delta p \ll n_0$ but also $\Delta n = \Delta p \gg p_0$, a fairly common situation, we see that E_{Fe} moves very little from E_F but E_{Fh} may move quite appreciably.

We no longer have a simple relationship like $n_0 p_0 = n_i^2$; however, we may note that

$$np = n_i^2 \exp [(E_{Fe} - E_{Fh})/\boldsymbol{k}T] \tag{32}$$

so that the difference $E_{Fe} - E_{Fh}$ gives a measure of the departure of the product np from its equilibrium value.

In the case of carrier depletion, when we have $n < n_0$ and $p < p_0$, the positions of the quasi-Fermi levels relative to the Fermi level are reversed as will be seen from equations (31) and (31a).

When we do not have a non-degenerate situation we may still define quasi-Fermi levels in terms of equations like (19) and (20) of § 4.2. They are not then so useful in calculations or in giving a picture of the electron concentration. Quasi-Fermi levels give no indication of the distribution of electrons among the levels in the conduction band or of holes in the valence band and are not very useful when these differ a great deal from their equilibrium form.

7.4 Electron–hole recombination

We must now consider the behaviour of a semiconductor in which a deviation given by Δn, Δp from the equilibrium carrier concentration has been created. Clearly, if the influence causing the deviation is removed, the concentrations will return after a time to their equilibrium values. Let Δn_0 and Δp_0 be the values of Δn and Δp at $t=0$ ($\Delta n_0 = \Delta p_0$), and let the (constant) rate of creation of hole–electron pairs be $\mathscr{R}_0$; we may then express the rate of change of n and p by means of the equation

$$\frac{dn}{dt} = \frac{d\Delta n}{dt} = \mathscr{R}_0 - \frac{n}{\tau_n}, \tag{33}$$

where τ_n is a quantity which may vary with p. In equilibrium $d\Delta n/dt = 0$ and $n = n_0$ so that $\mathscr{R}_0 = n_0/\tau_{n_0}$, where τ_{n_0} is the value of τ_n when $n = n_0$. Equation (33) may therefore be written in the form

$$\frac{d\Delta n}{dt} = \frac{n_0}{\tau_{n_0}} - \frac{n}{\tau_n}; \tag{34}$$

similarly we have

$$\frac{d\Delta p}{dt} = \frac{p_0}{\tau_{p_0}} - \frac{p}{\tau_p}. \tag{34a}$$

An important situation arises when $n \gg p$; p is then known as the minority carrier concentration. The variation of n will be small if $|\Delta p| \not> p$, since $|\Delta n| = |\Delta p| < p$; in this case τ_p will not depend appreciably on n and may be taken as a constant, so that we may then write

$$\frac{d\Delta p}{dt} = -\frac{\Delta p}{\tau_p} \tag{35}$$

with $\mathscr{R}_0 = p_0/\tau_p$. The solution of equation (35) is

$$\Delta p = \Delta p_0 \exp\left[-t/\tau_p\right] \tag{36}$$

and we see that the excess hole concentration decays to $1/e$ of its value at $t=0$ in a time τ_p, which for this reason is called the minority-carrier lifetime for holes. We note that we must also have

$$\Delta n = \Delta n_0 \exp\left[-t/\tau_p\right], \tag{36a}$$

so that τ_p represents the mean lifetime of excess electron–hole pairs. In equilibrium, the generation rate $\mathscr{R}$ is equal to p_0/τ_p, but when n is not much greater than p, a more complex situation occurs which we shall discuss later. Values of τ_p range from one or two milliseconds (for high purity Ge or Si) to less than 10^{-8} s. We may similarly define a quantity τ_n when $p \gg n$. There is a danger that τ_n and τ_p may be confused with the relaxation times τ_e, τ_h for electrons and holes and the space-charge relaxation time τ_0; we have used the above notation to try to avoid this confusion. The order of magnitude of the quantities is quite different; at room temperature τ_e, τ_h, τ_0 are of the order of 10^{-12} s, while τ_n, τ_p are generally very much greater than this.

7.5 Diffusion and conduction in extrinsic material ($n \gg p$ or $p \gg n$)

We are now in a position to discuss the flow of electrons and holes under the influence of electric and magnetic fields in circumstances in which n and p may depart appreciably from their equilibrium values; we shall first deal with one-dimensional motion in the x-direction under electric fields and concentration gradients directed along the x-axis. We shall suppose that we have strongly extrinsic n-type material, i.e. that $n \gg p$; the analysis for p-type material is quite analogous. We shall assume that there is no space-charge, i.e. that $\Delta n = \Delta p$, and that the material is homogeneous, and we shall show later that the conditions $n \gg p$ or $p \gg n$ are just those which make this a good approximation.

7.5.1 The equation of continuity

The equation of continuity in the theory of electric current flow expresses the condition that there is no accumulation of charge. It is generally written in the form

$$\frac{\partial \rho}{\partial t} = -\text{div}\,\mathbf{J}. \tag{37}$$

Here we must express the conditions that there should be no accumulation of electrons or of holes, so that an equation like (37) must be written down for both the hole and electron concentrations. We must also add a term to take account of creation and recombination of electron–hole pairs. For our purpose we need only the equation for holes, since if we know p we may obtain n from the equation

$$n = n_0 + p - p_0, \tag{38}$$

which follows from the condition $\Delta n = \Delta p$. The full equation of continuity for holes is

$$\frac{\partial p}{\partial t} = -\frac{(p - p_0)}{\tau_p} - \frac{1}{e}\,\text{div}\,\mathbf{J}_h, \tag{39}$$

where $\mathbf{J}_h$ is the hole current density, given by equation (6); inserting the value of $\mathbf{J}_h$, and restricting variation to the x-direction, this becomes

$$\frac{\partial p}{\partial t}=-\frac{(p-p_0)}{\tau_p}-p\mu_h\frac{\partial \mathscr{E}}{\partial x}-\mathscr{E}\mu_h\frac{\partial p}{\partial x}+D_h\frac{\partial^2 p}{\partial x^2}. \tag{40}$$

A similar equation may be used to express the rate of variation of n; some care is, however, required in writing down the term which comes from recombination; in this case we have

$$\frac{\partial n}{\partial t}=\frac{p_0}{\tau_p}-\frac{n}{\tau_n}+n\mu_e\frac{\partial \mathscr{E}}{\partial x}+\mathscr{E}\mu_e\frac{\partial n}{\partial x}+D_n\frac{\partial^2 n}{\partial x^2}. \tag{40a}$$

Although we shall not use this equation in the present section, we may note that it appears to be incompatible with equation (40) on substitution of the relationship (38). The reason lies in the approximation we have made of assuming that $\Delta n=\Delta p$: this approximation is quite valid for equation (40) when $n \gg p$, but not for equation (40*a*). We shall now solve equation (40) under a number of restricted conditions.

7.5.2 Small electric field

We shall first consider the condition when the electric field is so small that the conduction current due to minority carriers may be neglected; we then have a condition of almost pure diffusion, and equation (40) becomes

$$\frac{\partial \Delta p}{\partial t}=-\frac{\Delta p}{\tau_p}+D_h\frac{\partial^2 \Delta p}{\partial x^2}. \tag{41}$$

Let us consider the steady state in which $\partial\Delta p/\partial t=0$, when we have a source of holes at $x=0$ which maintains $\Delta p=\Delta p_0$ at $x=0$; the steady-state equation for Δp is then

$$D_h\frac{d^2\Delta p}{dx^2}-\frac{1}{\tau_p}\Delta p=0. \tag{42}$$

We shall assume that we have a semi-infinite bar of semiconducting material extending from $x=0$ to $x=\infty$. For this the appropriate solution is

$$\Delta p=\Delta p_0 \exp\left[-x/L_p\right], \tag{43}$$

Where $L_p^2=D_h\tau_p$; the equilibrium condition is therefore reached in a distance equal to a few times L_p. The length L_p is called the minority carrier diffusion length; for example, for $\mu_h = 10^3\ \text{cm}^2\ \text{V}^{-1}\ \text{s}^{-1}$, $D_h = 25\ \text{cm}^2\ \text{s}^{-1}$, $\tau_p = 10^{-4}$ s, we have $L_p = 0.05$ cm.

The steady state can only be maintained by means of a flow of holes into the material at $x = 0$; the hole current J_{h_0} at $x = 0$ is given by the equation

$$J_{h_0} = -D_h e \frac{\partial \Delta p}{\partial x}\bigg)_{x=0} = eD_h \Delta p_0 / L_p, \tag{44}$$

while the total current density $\mathbf{J}$ obeys equation (37) and so is constant. We now introduce a quantity γ called the injection ratio; the hole current at $x = 0$ is given in terms of γ by the equation

$$J_{h_0} = \gamma J \tag{45}$$

from which it follows that the electron current at $x = 0$, J_{e_0} is given by the equation

$$J_{e_0} = (1-\gamma)J. \tag{45a}$$

If $\gamma = 1$, *all* the current at $x = 0$ is carried by holes, while clearly for $x \gg L_p$ practically all the current is carried by electrons. In order that the electron current may be zero at $x = 0$ there must exist an electric field but, as we shall see, this can be quite small. We have from equations (44) and (45)

$$\Delta p_0 = \gamma J L_p / eD_h \tag{46}$$

$$= (\gamma J/e)(\tau_p/D_h)^{\frac{1}{2}} \tag{46a}$$

and we see that if $J > 0$, so that the total current flows *into* the material at $x = 0$, then $\Delta p_0 > 0$; the minority carrier concentration is therefore increased, unless $\gamma = 0$, i.e. unless *all* the current at $x = 0$ is carried by electrons. (There will be a small hole conduction current which we have neglected.) On the other hand, if $J < 0$, then $\Delta p < 0$, and the minority carrier concentration is decreased. Unless γ is small, the current carried by holes will be comparable in magnitude with that carried by electrons at $x = 0$; at a distance equal to a few times L_p, on the other hand, practically all the current will be carried by electrons, as may be seen as follows. We have

$$J_h = \gamma J \exp[-x/L_p], \tag{47}$$

so that

$$J_e = J - J_h = J\{1 - \gamma \exp[-x/L_p]\}. \tag{48}$$

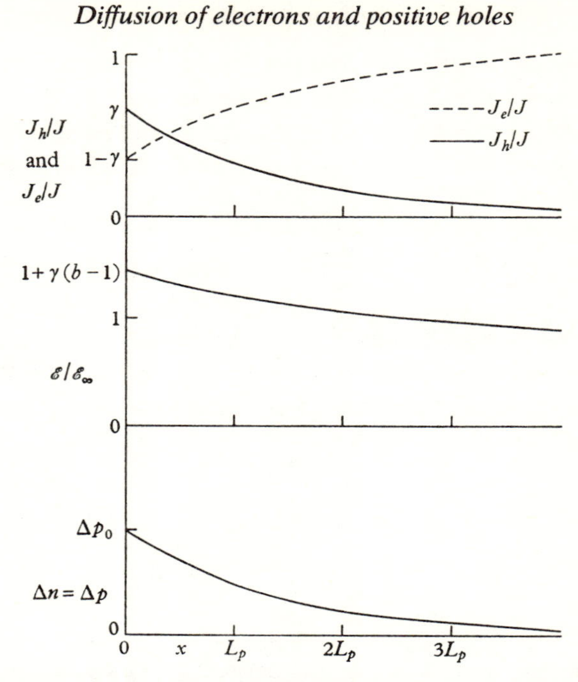

Fig. 7.3. Electron and hole currents J_e, J_h, electric field $\mathscr{E}$, and increase in carrier densities Δn, Δp ($\mathscr{E}$ small).

This is illustrated in Fig. 7.3. We may now calculate the electric field $\mathscr{E}$. The diffusion current due to electrons is equal to $eD_e\, \partial\Delta p/\partial x$, since $\Delta n = \Delta p$; also, since $D_e/D_h = \mu_e/\mu_h = b$, this is equal to $-bJ_h$. The conduction current due to electrons is approximately equal to $en_0\mu_e\mathscr{E}$, where $\mathscr{E}$ is the electric field, since we have assumed that $\Delta p \ll n$; this is also approximately equal to $\sigma\mathscr{E}$ since $n \gg p$. We then have for the total electron current J_e the equation

$$J_e = \sigma\mathscr{E} - bJ_{h_0} \exp[-x/L_p], \tag{49}$$

so that

$$J = \sigma\mathscr{E} - (b-1)\gamma J \exp[-x/L_p] \tag{50}$$

and

$$\mathscr{E} = \mathscr{E}_\infty\{1 + (b-1)\gamma \exp[-x/L_p]\}, \tag{51}$$

where $\mathscr{E}_\infty = J/\sigma$ is the value of the electric field when $x \gg L_p$; for $x = 0$ we have $\mathscr{E} = \mathscr{E}_0$, where

$$\mathscr{E}_0 = \mathscr{E}_\infty[1 + (b-1)\gamma]. \tag{52}$$

The variation of $\mathscr{E}$ with x is also shown in Fig. 7.3: the electric field is greatest at $x=0$ if, as is usual, $b>1$; *if* $\gamma=1$, $\mathscr{E}_0/\mathscr{E}_\infty=b$, and if $\gamma=0$, $\mathscr{E}$ is constant and equal to $\mathscr{E}_\infty$.

We may now examine the assumption that most of the hole current is due to diffusion. The conduction hole current will be equal to $\boldsymbol{e}(p+\Delta p)\mu_h\mathscr{E}$, and this will be small compared with the diffusion current at $x=0$, where $\mathscr{E}$ is greatest, provided

$$\mathscr{E}\mu_h \ll D_h/L_p, \tag{53}$$

i.e. provided

$$\boldsymbol{e}\mathscr{E}L_p \ll \boldsymbol{k}T. \tag{53a}$$

$\boldsymbol{e}\mathscr{E}L_p$ is the energy gained by an electron from the field in travelling a distance equal to the diffusion length. Let us define a critical field $\mathscr{E}_c$ by means of the equation

$$\mathscr{E}_c = \boldsymbol{k}T/\boldsymbol{e}L_p; \tag{54}$$

then the hole current is predominantly a diffusion current provided $\mathscr{E} \ll \mathscr{E}_c$. This will be so if $\mathscr{E}_\infty = J/\sigma \ll \mathscr{E}_c$; for example, if $L_p=0.05$ cm $=5\times10^{-4}$ m, $\mathscr{E}_c = 50$ V m^{-1}.

7.5.3 Carrier injection ($\mathscr{E}$ small)

It is important at this stage to get an impression of the magnitudes of the various quantities involved in these diffusion processes. Suppose we take $n=3\times10^{20}$ m^{-3}, $p=3\times10^{18}$ m^{-3} so that $n=100p$; with $\mu_h=0.10$ m^2 V^{-1} s^{-1} and $b=2$ we have $\sigma\simeq10\,\Omega^{-1}$ m^{-1}; let us also take $\mathscr{E}_\infty=1$ V m^{-1} and $\mathscr{E}_c=50$ V m^{-1}, so that $J=10$ A m$^{-2}=1$ mA cm^{-2}. We have for Δp_0

$$\Delta p_0 = \frac{\gamma J L_p}{\boldsymbol{e}D_h} = \frac{\gamma n_0 \mu_e L_p \mathscr{E}_\infty}{D_h} = \gamma b\left(\frac{\mathscr{E}_\infty}{\mathscr{E}_c}\right)n_0. \tag{55}$$

Since we have assumed that $\mathscr{E}_\infty \ll \mathscr{E}_c$ we have also $\mathscr{E}_0 \ll \mathscr{E}_c$, so that $\Delta p_0 \ll n_0$; on the other hand, Δp_0 is not necessarily small compared with p_0; taking $\gamma=0.5$ we have $\Delta p_0=0.02n_0=2p_0$. We thus see that with quite a small applied positive voltage we may get a large increase in minority carrier density; the increase in the total number of carriers is $2\Delta p_0=0.04n_0$ for the above example. This phenomenon is known as minority carrier injection, and takes place, for n-type material, at an electrode made positive with respect to the material and for which $\gamma\neq0$. The nearer γ approaches 1 the more efficient is the injection process, and we may note that unless γ is extremely small we shall always get *some* hole injection. It is in fact quite difficult to make γ less than about 0.01, but this is sometimes required; as we shall see, special precautions have to be taken to achieve this; normal soldered metal-to-semiconductor contacts usually have γ between about 0.5 and 0.1, for n-type material.

Two points should be particularly noted in connection with this injection process. For every hole injected an extra electron is also injected in order to ensure that $\Delta p = \Delta p$; the change $\Delta\sigma$ in the conductivity σ is therefore given by

$$\Delta\sigma = e\Delta p(1+b)\mu_h. \tag{56}$$

The second point is that, when injection is achieved with a small field such that $\mathscr{E}_\infty \ll \mathscr{E}_c$, the conductivity change $\Delta\sigma$ is restricted to a distance of the order of the diffusion length L_p from the injecting contact; as we shall see later, this limitation may be to some extent removed by using a higher field for the injection. Minority carrier injection is the basis on which transistor action depends.

7.5.4 Carrier extraction ($\mathscr{E}$ small)

We must now consider the corresponding phenomenon in which the hole concentration p is reduced; in this case the current flows *from* the electrode, so that both J and Δp, are negative; this phenomenon is known as carrier extraction. We see at once that there is one important difference between it and carrier injection, we have the restriction $|\Delta p| \leqslant p_0$, which is not required for the latter. The largest hole current that can be carried by diffusion is ep_0D_h/L_p, and when this is reached further increase in current will change the value of γ. In the above treatment we have assumed that γ is independent of J but this is true only for small values of J. We see, however, that for contacts for which γ is not far from unity considerable depletion of holes may be obtained.

For p-type material the above conditions are reversed, a contact with $\gamma \simeq 0$ giving good injection of electrons when held at a small negative potential relative to the adjacent material.

7.5.5 Large electric field ($\mathscr{E} > 0$)

Let us now consider the contrasting condition in which the electric field is so large that the diffusion currents may be neglected; this will occur when $\mathscr{E} \gg \mathscr{E}_c$. In this case equation (40) reduces, for the steady condition, to

$$\mathscr{E}\mu_h \frac{d\Delta p}{dx} + \frac{\Delta p}{\tau_p} = 0 \tag{57}$$

assuming $\mathscr{E}$ to be constant. Let us first suppose that $\mathscr{E}$ is positive; the solution appropriate to the problem under consideration is then

$$\Delta p = \Delta p_0 \exp\left[-x/L_{d0}\right], \tag{58}$$

where

$$L_{d0} = \mathscr{E}\mu_h\tau_p = (\mathscr{E}/\mathscr{E}_c)L_p. \tag{59}$$

The quantity L_{d0} is called the high-field down-stream diffusion length and is proportional to the electric field $\mathscr{E}$; it will be much greater than L_p provided $\mathscr{E} \gg \mathscr{E}_c$. The holes reach a mean distance L_{d0} from the injection point, which is just equal to the distance they would drift in the field in a time τ_p. The hole diffusion current will be small compared with the hole conduction current provided $D_h/L_{d0} \ll \mu_h\mathscr{E}$, i.e. provided $\mathscr{E} \gg \mathscr{E}_c$; the electron diffusion current will also be negligible compared with the conduction current.

In order that the ratio of hole current to total current should be equal to γ at $x = 0$ we must have

$$\gamma = \frac{p_0 + \Delta p_0}{b(n_0 + \Delta p_0) + p_0 + \Delta p_0}. \tag{60}$$

Since $p_0 \ll n_0$ this reduces to

$$\gamma = \frac{p_0 + \Delta p_0}{bn_0 + (1 + b)\Delta p_0}. \tag{61}$$

Unless, therefore, γ is small, say less than 0.1, we no longer have $n \gg p$ near $x = 0$, or $\mathscr{E}$ constant, and the assumption on which this simple treatment is based is violated. The strong field condition must therefore be treated without making this assumption, and this we shall do later. When γ is small we have approximately

$$\Delta p_0 = b\gamma n_0 - p_0 \tag{62}$$

and in this case the increased hole density extends a distance of the order of L_{d0} from the end of the rod. If, for the example previously considered, we take $\mathscr{E} = 10\mathscr{E}_c = 500\ \text{V m}^{-1}$ we have $(L_p = 0.05\ \text{cm})$ $L_{d0} = 0.5$ cm; the holes are therefore swept quite a long way into the semiconductor. When $\mathscr{E}$ is negative no solution similar to that given by equation (58) exists, and in this case we must consider both conduction and diffusion.

Before we proceed to the general solution for the condition $n \gg p$ we must consider the relative magnitude of the term involving the gradient of the electric field $\partial\mathscr{E}/\partial x$ in equation (40). For the small-field case we have from equation (51)

$$\frac{\partial\mathscr{E}}{\partial x} = -[(b-1)\gamma\mathscr{E}_\infty/L_p]\exp[-x/L_p], \tag{63}$$

and at $x = 0$ where $\partial\mathscr{E}/\partial x$ is greatest we have

$$p\mu_h\frac{\partial\mathscr{E}}{\partial x} = -(b-1)(p_0+\Delta p_0)\mu_h\gamma\mathscr{E}_\infty/L_p. \tag{64}$$

This we have to compare with the diffusion term $(D_h/L_p^2)\Delta p_0$; the latter will be large in comparison, provided $\mathscr{E}_\infty \ll D_h/\gamma\mu_h L_p = \mathscr{E}_c/\gamma$. Thus unless $\gamma \simeq 0$ the condition is the same as we have previously assumed. For the strong field case we have

$$\mathscr{E}_x = J/(\sigma+\Delta\sigma), \tag{65}$$

where $\Delta\sigma = \boldsymbol{e}(1+b)\Delta p\mu_h$, and at $x = 0$ we have approximately

$$\frac{1}{\mathscr{E}}\frac{\partial\mathscr{E}}{\partial x} = \frac{(b+1)\Delta p_0}{L_{d0}[bn_0+(b+1)\Delta p_0]}. \tag{66}$$

We have now to compare the magnitude of the terms $p\,\partial\mathscr{E}/\partial x$ and $\mathscr{E}(\partial p/\partial x)$; the former is small compared with the latter at $x = 0$ (where $\partial\mathscr{E}/\partial x$ is greatest) provided we have

$$\frac{(b+1)(p_0+\Delta p_0)}{bn_0+(b+1)\Delta p_0} \ll 1. \tag{67}$$

Since $p_0 \ll n_0$ this will hold provided $\Delta p_0 \ll n_0$, which is the same condition as we had before. Provided, therefore, we have $p_0 \ll n_0$ and restrict the value of Δp_0 to be sufficiently small so that the change in the *total* number of carriers is small, we may neglect the term involving $\partial\mathscr{E}/\partial x$ in equation (40). We shall assume that these conditions hold in the treatment given below of the situation in which we have comparable contributions from diffusion and conduction.

7.5.6 General solution ($p \ll n$)

The equation which we have to solve to determine the steady state is now

$$D_h\frac{d^2\Delta p}{dx^2} - \mathscr{E}\mu_h\frac{d\Delta p}{dx} - \frac{\Delta p}{\tau_p} = 0. \tag{68}$$

In the conditions discussed above the electric field $\mathscr{E}$ may be assumed to be constant and the general form of solution is

$$\Delta p = A \exp[-x/L_1] + B \exp[-x/L_2]; \tag{69}$$

$\lambda_1 = 1/L_1$ and $\lambda_2 = 1/L_2$ are the roots of the quadratic equation

$$\lambda^2 + \lambda(\mathscr{E}/\mathscr{E}_c)/L_p - 1/L_p^2 = 0, \tag{70}$$

where, as before, $\mathscr{E}_c = kT/eL_p$. We note that one of the roots of the equation (70) must be positive and one negative, both being real. For the problem of the semi-infinite rod only one is therefore permissible. The roots of equation (70) are

$$\lambda = -\frac{\mathscr{E}}{2\mathscr{E}_c}\frac{1}{L_p} \pm \frac{1}{2L_p}\left\{\left(\frac{\mathscr{E}}{\mathscr{E}_c}\right)^2 + 4\right\}^{\frac{1}{2}}. \tag{71}$$

If $\mathscr{E} \ll \mathscr{E}_c$ we have $\lambda = \pm L_p^{-1}$, the positive root being appropriate for a semi-infinite rod with $x \geqslant 0$; this gives the low-field condition. If $\mathscr{E} \gg \mathscr{E}_c$, the two roots are approximately given by the equations

$$\lambda_1 = -\frac{\mathscr{E}}{\mathscr{E}_c}\frac{1}{L_p} = -\frac{e\mathscr{E}}{kT}, \tag{72}$$

$$\lambda_2 = \frac{\mathscr{E}_c}{\mathscr{E}}\frac{1}{L_p} = \frac{1}{L_{d0}}. \tag{73}$$

The solution corresponding to λ_1 must be taken when $\mathscr{E} < 0$ giving

$$\Delta p = \Delta p_0 \exp\left[-\frac{e|\mathscr{E}|x}{kT}\right]; \tag{74}$$

this simply corresponds to a Boltzmann distribution of holes in a retarding field. The solution corresponding to λ_2 is appropriate when $\mathscr{E} > 0$ and is the same as that already obtained (equation (58)); this indicates that in equation (71) we take the negative sign when $\mathscr{E} < 0$ and the positive sign when $\mathscr{E} > 0$.

We may now define two quantities L_u, L_d given by the equations

$$\frac{1}{L_d} = \frac{1}{2L_p}\left\{\left(\frac{\mathscr{E}}{\mathscr{E}_c}\right)^2 + 4\right\}^{\frac{1}{2}} - \frac{|\mathscr{E}|}{2L_p\mathscr{E}_c}, \tag{75}$$

$$\frac{1}{L_u} = \frac{1}{2L_p}\left\{\left(\frac{\mathscr{E}}{\mathscr{E}_c}\right)^2 + 4\right\}^{\frac{1}{2}} + \frac{|\mathscr{E}|}{2L_p\mathscr{E}_c} \tag{76}$$

and these are known respectively as the down-stream and up-stream diffusion lengths; we have then

$$\Delta p = \Delta p_0 \exp[-x/L_d] \quad (\mathscr{E} > 0), \tag{77}$$

$$\Delta p = \Delta p_0 \exp[-x/L_u] \quad (\mathscr{E} < 0). \tag{78}$$

We may note that when $\mathscr{E} = \mathscr{E}_c$, we have $L_d = 2L_p/(\sqrt{5}-1)$, and $L_u = 2L_p/(\sqrt{5}+1)$. Let us now consider a long, uniform, thin rod of semiconducting material, and for the moment neglect surface effects. At the mid-point of the rod ($x = 0$) we shall suppose that we have an electrode from which holes are injected into the rod from a low-voltage source, so as to maintain at $x = 0$ an excess hole concentration Δp_0; a potential difference is applied between the ends of the rod so as to produce a field $\mathscr{E}$ along its length. Then clearly we have

$$\left.\begin{aligned} \Delta p &= \Delta p_0 \exp[-x/L_d] \quad (x \geqslant 0), \\ \Delta p &= \Delta p_0 \exp[x/L_u] \quad (x < 0). \end{aligned}\right\} \tag{79}$$

The distribution of holes along the bar is shown for various values of $\mathscr{E}$ in Fig. 7.4. When $\mathscr{E} \gg \mathscr{E}_c$ the holes are swept far down the bar if L_p is not too small, but do not penetrate far to the left ($x < 0$) of the injection point.

Suppose that a short pulse of holes is applied starting at $t = 0$ and ending at $t = \Delta t$; this pulse will be swept down the rod and will gradually decay in size as it moves away from the point of injection. For $\mathscr{E} \gg \mathscr{E}_c$, the holes move with velocity $\mathscr{E}\mu_h$ and we should expect the pulse to move with this speed as well. We note that the pulse of holes must carry with it a pulse of electrons so as to maintain approximate neutrality of space charge; we shall see later that this influences the velocity of the pulse only when $n \simeq p$, and that when $n \gg p$ the pulse moves with velocity $\mathscr{E}\mu_h$. This forms the basis of what is now a classical experiment, originally proposed and carried out by J. R. Haynes and W. Shockley,[2] to determine directly the mobility and lifetime of minority carriers in Ge. The pulse of holes may, as we shall see, be detected at any point of the rod, and its transit time and amplitude may be measured as a function of distance from the point of injection. We should point out that all the results obtained above for holes in n-type material apply equally well for electrons in p-type material; here we define a diffusion length L_n by means of the equation $L_n^2 = \tau_n D_e$. We must now consider

[2] *Phys. Rev.* (1949) **75**, 691.

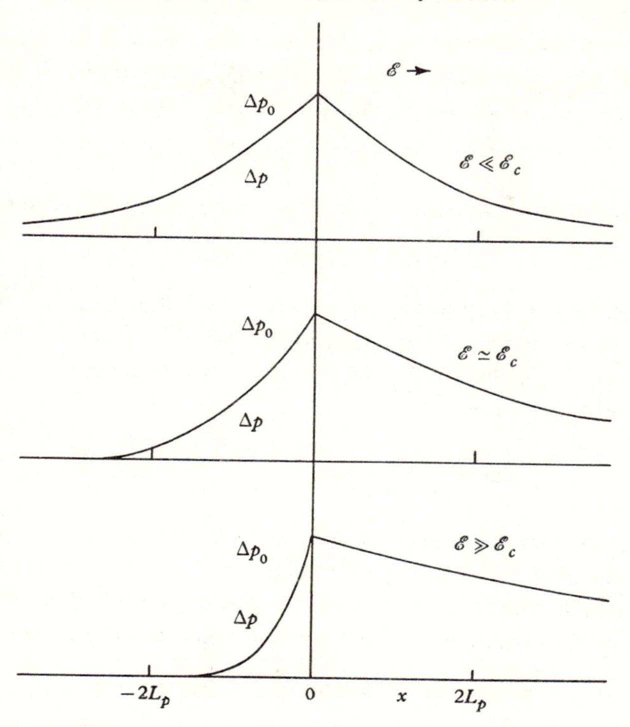

Fig. 7.4. Distribution of holes injected at a point on a thin uniform bar.

non-stationary processes in which the variation of Δp with time is considered.

7.6 Drift of a pulse of minority carriers in an electric field

If we neglect recombination of holes and electrons, the one-dimensional equation of continuity (equation (40)) for holes becomes, in the absence of an electric field,

$$\frac{\partial \Delta p}{\partial t} = D_h \frac{\partial^2 \Delta p}{\partial x^2}. \tag{80}$$

This equation occurs in the theory of heat flow and its solutions are well known; for example, if at $t = 0$ the excess holes are confined to a very narrow region near $x = 0$, the solution at a subsequent time t is of the form

$$\Delta p = Bt^{-\frac{1}{2}} \exp\left[-x^2/4D_h t\right]. \tag{81}$$

Thus we see that at time t the extent of the pulse is of the order of $4(D_h t)^{\frac{1}{2}}$. If a total number of excess holes N_p is contained in the pulse, integration of equation (81) from $x = -\infty$ to $x = +\infty$ gives

$$B = N_p/(4\pi D_h)^{\frac{1}{2}}.$$

In order to take recombination into account let us make the substitution

$$\Delta p = f(x, t) \exp [-t/\tau_p]; \tag{82}$$

on substituting in equation (40) with $\mathscr{E} = 0$ we find that we obtain an equation of the same form as equation (80) for the function $f(x, t)$. The solution of the problem discussed above, when account is taken of recombination, is then

$$\Delta p = \frac{N_p}{(4\pi D_h t)^{\frac{1}{2}}} \exp\left(-\frac{t}{\tau_p} - \frac{x^2}{4D_h t}\right), \tag{83}$$

which shows that the pulse spreads as before, but that its maximum value at time t is reduced by a factor $\exp [-t/\tau_p]$; the total number of excess holes at time t is now equal to $N_p \exp [-t/\tau_p]$. If instead of an infinitely narrow pulse at $t = 0$ we have a Gaussian pulse of the form

$$\Delta p = B \exp [-x^2/x_0^2], \tag{84}$$

we may use the same solution by replacing t by $t + t_0$, where $t_0 = x_0^2/4D_h$.

Now let us suppose that we have an electric field $\mathscr{E}$, and as before, we shall neglect the term in equation (40) which involves $\partial\mathscr{E}/\partial x$. Let us now make the substitution

$$x' = x - \mathscr{E}\mu_h t. \tag{85}$$

Again, the equation for $f(x, t)$ reduces to the form of (80) and we have the solution given by

$$\Delta p = \frac{N_p \exp [-t/\tau_p]}{(4\pi D_h t)^{\frac{1}{2}}} \exp [-(x - \mathscr{E}\mu_h t)^2/4D_h t]. \tag{86}$$

This proves the result mentioned above, that the pulse moves in the direction of the field with velocity $\mathscr{E}\mu_h$, spreading as it goes and having its peak amplitude reduced in the ratio $\exp [-t/\tau_p]$. If the pulse is detected at a subsequent time, and its shape observed, a direct measure of μ_h, τ_p, D_h may be obtained. The mobility obtained in this way is called the *drift mobility*; when $p \ll n$ and $n \ll p$ it is equal respectively to the conductivity mobility for holes and electrons, but when $n \simeq p$ we shall see that this is no longer so.

7.7 Near-intrinsic material

We have discussed in some detail the flow of minority carriers in an extrinsic semiconductor for which $p \ll n$ or $n \ll p$, since the analysis for these conditions is very much simpler than for the condition in which $n \simeq p$. We have assumed in this analysis that $\Delta n = \Delta p$ and that $\partial \mathscr{E}/\partial x$ may be neglected; these two assumptions are related since $\partial \mathscr{E}/\partial x$ and $\Delta n - \Delta p$ are connected by means of Poisson's equation. The object of the present section is to investigate the motion of carriers when these restrictions are removed. We have already had an indication that the previous analysis breaks down when injection of minority carriers in n-type material is such as to make Δp comparable with n (Δp may be much greater than p).

Let us return for a moment to the small-field condition for which we know $\mathscr{E}$ and so may calculate $\Delta n - \Delta p$. Using the value of $\mathscr{E}$ given by equation (51) we have from Poisson's equation (equation (25)) at $x = 0$

$$(\Delta n - \Delta p)/p = \epsilon(b-1)\gamma \mathscr{E}_\infty / eL_p p. \tag{87}$$

Taking $\epsilon = 10\epsilon_0$, and $L_p = 5 \times 10^{-4}$ m, $p = 3 \times 10^{18}$ m^{-3}, as before, we have $(\Delta n - \Delta p)/p \simeq 10^{-5}(\gamma \mathscr{E}_\infty / \mathscr{E}_c)$; we therefore see that in this case $\Delta n = \Delta p$ is a good approximation. For the high-field case we have seen that

$$\frac{\partial \mathscr{E}}{\partial x} \simeq \left(\frac{\Delta p_0}{n_0}\right)\frac{\mathscr{E}}{bL_{d0}} = \left(\frac{\Delta p_0}{n_0}\right)\frac{\mathscr{E}_c}{bL_p} \tag{88}$$

(cf. equation (66)) and we have therefore

$$\frac{\Delta n - \Delta p}{p} = \left(\frac{\Delta p_0}{n_0}\right)\frac{\epsilon \mathscr{E}_c}{ebpL_p} \tag{89}$$

$$= 10^{-5}(\Delta p_0 / n_0). \tag{89a}$$

We have already seen that when $\Delta p_0 \simeq n_0$ we *cannot* neglect the term involving $\partial \mathscr{E}/\partial x$ in equation (40); nevertheless, it appears that $(\Delta n - \Delta p)$ is very small compared with p, yet even this small space-charge is sufficient to give quite a strong field. It is therefore a good approximation to put $\Delta n = \Delta p$ in all terms in equations (40) and (40a) *except those involving* $\partial \mathscr{E}/\partial x$, and in treating the general case it will be desirable to eliminate such terms.

We must first consider rather more carefully the recombination terms. In the absence of electric fields or concentration gradients we have from equations (40), (40a) assuming we have no trapping of electrons and holes, so that they are generated and recombine in pairs,

$$\frac{\partial \Delta p}{\partial t} = \frac{\partial p}{\partial t} = \mathcal{R}_0 - \frac{p}{\tau_p'} = -\frac{\Delta p}{\tau_p}, \tag{90}$$

$$\frac{\partial \Delta n}{\partial t} = \frac{\partial n}{\partial t} = \mathcal{R}_0 - \frac{n}{\tau_n'} = -\frac{\Delta n}{\tau_n}. \tag{91}$$

$\mathcal{R}_0$ is the thermal generation rate, p/τ_p' the recombination rate for holes and n/τ_n' the recombination rate for electrons; the quantities τ_n, τ_p are *defined* by means of these equations. When $n \gg p$ we have seen that τ_p is the minority carrier lifetime and is independent of p. Denoting by τ_{p0}', τ_{n0}' the values of τ_p' and τ_n' corresponding to the equilibrium concentration n_0, p_0 we have when $n \gg p$, $\tau_p' = \tau_p = \tau_{p0}' = p_0/\mathcal{R}_0$; when we do not have this restriction, the recombination rate $\mathcal{R}_0$ is given by the equation

$$\mathcal{R}_0 = p_0/\tau_{p0}' = n_0/\tau_{n0}'. \tag{92}$$

From equations (91), (92) we have therefore

$$\frac{1}{\tau_p} = \left(\frac{p_0}{\tau_{p0}'} - \frac{p}{\tau_p'}\right) \Big/ (p_0 - p), \tag{93}$$

$$\frac{1}{\tau_n} = \left(\frac{n_0}{\tau_{n0}'} - \frac{n}{\tau_n'}\right) \Big/ (n_0 - n). \tag{94}$$

In general, τ_n and τ_p will vary with n and p, the exact form of this variation depending on which recombination mechanism predominates. For radiative recombination the rate will be proportional both to n and to p so that we obtain the relationships

$$\frac{n}{\tau_n'} = \frac{p}{\tau_p'} = Anp = \mathcal{R}_0 np/n_0 p_0; \tag{95}$$

thus we have

$$\frac{1}{\tau_n} = \frac{\mathcal{R}_0(np - n_0 p_0)}{n_0 p_0 (n - n_0)}, \tag{96}$$

$$\frac{1}{\tau_p} = \frac{\mathcal{R}_0(np - n_0 p_0)}{n_0 p_0 (p - p_0)}. \tag{97}$$

In order that $\partial n/\partial t = \partial p/\partial t$ we must have $(p - p_0)/\tau_p = (n - n_0)/\tau_n$; the values given in equations (96), (97) satisfy this condition. We may write equation (97) in the form

$$\frac{1}{\tau_p} = \frac{\mathcal{R}_0}{n_0 p_0}\left(n_0 + p_0 \frac{\Delta n}{\Delta p} + \Delta n\right), \tag{98}$$

and when $p_0 \ll n_0$ and also $\Delta n \ll n_0$ we have

$$\frac{1}{\tau_p}=\frac{\mathcal{R}_0}{p_0}, \tag{99}$$

so that τ_p is constant; similarly we have

$$\frac{1}{\tau_n}=\frac{\mathcal{R}_0}{n_0 p_0}\left(p_0+n_0\frac{\Delta p}{\Delta n}+\Delta p\right). \tag{100}$$

If we may take (as we shall show) $\Delta n=\Delta p$ in equations (98) and (100) we have

$$\frac{1}{\tau_n}=\frac{1}{\tau_p}=\frac{\mathcal{R}_0}{n_0 p_0}(n_0+p_0+\Delta p). \tag{101}$$

If Δp is not small compared with n_0, τ_p is no longer constant and the decay to equilibrium is not exponential. The result that $\tau_n=\tau_p$ if $\Delta n=\Delta p$ is, however, quite general and follows from the condition that $\partial n/\partial t=\partial p/\partial t$.

Let us now return to the equations of continuity and write them in the form

$$\frac{\partial \Delta p}{\partial t}=-\frac{\Delta p}{\tau_p}+D_h\frac{\partial^2 \Delta p}{\partial x^2}-\mu_h p\frac{\partial \mathscr{E}}{\partial x}-\mu_h\mathscr{E}\frac{\partial \Delta p}{\partial x}, \tag{102}$$

$$\frac{\partial \Delta n}{\partial t}=-\frac{\Delta n}{\tau_n}+D_e\frac{\partial^2 \Delta n}{\partial x^2}+\mu_e n\frac{\partial \mathscr{E}}{\partial x}+\mu_e\mathscr{E}\frac{\partial \Delta n}{\partial x}. \tag{103}$$

We also have a third equation, namely Poisson's equation

$$\frac{\partial \mathscr{E}}{\partial x}=e\frac{(\Delta p-\Delta n)}{\epsilon} \tag{104}$$

to determine Δn, Δp, $\mathscr{E}$. Let us substitute for $\partial\mathscr{E}/\partial x$ in equations (102) and (103), obtaining the equations

$$\frac{\partial \Delta p}{\partial t}=-\frac{\Delta p}{\tau_p}+D_h\frac{\partial^2 \Delta p}{\partial x^2}+\frac{\sigma_h}{\epsilon}(\Delta n-\Delta p)-\mu_h\mathscr{E}\frac{\partial \Delta p}{\partial x}, \tag{105}$$

$$\frac{\partial \Delta p}{\partial t}=-\frac{\Delta n}{\tau_n}+D_e\frac{\partial^2 \Delta n}{\partial x^2}-\frac{\sigma_e}{\epsilon}(\Delta n-\Delta p)+\mu_e\mathscr{E}\frac{\partial \Delta n}{\partial x}, \tag{106}$$

where σ_e and σ_h are respectively equal to the partial electron and hole conductivities. Now, from our previous considerations we have seen that $|\Delta n-\Delta p|$ is small mainly in view of the large factors σ_n/ϵ or σ_p/ϵ. It is in the terms involving these large factors that it is a poor approximation to put $\Delta n = \Delta p$, i.e. to omit space-charge; fortunately we can eliminate these terms. Let us multiply equation (105) by σ_e and equation (106) by σ_h and add; we obtain the equation

$$\sigma_e \frac{\partial \Delta p}{\partial t} + \sigma_h \frac{\partial \Delta n}{\partial t} = -\left(\frac{\sigma_e \Delta p}{\tau_p} + \frac{\sigma_h \Delta n}{\tau_n}\right) + \left(\sigma_e D_h \frac{\partial^2 \Delta p}{\partial x^2} + \sigma_h D_e \frac{\partial^2 \Delta n}{\partial x^2}\right)$$

$$+ \sigma_h \mu_e \mathscr{E} \frac{\partial \Delta n}{\partial x} - \sigma_e \mu_h \mathscr{E} \frac{\partial \Delta p}{\partial x}. \tag{107}$$

Now that we have eliminated the terms with a large coefficient of $(\Delta n - \Delta p)$ we may more reasonably make the approximation $\Delta n = \Delta p$; we shall later calculate $(\Delta n - \Delta p)$ and show it to be small, but that $(\sigma/\epsilon)(\Delta n - \Delta p)$ is small only under certain conditions. Thus, putting $\Delta n = \Delta p$ in equation (107) we have, on dividing by $(\sigma_e + \sigma_h)$,

$$\frac{\partial \Delta p}{\partial t} = -\frac{\Delta p}{\tau_p} + D \frac{\partial^2 \Delta p}{\partial x^2} - \mu \mathscr{E} \frac{\partial \Delta p}{\partial x}, \tag{108}$$

where
$$D = \frac{\sigma_e D_h + \sigma_h D_e}{\sigma_e + \sigma_h} \tag{109}$$

$$= \frac{n\mu_e D_h + p\mu_h D_e}{n\mu_e + p\mu_h}. \tag{109a}$$

Using the Einstein relationships equation (109*a*) reduces to the form

$$D = (n+p)\Big/\left(\frac{n}{D_h} + \frac{p}{D_e}\right). \tag{109b}$$

Also we have
$$\mu = \frac{\sigma_e \mu_h - \sigma_h \mu_e}{\sigma_e + \sigma_h} \tag{110}$$

$$= (n-p)\Big/\left(\frac{n}{\mu_h} + \frac{p}{\mu_e}\right). \tag{110a}$$

D is an effective diffusion coefficient and μ an effective mobility for the motion of the quantity Δp; this follows from the fact that equation (108) is of the same form as we have discussed in detail in § 7.5.6, but with the new values for the diffusion coefficient and mobility. Only when $p \ll n$ does μ reduce to μ_h and D to D_h; similarly, when $n \ll p$, μ reduces to $-\mu_e$ and D to D_e. Thus we see that the previous use of equation (68) with μ_e, D_e, or μ_h, D_h is justified, i.e. we were then justified in neglecting the term including $\partial\mathscr{E}/\partial x$ in equation (40). The above treatment follows closely that given by W. van Roosbroeck;[3] similar equations have also been discussed by J. B. Gunn,[4] who uses as variables the quantities $N=(bn+p)/n_i$, $Q=(p-n)/n_i$ and by this means reduces the equations to a form in which the significance of the quantity $(\Delta p-\Delta n)$ is rather more apparent.

The use of near-intrinsic material has recently become of importance in the development of photo-conductive infra-red detectors and has been discussed in this connection by E. S. Rittner.[5] In the treatment which we have given, the trapping of electrons at deep-lying levels has been neglected; the analysis has been extended by Rittner to include conditions when both excitation of electrons from traps and capture of electrons into traps are of importance, and we shall later include trapping levels in our discussion of photo-conductivity (see § 10.9.2). Near-intrinsic material is also of great importance in certain types of transistor and this aspect has been discussed by J. B. Gunn (*loc. cit.*). When $n=p=n_i$, D has a constant value given by the equation

$$D=\frac{2D_eD_h}{D_e+D_h}=\frac{2bD_h}{b+1}. \tag{111}$$

When b is large, as in InSb, the effective diffusion coefficient for intrinsic material is similar to that for holes in spite of the much larger value for electrons.

For intrinsic material we also get the somewhat surprising result that $\mu=0$. The mobility μ does not represent the drift velocity of *particles* in an electric field, it represents the drift velocity of a disturbance. If the material is n-type then μ is positive and the disturbance moves in the direction in which a *positive* charge would move, i.e. in the direction in which excess *holes* would move; if the material is p-type the disturbance moves in the opposite direction, i.e. in the direction in which excess *electrons* would move. We thus see that a disturbance moves in an electric field in the direction in which the *minority* carriers would move.[6] If $n\gg p$ or $p\gg n$, the drift velocity of the disturbance is just equal to that of the minority carriers themselves but as $n\to p$ the velocity becomes

[3] *Bell. Syst. Tech. J.* (1950) **29**, 560; *Phys. Rev.* (1953) **91**, 282.
[4] *J. Electron. Control* (1958) **4**, 17.
[5] *Proceedings of Atlantic City Photoconductivity Conference* (Chapman and Hall; John Wiley and Sons, 1956), p. 215.
[6] This result was obtained by C. Herring, *Bell Syst. Tech. J.* (1949) **28**, 401, when recombination may be neglected; see also, W. Shockley, *Electrons and Holes in Semiconductors* (Van Nostrand, 1950), p. 238.

less for a given field, so that finally when $n = p$ the drift velocity is zero. It is this variation of μ with n and p that makes it necessary to distinguish between drift mobility μ and the conductivity mobilities μ_e, μ_h.

When $n_0 \simeq p_0$ and $\Delta n \simeq \Delta p \ll n_0$, n and p may be replaced in equations (109) and (110) by their equilibrium values. Thus, for small injection, μ and D are constant and our previous analysis applies, provided we replace μ_h by μ and D_h by D, etc. We can again define a diffusion length L by means of the equation

$$L^2 = D\tau, \tag{112}$$

where to this approximation $\tau_n = \tau_p = \tau$. D and L are, in this case, known as the anbipolar diffusion constant and diffusion length. The steady-state solution of equation (108) will be in the form

$$\Delta n = \Delta p = \Delta p_0 \exp[-x/L'], \tag{113}$$

where L' is the 'up-stream' or 'down-stream' diffusion length in the field $\mathscr{E}$.

As a second approximation let us write

$$\Delta n = \Delta p + \eta \tag{114}$$

and return to equation (105) to estimate the magnitude of the space-charge term. To make the analysis simpler let us assume a small-field condition and consider diffusion only; we may then take $L' = L = (D\tau)^{\frac{1}{2}}$, and we obtain

$$\begin{aligned} \frac{\sigma_h \eta}{\epsilon} &= \frac{\Delta p}{\tau_p}\left(1 - \frac{D_h}{D}\right) \\ &= \frac{\Delta p}{\tau_p} \frac{(1+b)}{b} \frac{p}{(n+p)}. \end{aligned} \tag{115}$$

Thus again we see that the space-charge term in equation (105) is negligible only if $p \ll n$. We note that the similar term in equation (106) is by no means negligible under this condition since

$$\frac{\sigma_e}{\epsilon}\eta = \frac{\Delta p}{\tau_p}(1+b)\frac{n}{(n+p)} \tag{116}$$

and this is comparable with $\Delta p/\tau_p$. However, when $p \ll n$ we need only equation (105), neglecting the term in $|\Delta n - \Delta p|$, and do not use (106), but simply the condition $\Delta n = \Delta p$; similarly, when $n \ll p$ we use only equation (106). As before, we may readily show by means of a typical numerical example that $\eta/\Delta p \sim 10^{-5}$, and we conclude that unless there are very strong fields, of the order $10^7\ \mathrm{V\,m^{-1}}$, $\Delta n \simeq \Delta p$. The small space-charge created when n and p are of comparable magnitude is, however, sufficient to give fields large enough to affect the conduction.

7.7.1 Small-field condition

We shall now suppose that n_0, p_0, $n_0 - p_0$ are comparable in magnitude, with $n_0 > p_0$. When we have a small applied field so that $\Delta p \ll n$ or p, the quantities D and μ are constant, but differ from D_e, D_h, and μ_e, μ_h. The solution with $\mathscr{E} > 0$ for a semi-infinite rod is given by equation (113) with $L' = L'_d$, where L'_d is a down-stream diffusion length obtained by replacing D_h by D in equation (75), so that the hole current is given by

$$J_h = \gamma J = e\mu_h(p_0 + \Delta p)\mathscr{E} + \frac{eD_h}{L'_d}\Delta p. \tag{117}$$

Near $x = 0$ we may write approximately

$$\mathscr{E} = J/e\mu_h[bn_0 + p_0 + (b+1)\Delta p_0] \tag{118}$$

and at $x = 0$ we have

$$\gamma J = J\frac{p_0 + \Delta p_0}{bn_0 + p_0 + (b+1)\Delta p_0} + \frac{eD_h}{L'_d}\Delta p_0. \tag{119}$$

When Δp_0 is small we may write this equation in the form

$$(\gamma - \gamma_0)J = \frac{eD_h}{L'_d}\Delta p_0, \tag{120}$$

where

$$\gamma_0 = \frac{p_0}{bn_0 + p_0}. \tag{121}$$

When $\gamma = \gamma_0$ we have $\Delta p = 0$, and the contact is then said to be 'ohmic'. If $\gamma > \gamma_0$ then $\Delta p > 0$ and we get hole injection; this condition always applies for strongly extrinsic n-type material unless γ is extremely small. For near-intrinsic material we can quite easily have $\gamma < \gamma_0$, and we see that in this case we get minority carrier depletion. An extremely interesting situation arises when γ is very small and a strong positive field is applied. If we neglect the diffusion term in equation (117) we see that as $\gamma \to 0$ we have $\Delta p \to -p_0$; this means that all the minority carriers are removed near $x = 0$. For a strong field L'_d will be large and the depletion of carriers may extend throughout a sample whose length is short compared with L'_d; this phenomenon is known as *carrier exclusion*.

7.7.2 Carrier exclusion

Carrier exclusion was first reported by A. F. Gibson.[7] It has been studied in Ge by J. B. Arthur, W. Bardsley, M. A. C. S. Brown and A. F. Gibson[8] and also by R. Bray.[9] (In the original paper it is called carrier extraction but this has caused some confusion with the low-field depletion effect.) The effect is a very striking one; for example, if $n_0 = 1.1n_i$ and $p_0 = 0.9n_i$, the residual number of carriers after extraction is $0.2n_i$ and the conductivity will have been reduced by at least a factor of 10 if $b \simeq 2$; even larger reductions than this have been observed, but there are a number of conditions which must be fulfilled if this is to be achieved.

Suppose we have a sample of length l; then if the down-stream diffusion length L'_d is considerably less than l, the carrier depletion will not extend throughout the sample. For the high-field condition we have $L'_d = \mu\mathscr{E}\tau_p$, and the condition for complete extraction is

$$\mu\mathscr{E}\tau_p \gg l. \tag{122}$$

Also, in the depleted condition, after most of the holes have been extracted we shall have $n \gg p$ so that $\mu \simeq \mu_h$. Now let V_e be the applied voltage required to make $\mu_h\mathscr{E}\tau_p = l$ then, using the Einstein relationship we see that we must have

$$V_e = \frac{kT}{be}\left(\frac{l}{L_p}\right)^2. \tag{123}$$

An applied voltage somewhat greater than V_e is therefore required to achieve almost complete extraction.

When L_d is comparable with l we should really use a solution of the form

$$\Delta p = \Delta p_0 \exp\left[-x/L_d\right] + \Delta p_l \exp\left[(x-l)/L_u\right]. \tag{124}$$

In the strong-field condition $L_u \ll L_d$ so the concentration is affected only near $x = l$. As we shall see below, this will have an appreciable effect on the value of Δp only when γ is small for the contact at $x = l$; we shall therefore assume that its effect is negligible and write Δp in the

[7] *Physica* (1954) **20**, 1058. [8] *Proc. Phys. Soc.* B (1955) **68**, 43.
[9] *Phys. Rev.* (1955) **100**, 1047.

form
$$\Delta p = \Delta p_0 \exp[-x/L_d]; \tag{125}$$
we shall also assume that $L_d \gg l$. Let us define the transit time t as equal to $l/\mathscr{E}\mu$ so that we have $L_d = l\tau_p/t$; our assumption is therefore equivalent to the condition $t \ll \tau_p$. From equation (119), neglecting the diffusion term, we have when $\gamma \ll 1$,

$$\Delta p_0 = -p_0 + \gamma b(n_0 - p_0), \tag{126}$$

so that we write equation (125) in the form

$$p(x) = p_0 - [p_0 - \gamma b(n_0 - p_0)] \exp[-x/L_d]. \tag{127}$$

At $x = l$ we have approximately

$$p(l) = \gamma b(n_0 - p_0) + p_0 t/\tau_p, \tag{128}$$

and in order that $p(l) \ll p_0$ we must have both

$$\gamma \ll b(n_0 - p_0)/p_0 \quad \text{and} \quad t \ll \tau_p;$$

the second condition is equivalent to $V \gg V_e$, where V is the applied voltage. We shall later discuss methods of obtaining a small value of γ.

It may readily be seen why severe carrier depletion takes place in a strong positive field when γ is small; holes are swept out of the sample by the field more rapidly than they can be replaced by thermal generation or by injection at the positive electrode; for larger values of γ sufficient holes are injected to maintain a concentration in the material. A more complete theoretical discussion of carrier exclusion than that given above has been given by W. T. Read.[10]

An interesting application of carrier exclusion has been made by J. B. Arthur, W. Bardsley, M. A. C. S. Brown and A. F. Gibson[11] who have used it to study the variation of the drift mobility μ as the intrinsic condition is approached. They were able in this way to verify equation (110*a*).

7.7.3 Carrier accumulation

When γ is small, and the direction of the field is such as to drive the minority carriers (holes) towards the electrode at $x = 0$, a phenomenon which is the counterpart of carrier exclusion may be observed. Since γ is small the holes cannot flow sufficiently rapidly into the electrode when

[10] *Bell Syst. Tech. J.* (1956) **35**, 1239.

[11] *Proc. Phys. Soc.* B (1955) **68**, 43.

their concentration is normal; they therefore tend to accumulate, and a high density of holes is produced. Here the appropriate solution is given by

$$\Delta p = \Delta p_0 \exp[-x/L_u], \tag{129}$$

where for strong fields, as we have seen, $L_u = kT/e|\mathscr{E}|$. We now have near $x = 0$

$$\begin{aligned} \gamma J = J_h &= -(p_0 + \Delta p_0)\mu_h|\mathscr{E}| + \Delta p_0 D_h/L_u \\ &= -p_0\mu_h|\mathscr{E}| \end{aligned} \tag{130}$$

and

$$\begin{aligned} (1-\gamma)J = J_e &= -(n_0 + \Delta p_0)\mu_e|\mathscr{E}| - \Delta p_0 D_e/L_u \\ &= -(n_0 + 2\Delta p_0)b\mu_h|\mathscr{E}|, \end{aligned} \tag{131}$$

so that

$$\gamma = \frac{p_0}{bn_0 + p_0 + 2b\Delta p_0} \tag{132}$$

and

$$\Delta p_0 = \frac{p_0(1-\gamma) - \gamma b n_0}{2b\gamma}. \tag{133}$$

From equation (133) we see that when γ is small Δp_0 may become very large. Carrier accumulation has been studied in Ge by J. B. Arthur, A. F. Gibson and J. B. Gunn.[12] Instead of using the injection ratio γ to describe the properties of the contact these authors have introduced a quantity v_l called the leakage velocity of the contact. It is defined by means of the equation

$$ev_l\Delta p_0 = J_h = \gamma J, \tag{134}$$

so that we have $\gamma = ev_l\Delta p_0/J$; neither γ nor v_l are independent of J, but when v_l is small γ also is small. We have so far considered only one-dimensional flow; the problem of radial flow, which is more appropriate to point contacts has been treated by J. B. Gunn.[13]

7.8 Comparison of contact phenomena

At this stage it is interesting to compare various phenomena associated with semiconductor contacts; four effects have been listed and discussed

[12] *Proc. Phys. Soc.* B (1956) **69**, 697. [13] *J. Electron. Control* (1958) **4**, 17.

by G. G. E. Low[14] corresponding to those already discussed. Minority carrier injection and extraction are low-field phenomena depending mainly on diffusion; the former occurs at a positive contact and the latter at a negative contact on n-type material, and vice-versa on p-type material. They are associated with values of γ nearly equal to unity for n-type material and nearly equal to zero for p-type material, and are mainly associated with extrinsic material. Carrier exclusion and accumulation are associated with relatively high fields. The former occurs when a contact on slightly n-type material, with γ nearly equal to zero, is made positive and the latter when it is made negative; for slightly p-type material, exclusion occurs when a contact with γ nearly equal to unity is made negative and accumulation when it is made positive. Of the four effects, that of minority carrier injection is by far the most important technologically, as on it is based the operation of most types of transistor and many other devices. The other effects have, however, important technological consequences in the detailed operation of these devices and have also found special applications.

All these effects have been studied experimentally by N. J. Harrick[15] using a rather beautiful experimental technique. A very fine beam of infra-red radiation is passed through the specimen, of wavelength such that most of the absorption is due to free carriers (see § 10.2). By measuring the absorption, the carrier concentration and its variation throughout the specimen may be determined. Harrick has, in this way, been able to observe most of the phenomena discussed in the previous sections, associated with fairly low electric fields; he has also been able to verify the equations derived for near-intrinsic material.

7.9 The *p–n* junction

We must now deal with a number of problems in which the material with which we are concerned is no longer homogeneous; in this case, as we have seen in § 7.1, appreciable space-charges may exist, leading to high electric fields. This is in strong contrast to the situation in homogeneous material in which the space charge in the bulk material must be small. An extreme example of inhomogeneity occurs in a semiconductor in which the donor and acceptor concentrations N_d, N_a vary in such a way that the material changes from n-type to p-type; we should stress that

[14] *Proc. Phys. Soc.* B (1955) **68**, 310.
[15] *Phys. Rev.* (1956) **101**, 491; *ibid.* (1956) **103**, 1173.

we are dealing here with single crystals and not with contacts between two separate crystals of different type.

For example, suppose we have linear gradients in the donor and acceptor concentrations over a certain range as shown in Fig. 7.5 such

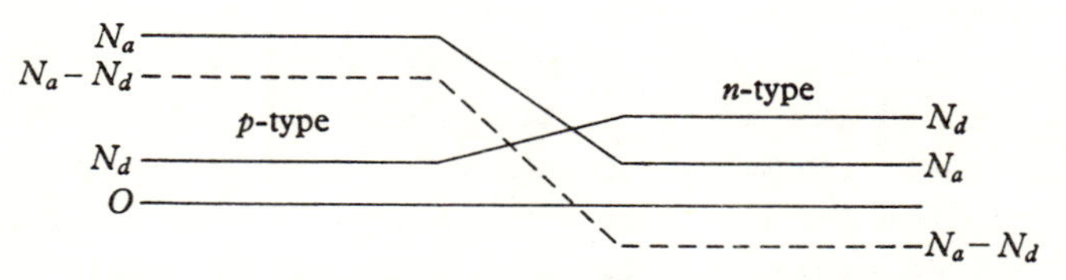

Fig. 7.5. Transition from n-type region to p-type region.

that to the left of the transition region we have $N_a > N_d$, and on the right of the transition region we have $N_d > N_a$. To the left of the transition region the semiconductor is p-type with an effective acceptor concentration equal to $(N_a - N_d)$ and to the right it is n-type with an effective donor concentration equal to $(N_d - N_a)$ (see § 4.3). Such junctions between p-type and n-type material are known as p–n junctions and are of supreme technological importance in many semiconductor applications. They seem to have been discussed as long ago as 1938 by B. Davidov[16] who considered their rectifying and photo-voltaic properties, although their technological significance appears to have been forgotten for many years. Before the advent of the transistor some measurements were made on p–n junctions in Ge, for example, by S. Benzer[17] and by M. Becker and H. Y. Fan;[18] the effect of p–n junctions on the photo-voltaic and photo-conduction properties of PbS films was also discussed by L. Sosnowski, J. Starkiewicz and O. Simpson.[19] The mathematical theory of junctions of the type illustrated in Fig. 7.5 was given by W. Shockley[20] in 1949; as a result he was able to propose in considerable detail a scheme for the junction transistor which worked remarkably well, as predicted. This brought the p–n junction to the forefront of transistor technology, a place which it has continued to occupy.

The mathematical treatment of a junction with a linear concentration gradient of donors and acceptors is rather complex; also, it turns out that the junctions of greatest technological importance have very narrow transition regions: we shall later see by what standards a junction is

[16] *J. Tech. Phys., Moscow* (1938) **5**, 79, 87.
[17] *Phys. Rev.* (1947) **72**, 1267.
[18] *Phys. Rev.* (1949) **75**, 1631.
[19] *Nature, Lond.* (1947) **159**, 818.
[20] *Bell Syst. Tech. J.* (1949) **28**, 435.

considered narrow. For many practical purposes the junction may be regarded as arising from a sudden transition from n-type material to p-type material such as is illustrated in Fig. 7.6. We have shown only donor centres in the n-type region and only acceptors in the p-type; if

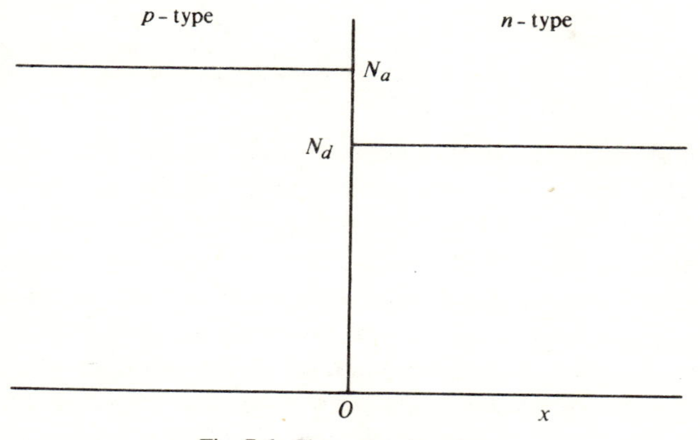

Fig. 7.6. Sharp p–n junction.

both types exist then we simply replace N_a and N_d by the *effective* number of donors and acceptors. We shall assume that the donors and acceptors, respectively, have their energy levels so near the conduction and valence bands that they are all ionized. We shall also assume that $N_a \gg n_i$ and $N_d \gg n_i$. To the right of the junction let $n = n_n$ and $p = p_n$ and to the left of the junction $n = n_p$ and $p = p_p$. Let us suppose, for the moment, that the junction had not been formed; to the right we should have approximately $n_n = N_d$, $n_n \gg p_n = n_i^2/N_d$, while to the left we have $p_p = N_a$, $p_p \gg n_p = n_i^2/N_a$; we also have $n_n \gg n_p$ and $p_n \ll p_p$. If now we imagine the junction to be formed, there will exist a high concentration gradient of both electrons and holes at the junction, and electrons will diffuse from the n-type material to the p-type and holes from the p-type to the n-type; this is illustrated in Fig. 7.7(a). Another way of looking at this is to consider the energy level diagram, such as is shown in Fig. 7.7(b); on formation of the junction, electrons in the conduction band of the n-type material will fall into the lower states represented by the holes in the p-type material. Thus holes will disappear from the p-type material and appear as though they had been transferred to the n-type material, and this would even happen at very low temperatures when the donors and acceptors are un-ionized, the electrons from the full donors going down to fill the empty acceptors. This process cannot go on

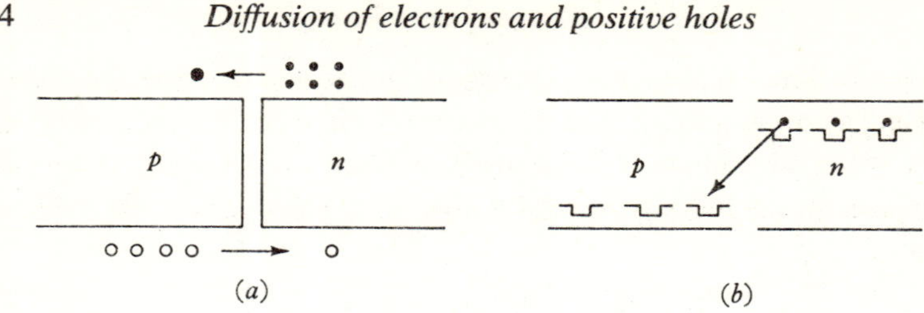

Fig. 7.7. Diffusion of electrons and holes across a junction.

indefinitely, for as soon as electrons begin to flow *from* the n-type material and holes *into* it a space-charge will be set up; a similar space-charge will be set up in the p-type material. This space-charge will set up a strong electric field in a direction such as to oppose the flow of both electrons and holes and an equilibrium will finally be reached.

In the condition of equilibrium the electron energy levels of the p-type material will be raised, because of the negative charge it has acquired. We may readily calculate the amount by which the energy levels must be raised, by noting that the equilibrium condition is simply that the Fermi level should be constant throughout the material. If we denote the p-type material by suffix a and the n-type by suffix b then the energy of the former must be raised by an amount $(E_{Fa} - E_{Fb})$; this is illustrated in Fig. 7.8.

We have seen in §7.1 that the electrostatic potential ψ may be taken as differing by only a constant from the energy $-E_c/e$; then, if we denote

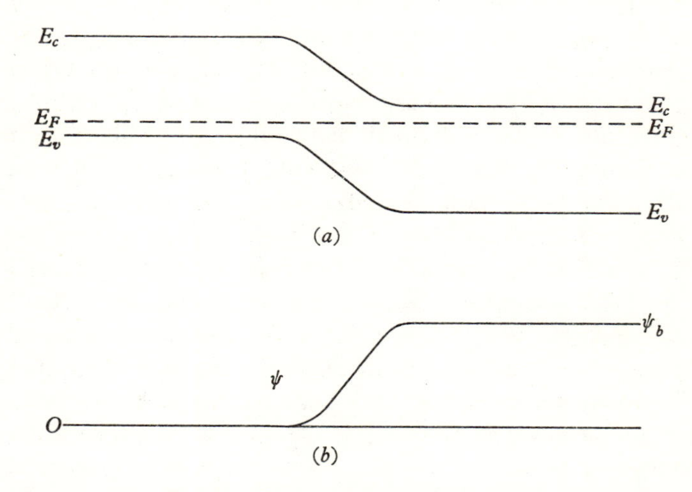

Fig. 7.8. Energy bands and electrostatic potential ψ for a p–n junction.

by the suffixes a, b the values of various quantities far from the junction, we have

$$E_{ca}-E_{cb}=E_{va}-E_{vb}=\boldsymbol{e}(\psi_b-\psi_a). \tag{135}$$

Also we have (§ 4.3) for non-degenerate conditions

$$\left.\begin{aligned} n_n&=N_c\exp[(E_F-E_{cb})/\boldsymbol{k}T],\\ p_n&=N_v\exp[(E_{vb}-E_F)/\boldsymbol{k}T],\\ p_p&=N_v\exp[(E_{va}-E_F)/\boldsymbol{k}T],\\ n_p&=N_c\exp[(E_F-E_{ca})/\boldsymbol{k}T],\end{aligned}\right\} \tag{136}$$

$$\begin{aligned} n_i^2&=N_cN_v\exp[-(E_{ca}-E_{va})/\boldsymbol{k}T]\\ &=N_cN_v\exp[-\Delta E/\boldsymbol{k}T],\end{aligned} \tag{137}$$

N_c, N_v, being the quantities defined in § 4.2. In terms of the electrostatic potential we have (from equations (135) and (136))

$$n_p/n_n=p_n/p_p=\exp[\boldsymbol{e}(\psi_a-\psi_b)/\boldsymbol{k}T]; \tag{138}$$

we also have

$$E_{cb}-E_{va}=-\boldsymbol{k}T\ln(n_n/N_c)-\boldsymbol{k}T\ln(p_p/N_v), \tag{139}$$

or, using equation (135),

$$\boldsymbol{e}(\psi_b-\psi_a)=E_{ca}-E_{cb}=\Delta E+\boldsymbol{k}T\ln(n_np_p/N_cN_v). \tag{140}$$

When both n_n and p_p are very much larger than n_i we have approximately $\boldsymbol{e}(\psi_b-\psi_a)=\Delta E$. Until we know the distance on either side of the junction over which this change takes place we cannot estimate the magnitude of the electric field at the junction.

The space-charge ρ produced by the flow of electrons across the junction is given by the equation

$$\rho=\boldsymbol{e}(p-n+N_d-N_a). \tag{141}$$

If we take the zero of the electrostatic potential such that $\psi=0$ when an electron has energy E_{ca}, i.e. $\psi_a=0$, we may express the hole and electron concentrations in the form

$$p=N_a\exp[-\boldsymbol{e}\psi/\boldsymbol{k}T], \tag{142}$$

$$n=N_d\exp[\boldsymbol{e}(\psi-\psi_b)/\boldsymbol{k}T]. \tag{143}$$

Combining equations (141), (142), (143) we get an expression for ρ in terms of ψ. Poisson's equation may then be written in the

one-dimensional form
$$\epsilon \frac{d^2\psi}{dx^2} = -\rho(\psi). \tag{144}$$

With the form of $\rho(\psi)$ obtained, this is a difficult equation to solve exactly; an approximate solution may, however, be found as follows.[21] We note that the carrier density n varies exponentially with ψ so that n will drop very rapidly as $(\psi - \psi_b)$ becomes negative; at the same time p will increase rapidly. In this region, the charge density ρ quickly reaches a value eN_d to the right of the junction and $-eN_a$ to the left; this situation is illustrated in Fig. 7.9 which shows the variation of ρ and ψ. It

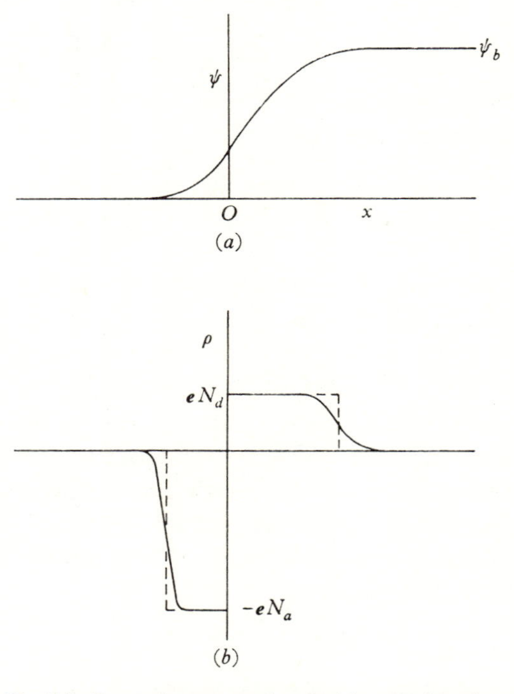

Fig. 7.9. Space charge associated with a p–n junction.

will be seen that it is quite a good approximation to take ρ as constant and equal to $-eN_a$ from $x = -d_a$ to $x = 0$, constant and equal to eN_d from $x = 0$ to $x = d_b$, and zero elsewhere. The lengths d_a, d_b are still to

[21] See, for example, R. J. Elliott and A. F. Gibson, *Solid State Physics* (Macmillan, 1974).

be determined from Poisson's equation, which may now be written in the form

$$\left.\begin{aligned}(d^2\psi/dx^2) &= 0 && (-\infty < x < -d_a)\\ &= (e/\epsilon)N_a && (-d_a \leqslant x < 0)\\ &= -(e/\epsilon)N_d && (0 \leqslant x < d_b)\\ &= 0 && (d_b \leqslant x < \infty).\end{aligned}\right\} \tag{145}$$

The boundary conditions which must be satisfied are

$$\begin{aligned}\psi &= 0 && (x = -d_a),\\ (\partial\psi/\partial x) &= 0 && (x = -d_a),\\ \psi &= \psi_b && (x = d_b),\\ (\partial\psi/\partial x) &= 0 && (x = d_b)\end{aligned}$$

and $(\partial\psi/\partial x)$ continuous at $x = 0$ (see Fig. 7.10). Clearly we must have

$$d_aN_a = d_bN_d \tag{146}$$

in order to preserve electrical neutrality. We have for the electric field $\mathscr{E}$

$$\left.\begin{aligned}\mathscr{E} = -\frac{\partial\psi}{\partial x} &= -\frac{e}{\epsilon}N_a(d_a + x) \quad (-d_a < x \leqslant 0)\\ &= (e/\epsilon)N_d(x - d_b) \quad (d_b \geqslant x > 0).\\ &= 0 \quad \text{elsewhere.}\end{aligned}\right\} \tag{147}$$

Continuity of $\mathscr{E}$ at $x = 0$ is equivalent to equation (146). On integrating equations (147) we have, using the boundary conditions at $x = -d_a$ and $x = d_b$,

$$\left.\begin{aligned}\psi &= (e/2\epsilon)N_a(d_a + x)^2 && (-d_a \leqslant x < 0)\\ &= \psi_b - (e/2\epsilon)N_d(d_b - x)^2 && (0 \leqslant x \leqslant d_b).\end{aligned}\right\} \tag{148}$$

Continuity of ψ at $x = 0$ then gives

$$\psi_b = (e/2\epsilon)(N_dd_b^2 + N_ad_a^2), \tag{149}$$

and, since we know ψ_b, equations (146) and (149) determine d_a and d_b.

If one side of the junction is very impure compared with the other these equations may be somewhat simplified; for example, let us suppose $N_a \gg N_d$. Such a junction is usually referred to as a p^+–n junction, and from equation (146) we see that $d_a \ll d_b$ and may be taken

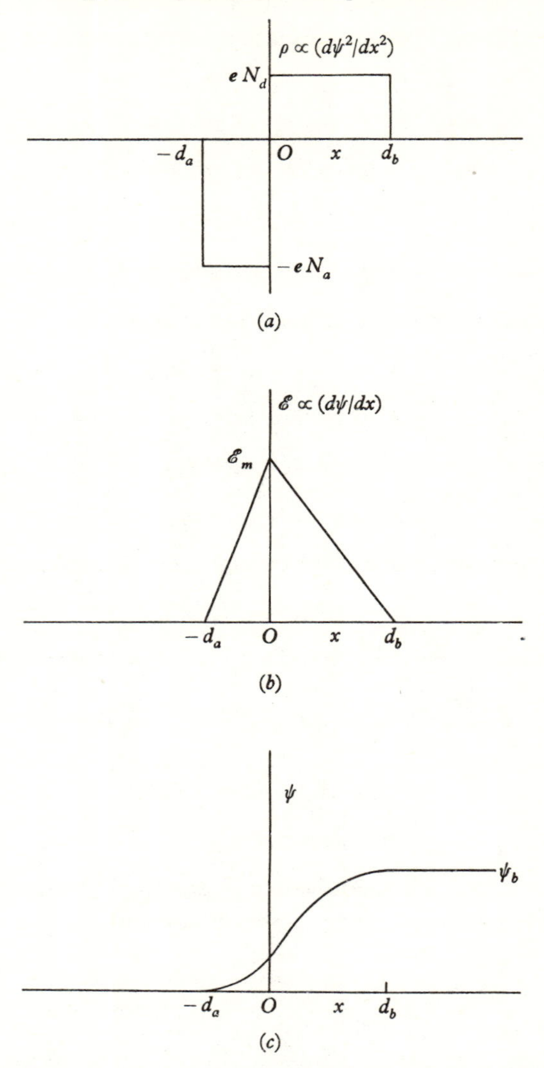

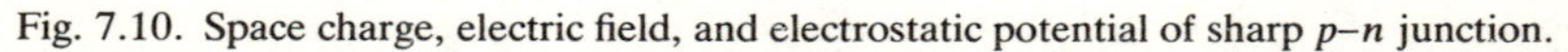
Fig. 7.10. Space charge, electric field, and electrostatic potential of sharp *p–n* junction.

equal to zero. We then have

$$d_b^2 = \frac{2\epsilon}{eN_d}\psi_b \simeq \frac{2\epsilon\Delta E}{e^2 N_d} \tag{150}$$

(note that $\psi_b > 0$). For example, if we take $\Delta E = 1$ eV, $N_d = 10^{21}$ m^{-3}, $\epsilon/\epsilon_0 = 10$ we have $d_b \simeq 1$ μm; the maximum value of the field $\mathscr{E}$ is then

2×10^6 V m^{-1}. It will thus be seen that quite large fields are encountered near the junction.

We may now state more precisely what we mean by a narrow junction: a junction will approximate to the ideal condition discussed and may be termed narrow if the change from n- to p-type takes place in a distance d small compared with the distance d_a+d_b which would be calculated for an ideal junction. A junction for which this is not so is generally called a graded junction and requires a more elaborate treatment.[22]

7.9.1 Barrier-layer capacity

It will be clear, from the above considerations, that the resistance of a p–n junction will be high as compared with that of a sheet of the n- or p-type material of the same area and thickness as the space-charge layer. The junction will, moreover, behave as a condenser, since it consists of two 'plates' of high conductivity material separated by a region of low conductivity. The capacity C_b per unit area, known as the barrier-layer capacity, is given by the equation

$$C_b=\frac{\epsilon}{d_a+d_b}. \tag{151}$$

When $N_a \gg N_d$ we have

$$C_b=\frac{\epsilon^{\frac{1}{2}}e^{\frac{1}{2}}N_d^{\frac{1}{2}}}{(2\psi_b)^{\frac{1}{2}}}\simeq\frac{\epsilon^{\frac{1}{2}}eN_d^{\frac{1}{2}}}{(2\Delta E)^{\frac{1}{2}}}, \tag{152}$$

and for the values considered above we have $C_b=0.01\ \mu$F cm^{-2}.

7.9.2 Current–voltage characteristic

Suppose we now apply an external voltage V_0 across the junction, positive values of V_0 being such as to make the p-side positive; because of the high resistivity of the space-charge layer, extending from $x=-d_a$ to $x=d_b$, the voltage drop will take place almost entirely within this region. It is quite a good approximation then to take for the electrostatic potential ψ the values (see Fig. 7.11)

$$\left.\begin{aligned}\psi&=0 & (x\leqslant -d_a)\\ &=\psi_b-V_0 & (x\geqslant d_b).\end{aligned}\right\} \tag{153}$$

[22] See W. Shockley, *Bell Syst. Tech. J.* (1949) **28**, 435.

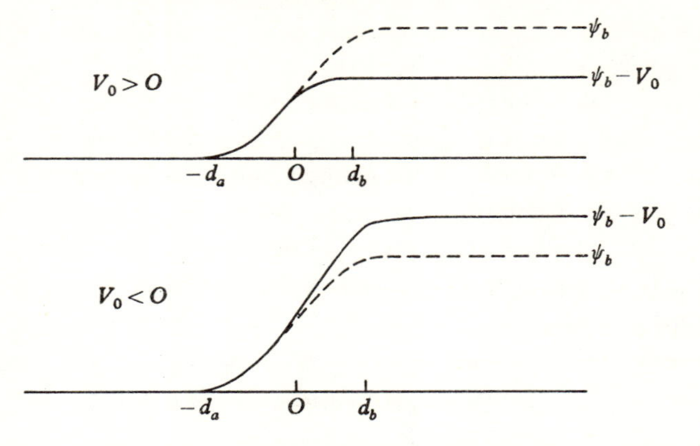

Fig. 7.11. Effect of applied voltage on potential barrier of *p–n* junction.

Since we no longer have equilibrium, the Fermi level will no longer have a constant value. For a non-degenerate condition, in the steady state we have to replace ψ by $\psi - V_0$, so that we have (cf. equation (138), with $\psi_a = 0$)

$$\frac{p_n}{p_p} = \frac{n_p}{n_n} = \exp\left[-e(\psi_b - V_0)/kT\right]. \tag{154}$$

The excess hole concentration Δp_0 at $x = d_b$ is then given by

$$\Delta p_0 = p_n[\exp(eV_0/kT) - 1], \tag{155}$$

and similarly, the extra electron concentration Δn_0 at $x = -d_a$ by

$$\Delta n_0 = n_p[\exp(eV_0/kT) - 1]. \tag{156}$$

These excess concentrations will be maintained only for a distance of the order of L_p and L_n on the right- and left-hand sides of the junction if, as we shall suppose, the applied field is not too large. We have seen, in any case, that most of the voltage drop will take place in the space-charge region, so that current near the junction will flow mainly by diffusion. We shall assume that L_p and L_n are much greater than d_a or d_b and shall take Δn_0 and Δp_0 to be the excess concentrations of *minority* carriers on the two sides of the junction; we note that for $V_0 > 0$ they both correspond to increases in the minority carrier concentration.

By means of equations (155) and (156) we may readily calculate the current carried by holes in the right-hand side of the junction and by electrons in the left-hand side; we have then from equations (155), (156)

$$J_e = \frac{en_pD_e}{L_n}[\exp(eV_0/kT)-1] \quad (x<0), \tag{157}$$

$$J_h = \frac{ep_nD_h}{L_p}[\exp(eV_0/kT)-1] \quad (x>0). \tag{158}$$

Now, the space-charge region is so narrow that we may reasonably assume that no appreciable recombination takes place within it; if this is not so the treatment must be modified. The hole current and electron current may therefore each be taken as continuous at the junction and we may write for the total current density J

$$J = J_s[\exp(eV_0/kT)-1], \tag{159}$$

where

$$J_s = e\left(\frac{p_nD_h}{L_p}+\frac{n_pD_e}{L_n}\right) \tag{160}$$

and L_p, D_h are the diffusion length and diffusion coefficient for holes in the n-type material; D_e and L_n are the corresponding quantities for electrons in the p-type material. When $V_0>0$, this current increases rapidly with V_0; when, on the other hand, $V_0<0$ the current density saturates at a value $-J_s$. Equation (159), derived originally by W. Shockley[23] is frequently referred to as Shockley's equation. It shows that the junction gives a strong rectifying action and in fact behaves like an ideal rectifier. (Most rectifiers may have their current voltage characteristics represented by an equation of the form

$$J = A[\exp(\beta eV_0/kT)-1], \tag{161}$$

where A and β are constants; a rectifier is called *ideal* when $\beta=1$.) Equation (159) has been verified experimentally by many workers, and provided the transition region is narrow in the sense discussed above, the agreement between theory and experiment is excellent. Equations (159), (160) may also be expressed in terms of the conductivities σ_n, σ_p of the n and p sections of the junction and σ_i the conductivity of intrinsic material; we obtain in this way

$$\left.\begin{aligned}\sigma_n &= e(bn_n+p_n)\mu_h \simeq ebn_n\mu_h = ebn_i^2\mu_h/p_n,\\ \sigma_p &= e(bn_p+p_p)\mu_h \simeq ep_p\mu_h = en_i^2\mu_h/n_p,\end{aligned}\right\} \tag{162}$$

$$\sigma_i = en_i(b+1)\mu_h. \tag{163}$$

[23] *Bell. Syst. Tech. J.* (1949) **28**, 435.

Using Einstein's relationship we then have, for the saturation current J_s,

$$J_s = \frac{b\sigma_i^2}{(1+b)^2}\left(\frac{1}{\sigma_n L_p} + \frac{1}{\sigma_p L_n}\right)\frac{kT}{e}. \tag{164}$$

A particularly interesting case occurs when $N_a \gg N_d$ and $N_d \gg n_i$. When this is so we have $\sigma_p \gg \sigma_n$ and $J_e' \ll J_h$; the junction is then called a p^+-n junction and makes a very efficient injecting contact for holes since $\gamma \simeq 1$. We should note, however, that the condition for $\gamma \simeq 1$ is $\sigma_p L_n \gg \sigma_n L_p$, i.e. the diffusion length L_n in the heavily doped p-type section should not be small; this generally means in practice that the heavy doping should not decrease τ_n too much. It will be seen that the saturation current J_s is now given by

$$J_s = \frac{bkT\sigma_i^2}{e(1+b)^2\sigma_n L_p} \tag{165}$$

and the properties of the junction are determined entirely by the n-type section.

The injection ratio γ is given by the equation

$$\gamma = J_h/J = \sigma_p L_n/(\sigma_p L_n + \sigma_n L_p). \tag{166}$$

If $\sigma_p L_n \ll \sigma_n L_p$, then $\gamma \simeq 0$, and the junction becomes an efficient contact for injecting electrons into p-type material, the injecting electrode being the n-type section, which is negatively biased with respect to the p-type section. A junction of this type is known as an n^+-p junction and has properties similar to those of a p^+-n junction.

It is of interest to discuss equations (157)–(164) in a rather more physical manner. The forward current is produced by injection of holes into the n-type material and of electrons into the p-type material; how this comes about may be readily seen from Fig. 7.11. The potential barrier set up in equilibrium is shown by the full line; this contains most of the electrons in the n-type section and most of the holes in the p-type section. On application of a positive voltage V_0 the height of the barrier is lowered by an amount V_0, as shown by the broken line in Fig. 7.11. Thus, more electrons can pass from the n-type to the p-type section, giving the forward electron current; similarly, more holes pass from the p-type section to the n-type section giving the forward hole current. When $V_0 < 0$, on the other hand, the height of the barrier is raised and the flow in these directions is decreased leaving only the backward current. When V_0 is large and negative no electrons flow from the n-type region to the p-type region, the only charge flow being due to the diffusion of electrons from the p-type region towards the potential barrier, over which they fall into the n-type region, and the similar diffusion of holes from the n-type region. The order of magnitude of the saturation current density may readily be estimated by considering the magnitude of these diffusion currents. Holes are created thermally in the n-type region at a rate p_n/τ_p per unit volume, and they diffuse towards the barrier from a distance of the order of L_p; the hole current will therefore be of the order of $eL_p p_n/\tau_p$, which is equivalent to $ep_n D_h/L_p$ as given in equation (160); similarly, the saturation electron current is $en_p D_e/L_n$.

For a p^+–n junction we note that the saturation current in the backward direction is proportional to p_n, the minority carrier density in the n-type section, so that the saturation current gives a measure of the minority carrier density. If this is increased artificially, for example, by illumination or by injection from another contact, the saturation current will increase accordingly. A contact made in the form of a small p^+–n junction on n-type material, and biased negatively, is called a *collector* contact as it collects the minority carriers in its neighbourhood, and the saturation current gives a measure of their concentration. Such a contact may be used, for example, to detect a 'pulse' of holes passing it, as explained in § 7.6; when illuminated, a contact of this type will show a marked photo-conductive effect, and used in this way is known as a photo-diode.

The backward current does not, in practice, remain constant as the negative voltage is increased indefinitely, but increases quite slowly till a critical voltage $-V_b$ is reached; at this voltage the current increases rapidly. This voltage is sometimes referred to as the Zener voltage, since it was thought to represent the voltage at which electron-hole pairs can be produced by an electric field through excitation of electrons from the full-band to the conduction band – a process known as the Zener effect. The large increase in current is now thought to be due, in most cases, to impact ionization and the formation of an avalanche (see § 12.3). The breakdown characteristic is quite reversible and the effect is used in a variety of ways to give voltage stabilizers, etc.; a typical current–voltage characteristic is shown in Fig. 7.12. The value of V_b may be varied by changing the composition of the more lightly doped section.

The maximum field $\mathscr{E}_m$ occurs at the junction and is given by equation (147), on insertion of the value of d_b; we have

$$\mathscr{E}_m^2 = \frac{2e|V_t|N_d}{\epsilon} = \frac{2|V_t|\sigma_n}{\epsilon\mu_n}, \tag{167}$$

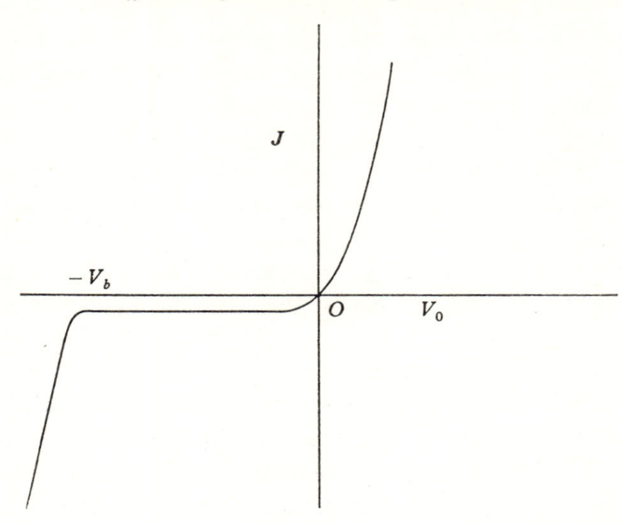

Fig. 7.12. Current–voltage characteristic of *p–n* junction diode.

where V_t is the total change in potential across the junction, approximately equal to V_0 when $|V_0|$ is large compared with ΔE. If $\mathscr{E}_a$ is the avalanche voltage we have the maximum voltage V_m given by the equation

$$V_m = \epsilon \mathscr{E}_a^2 \mu_n / 2\sigma_n \tag{168}$$

from which we note that $V_m \propto \sigma_n^{-1}$; if we take $\epsilon = 10\epsilon_0$, $\mathscr{E}_a = 10^7\ \mathrm{V\ m^{-1}}$, $\sigma_n = 10^2\ \Omega^{-1}\ \mathrm{m^{-1}}$, $\mu_n = 0.2\ \mathrm{m^2\ V^{-1}\ s^{-1}}$, then $V_m = 10\ \mathrm{V}$.

The saturation current in the backward direction is proportional to σ_i^2 and is small for material with a large energy gap, with correspondingly small value of σ_i; for example, it is very much smaller for Si than for Ge, and this makes the former a very attractive material for the manufacture of rectifiers. The *p–n* junction is now rapidly superseding all other forms of rectifier.

Using equations (159), (164) we may calculate the conductance of the junction. If i is the current and A the area of the junction we have

$$i = AJ_s[\exp(eV_0/kT) - 1], \tag{169}$$

$$= i_s[\exp(eV_0/kT) - 1], \tag{169a}$$

where i_s is the saturation current. The differential conductance G is given by the equation

$$G = \frac{\partial i}{\partial V_0} = \frac{ei_s}{kT} \exp[eV_0/kT]. \tag{170}$$

The low-current value G_0 is obtained by letting $V_0 \to 0$ and we have

$$G_0 = ei_s/kT \tag{171}$$

$$= \frac{Ab\sigma_i^2}{(1+b)^2}\left(\frac{1}{\sigma_n L_p} + \frac{1}{\sigma_p L_n}\right). \tag{172}$$

For a p^+–n junction we have

$$G_0 = \frac{Ab}{(1+b)^2}\left(\frac{\sigma_i}{L_p}\right)\left(\frac{\sigma_i}{\sigma_n}\right); \tag{173}$$

this is smaller by a factor of the order of (σ_i/σ_n) than the conductance of a length L_p of intrinsic material

7.9.3 High-frequency behaviour of a *p–n* junction

We have seen in § 7.9.1 that a p–n junction has a barrier-layer capacity given by equation (152), which will be in shunt with the conductance G; equation (152) must, however, be modified in one respect, the change in electrostatic potential across the junction is $\psi - V_0$ when a voltage V_0 is applied, so that for a p^+–n junction we have now

$$C_b = \frac{A\epsilon^{\frac{1}{2}} e^{\frac{1}{2}} N_d^{\frac{1}{2}}}{2^{\frac{1}{2}}(\psi_b - V_0)^{\frac{1}{2}}}. \tag{174}$$

The barrier-layer capacity is thus reduced for backward bias; in this condition G is small and the capacity is more important than for the forward direction.

There is, moreover, another effect which introduces a reactance term into the impedance associated with a p–n junction; the diffusion of electrons and holes takes time, and there is a storage effect which, as we shall see, is also equivalent to the junction having a capacity which will be in parallel with C_b. If the applied voltage is of the form $V = V_0\, e^{i\omega t}$, we may assume a solution of the diffusion equation periodic in time; we have therefore

$$D_h \frac{d^2 \Delta p}{dx^2} = \left(\frac{1}{\tau_p} + i\omega\right)\Delta p. \tag{175}$$

The solution of this equation is of the same form as we had previously, provided we replace L_p by $L_p/(1+i\omega\tau_p)^{\frac{1}{2}}$; we have, therefore, for the current i_j the equation

$$i_j = i_s[\exp(eV_0\, e^{i\omega t}/kT) - 1](1 + i\omega\tau_p)^{\frac{1}{2}}, \tag{176}$$

and if $eV_0 \ll kT$ and $\omega\tau_p \ll 1$ we have approximately

$$i_j = G_0 V_0 e^{i\omega t}(1 + \tfrac{1}{2}i\omega\tau_p). \tag{177}$$

If we let C_d represent the diffusion capacity we have, therefore,

$$C_d = \tfrac{1}{2}G_0\tau_p \tag{178}$$

and the admittance of the junction for small currents is equal to

$$G_0 + i\omega(C_b + C_d). \tag{178a}$$

We have, so far, considered only planar junctions. For a junction in the form of a point contact the situation is more complex, but an approximation to such a junction may be made by assuming spherical symmetry, and the diffusion current calculated using the appropriate form of the diffusion equation. In this case the resistance of the material near the junction is of greater importance, as this determines the so-called spreading resistance, which is equal to $\frac{1}{2}\pi\sigma r_0$ for a hemispherical contact of radius r_0. This will change appreciably, since σ is changed by injection when the current is in the forward direction, and this in turn will change the voltage drop across the space-charge section, with the result that the voltage–current characteristic may depart somewhat from that given by equation (159). A variety of conditions are discussed by H. K. Henisch[24] in a comprehensive treatment of rectifying contacts. There also exists a very extensive literature on p–n junctions in connection with transistor development.

7.10 The n^+–n and p^+–p junctions

Another type of junction of great technical importance is that which occurs between two sections of a semiconductor which are both of the same type but have very different values for the effective donor or acceptor concentrations. When one section is much more heavily doped than the other, the junction is referred to as an n^+–n junction or a p^+–p junction. The form of the energy-level diagram representing the equilibrium of such an n-type junction is shown in Fig. 7.13. As before, we shall denote the left- and right-hand sides of the junction by the letters a, b, and we shall suppose that $N_{da} \gg N_{db}$. The change in electrostatic potential ψ across the junction is obtained as for the p–n junction. In

[24] *Rectifying Semiconductor Contacts* (Oxford University Press, 1957).

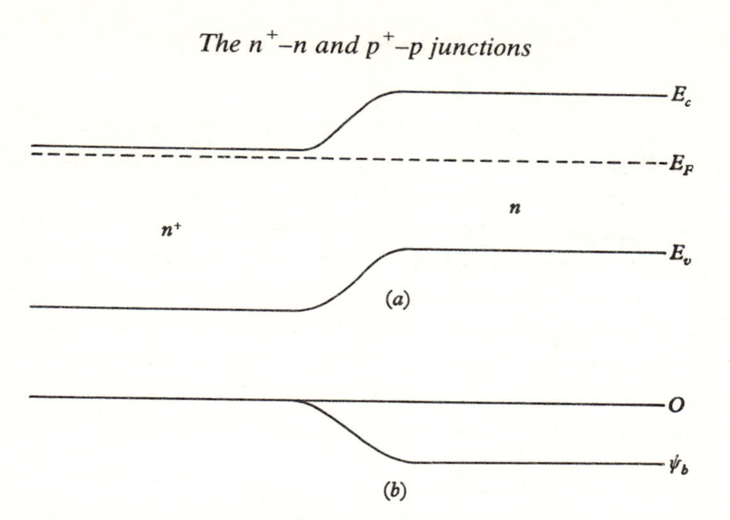

Fig. 7.13. Energy bands and electrostatic potential for an n^+–n junction.

this case we have

$$\left.\begin{aligned} N_{da} &= N_c \exp\left[-(E_{ca}-E_F)/kT\right], \\ N_{db} &= N_c \exp\left[-(E_{cb}-E_F)/kT\right], \end{aligned}\right\} \tag{179}$$

so that

$$\psi_b - \psi_a = -(1/e)(E_{cb}-E_{ca}) = -(kT/e)\ln(N_{da}/N_{db}), \tag{180}$$

and if, as before, we take $\psi_a = 0$ we obtain for ψ_b the value

$$\psi_b = -(kT/e)\ln(N_{da}/N_{db}). \tag{181}$$

The n^+–n junction differs from the p–n junction in several important respects; there is, for example, no high-resistance region at the junction, the majority carrier density passing smoothly from N_{da} to N_{db}. The potential drop, when an external voltage is applied, does not, therefore, take place mainly over a narrow region as for the p–n junction. When a positive voltage is applied to the junction (n^+ side positive) the potential barrier tends to be increased in this case and holes are driven away from the junction on the right-hand side; the hole current then consists of a diffusion of holes in the left-hand side towards the junction. (The field in the high conductivity n^+ section will be very small.) The hole current density may be obtained as for the p–n junction and is given by

$$J_h = \frac{ep_a L_{pa}}{\tau_{pa}} = \frac{en_i^2 L_{pa}}{N_{da}\tau_{pa}}. \tag{182}$$

This will be very small when N_{da} is large so that for such a junction $\gamma \simeq 0$. A junction of this type is therefore very useful as a contact when it is particularly desired not to inject holes into n-type material; for example, such junctions are frequently used as base contacts for diodes.

When an n^+–n junction is biased so that the n^+ section is negative, holes will be drawn towards it, but the hole current will be unable to flow in the n^+ section unless there is a large excess hole density to enable current to flow by diffusion. We have already seen how such a large density can arise for a contact with $\gamma \simeq 0$ in our discussion of carrier accumulation. If Δp_a is the excess hole concentration to the left of the junction and Δp_b the excess concentration to the right we must have

$$\frac{p_a + \Delta p_a}{p_b + \Delta p_b} = \frac{p_a}{p_b} = \exp\left[e\psi_b/kT\right], \tag{183}$$

since there is no marked voltage drop across the junction; it follows that

$$\frac{\Delta p_a}{\Delta p_b} = \frac{p_a}{p_b} = \frac{N_{db}}{N_{da}}. \tag{184}$$

The hole current density J_h will thus be given by

$$J_h = -\frac{e\Delta p_a D_{ha}}{L_{pa}} = -\frac{e\Delta p_a L_{pa}}{\tau_{pa}}. \tag{185}$$

We have therefore

$$\Delta p_a = |J_h|\tau_{pa}/eL_{pa} \tag{186}$$

and

$$\Delta p_b = (|J_h|\tau_{pa}/eL_{pa})(N_{da}/N_{db}), \tag{187}$$

and we see that Δp_b can be quite large if $N_{da} \gg N_{db}$. If we introduce the leakage velocity v_l defined by equation (134) we have

$$v_{la}\Delta p_a = v_{lb}\Delta p_b, \tag{188}$$

and from equation (185) we see that $v_{la} = D_{ha}/L_{pa} = (D_{ha}/\tau_{pa})^{\frac{1}{2}}$; for example, if $D_{ha} = 25\ \text{cm}^2\ \text{s}^{-1}$, $\tau_{pa} = 10^{-4}$ s we have $v_{la} = 500\ \text{cm s}^{-1}$. If v_d is the drift velocity for holes towards the junction in the electric field we have

$$J_h = ev_d p_b = ev_{la}\Delta_{pa} = ev_{la}\Delta p_b(N_{db}/N_{da}), \tag{189}$$

so that

$$\frac{\Delta p_b}{p_b} = \left(\frac{v_d}{v_{la}}\right)\left(\frac{N_{da}}{N_{db}}\right). \tag{190}$$

The above value of D_{ha} corresponds to $\mu_h = 10^3\ \mathrm{cm^2\ V^{-1}\ s^{-1}}$ so that for $\mathscr{E}_b = 0.5\ \mathrm{V\ cm^{-1}}$, $v_d = v_{la}$; for this condition we see that $\Delta p_b/p_b = N_{da}/N_{db}$ and so can be quite large.

A full mathematical discussion of n^+–n and p^+–p junctions has been given by J. B. Gunn.[25]

7.11 Surface effects

We shall now discuss contacts between metals and semiconductors. The properties of such contacts are found to depend very much on the condition of the surface of the semiconductor; this is not surprising for pressure contacts, since oxide layers and other surface contamination are likely to affect the contact; surface effects are, however, much more important for all contacts between metals and semiconductors than for contacts which occur between two metals, and have a very marked effect on the rectifying and injection properties of a contact. When we come to consider welded and soldered contacts the situation is even more complex, as the process of making the contact almost always changes the properties of the semiconductor in its vicinity; p^+–n or n^+–n junction contacts are much more controllable, being part of a single crystal.

Before proceeding to discuss the contact between a metal and a semiconductor we must therefore consider some of the properties of a semiconductor surface. The most significant property of the surface is that it generally introduces energy levels lying in the forbidden energy gap between the valence band and the conduction band; such surface states were envisaged even in a perfect crystal by I. Tamm,[26] and arise from the effect of the boundary of the crystal. The surface states met with in practice, however, are generally due to the adsorption of impurities on the surface, electronegative impurities such as oxygen acting as electron acceptors. There is also ample evidence that traps for holes may also exist at the surface. These surface states generally have energy levels significantly below the conduction band or above the valence band; the levels may even be spread over a band.

The theory of such surface states has been studied by a number of authors, together with the effect of impurities adsorbed on the surface of a semiconductor. An extensive review of the subject has been given by S. G. Davison and J. D. Levine[27] who have discussed the various

[25] *J. Electron. Control.* (1958) **4**, 17. [26] *Z. Phys.* (1932) **76**, 849.
[27] *Solid State Physics* (Academic Press, 1970) **25**, 1.

methods for examining the modification to the crystal structure by the presence of the surface.

Let us consider what will happen in an n-type semiconductor whose donor levels lie very close to the conduction band when a number of acceptor states on the surface are below the Fermi level in the bulk material; the situation is illustrated in Fig. 7.14. Electrons from the

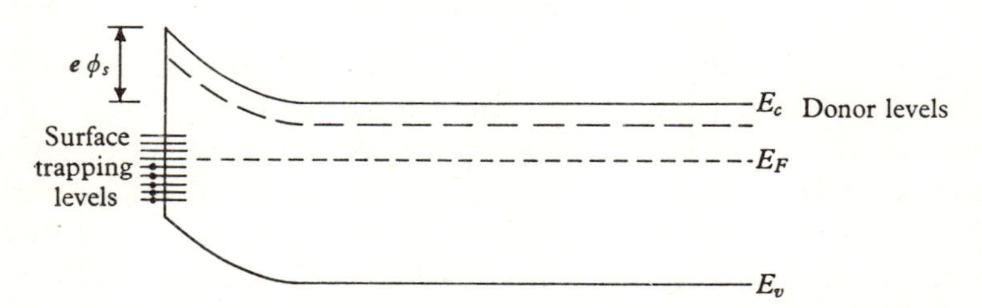

Fig. 7.14. Effect of electron traps on the surface of an n-type semiconductor.

conduction band or from the donor levels, if the temperature is so low that they are not all ionized, will tend to fill up the surface states which lie below the Fermi level. Let N_s represent the number of available surface states, per unit area, below the Fermi level; then the surface will acquire a negative surface charge of amount $-eN_s$ per unit area. As we have already seen for a p–n junction, this will cause a depletion region from which almost all the electrons are removed and which will have a positive space-charge approximately equal to eN_d, where N_d is the donor concentration. The result will be a strong electric field directed towards the surface so as to oppose the flow of more electrons into the surface states. The result will be that the energy of the surface is raised with respect to the bulk material as shown in Fig. 7.14. If the amount by which the conduction band is raised at the surface is denoted by $e\phi_s$, then ϕ_s is called the surface potential. In order to calculate ϕ_s we have to solve Poisson's equation with the appropriate boundary conditions. We may here give an approximate treatment as for the p–n junction by assuming the space-charge ρ to be equal to eN_d from $x=0$ (the surface) to $x=d$, where $d=N_s/N_d$; we then solve Poisson's equation with the boundary conditions $\mathscr{E}=0$ at $x=d$ and $\mathscr{E}=-eN_s/\epsilon$ at $x=0$. The solution is quite straightforward and gives

$$\mathscr{E}=-\frac{eN_d}{\epsilon}(d-x)=-\frac{eN_s}{d\epsilon}(d-x). \tag{191}$$

For the electrostatic potential ψ we have, if we take $\psi = 0$ at $x = d$,

$$\psi = -\frac{eN_d}{2\epsilon}(d-x)^2. \tag{192}$$

The surface potential ϕ_s is clearly equal to the value of $-\psi$ when $x = 0$; we have therefore

$$\phi_s = \frac{eN_s d}{2\epsilon} = \frac{eN_s^2}{2\epsilon N_d}. \tag{193}$$

The energy E_s by which the energy bands at the surface are raised is given by the equation

$$E_s = e\phi_s = \frac{e^2 N_s^2}{2\epsilon N_d}. \tag{194}$$

For example, if $\epsilon = 10\epsilon_0 \simeq 10^{-10}$ F m^{-1}, $N_d = 2 \times 10^{21}$ m^{-3}, and $N_s = 10^{15}$ m$^{-2} = 10^{11}$ cm^{-2} we have $\phi_s \simeq 0.5$ V.

A more elaborate treatment of the problem, involving the use of Poisson's equation to determine the space-charge distribution near the surface, has been given by R. H. Kingston and S. F. Neustadter,[28] the only assumptions being that the donors are fully ionized in the n-type region and that Boltzmann's equation may be used to give the electron concentration; the results are presented in numerical form.

If the surface states are capable of trapping holes, a *decrease* in the electron energy corresponding to the band edges will result. Both electron and hole traps are required to explain the effects found when the surface is exposed to different kinds of atmosphere. For p-type material, the bands will also be lowered when holes are trapped at the surface and raised when electrons are trapped. It is very difficult to control the surface when exposed to gaseous ambients, so that studies have frequently been made with surfaces exposed to electrolytes. W. H. Brattain and C. G. B. Garrett[29] have carried out extensive investigations with Ge in this way and have shown, by means of suitable electrolytes, that either a p-type surface or an n-type surface may be obtained. W. H. Brattain and J. Bardeen[30] have investigated the contact potential at a Ge surface using the classical Kelvin method; by this means changes in the value of ϕ_s may readily be observed. Changes with atmosphere were studied and found to be quite large, the contact potential being changed

[28] *J. Appl. Phys.* (1955) **26**, 718.
[29] *Physica* (1954) **20**, 885; *Bell Syst. Tech. J.* (1955) **34**, 129.
[30] *Bell Syst. Tech. J.* (1953) **32**, 1.

successively from a low value to a relatively high value by cyclic changes in atmosphere; the effect of illumination of the surface with light was also studied.

If, for an *n*-type semiconductor, ϕ_s is great enough to raise the valence and conduction bands sufficiently to bring the valence band nearer to the Fermi level than the conduction band, we shall have what is effectively a *p*-type layer near the surface; if the valence band were close to the Fermi level the hole density would be quite high. A layer of this type is called an inversion layer and is illustrated in Fig. 7.15; such

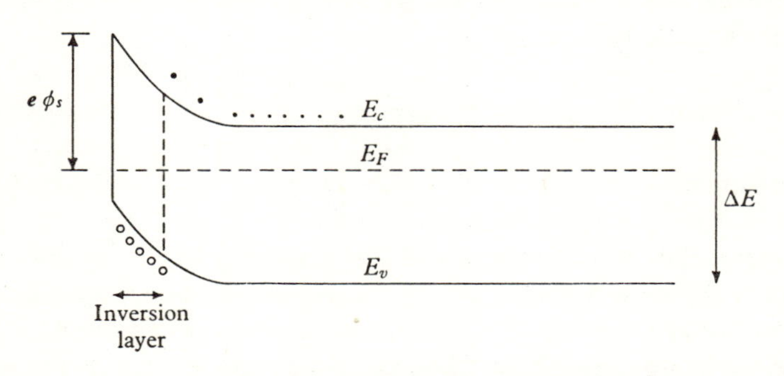

Fig. 7.15. *p*-type inversion layer on *n*-type semiconductor.

layers play an important part in semiconductor technology. For example, a *p*-type layer on the surface of a bar in which there is a *p–n* junction would modify considerably the properties of the junction.

It is interesting to note that it was the study of surface states by J. Bardeen[31] that led to the discovery of transistor action. Bardeen found that it was necessary to assume the existence of such states in order to explain the observation that the rectifying properties of a metal–semiconductor contact hardly depend at all on the properties of the metal, but depend markedly on the state of the surface of the semiconductor; he also found that such states explain observations on a phenomenon known as the field effect which we shall discuss below. These considerations led to the idea of an inversion layer and then to the concept of hole injection.

7.11.1 The field effect

If one of the plates of a parallel-plate condenser consists of a layer of *n*-type semiconductor and the other of a metal, and the metal plate is charged positively, the semiconductor should have a negative charge

[31] *Phys. Rev.* (1947) **71**, 717.

and so have excess electrons over the equilibrium value. These electrons might be expected to take part in conduction along with the electrons which would normally occupy the conduction band, so that, on charging the condenser, an increase in the conductivity of the semiconductor should be observed. An experiment of this kind was carried out in the Bell Telephone Laboratories, U.S.A., in 1948 by W. Shockley and G. L. Pearson[32] and proved to be a crucial step in the invention of the transistor. The expected effect was found, but the magnitude was only one-tenth of that anticipated. To explain this J. Bardeen proposed that 90% of the excess electrons are trapped in surface states and are not free to take part in conduction.[33]

The field effect has been developed into a powerful tool for investigating the surfaces of semiconductors. It is found that the change in conductivity on application of a field normal to the surface depends markedly on the surface potential ϕ_s and the change may be either positive or negative; it is found, moreover, that the change is not permanent but that the conductivity returns to its original value, the field being still applied, with a time-constant τ_f. A great deal of work has been published on this effect, mainly on observations of Ge and Si surfaces; a comprehensive discussion of its use for evaluating the various surface parameters has been given by G. G. E. Low.[34] A series of review papers covering the subject has been edited by R. H. Kingston.[35] A review of work on the field effect has also been given by L. R. Godefroy.[36] The main result of these studies is that it appears that two distinct types of surface state generally occur in semiconductors, and are distinguished by their very different values of τ_f. States, known as 'slow' states, have values of τ_f from a few seconds to several minutes, and are thought to be situated at the outside surface of an oxide layer on the semiconductor, the high value of τ_f being associated with the time taken by electrons to penetrate the layer.[37] These states have been extensively investigated by S. R. Morrison.[38] The other type of states known as 'fast' states have values of τ_f ranging from a few microseconds to a few

[32] *Phys. Rev.* (1948) **74**, 223.

[33] An interesting account of the part played by the field effect in the early days of transistor research is given by W. Shockley in his book *Electrons and Holes in Semiconductors* (Van Nostrand, 1950), § 2.1.

[34] *Proc. Phys. Soc.* B (1956) **69**, 1331.

[35] *Semiconductor Surface Physics*, ed. R. H. Kingston (Pennsylvania University Press; Oxford University Press, 1957).

[36] *Progress in Semiconductors* (Heywood, 1956) **1**, 195.

[37] See, for example, M. Lasser, C. Wysocksi and B. Bernstein, *Phys. Rev.* (1957) **105**, 491.

[38] *Semiconductor Surface Physics*, ed. R. H. Kingston (Pennsylvania University Press; Oxford University Press, 1957), p. 169.

milliseconds, and are thought to exist at the surface between the oxide layer and the semiconductor; they are the states mainly responsible for recombination of electrons and holes at the surface[39] (see § 9.8). The electrochemistry of surfaces and the effect of etching on surface states has been treated in a series of papers by various authors.[40]

Inversion layers on the surface of semiconductors play an important part in determining their breakdown properties under high electric fields. This aspect has been studied by C. G. B. Garrett and W. H. Brattain[41] who have shown that breakdown is due to an avalanche multiplication process and is strongly influenced by the value of the surface potential.

The electrical properties of semiconductor surfaces have been treated in an extensive review by T. B. Watkins,[42] including the field effect which provides the physical basis of one of the most widely used types of transistor (F.E.T.).

7.12 Metal–semiconductor contacts

From the above discussion we should not be surprised to find that the state of the surface of a semiconductor has a marked effect on metal–semiconductor contacts – particularly on small-area contacts. It is, however, of interest to see what we should expect from a clean contact between a metal and semiconductor. For an extrinsic *n*-type semiconductor, whose donor levels lie near the conduction band, electrons would flow into the metal on making contact, provided the work-function W_s of the semiconductor is less than W_m, that of the metal; this is illustrated in Fig. 7.16. It appears that this condition generally is satisfied and so we should expect the energy levels of the semiconductor to be raised near the contact, as shown in Fig. 7.16. The amount by which the levels are raised should be determined by the difference in work-function between the semiconductor and metal. A surface potential ϕ_s may now be defined as before, and we should expect ϕ_s to be approximately equal to $W_m - W_s$; experiment shows, however, that this is not generally so, the value of ϕ_s being almost independent of W_m. The value of ϕ_s is in this case determined mainly by surface states and hardly at all by the metal. There must, therefore, exist an electric field across an insulating layer at the surface in order to bring the Fermi level of the metal to the same value as for the semiconductor.

[39] See, for example, A. Many and D. Gerlich, *Phys. Rev.* (1957) **107**, 404.
[40] *The Electrochemistry of Semiconductors*, ed. P. J. Holmes (Academic Press, 1973).
[41] *J. Appl. Phys*, (1956) **27**, 299.
[42] *Progress in Semiconductors* (1960) **5**, 1.

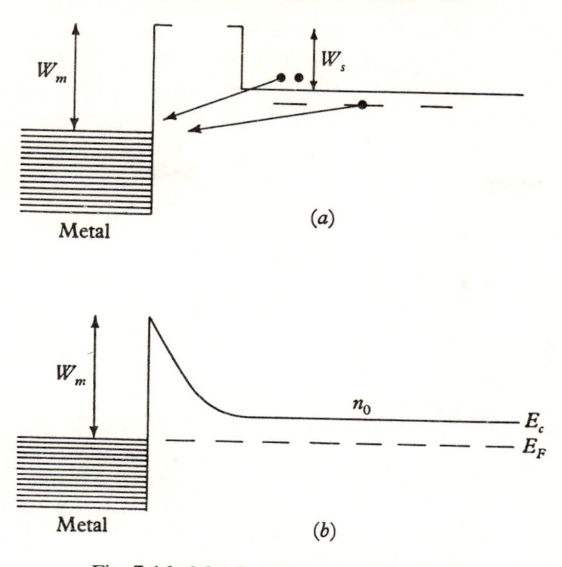

Fig. 7.16. Metal–semiconductor contact.

On application of a potential which makes the metal positive with respect to the semiconductor (n-type), the potential barrier preventing flow of electrons from the semiconductor to the metal is lowered and electrons flow readily to the metal.[43] (Very little potential drop takes place in the metal.) When the metal is made negative with respect to the semiconductor the height of the barrier, which electrons have to surmount to reach the metal, is raised and this flow is substantially cut off leaving only the small flow of electrons from the metal to the semiconductor. This may be expressed quantitatively as follows. Let $-J_s$ represent the current flowing from metal to semiconductor; this is independent of applied voltage and depends only on the height of the top of the barrier above the Fermi level. If we assume that most of the voltage drop V_0 occurs across the space-charge region (see Fig. 7.17) the effective height of the barrier will be $\boldsymbol{e}(\phi_s - V_0)$, and we may express the electron-current density $-J_e$ from the semiconductor to the metal in the form

$$-J_e = A \exp\left[-\boldsymbol{e}(\phi_s - V_0)/\boldsymbol{k}T\right].$$

When $V_0 = 0$ we have $J_e = J_s$ so that

$$-J_s = A \exp\left(-\boldsymbol{e}\phi_s/\boldsymbol{k}T\right). \tag{195}$$

[43] W. Schottky and H. Hartmann, *Zeits. f. Tech. Phys.* (1935) **16**, 512; W. Schottky, *Naturwiss.* (1938) **26**, 843.

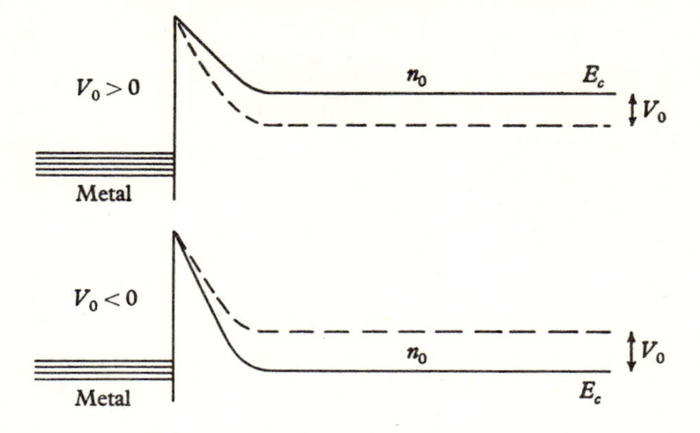

Fig. 7.17. Effect of voltage applied to metal–semiconductor contact.

The total current density J is therefore given by the equation

$$J = -J_s - J_e = J_s[\exp(V_0/\boldsymbol{k}T) - 1], \tag{196}$$

which is of the same form as for a p–n junction; the value of the saturation current J_s is, however, not so readily obtained.

Various theories have been given to enable J_s to be calculated; their limitations and conditions of applicability have been discussed by H. K. Henisch.[44] The main difference between the theories concerns the width of the space-charge layer and the probability of electrons and holes making collisions while passing through this layer. When the layer is thin, the so-called diode theory applies and gives for A the value

$$A = ne(\boldsymbol{k}T/2\pi m_e)^{\frac{1}{2}} \tag{197}$$

for electrons of effective mass m_e. Unless great care is taken in forming the metal–semiconductor contact, under conditions of very high vacuum, a thin oxide layer inevitably forms and the contact becomes metal oxide–semiconductor (M.O.S.). This has a marked effect on the current–voltage characteristic. The theory has been extended to take this condition into account and also to take account of the image force on the electrons in the semiconductor, and in various other ways. The effect of electrons tunnelling through the potential barrier has also been considered.

The problem of calculating the voltage–current characteristic of such rectifying contacts is extremely complex. For example, in the simple treatment given above, minority carriers were assumed to play no

[44] *Rectifying Semiconductor Contacts* (Oxford University Press, 1957), pp. 68–220.

significant part; on the other hand, we have seen how inversion layers can be formed, and it is well known that certain metal–semiconductor (p-type) contacts have quite a high value of the injection ratio γ. Very few metal–semiconductor contacts have current–voltage characteristics of the form given by equation (196); the characteristic may frequently be fitted with an equation of this form with e/kT replaced by $\beta e/kT$, the factor β generally differing quite appreciably from unity. This complexity is in strong contrast to the elegant simplicity of the p–n junction type of rectifier. For this reason, in making welded or fused metal–semiconductor contacts, the tendency is to make them in the form of p^+–n or n^+–n junctions as required; the contact is really in the form metal–p^+–n or metal–n^+–n, the p^+ or n^+ section in contact with the metal having a very low resistance. Such contacts are made by alloying with the metal a small amount of a group III or group V element; for example, if a contact to n-type Ge, using gold wire, is required to have a high value of γ, the gold is alloyed with In or Ga; on welding or fusing the wire to the Ge, the In or Ga diffuses for a short distance into the Ge and a p^+–n junction is formed, for which we have seen that $\gamma \simeq 1$; if a non-injecting contact is required, a gold–antimony alloy might be used, giving an n^+–n junction with $\gamma \simeq 0$. Many of the metal–semiconductor contacts which have been studied in the past have been unintentionally made in this way and their properties are now better understood in terms of p^+–n and n^+–n junctions. Contacts of this kind can be made having a nearly linear current–voltage characteristic showing little or no rectification. They are known as 'ohmic' contacts. For a rectifier, for example, one contact will be of the rectifying type and the other 'ohmic'.[45]

Most of the earlier work on rectifying contacts was carried out with Si and Ge. Contacts between various metals and a variety of semiconductors that have been studied include some ternary compounds[46] such as $AgGaSe_2$. These show some interesting differences in behaviour as compared with contacts with Si and Ge, but even they have voltage–current characteristics very similar to that given by equation (196).

Although in many instances the metal used for the rectifying contact makes very little difference, large variations in work-function being compensated by the effect of surface states, there are some exceptions. It has been shown, for example, by V. Heine[47] that the band structure of

[45] *Ohmic Contacts to Semiconductors*, ed. B. Schwarz (Electrochem. Soc. N.Y., 1969).

[46] P. Robinson and J. I. B. Wilson, *Ternary Compounds*, ed. G. Holah (Inst. of Phys., London, 1977), p. 229.

[47] *Phys. Rev.* (1965) **138**, A1, 689.

the metal may in some instances have an effect. This is particularly so for transition metals such as Ni.

An account of some of the more recent work on semiconductor–metal contacts is given in the report of a conference on the subject.[48]

The injection of minority carriers by metal–semiconductor contacts was extensively studied in the early days of transistor development, when such contacts were widely used. Methods of measuring the injection ratio γ have been developed by W. Shockley, G. L. Pearson and J. R. Haynes,[49] and by B. Valdes.[50] Theories of injection and rectification by metal–semiconductor contacts have been given by J. B. Gunn[51] and by P. C. Banbury;[52] an experimental study of the variation of γ with surface conditions and with illumination has been made by C. A. Hogarth[53] for small-area contacts, and large-area contacts have been studied by H. K. Henisch.[54]

If we assume that γ is given by the ratio $p/(nb+p)$, where n and p are the carrier densities near the contact, we may readily show from the values of n and p, calculated by using Boltzmann's equation, that

$$\gamma = \frac{1}{1+b\exp\{(\Delta E - 2e\phi_s)/kT\}}. \tag{198}$$

An expression of this form has been given by Hogarth (*loc. cit.*), but would imply that γ is almost always nearly equal to 1 or to 0, except for a very restricted range of values of ϕ_s. A similar expression given by Henisch (*loc. cit.* p. 226), but involving the diffusion constant and minority carrier lifetime, leads essentially to the same conclusion, and in particular appears to show that γ is very small indeed, unless $\phi_s < \frac{1}{2}\Delta E$. No such rapid variation of γ with surface conditions is found, and both formulae indicate a much too rapid variation of γ with ϕ_s. It is very unusual to find a value of γ less than about 0.1 for a true metal–semiconductor contact and values between 0.1 and 0.5 are common for n-type material, while for p-type material γ lies usually between 0.5 and 0.9. For a hole-injecting contact an increase in γ from 0.5 to 1.0 means a fourfold increase in total excess carrier concentration for a given current, since each excess hole takes with it an excess electron; alloy contacts as discussed above, giving values of γ nearly equal to 1, have

[48] *Metal–Semiconductor Contacts*, ed. M. Pepper (Inst. of Physics Conf. Series, Lond., 1974).

[49] *Bell. Syst. Tech. J.* (1949) **28**, 344.

[50] *Proc. Inst. Radio Engrs.* (1952) **40**, 1420.

[51] *Proc. Phys. Soc.* B (1954) **67**, 575.

[52] *Proc. Phys. Soc.* B (1953) **66**, 833.

[53] *Proc. Phys. Soc.* B (1953) **66**, 845.

[54] *Proc. Phys. Soc.* B (1953) **66**, 841.

therefore a considerable advantage over pure metal–semiconductor contacts when used for injection.

The subject of current injection has been treated in detail by M. A. Lampert and P. Mark.[55]

7.13 Drift mobility of electrons and holes

The most direct method of measuring the drift mobility μ_d of electrons and holes is that derived from the drift experiment of J. R. Haynes and W. Shockley[56] as already discussed in § 7.6, and further developed by W. Shockley, G. L. Pearson and J. R. Haynes.[57] The method consists essentially of injection of a narrow pulse of minority carriers into a filament and sweeping the pulse along the filament by means of an electric field; one form of the arrangement commonly used is shown in Fig. 7.18. The passage of the pulse past a collector contact (see § 7.9.2)

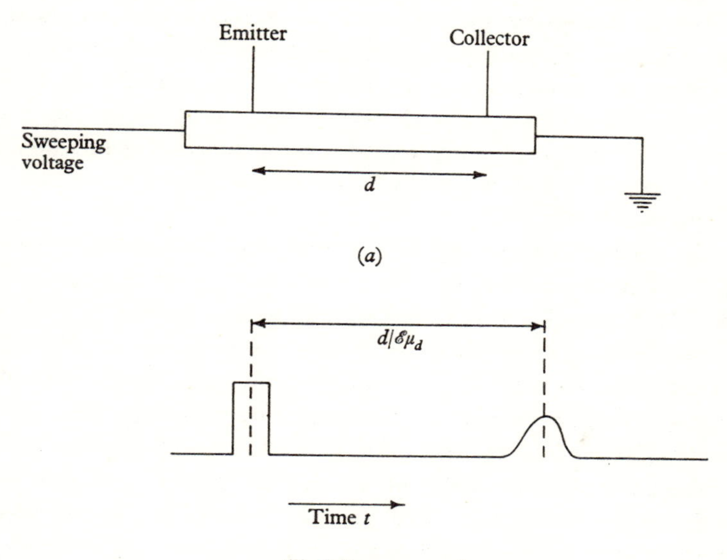

Fig. 7.18. Measurement of drift mobility.

is observed by means of a cathode-ray oscilloscope. When the positive voltage-pulse producing an electric field $\mathscr{E}$ is applied to the emitting contact, a similar pulse appears almost at once at the collector contact,

[55] *Current Injection in Solids* (Academic Press, 1970).
[56] *Phys. Rev.* (1949) **75**, 691. [57] *Bell Syst. Tech. J.* (1949) **28**, 344.

the time delay corresponding to the propagation time of an electromagnetic pulse down the filament; at a later time of the order of $l/\mathscr{E}\mu_d$, where l is the distance between the emitter and collector contacts, the pulse of minority carriers arrives beneath the collector contact and gives rise to a voltage pulse at the output, since the holes change the collector current. The first pulse will be a good replica of the pulse applied to the injecting electrode, provided the amplifiers associated with the display equipment have sufficient bandwidth to reproduce the pulse without appreciable distortion; the minority carrier pulse will, however, be somewhat spread out due to diffusion, as illustrated in Fig. 7.18. The time t between the mid-points of the two pulses is equal to $l/\mathscr{E}\mu_d$ and gives a direct measure of μ_d (this assumes that $\mathscr{E} \gg \mathscr{E}_c$, see § 7.5.5). For n-type Ge, $\mu_{dh} = 1900\ \text{cm}^2\ \text{V}^{-1}\ \text{s}^{-1}$ so that if $\mathscr{E} = 10\ \text{V cm}^{-1}$, which is a typical value for such an experiment, we have $t = 30\ \mu\text{s}$; if the minority carrier lifetime is much greater than this, the decay in the excess carrier concentration due to recombination will be small. If a short injection pulse of the order of 1 μs in duration is used, as is common in modern practice, the two pulses will be well separated in time, but if a considerably longer pulse is used, as in the original experiments, a more complex wave-form will be observed at the collector; this may be analysed to give the transit time t of the pulse of minority carriers.

If the number of holes injected is large, the field in which the pulse moves will not be uniform and the pulse may become unsymmetrical in shape. This may lead to errors in timing the arrival of the pulse; however, by using short pulses and low minority carrier concentrations, quite accurate values of the drift mobility may be obtained. As we have seen in § 7.7, the drift mobility μ_d is equal to the conductivity mobility μ_c in strongly extrinsic material, but may differ from it appreciably when the material is near-intrinsic; by varying the impurity content of the material of the filament μ_{ce} and μ_{ch} may be obtained in this way and their variation with temperature and with impurity concentration may also be obtained.

7.14 Heterojunctions

A form of junction which has become of increasing technological importance, especially with the expansion of the opto-electronic device development in the form of photo-diodes and lasers (see § 14.6), is the heterojunction. This consists of a junction between two different semiconductors. These may be quite different materials such as Ge and GaAs, although there are difficulties in forming the junction if the crystal lattice parameters differ too much from each other. More frequently one material is a particular compound such as GaAs while the other is an alloy of this compound with a related compound such as $GaAs_xP_{1-x}$. The latter has a larger forbidden energy gap than the former and this is the characteristic feature of heterojunctions. The properties of heterojunctions were first studied by R. L. Anderson[58] who realised their potential uses in semiconductor technology and who gave their basic principles of operation.

There are two distinct types of heterojunction. In the first, one semiconductor is *p*-type and the other *n*-type, thus providing an extension of the *p–n* junction. In the second both materials are *n*-type or both *p*-type and are somewhat similar to n^+–n and p^+–n junctions. We shall consider first the *p–n* type.

We shall suppose that the *p*-type material has the larger forbidden energy gap ΔE_p. The *n*-type has gap ΔE_n with $E_n < E_p$. Because of this we must modify Fig. 7.8 for the band shapes, in equilibrium, as shown in Fig. 7.19, the Fermi level being the same at both sides of the junction.

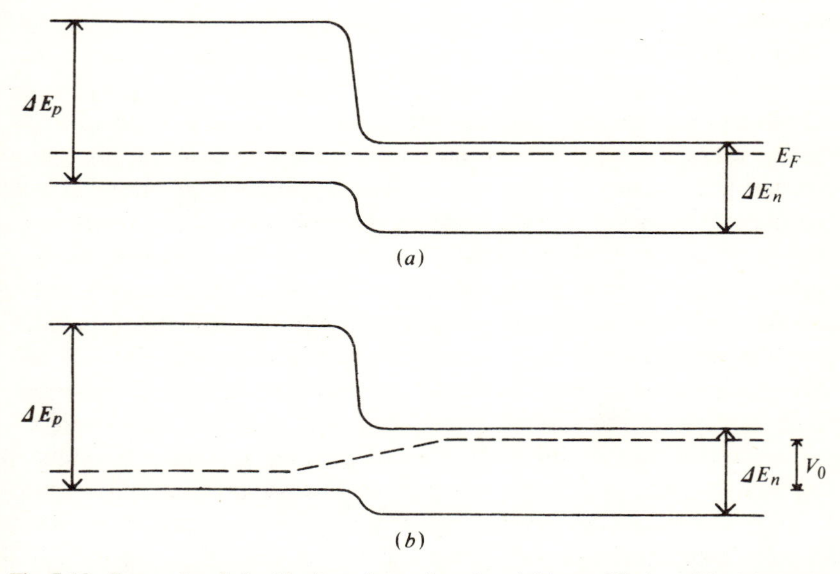

Fig. 7.19. Energy bands for ideal *p–n* heterojunction, (*a*) in equilibrium, (*b*) with applied voltage V_0.

[58] *Solid State Electron.* (1962) **5**, 341.

Fig. 7.19 also shows the effect of applying a voltage V_0 in the 'forward' direction, i.e. so as to drive holes from left to right. It will be seen that the potential barrier which holes must pass over to get from the left- to the right-hand side of the junction is lowered and if V_0 is large enough will be reduced to zero as for a normal p–n junction, so that the hole current in the 'forward' direction is very much the same as before. On the other hand the potential barrier which electrons must cross in passing from the right- to the left-hand side of the junction will still be appreciable if ΔE_p is somewhat greater than ΔE_n. The electron current will therefore be quite small even when the hole current is large. The junction therefore behaves rather like a p^+–n junction with $\gamma \simeq 1$.

One of the important uses of p–n heterojunctions is indeed as efficient injectors of holes into n-type material without the use of a heavily-doped p-type section. For use of junctions in devices from which it is desired to extract recombination radiation arising from the injected holes the heterojunction has some advantages. The p-type section, having the wider forbidden energy gap, is transparent to the recombination radiation from the n-type section. Also free-carrier absorption which would be obtained in a heavily doped p^+ section is avoided. The roles of the n-type and p-type sections can, of course, be reversed giving an electron injecting junction with $\gamma \simeq 0$.

The form of the forward current is very similar to that for a normal p–n junction, as we should expect. It is given as a function of V_0 by an equation of the form of (159) but with a somewhat modified value for J_s. On the other hand the current in the backward direction, though very much smaller, does not saturate as for a normal p–n junction but continues to increase as $-V_0$ increases. The reason for these differences lies mainly in the properties of the transition layer joining the two semiconductors in the heterojunction. Because of differences in their lattice parameters some strain must be present and this produces both hole and electron traps at the interface, as for a surface (see § 7.11). These traps tie the Fermi level at the interface to a position about midway between the valence and conduction bands of both semiconductors and so cause a rise in the bands on the n-type side and a fall of the bands on the p-type side as for a metal–semiconductor contact (see § 7.12). As a result, the energy bands are modified from the form shown in Fig. 7.19(a) to that shown in Fig. 7.20.

The effect of interfacial states has been discussed by W. G. Oldham and A. G. Milnes[59] and the properties of GeGaAs junctions as affected

[59] *Solid State Electronics* (1964) **7**, 153.

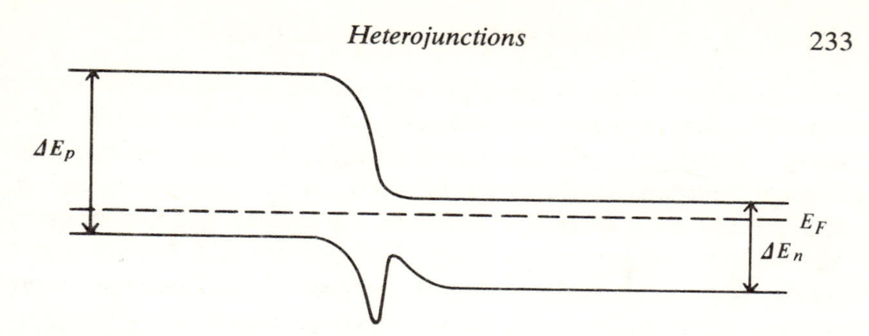

Fig. 7.20. Energy bands for *p–n* heterojunction showing the effect of interfacial traps.

by these states have been described by S. Perlman and D. L. Feucht.[60] It will be seen from Fig. 7.20 that a potential barrier has to be overcome in the 'forward' direction as well as in the 'backward'. The current through this barrier is thought to be mainly due to hole tunnelling because of the small variation with temperature of the quantity J_s in the 'forward' characteristic (see § 14.4.3).

For an *n–n* heterojunction the form of the energy bands is as shown in Fig. 7.21. Here we have included the effect of interfacial states. It will

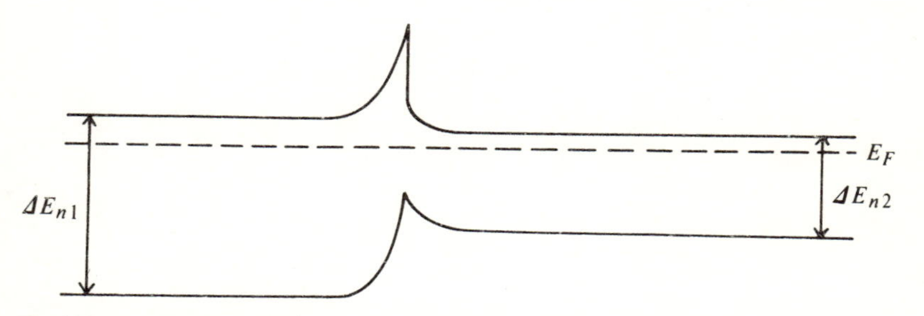

Fig. 7.21. Energy bands for *n–n* heterojunction showing the effect of interfacial traps.

be seen that electrons are faced with a Schottky-type barrier rather similar to that for a metal–semiconductor contact, but usually more complex in form. Indeed the valence band edge may have a rise followed by a dip. The characteristics of such junctions have been discussed by E. D. Hinkley, R. H. Rediker and D. K. Jadus[61] who have shown that the 'forward' characteristics are given by an equation of the form

$$J = AT^2 \exp[-\phi kT]\{\exp[eV_0/\beta kT]-1\}. \tag{199}$$

[60] *Solid State Electronics* (1964) **7**, 911.

[61] *Appl. Phys. Lett.* (1965) **6**, 144.

A is a constant which depends on the materials of the junction, ϕ is a potential, again dependent on the materials, and β a constant of order unity. This is very like equation (196) for a metal–semiconductor contact. Junctions of this form between GaAs and InSb for which Hinkley and his colleagues (*loc. cit.*) have measured the characteristics have shown very good rectification, of the order of 10^8. The value of ϕ was found to be about 0.8 V, similar to values of metal–semiconductor contacts. Again the reverse current does not saturate.

Heterojunctions have been treated in detail by T. L. Tansley[62] who has described the various complex profiles which under certain circumstances the shape of the energy bands may take through the effect of different kinds of interfacial states. He has also examined a large number of semiconductor combinations with a view to seeing which would best seem to fit together as regards similarity of lattice parameters. Although such a fit is of importance in obtaining a simpler structure without too much influence of interfacial states, it is surprising how well some junctions work with a large discrepancy between lattice parameters. For example Hinkley and his colleagues used GaAs and InSb which have an 18% difference in lattice parameter and 9.5 ratio in forbidden energy gap (1.4/0.18).

[62] *Semiconductors and Semimetals* (Academic Press, 1971) **7A**, 293.

8

Scattering of electrons and holes

8.1 Change of state

We have introduced in § 5.1 the relaxation times τ_e, τ_h which determine the rate at which electrons and holes are caused to change their **k**-vectors and which determine their mobility in electric fields. For scalar effective masses m_e and m_h the mobilities are simply given by $e\langle\tau_e\rangle/m_e$ and $e\langle\tau_e\rangle/m_h$, the symbol $\langle\,\rangle$ representing a particular average over the energy distribution. In § 7.4 we introduced the quantities τ_n and τ_p called the minority-carrier lifetimes of electrons and holes. These determine the rate at which electrons and holes change their state by recombination or trapping and are related to the generation rate of hole–electron pairs and to the excitation of electrons and holes from impurity centres. In this chapter we shall discuss various mechanisms which determine the magnitude of τ_e and τ_h and quantities related to them, and in the next chapter we shall discuss the various processes which determine τ_n and τ_p.

8.2 Scattering mechanisms

The problem of determining τ_e and τ_h as functions of the wave-vector **k** is fundamentally one involving the quantum theory of collision processes. This requires a rather advanced wave-mechanical treatment and will not be given in detail here; instead reference will be made to a number of theoretical papers dealing with the wave-mechanics of various types of scattering. Although the full wave-mechanical treatment is necessary to obtain the absolute value of τ associated with each type of scattering it is possible, in some instances, to determine the *variation* of τ with the energy E by fairly simple arguments when τ may be expressed as a function of E, i.e. when we have a relaxation time which does not depend on the direction of motion of an electron in a crystal. As we have seen in § 5.3.2 directional variation of τ may, in certain cases, be taken up by an adjustment of the values of the effective mass. Apart from the expression for the electrical conductivity which involves $\langle\tau\rangle$ (or $\langle\tau^{-1}\rangle$ for high magnetic fields) all the formulae of § 5.3 depend only on ratios of the form

$$\langle\tau^2\rangle/\langle\tau\rangle^2, \quad \langle\tau^3\rangle\langle\tau\rangle/\langle\tau^2\rangle^2, \quad \text{etc.},$$

so that they do not depend on the absolute value of τ but only on its variation with E. Our main problem will therefore be to determine this variation.

8.3 Scattering by lattice vibrations

As we have seen, the most fundamental process whereby electrons in a crystal are scattered is their interaction with the thermally generated vibrations of the atoms of the crystal. These are present even in a perfect crystal. Other scattering mechanisms depend on the presence of imperfections of the various kinds which we have discussed in § 3.1.

The atoms of a perfect crystal are often said to occupy sites which form a periodic lattice. We must, however, be more explicit as to what is meant by the saying that they 'occupy' such sites. It certainly does *not* mean that at any instant the positions of the atoms of the crystal form a perfect periodic lattice. At ordinary temperatures they execute oscillations, due to thermal agitation, about their positions of equilibrium, and the mean energy of the oscillations is given by the principle of equipartition of energy. According to classical mechanics they would occupy their positions of equilibrium at the absolute zero of temperature and so might then be regarded as being at the points of a perfect lattice. According to quantum theory, however, they still execute small vibrations about their positions of equilibrium at the absolute zero of temperature as do the atoms of a molecule – the so-called zero-point vibration (see equation (4)). Such zero-point vibrations are, however, *ordered* as between the different atoms of the crystal and do not contribute to scattering. One may therefore define the term 'occupy' by saying that the mean position of each atom of a perfect crystal averaged over a time long compared with their periods of vibration, is at a point of a perfect periodic lattice. At the absolute zero of temperature it is not at all a bad approximation to consider the atoms as at rest at their lattice sites, and to regard their random displacements at higher temperatures as random vibrations about the lattice points. Such vibrations are now commonly called 'lattice' vibrations, though the term is in some ways unfortunate, since the lattice itself is of course fixed and does not vibrate, being a geometrical configuration of points fixed in space. The use of the term has, however, now been accepted in the literature. The average amplitude of such vibrations, except at very low temperatures, is proportional to $T^{\frac{1}{2}}$, T being the absolute temperature, since the energy of each atom to a first approximation is of the order $3\boldsymbol{k}T$. This leads to the well-known law of Dulong and Petit that the atomic heat of a monatomic crystalline solid at ordinary temperatures is of the order of $3R$. At lower temperatures quantum theory makes a major correction and leads to values of the specific heat proportional to T^3 at low temperatures.

There are two equivalent approaches to the theory of vibrations of the atoms of a crystal about their positions of equilibrium. The first is the classical method of small oscillations. A crystal containing N atoms may be regarded as a system having $3N$ degrees of freedom. It may therefore be described in terms of $3N$ normal co-ordinates Q_r $(r = 1, \ldots, 3N)$ which are functions of the positional co-ordinates (x_s, y_s, z_s) $(s = 1, \ldots, N)$ giving the displacements of the N atoms from their positions of equilibrium. Each normal co-ordinate represents a configuration which oscillates with simple harmonic motion with period $T_s = 2\pi/\omega_s$. Such an oscillation is known as a normal mode and the period T_s as a normal period. The co-ordinate Q_s then satisfies the simple equation

$$\ddot{Q}_s + \omega_s^2 Q_s = 0 \quad (s = 1, \ldots, 3N). \tag{1}$$

These normal co-ordinates enable quantum theory to be applied to the problem in a very simple manner, since each co-ordinate Q_s is independent. The general motion, for small oscillations, is a linear superposition of the motions due to the separate normal modes. Equation (1), being that of a simple harmonic oscillator, has its counterpart in quantum mechanics for which the solution is very well known.[1] Only certain energies E_{ns} are allowed, given by

$$E_{ns} = \hbar(n + \tfrac{1}{2})\omega_s = \boldsymbol{h}(n + \tfrac{1}{2})\nu_s, \tag{2}$$

where n is an integer, ν_s is the frequency of the sth normal oscillation and $\omega_s = \nu_s/2\pi$. If $n_1, n_2, \ldots, n_s, \ldots$ are a series of integers, the total vibrational energy E_v of the crystal may be expressed as

$$E_v = \boldsymbol{h}\Sigma(n_s + \tfrac{1}{2})\nu_s. \tag{3}$$

[1] *W.M.C.S.*, ch. 8.

We may note that the lowest value of the vibrational energy due to the sth normal mode is $\frac{1}{2}h\nu_s$; this is what we have called the zero-point energy for the particular mode. The zero-point energy W_{0v} for the whole crystal is given by

$$W_{0v} = \tfrac{1}{2}h\Sigma\nu_s. \tag{4}$$

The wave-function corresponding to the zero-point energy is spread over a small region around the equilibrium configuration and represents a vibration of small but finite amplitude. Thus in quantum theory the state in which each atom is at rest at its position of equilibrium is not allowed.

The other approach to the problem of small oscillations is to use the methods of wave-motion and to express the motion of each atom of the crystal in terms of displacement waves. In terms of the displacement **R** of an atom whose equilibrium position is at **r**, such a wave may be expressed in the form

$$\mathbf{R} = \mathbf{A} \exp [i(\mathbf{K} \cdot \mathbf{r} - \omega t)], \tag{5}$$

where **K** is a wave-vector. The relationship between K and ω is determined by the crystalline configuration and leads to the so-called dispersion law for the crystal. By applying appropriate boundary conditions it is found that only certain values of ω are allowed and these, as might be expected, turn out to be equal to the quantities ω_s associated with the normal modes. Moreover, the quantities A_s in equation (5) corresponding to ω_s are closely related to the normal co-ordinates Q_s.[2] The wavelength λ_s associated with each value of ω_s is equal to $2\pi/K_s$; for small values of K, corresponding to long wavelengths, the waves represented by equation (5) are the ordinary acoustic waves in the crystal. The form of relationship between ω and **K** is very similar to the E–k relationship for electrons in crystals. Due to the periodicity of the lattice, **K** is subject to the same limitations as the electron wave-vector **k**, and its values may be restricted to the first Brillouin zone. Its allowed values, which then determine the allowed values of ω are, for an infinite crystal, determined in the same way as the allowed values of k (see § 2.2). For a crystal with N_a atoms per unit cell the ω–K curves break up into $3N_a$ bands, and not into an infinite number as for electrons. This is illustrated for one direction in the crystal in Fig. 8.1 for a crystal (such as Ge or PbS) with two atoms per unit cell; two bands, A, O, are shown here, and it will be seen that they are separated by a band of frequen-

[2] *W.M.C.S.*, ch. 8.

cies, lying between the two angular frequencies marked $\omega^a_{\text{max.}}$ and $\omega^0_{\text{min.}}$. When $K=K_{xm}$, the value of K corresponding to the edge of the first zone in the particular direction (K_x in Fig. 8.1), the curves have tangents

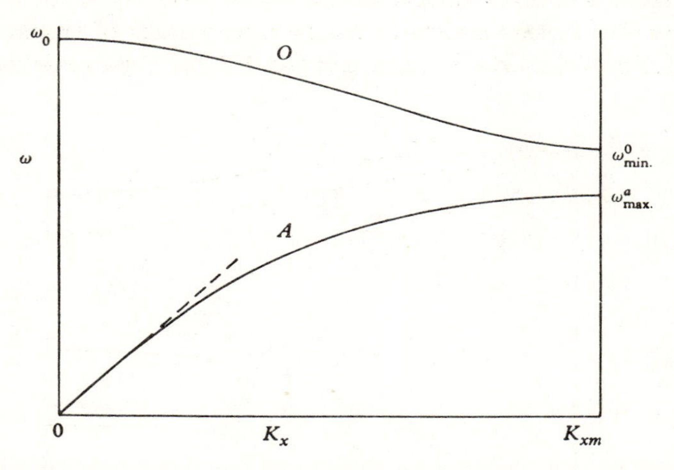

Fig. 8.1. Dispersion curves for lattice vibrations of crystal with two atoms per unit cell.

parallel to the K-axis, and this is also true for the O branch of the curve at $K=0$; for the A branch the curve tends to a straight line of slope u_a at $K=0$. The phase velocity of the displacement waves is given by ω/K and tends to u_a at $K=0$ for the A branch; the group velocity is equal to $d\omega/dK$ and also tends to u_a for small values of K for the A branch. The A branch therefore corresponds to ordinary acoustic waves for small values of K, the wave-velocity being just equal to the velocity of sound u_a in the crystal. From the macroscopic theory of acoustic waves in solid media it is known that in an isotropic medium we have two types of waves, those corresponding to compression and expansion of the solid and those corresponding to propagation of a shear strain. The former are longitudinal waves, i.e. the motion of the particles of the solid is along the direction of propagation, while the latter are transverse waves, the motion of the particles being at right angles to the direction of propagation. The latter may be resolved along two directions at right angles giving two independent sets of transverse shear waves. The general form of the equation for displacements $\mathbf{R}$ may therefore be written in the form

$$\mathbf{R}=\Sigma R_i=\sum_i\sum_s A_{is}\exp\left[i(\mathbf{K}_{is}\cdot\mathbf{r}-\omega_{is}t)\right], \tag{6}$$

where i takes the values 1, 2, 3 corresponding to the longitudinal and two transverse modes of oscillation. In general, the ω–K curves are different for the three modes, but for an isotropic solid the two transverse shear modes are degenerate having the same values of ω. This is also true for certain directions having a high order of symmetry in a crystal. This situation is illustrated in Fig. 8.2, the curves marked A_l, O_l

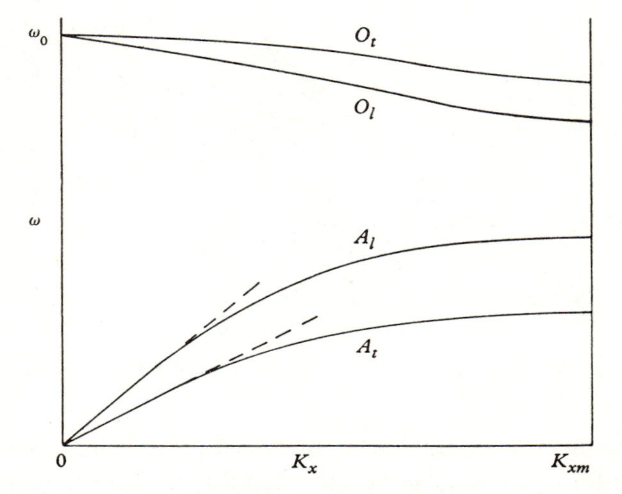

Fig. 8.2. Dispersion curves for longitudinal and degenerate shear modes.

corresponding to the longitudinal modes and those marked A_t, O_t to the transverse shear modes. The two sets of curves generally meet at $K = 0$ for non-polar crystals. It is clear why the A branches are called acoustic modes; they correspond, as we have seen, to acoustic waves for small values of K. For long waves, in this case, both atoms in a unit cell move in the same direction, as do atoms in quite a number of neighbouring cells. The O branches correspond to modes in which the two atoms in each unit cell move in opposite directions. The situation with $K = 0$ corresponds to an oscillation in which all the atoms of the crystal of type 1 in the unit cell move in phase and all the atoms of type 2 move in phase, the movement of the two types being in opposite directions. This is clearly a standing wave and it is readily verified from Figs. 8.1 or 8.2 that for the O branches the group velocity, given by $d\omega/dK$, is equal to zero when $K = 0$. This type of oscillation is called an optical mode, since it couples strongly with electromagnetic waves (infra-red) when the two atoms of the unit cell are charged in the opposite sense as in a polar crystal; as they are displaced in opposite directions a strong dipole moment is created which couples with the electric field.[3] The frequency ω_0 at $K = 0$ generally lies in the far infra-red and so corresponds to a wavelength which is very large compared with the atomic spacing. This coupling is therefore strong only for small values of K, being averaged out when $2\pi/K$ is small compared with the wavelength of the infra-red radiation; there is therefore a strong absorption in polar crystals at a frequency $\nu_0 = \omega_0/2\pi$, called the *reststrahl* frequency. The distinction between the two forms of oscillations becomes less distinct as the value of K_x approaches K_{xm}, but the optical and acoustic branches may be distinguished by their behaviour for small values of K. We note that as the value of $\mathbf{K}$ approaches that corresponding to the zone edge we again have, as for electrons, under certain symmetry conditions, a zero group velocity ($d\omega/dK = 0$), showing that here again we have only standing waves. The values of ω corresponding to the zone edge, which we have denoted by $\omega^0_{\text{max.}}$, etc., play an important part in the theory of absorption of infra-red radiation by electrons in certain types of semiconductor.

In Fig. 8.2 we have shown the two optical branches having the same frequency at $\mathbf{K} = 0$. This would be so for crystals like Si and Ge which are non-polar. For polar compounds such as the III–V semiconductors the two optical branches will not be coincident at $\mathbf{K} = 0$ and we have two zero-$\mathbf{K}$ optical frequencies ω_{L0} and ω_{T0}. When we take into account the interaction of the electromagnetic waves of frequency $\omega/2\pi$ with the lattice vibrations we find that there is a relationship connecting ω_{L0} and ω_{T0} with the static permittivity ϵ_s and the high-frequency permittivity ϵ_h. This relationship $\omega^2_{L0}/\omega^2_{T0} = \epsilon_s/\epsilon_h$ is known as the Lyddane–Sachs–Teller relationship. Since ϵ_s includes polarization due to ionic movement while ϵ_h does not, in general, $\epsilon_s > \epsilon_h$ for polar compounds so that for these $\omega_{L0} > \omega_{T0}$. When damping of the lattice vibrations is neglected it is found that $\epsilon \to \infty$ at $\omega = \omega_{T0}$ and $\epsilon = 0$ at $\omega = \omega_{L0}$. For $\omega_{T0} < \omega < \omega_{L0}$, ϵ is negative and this leads to strong reflection over this band of frequencies. On the other hand when damping is taken into account there is also very strong absorption at $\omega = \omega_{T0}$. This is the *reststrahl* absorption. To obtain ω_{L0} and ω_{T0} both absorption and reflectivity or emissivity measurements are necessary. (See § 10.8 and § 14.3.)

Since the difference between ω_{T0} and ω_{L0} depends on the effective ionic charge e^*, such measurements provide a means for determining

[3] *W.M.C.S.*, § 8.18.

e^*. The polarization per ion pair is at low frequencies given by $e^{*2}/M_r\omega_{TO}^2$ where M_r is the reduced mass of the ion pair.[4] (See §§ 13.5, 14.3.)

Quantization of the oscillations described by the travelling waves is carried out as before. For each suffix i it turns out that we have N allowed frequencies giving as before $3N$ in all. The energies are given, as before, by equations such as (2). When energy is exchanged between the vibrational states of the crystal and an external radiation field this can only take place in units of $h\nu_s$, where ν_s is one of the allowed frequencies. A similar condition holds for exchange of energy between electrons and the lattice vibrations. This is reminiscent of the exchange of energy with a radiation field in which single photons are created or destroyed. We shall see in the next section that there is a very close analogy between photons and lattice vibrations.

8.4 Phonons

Let us now consider the effect of a travelling wave such as that given by equation (5) on the motion of an electron in the conduction band. Let us first consider a compressional wave. Such a wave will cause the lattice spacing to vary periodically and, as a consequence, we shall have a small periodic perturbation of the crystal potential. The velocity with which this disturbance moves through the crystal will be equal to the speed of sound ($\sim 10^5$ cm s^{-1}) for small values of K. An average electron at ordinary temperatures will be moving with a velocity of the order of 10^7 cm s^{-1}, so the periodic potential from which the electron may be scattered is moving slowly with respect to the electron wave. On scattering, the latter will suffer a small Doppler shift in frequency corresponding to a small change in energy which we shall now calculate.

Consider an electron whose wave-vector before scattering is $\mathbf{k}$ and so is represented by a wave-function of the form

$$\Psi_i = u(r)\exp\left[i(\mathbf{k}\cdot\mathbf{r}-\omega t)\right], \tag{7}$$

its energy E being equal to $\hbar\omega$. As we shall see, we are mainly concerned with long displacement waves and we may take an average value of $u(r)$ and regard it as a constant. Let the scattered wave have wave-vector $\mathbf{k}'$ and so be represented by

$$\Psi_s = u'(r)\exp\left[i(\mathbf{k}'\cdot\mathbf{r}-\omega' t)\right]. \tag{8}$$

[4] *W.M.C.S.*, § 8.18.

The compressions and rarefactions due to the displacement waves (5) have their maximum values on a set of planes spaced by an amount $\lambda_s = 2\pi/K_s$ apart. The condition for strong reflection is therefore given by the well-known Bragg condition

$$\mathbf{k} - \mathbf{k}' = \mathbf{K}_s. \tag{9}$$

This may be seen as follows: let the direction of motion of the electron before scattering make an angle ϕ with the vector $\mathbf{K}_s$, i.e. with the normal to the scattering planes. We have then $\mathbf{k} \cdot \mathbf{K}_s = kK_s \cos\phi$. Similarly, let the direction of motion of the scattered electrons make an angle $\pi - \psi$ with the vector $\mathbf{K}_s$, so that $\mathbf{k}' \cdot \mathbf{K}_s = -k'K_s \cos\psi$, where k, k', K_s are respectively the magnitudes of the vectors $\mathbf{k}$, $\mathbf{k}'$, $\mathbf{K}_s$ (see Fig. 8.3). In order that the phase of all wavelets scattered from points in *one*

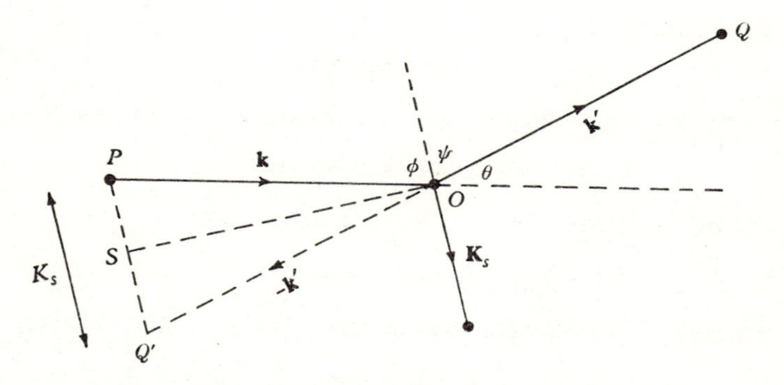

Fig. 8.3. Scattering of waves from a set of parallel planes.

plane be in phase we must have $k \sin\phi = k' \sin\psi$. If in Fig. 8.3 *PO* represents the vector $\mathbf{k}$, *OQ* the vector $\mathbf{k}'$, *OQ'* the vector $-\mathbf{k}'$, and *OS* is perpendicular to $\mathbf{K}_s$ in the plane *OPQ'*, then from the above relationship *PQ'* is perpendicular to *OS*. We may therefore write

$$\mathbf{k} - \mathbf{k}' = A\mathbf{K}_s, \tag{10}$$

where A is a constant. Now, in order that reflections from successive planes spaced λ_s apart should be in phase, we must have $A = 1$, as may readily be seen by taking $\phi = 0$, since then we must have

$$(k + k') = 2\pi/\lambda_s = K_s,$$

so that $A = 1$. The scattering condition therefore is given by equation (9). The angle θ through which the electron is scattered is $(\pi - \phi - \psi)$.

Since the Doppler shift is small we have $k' \simeq k$, $\phi \simeq \psi$, $\theta \simeq \pi - 2\phi$, so that

$$K_s = |k' - k| = 2k \sin \tfrac{1}{2}\theta. \tag{11}$$

In order that the phase of the incident and reflected waves at a point moving with the scattering wave should remain constant (a condition for reflection) we must have the same phase term in (7) and (8) on replacing $\mathbf{r}$ by $(\mathbf{K}_s/K_s)ut$ where $u = \omega_s/K_s$. Thus we have

$$\frac{(\mathbf{k} \cdot \mathbf{K}_s)u}{K_s} - \omega = \frac{(\mathbf{k}' \cdot \mathbf{K}_s)u}{K_s} - \omega'. \tag{12}$$

This is simply the equation which determines the Doppler shift in the reflected wave, giving

$$\omega - \omega' = [(\mathbf{k} - \mathbf{k}') \cdot \mathbf{K}_s]u/K_s, \tag{13}$$

or, using equation (11),

$$\omega - \omega' = K_s u = \omega_s. \tag{14}$$

Similarly, for a travelling wave in the opposite direction we have

$$\omega - \omega' = -K_s u = -\omega_s. \tag{15}$$

Thus the two possible values of $(\omega - \omega')$ are given by

$$\omega - \omega' = \pm\omega_s. \tag{16}$$

Multiplying by $\hbar$ we have for the change in energy of the electron

$$E - E' = \pm h\nu_s, \tag{17}$$

and we thus see that in being scattered the electron may either lose or gain energy of amount $h\nu_s$, i.e. one of the units of energy by which the total energy of the crystal may change.

It remains to verify that the change of energy on scattering is small compared with the average energy of an electron, so that $k \simeq k'$. By combining equation (11) with equation (17) we see that the gain in energy $\delta E = E' - E$ when an electron is scattered through an angle θ is given approximately by

$$\delta E = \pm 2mvu \sin(\tfrac{1}{2}\theta), \tag{18}$$

where v is the velocity of the electron, the positive sign being taken for absorption and the negative sign for emission of energy. It will be seen that the biggest change in energy will take place when $\theta = \pi$ and that then we have $|\delta E|/E = 4u/v$. Since u/v is small except at low temperatures this ratio is in general quite small.

Let us now consider the change in crystal momentum **P** of the electron. We have from equation (10)

$$\mathbf{P}-\mathbf{P}'=\hbar(\mathbf{k}-\mathbf{k}')=\hbar\mathbf{K}_s. \tag{19}$$

We may therefore interpret $\hbar K_s$ as the momentum gained by the lattice as a result of scattering of the electron, the gain in energy being $h\nu_s$. The momentum gain may also be written as h/λ_s. This is exactly analogous with the creation of a photon of energy $h\nu$ and momentum h/λ. The quantum of lattice vibration which may be regarded as being created in interactions with the atoms of the crystal is termed a *phonon*, and a process such as we have described is termed a one-phonon process. It is actually possible for more than one phonon to be absorbed or emitted in a scattering process, since the vibrations are not completely described by simple harmonic waves, anharmonic effects introducing terms with frequencies $2\nu_s$, etc. The probability of such processes is, however, generally small compared with that of the one-phonon processes and we shall not consider them for the present.

We may now see why the zero-point vibrations cannot scatter electrons. Near the absolute zero of temperature the lattice vibrations will be in their lowest state and so will be unable to give a quantum $h\nu_s$ of energy to the electron. The electron also will not have enough energy to create a phonon and so cannot change its value of **k**. We shall later consider the important part played by phonons in the absorption of infra-red radiation through exciting electrons from the full band to the conduction band of a semiconductor, and also the absorption and emission of infra-red radiation by the lattice vibrations (see §§ 10.5.3 and 10.8). We shall now consider in more detail the scattering of electrons by the lattice vibrations, in order to find the variation of the relaxation time with the energy of the electron and with the temperature of the crystal.

8.5 Relaxation time for lattice scattering

An electron incident on a displacement wave in a direction making an angle ϕ with the normal, is either unaffected or is scattered by the wave. If it is scattered there is a very high probability that it will be scattered through an angle $\pi-2\phi$. Not all displacement waves travelling in the same direction will cause such scattering – only those for which the frequency ν_s is such that equation (17) is satisfied. The largest value of ν_s involved is given by $h\nu_s/E_e=4u/v$, where E_e is the energy of the electron. For $E_e = kT$, with $T = 300\,°K$, $h\nu_s \simeq kT/10 = k\theta_s$, where $\theta_s \simeq 30\,°K$. The upper limits of the acoustic branches of the lattice vibration spectrum generally correspond to energies $k\theta$, where θ is of the order of 100–300 °K, so that the vibrations concerned with this kind of scattering lie well down the acoustic branch of the spectrum and have wavelengths long compared with the lattice spacing.

Since displacement waves are travelling in all directions we see that the scattering will be isotropic provided that there is an equal chance of finding the necessary phonons. By the principle of detailed balancing the probabilities of absorption and emission of a phonon are equal in thermal equilibrium; also the probability of absorption of a phonon is proportional to the number present. We shall require phonons whose energy ranges from nearly zero to the maximum value E_m and the ratio of the number present with the maximum energy to the number with energy nearly equal to zero is $\exp[-E_m/kT]$. Except at low temperatures, $E_m/kT \ll 1$, and there is little variation in the density of phonons over the range of energy required for scattering through various angles; thus in these circumstances the scattering is isotropic. At very low temperatures, however, some falling off in the large-angle scattering may occur owing to the lack of sufficiently energetic phonons, but at such temperatures impurity scattering is in any case likely to predominate, as we shall see.

The amplitude of the electron wave scattered from the small periodic potential set up by the displacement wave will be proportional to the potential, and this in turn will be proportional to the amplitude of the displacement wave. The probability of scattering is proportional to the square of the amplitude of the electron wave and therefore also to the square of the amplitude of the displacement wave; this is proportional to the energy carried by the displacement wave which, for the long acoustic waves, is of the order of kT. The scattering cross-section σ_l is therefore proportional to T and the mean free path l for scattering by the longitudinal acoustic modes is therefore given by an equation of the form

$$1/l = B_l T, \tag{20}$$

where B_l is a constant.

To calculate the constant B_l requires a full wave-mechanical treatment; this has been given by W. Shockley and J. Bardeen[5] and substantiates the conclusions which we have reached by elementary arguments.

[5] *Phys. Rev.* (1950) **77**, 407; *ibid.* (1950) **80**, 72.

The value of the mean free path l for a semiconductor with spherical constant-energy surfaces is found to be[6]

$$l = \frac{\rho u_l^2 \boldsymbol{h}^4}{16\pi^3 m_e^2 E_1^2 \boldsymbol{k}T}, \tag{21}$$

where ρ is the density, u_l is the velocity of long compressional waves in the crystal and E_1 is an energy defined by means of the equation

$$\Delta E_c = E_1 \Delta V / V_0. \tag{22}$$

Here ΔE_c is the magnitude of the change in the energy corresponding to the bottom of the conduction band due to a small change ΔV of the original volume V_0. E_1 is known as a deformation potential. The relaxation time $\tau = l/v$ is then given by the equation

$$\tau = \frac{\rho u_l^2 \boldsymbol{h}^4}{8\pi^3 (2m_e)^{\frac{3}{2}} \boldsymbol{k}TE_1^2 E^{\frac{1}{2}}}. \tag{23}$$

We thus see that τ has the form

$$\tau = aE^{-\frac{1}{2}} T^{-1}, \tag{24}$$

which we have already used for the relaxation time due to lattice scattering, a being a constant (§ 5.3.2).

When the constant-energy surfaces are no longer spherical the effective mass m_e must be replaced by an appropriate averaged value. The factor $E^{\frac{1}{2}} m_e^{\frac{3}{2}}$ in equation (23) arises simply from the density of states into which an electron may be scattered. We have seen that for a semiconductor with a single minimum in the conduction band and constant-energy surfaces in the form of ellipsoids, this factor is replaced by $E^{\frac{1}{2}}(m_1 m_2 m_3)^{\frac{1}{2}}$ (see § 4.2). For a semiconductor with cubic symmetry with spheroidal constant-energy surfaces the factor is $E^{\frac{1}{2}} m_T m_L^{\frac{1}{2}}$.

Equation (24) leads at once to the result that the lattice mobility μ_l has the form $\mu_l = \mu_0 (T_0/T)^{\frac{3}{2}}$, where μ_0, T_0 are constants. Using equation (97) of § 5.3.2 for the average value $\langle \tau \rangle$ of τ we have

$$\mu_l = \frac{2^{\frac{3}{2}} \pi^{\frac{1}{2}} e \hbar^4 \rho u_l^2}{3 m_e^{\frac{5}{2}} E_1^2 (\boldsymbol{k}T)^{\frac{3}{2}}}. \tag{25}$$

A similar treatment may be given for the scattering of positive holes. The formula for μ_l for holes is the same as equation (25), except that the effective mass m_h is that appropriate to holes and E_1 is replaced by the

[6] *W.M.C.S.*, § 13.4.4.

deformation potential for the valence band. For a semiconductor with spherical constant-energy surfaces in both the valence and conduction bands we have

$$\frac{\mu_{el}}{\mu_{hl}}=\left(\frac{m_h}{m_e}\right)^{\frac{5}{2}}\left(\frac{E_{1h}}{E_{1e}}\right)^2. \tag{26}$$

If $E_{1h}\simeq E_{1e}$ we have

$$(\mu_{el}/\mu_{hl})\simeq(m_h/m_e)^{\frac{5}{2}}. \tag{27}$$

This relationship is frequently used in the literature without any indication of its limitations.

We have seen that for ellipsoidal constant-energy surfaces we must replace $1/m_e$ by

$$\frac{1}{3}\left(\frac{1}{m_1}+\frac{1}{m_2}+\frac{1}{m_3}\right)$$

in expressing μ in terms of τ (§ 5.2.1). In this case we must replace $m_e^{\frac{5}{2}}$ in equation (25) by

$$\frac{1}{m_e^{\frac{5}{2}}}=\frac{1}{3(m_1m_2m_3)^{\frac{1}{2}}}\left[\frac{1}{m_1}+\frac{1}{m_2}+\frac{1}{m_3}\right]. \tag{28}$$

For cubic symmetry and spheroidal constant-energy surfaces this becomes

$$\frac{1}{m_e^{\frac{5}{2}}}=\frac{1}{3m_Tm_L^{\frac{1}{2}}}\left[\frac{2}{m_T}+\frac{1}{m_L}\right]. \tag{28a}$$

Scattering due to the transverse shear modes should also be taken into account. The variation with temperature and energy is similar to that for the longitudinal modes, but the expression for the absolute value of τ_s is somewhat complex. This form of scattering has been discussed theoretically by C. Herring[7] who has shown that scattering by transverse modes in Si and Ge may be quite appreciable as compared with the scattering by the longitudinal modes. For semiconductors with spherical constant-energy surfaces, the relative contribution should, however, be a good deal smaller.

When we have a degenerate valence band at $\mathbf{k}=0$, as for Si and Ge, scattering of holes by the acoustic modes is a much more complex process. Even if we approximate to the band structure by assuming that we have two scalar hole masses m_{h1}, m_{h2} (see § 2.3), which because of

[7] *Bell Syst. Tech. J.* (1955) **34**, 237.

their differences in magnitude are generally referred to as light holes and heavy holes, we have the possibility of light holes being scattered into the heavy hole band and vice versa. If τ_{h1} and τ_{h2} are averaged relaxation times for the two types of holes it may be shown that[8]

$$\frac{1}{\tau_1}=\frac{1}{\tau_2}=\frac{1}{\tau_0}\left(\frac{m_{h1}}{m_0}\right)^{\frac{3}{2}}+\frac{1}{\tau_0}\left(\frac{m_{h2}}{m_0}\right)^{\frac{3}{2}}, \tag{29}$$

τ_0 being the relaxation time for a hole of scalar effective mass m_0 in a 'spherical' band having the same deformation potential. The mobilities μ_{h1}, μ_{h2} of the two types of holes are then given by

$$\frac{e}{\mu_{h1}}=\frac{m_0}{\tau_0}\left\{\left(\frac{m_{h1}}{m_0}\right)^{\frac{5}{2}}+\left(\frac{m_{h2}}{m_0}\right)^{\frac{3}{2}}\left(\frac{m_{h1}}{m_0}\right)\right\} \tag{30}$$

$$\frac{e}{\mu_{h2}}=\frac{m_0}{\tau_0}\left\{\left(\frac{m_{h2}}{m_0}\right)^{\frac{5}{2}}+\left(\frac{m_{h1}}{m_0}\right)^{\frac{3}{2}}\left(\frac{m_{h2}}{m_0}\right)\right\}. \tag{30a}$$

If $m_{h2} \ll m_{h1}$ (light hole) it will be seen that the relaxation time for light holes to change to heavy holes is almost equal to the relaxation time for scattering of heavy holes. They are therefore short lived.

Separate deformation potentials are naturally required for the transverse shear modes and they depend on various components of the elastic deformation tensor and are quite complex. With so many disposable parameters it is somewhat difficult to separate the various components of the lattice scattering but a great deal of combined experimental and theoretical work has been devoted to this so that in the well-studied semiconductors these processes are now fairly well understood.

Scattering from the long-wave optical modes with frequencies nearly equal to the *reststrahl* frequency will also take place. While it has long been appreciated that these modes would be of great importance in polar semiconductors it was thought that for the element semiconductors their contribution would be small. This one should expect at low temperatures since there will be few optical phonons present to cause scattering by *absorption* of a phonon. Also there will be too few electrons present with sufficient energy to *excite* an optical phonon. At higher temperatures, however, C. Herring (*loc. cit.*) has shown that for Si and Ge the optical phonons can make an appreciable contribution.

The complexity introduced by non-spherical constant-energy surfaces in the conduction band of semiconductors such as Si and Ge is quite severe, complicating the deformation potentials and causing 'inter-

[8] *W.M.C.S.*, § 13.4.4.

valley' scattering (see § 8.7). The problem of determining the relative magnitudes of the contributions from various types of lattice vibrations is very complex and a great deal has been written on the subject. Extensive accounts have been given by F. J. Blatt,[9] M. S. Sodha,[10] J. L. Moll,[11] and by Esther M. Conwell[12] who has also considered the condition in which the electrons are no longer even approximately in thermal equilibrium with the crystal lattice.

Values of the various parameters for Si, Ge and GaAs have been calculated more recently using pseudopotential theory and a critical review given of previous work by W. Fawcett and colleagues.[13]

Forms of the deformation potential for even more complex bands have been discussed by G. E. Picus and G. L. Bir.[14] The effect of more general deviations from the quadratic form of the $E/\boldsymbol{k}$ relationship has been treated by a number of authors and discussed in some detail by L. Sosnowski.[15] This is particularly marked for semiconductors of the III–V compound group such as InSb, GaAs etc., for which we have constant-energy surfaces which are nearly 'spherical', i.e. E is a function of k but not a *quadratic* function. This kind of variation may be represented by an effective mass which varies with the average energy of the carriers, and hence with the added impurity concentration. Experimental evidence for the effect has been given by V. V. Galavanov *et al.*[16] and by B. Byszewski *et al.*[17]

8.6 Optical mode scattering in polar semiconductors

Because of the strong dipole moment set up by the optical modes in polar crystals, the coupling between an electron and the optical modes is likely to be much stronger than in non-polar crystals. Hence, the optical-mode scattering is likely to be much more important for polar crystals. Approximate formulae have been given for the contribution to the mobility from optical-mode scattering by H. Fröhlich and N. F.

[9] *Solid State Physics* (Academic Press, 1957) **4**, 199.
[10] *Progress in Semiconductors* (Heywood, 1958) **3**, 153.
[11] *Physics of Semiconductors* (McGraw-Hill, 1964).
[12] *Solid State Physics* (Academic Press, 1967), suppl. 9.
[13] W. Fawcett and E. G. S. Paige, *J. Phys. C: Solid State Phys.* (1971) **4**, 1801; D. C. Herbert, W. Fawcett, A. H. Lettington and D. Jones, *Proc. XIIth Int. Conf. on Phys. of Semiconductors* (Polish Sci. Publ., 1972), p. 1221.
[14] *Proc. Vth Int. Conf. on Phys. of Semiconductors* (Czechoslovak Acad. Sci., 1960), p. 89.
[15] *Proc. VIIth Int. Conf. on Phys. of Semiconductors* (Dunod, 1964), p. 341.
[16] *Fiz. Tverd. Tela* (1962) **4**, 546.
[17] *Phys. Stat. Solidi* (1963) **3**, 1880.

Mott,[18] D. Howarth and E. H. Sondheimer[19] and also by F. E. Low and D. Pines,[20] and the application of these formulae to polar semiconductors has been discussed by R. L. Petritz and W. W. Scanlon,[21] who have also suggested some modifications. The strength of the interaction between an electron and the optical modes is characterized by a coupling constant α defined by means of the equation

$$\alpha = \frac{e^2}{2\boldsymbol{h}}\left(\frac{m}{2\boldsymbol{h}\nu_e}\right)^{\frac{1}{2}}\left(\frac{m_e}{m}\right)^{\frac{1}{2}}\frac{(\epsilon-\epsilon')}{\epsilon\epsilon'}. \tag{31}$$

The frequency ν_e is equal to $(\epsilon/\epsilon')^{\frac{1}{2}}\nu_0$, where ν_0 is the *reststrahl* frequency corresponding to the transverse optical mode with $k=0$, and ϵ' is the high-frequency permittivity. The difference between ϵ and ϵ' is due to the fact that ϵ' does not include the polarization due to the motion of the ions. The expression for α can also be written as

$$\alpha = \frac{l_e}{a_0}\left(\frac{m_e}{m}\right)\frac{\epsilon_0(\epsilon-\epsilon')}{\epsilon\epsilon'} \tag{31a}$$

where $a_0 = 4\pi\epsilon_0\hbar^2/me^2$ is the Bohr radius and l_e is a length equal to $\hbar/(2m_e\hbar\nu_e)^{\frac{1}{2}}$. In a non-polar crystal $\epsilon-\epsilon'$ is small, so α is small, but for polar crystals the value of α may be just less than unity. For example for InSb $\alpha=0.02$, for GaAs $\alpha=0.06$, while for PbS $\alpha=0.28$ and for CdTe $\alpha=0.35$. The theory due to Fröhlich and Mott, and also its development by Howarth and Sondheimer, are valid only for $\alpha<1$; the rather more elaborate treatment of Low and Pines is valid when $\alpha>1$, but does not give explicit expressions for the relaxation time except at low temperatures. The expression given by Petritz and Scanlon for the contribution μ_0 of the optical-mode scattering to the mobility is given by

$$\frac{1}{\mu_0} = \frac{3\pi^{\frac{3}{2}}\alpha m_e\nu_e}{2e}[f(z)]^{-1} \tag{32}$$

where

$$f(z) = \chi(z)[e^z-1]/z^{\frac{1}{2}}$$

with

$$z = \boldsymbol{h}\nu_e/\boldsymbol{k}T.$$

$\chi(z)$ is a slowly varying function of z having the value 1.0 for $z=0$ and $z\simeq 3$ and falling to about 0.6 for $z=1$. Because of the exponential factor in $f(z)$ we see that $1/\mu_0$ becomes very small at low temperatures; the reason for this is simply that the excitation of the optical modes requires an energy of the order of $\boldsymbol{h}\nu_e$ and this is not available at low

[18] *Proc. Roy. Soc.* A (1939) **171**, 496.
[19] *Proc. Roy Soc.* A (1953) **219**, 53.
[20] *Phys. Rev.* (1953) **91**, 193.
[21] *Phys. Rev.* (1955) **97**, 1620.

temperatures. Petritz and Scanlon (*loc. cit.*) have used this expression (or an equivalent one) to discuss the mobility of electrons in the lead chalcogenides, especially PbS, and have shown that at room temperature optical-mode scattering mainly determines the mobility.

8.7 Inter-valley scattering

Another form of scattering which is of importance in semiconductors with more than a single minimum in the conduction band is that known as 'inter-valley' scattering. An electron with wave-vector $\mathbf{k}$ nearly equal to $\mathbf{k}_1$, the value corresponding to one of the minima, may change its wave-vector to a value $\mathbf{k}'$ nearly equal to $\mathbf{k}_2$ the value corresponding to a second minimum. However, if $\mathbf{k}_1$ and $\mathbf{k}_2$ are not nearly equal, quite a large change in crystal momentum will be involved. This momentum change will be approximately equal to $\hbar(\mathbf{k}_1-\mathbf{k}_2)$, and to make the change the emission or absorption of an energetic phonon will be involved. The scattering will therefore be highly inelastic and is unlikely to take place at low temperatures, since either there are very few energetic phonons present to supply momentum or the electron will have insufficient energy to create a phonon. The effect of this kind of scattering on the mobility of electrons in n-type Ge and Si has been discussed in some detail by C. Herring.[22] Herring writes the expression for τ in the form

$$\tau^{-1} = W_a + W_e, \tag{33}$$

where
$$W_a = W\frac{(1+E/\hbar\omega)^{\frac{1}{2}}}{\exp\,[\hbar\omega/kT]-1}$$

and
$$W_e = W\frac{(E/\hbar\omega-1)^{\frac{1}{2}}\exp\,[\hbar\omega/kT]}{\exp\,[\hbar\omega/kT]-1} \qquad (E \geqslant \hbar\omega)$$

$$= 0 \qquad (E < \hbar\omega).$$

W_a corresponds to absorption of a phonon, and the factor

$$[\exp\,(\hbar\omega/kT)-1]^{-1}$$

is proportional to N_p the number of phonons of energy $\hbar\omega$ at temperature T. The corresponding factor $[\exp\,(-\hbar\omega/kT)-1]^{-1}$ in W_e, which corresponds to emission of a phonon, is proportional to $(1+N_p)$ (see § 14.6). The frequency $\omega/2\pi$ is such that $\hbar\omega$ corresponds to the

[22] *Bell. Syst. Tech. J.* (1955) **34**, 237.

difference in vibrational energy involved in a change of wave-vector between two extrema. It will be seen that W_a becomes very small when $\hbar\omega \gg \boldsymbol{k}T$, and that under this condition the contribution from W_e will also be small, since we have $W_e = 0$ when $E < \hbar\omega$, which is much greater than $\boldsymbol{k}T$. We may note that no simple power law holds for the variation of τ with E in this case.

8.8 Inter-electron scattering

Collisions between electrons have generally been considered to be unimportant in determining the mobility except under highly degenerate conditions. Even then the contribution must be small as we may deduce from the fact that such collisions have very little effect in determining the resistance of metals. Collisions between electrons and between electrons and holes may, however, play an important part in establishing a steady state under high-field conditions. Because of the similarity of the masses of the colliding particles, energy is exchanged much more rapidly between them than between electrons and impurity centres or lattice vibrations. This has been discussed by H. Fröhlich and B. V. Paranjape[23] in connection with a theory of breakdown in dielectrics. The effect of inter-electron collisions on mobility has been discussed by H. Fröhlich, B. V. Paranjape, C. G. Kuper and S. Nakajima[24] and also by M. S. Sodha and P. C. Eastman.[25] Experiments are discussed in which the inter-electron scattering could, in principle, be separated from the other scattering; so far, however, there is little experimental evidence on the magnitude of the inter-electron scattering, except that it is generally small.

8.9 Ionized impurity scattering

The scattering of an electron in the Coulomb field of an ionized impurity atom may be discussed in terms of classical mechanics or by means of wave-mechanics – both methods give the same result.[26] The cross-section for scattering into a solid angle $d\omega$ through an angle θ is given by

$$\sigma(\theta)\,d\omega = \tfrac{1}{4}R^2 \operatorname{cosec}^4(\tfrac{1}{2}\theta)\,d\omega, \tag{34}$$

where

$$R = Ze^2/4\pi\epsilon m_e v^2$$

[23] *Proc. Phys. Soc.* B (1956) **69**, 21.
[24] *Proc. Phys. Soc.* B (1956) **69**, 842. [25] *Z. Phys.* (1958) **150**, 242.
[26] See, for example, N. F. Mott and H. S. W. Massey, *Theory of Atomic Collisions* (Oxford University Press, 1933), ch. 3.

and ϵ the permittivity.[27] We thus see that the scattering is highly anisotropic, small angles of scattering being strongly favoured. To obtain the relaxation time τ we must average over the various angles of scattering in the manner indicated by equation (8) of § 5.1. If, however, we substitute the above expression for $\sigma(\theta)$ in equation (8) and integrate with respect to θ to obtain σ_c we get an infinite conduction cross-section, which clearly does not represent the situation in a crystal. The reason is that the Coulomb field of the ionized impurity centre does not extend unmodified to large distances from the centre. It is collisions with large impact parameter (distance of closest approach if the particle were undeflected by the field of the centre) that cause the divergence of the integral for σ_c. In a crystal, the Coulomb field due to an impurity centre is modified by neighbouring ionized impurities, and also by the presence of free electrons or holes; two approaches to the problem of finding the right way to modify the Coulomb field have been made. Esther M. Conwell and V. F. Weisskopf[28] have assumed that the Coulomb field ceases to be effective at a radius r_m given by $(2r_m)^{-3} = N_I$, where N_I is the number of ionized impurities per unit volume, so that r_m is equal to half the mean distance between impurities. The deflection θ_m corresponding to impact parameter r_m is given by the well-known relationship

$$\tan(\theta_m/2) = R/r_m. \tag{35}$$

To obtain the conductivity cross-section σ_c we therefore insert the value for $\sigma(\theta)$ given by equation (34) in equation (8) of § 5.1 and integrate from $\theta = \theta_m$ to $\theta = \pi$. Thus we have

$$\begin{aligned}\sigma_c &= 2\pi \int_{\theta_m}^{\pi} (1-\cos\theta)\sigma(\theta)\sin\theta\, d\theta \\ &= -4\pi R^2 \ln\sin(\tfrac{1}{2}\theta_m) \\ &= 2\pi R^2 \ln(1 + r_m^2/R^2).\end{aligned} \tag{36}$$

We may write $r_m/R = 2E/E_m$, where $E_m = Ze^2/4\pi\epsilon r_m$ and is the potential energy of an electron in the field of the impurity centre at a distance equal to r_m. Let us estimate the value of E_m. For

$$N = 10^{15}\ \text{cm}^{-3}, \quad 2r_m = 10^{-5}\ \text{cm};$$

if we take $\epsilon/\epsilon_0 = 10$ we have $E_m = 1.35 \times 10^{-3}$ eV. For $T = 300$ °K,

[27] N.B. The factor 4π in the denominator comes from the use of M.K.S. units.
[28] *Phys. Rev.* (1950) **77**, 388.

$E = \frac{3}{2}kT = 0.037$ eV so that $r_m/R \simeq 50$, and we see that for ordinary temperatures r_m^2/R^2 is large. To obtain the relaxation time τ we use the relationship $\tau_I^{-1} = N_I\sigma_c v$ (see § 5.1). Thus we have

$$\frac{1}{\tau_I} = \frac{Z^2 e^4 N_I}{16\pi(2m_e)^{\frac{1}{2}}\epsilon^2 E^{\frac{3}{2}}} \ln\left[1 + (2E/E_m)^2\right], \tag{37}$$

where E_m, as before, is equal to minus the potential energy of an electron at a distance r_m from an impurity.

To obtain the contribution of ionized impurity scattering to the mobility we must calculate $\langle\tau_I\rangle$, which is given by

$$\langle\tau_I\rangle = \frac{16\pi(2m_e)^{\frac{1}{2}}\epsilon^2}{Z^2 e^4 N_I}\int_0^\infty E^3 \exp\left[-E/kT\right]$$

$$\times\left[\ln\{1 + (2E/E_m)^2\}\right]^{-1} dE \bigg/ \int_0^\infty E^{\frac{3}{2}} \exp\left[-E/kT\right] dE. \tag{38}$$

The logarithmic term varies slowly, and it is quite a good approximation to take it outside the integral, replacing E by its value $3kT$ at the maximum of the remainder of the integrand. In this way we obtain the equation

$$\mu_I = \frac{e}{m_e}\langle\tau_I\rangle = \frac{64\pi^{\frac{1}{2}}\epsilon^2(2kT)^{\frac{3}{2}}}{N_I Z^2 e^3 m_e^{\frac{1}{2}}}\left[\ln\left\{1 + \left(\frac{12\pi\epsilon kT}{Ze^2 N_I^{\frac{1}{3}}}\right)^2\right\}\right]^{-1}, \tag{39}$$

which is known generally as the Conwell–Weisskopf formula.[29]

Another approach to the problem of how to cut off the Coulomb field has been made by R. B. Dingle[30] and by Harvey Brooks.[31] This takes account of the screening of the Coulomb field of the ionized impurity centres by the free electrons and holes. The field of the charged centre modifies the average carrier density in its neighbourhood; let n' be the density at distance r from the centre, n being the density for large values of r. The electrostatic potential ϕ near the centre, assuming spherical symmetry, will satisfy Poisson's equation in the form

$$\frac{1}{r}\frac{d^2(r\phi)}{dr^2} = e(n' - n)/\epsilon. \tag{40}$$

[29] This differs from the formula usually quoted, due to the use of M.K.S. units, i.e. ϵ is replaced by $4\pi\epsilon$ since the Coulomb force is $Ze^2/4\pi\epsilon r^2$.

[30] *Phil. Mag.* (1955) **46**, 831.

[31] *Advances in Electronics and Electron Physics* (1955) **7**, 85. A similar treatment has been applied to impurities in metals by N. F. Mott, *Proc. Camb. Phil. Soc.* (1936) **32**, 281.

From Boltzmann's equation, assuming no degeneracy, we shall also have

$$n' = n \exp [e\phi/kT]. \tag{41}$$

Substituting in equation (40) we obtain

$$\frac{1}{r}\frac{d^2(r\phi)}{dr^2} = ne\epsilon^{-1}(\exp [e\phi/kT]-1). \tag{42}$$

We have seen that at distances of the order of r_m for which screening becomes important $e\phi \ll kT$, and if this approximation is made (42) becomes

$$\frac{d^2(r\phi)}{dr^2} = \frac{e^2 n}{\epsilon kT}(r\phi). \tag{43}$$

The appropriate solution is

$$\phi = \phi_0 \, e^{-r/d}/r,$$

where $d^2 = \epsilon kT/e^2 n$. As $r \to 0$, $\phi \to Ze/4\pi\epsilon r$ so we have

$$\phi = \frac{Ze}{4\pi\epsilon r} e^{-r/d}. \tag{44}$$

The length d is known as the Debye length. (A potential of the form given by equation (44) also gives the field near an ion in an electrolyte.)

The differential scattering cross-section $\sigma(\theta)$ for a field derived from the potential ϕ is given by[32]

$$\sigma(\theta) = \tfrac{1}{4}R^2[\sin^2 \tfrac{1}{2}\theta + \hbar^2/(2m_e vd)^2]^{-2}, \tag{45}$$

where R is defined as in equation (34). With this form of $\sigma(\theta)$ we may integrate from $\theta = 0$ to $\theta = \pi$ to obtain σ_c. We have

$$\sigma_c = 4\pi R^2 \int_0^{\pi} \frac{\sin^3 (\tfrac{1}{2}\theta)\, d[\sin (\tfrac{1}{2}\theta)]}{[\sin^2 (\tfrac{1}{2}\theta) + (\hbar^2/2m_e vd)^2]^2}$$

$$= 2\pi R^2[\ln (1+\beta^2) - \beta^2/(1+\beta^2)], \tag{46}$$

where

$$\beta^{-2} = \hbar^2/4m_e^2 v^2 d^2$$

$$= \hbar^2 e^2 n/8m_e \epsilon kTE;$$

β^2 is in general rather large and (46) may be written approximately as

$$\sigma_c = 2\pi R^2 \ln (\beta^2). \tag{46a}$$

[32] L. I. Schiff, *Quantum Mechanics* (McGraw-Hill, 2nd edition, 1955), p. 170.

The relaxation time τ_I is then given by

$$\frac{1}{\tau_I}=\frac{Z^2e^4N_I\ln(\beta^2)}{16\pi(2m_e)^{\frac{1}{2}}E^{\frac{3}{2}}\epsilon^2}, \tag{47}$$

assuming that $n=N_I$, and on averaging over E we obtain

$$\mu_I=\frac{e}{m_e}\langle\tau_I\rangle=\frac{64\pi^{\frac{1}{2}}\epsilon^2(2kT)^{\frac{3}{2}}}{N_IZ^2e^3m_e^{\frac{1}{2}}}\left[\ln\left\{\frac{24m_ek^2T^2\epsilon}{e^2\hbar^2N_I}\right\}\right]^{-1}. \tag{47a}$$

When we have both ionized donor and ionized acceptor impurities present in comparable numbers, N_I outside the square brackets must be replaced by the total number of ionized impurities weighted with the appropriate charge ($N_{ID}+N_{IA}$ if $Z=1$), while inside the brackets N_I is replaced by the total number of free carriers. Ionized impurity scattering under conditions of degeneracy has been discussed by R. Mansfield,[33] who has also dealt with the problem of combining ionized impurity scattering with lattice scattering under these conditions.

The expressions given by equations (39) and (47*a*) for μ_I differ mainly in their logarithmic factors. If we write $r'_m=R\beta$, then r'_m is the equivalent 'cut-off' distance for the second method of calculation. We have, if $E=3kT$,

$$r'^2_m=Z^2e^2m_e/24\pi^2\epsilon\hbar^2N_I$$
$$=Z^2(m_e/m)/6\pi(\epsilon/\epsilon_0)a_0N_I \quad (a_0=0.53\times10^{-8}\text{ cm}).$$

Taking $Z=1$ and $m_e=m$, $\epsilon/\epsilon_0=10$ we have if N_I is in cm^{-3}

$$r'^2_m\simeq10^6/N_I\text{ cm}^2.$$

With values of N_I of the order of 10^{15} there is thus not a great deal of difference between $\ln(r'^2_m/R^2)$ and $\ln(r^2_m/R^2)$. In general, the expression for μ_I given by equation (47*a*) is to be preferred, since it is based on a rather better theoretical foundation and may be suitably modified to take account of the fact that many of the impurities may not be ionized at low temperatures, for which the impurity scattering is most important.

When the constant-energy surfaces are not spherical it is difficult to give an exact calculation of the relaxation time τ_I for ionized impurity scattering. Since a large contribution comes from small angles of scattering we should expect that if we have ellipsoidal constant-energy surfaces the factor $(m_e)^{-\frac{1}{2}}$ in equation (47*a*) would be replaced by $(m_1^{-\frac{1}{2}})$,

[33] *Proc. Phys. Soc.* B (1956) **69**, 76.

$(m_2^{-\frac{1}{2}})$, etc., for motion in the principal directions. It is not altogether clear how to average in order to obtain the mobility in a semiconductor with cubic symmetry, but an effective mass m^* given

$$\frac{3}{(m^*)^{\frac{1}{2}}} = \frac{1}{m_L^{\frac{1}{2}}} + \frac{2}{m_T^{\frac{1}{2}}} \tag{48}$$

is not likely to be greatly in error when used in equations (39) or (47*a*). Actually a tensor form of τ_I should be used, if full account is to be taken of the anisotropy of the scattering; F. Ham[34] has suggested that τ_I should be a diagonal tensor in the same representation in which the effective-mass tensor is diagonal. That such a refinement may be necessary to interpret the effect of ionized impurity scattering in *n*-type Ge has been suggested by M. Glicksman.[35] (See § 5.3.2.)

8.10 Neutral impurity scattering

So far, we have only taken account of the effect of ionized impurities. Although neutral impurities may be expected to contribute less to the scattering than ionized impurities, their effect may not be by any means negligible, particularly at low temperatures. At such temperatures they may far outnumber the ionized impurities. As we have seen in § 3.4.1 the orbit of an electron bound to an impurity centre will extend over a large number of lattice spacings, so that it is only at quite large distances that the electron will screen the field of the impurity centre. The effect of scattering by neutral impurities has been discussed by J. Bardeen and G. L. Pearson[36] who show that the scattering may be calculated as for electrons scattered by neutral hydrogen atoms. Two processes are of importance, direct elastic scattering and exchange scattering, in which the incident electron changes places with the electron of the impurity centre. The approximate methods normally used for the calculation of atomic scattering cross-sections are not really valid under the present conditions. A treatment of collisions of slow electrons with atoms by H. S. W. Massey and B. L. Moiseiwitsch[37] has been applied to this problem by C. Erginsoy[38] who has shown that provided $E < \frac{1}{4}E_i$, where E_i is the ionization energy of the impurity, the cross-section σ_c is given by

$$\sigma_c = 20a_1/k, \tag{49}$$

[34] *Phys. Rev.* (1955) **100** 1251.
[35] *Phys. Rev.* (1957) **108**, 264.
[36] *Phys. Rev.* (1949) **75**, 865.
[37] *Phys. Rev.* (1950) **78**, 180.
[38] *Phys. Rev.* (1950) **79**, 1013.

where k is the magnitude of the wave-number of the electron being scattered and a_1 is the radius of the orbit of the outer electron in the neutral impurity centre. The relaxation time τ_N due to neutral impurity scattering is then given by

$$\frac{1}{\tau_N}=\frac{20a_1 vN_n}{k}=\frac{20a_1\hbar N_n}{m_e}, \tag{50}$$

where N_n is the number of neutral impurities per unit volume. For impurities for which the 'hydrogenic' model is applicable the quantity a_1 is given by equation (4) of § 3.4.1. In terms of the binding energy W_n of the neutral impurity, given by equation (3) of § 3.4.1 we may express τ_N in the form

$$\frac{1}{\tau_N}=\frac{5N_n\hbar e^2}{2\pi\epsilon W_n}. \tag{50a}$$

Since τ_N does not vary with the energy E the average value $\langle\tau_N\rangle$ is simply equal to τ_N and the mobility μ_N is given by $e\tau_N/m_e$. Thus we see that the neutral impurity relaxation time is independent of the temperature and also does not vary with the energy of the incident electron or hole, at least to this approximation; hence its importance at very low temperatures. In computing the quantity a_1 for semiconductors with spheroidal constant-energy surfaces the effective mass m_e is replaced by $(m_L m_T^2)^{\frac{1}{3}}$, whereas in the other factor the mobility effective mass is used.

Erginsoy's calculation of the relaxation time due to scattering by neutral impurities used only the phase-shift in the s-state scattered waves. Higher-order terms are available and these have been used by T. C. McGill and R. Baron[39] to obtain a better approximation. It is found that while over a considerable range of energies of the electron being scattered the relaxation time is nearly constant, as given by Erginsoy's formula, for values of the energy E such that $E/W_n<0.01$ this is no longer so and τ_N shows a substantial increase. This is illustrated in Fig. 8.4 in which the ratio τ_N/τ_{NE} is shown as a function of E/W_n, τ_{NE} being the constant value given by Erginsoy's formula.

Since τ_N is now a function of the energy E, to obtain μ_N we must calculate the average $\langle\tau_N\rangle$ as indicated in § 5.3.1. In this way the mobility μ_N obtained as a function of temperature T is shown in Fig. 8.5 as a ratio to the value τ_{NE} given by Erginsoy's formula and plotted against the quantity kT/W_n. It will be seen that provided $kT/W_n>0.01$ the difference from Erginsoy's value is small but if kT/W_n is of the order of

[39] *Phys. Rev.* (1975) **11**, 5208.

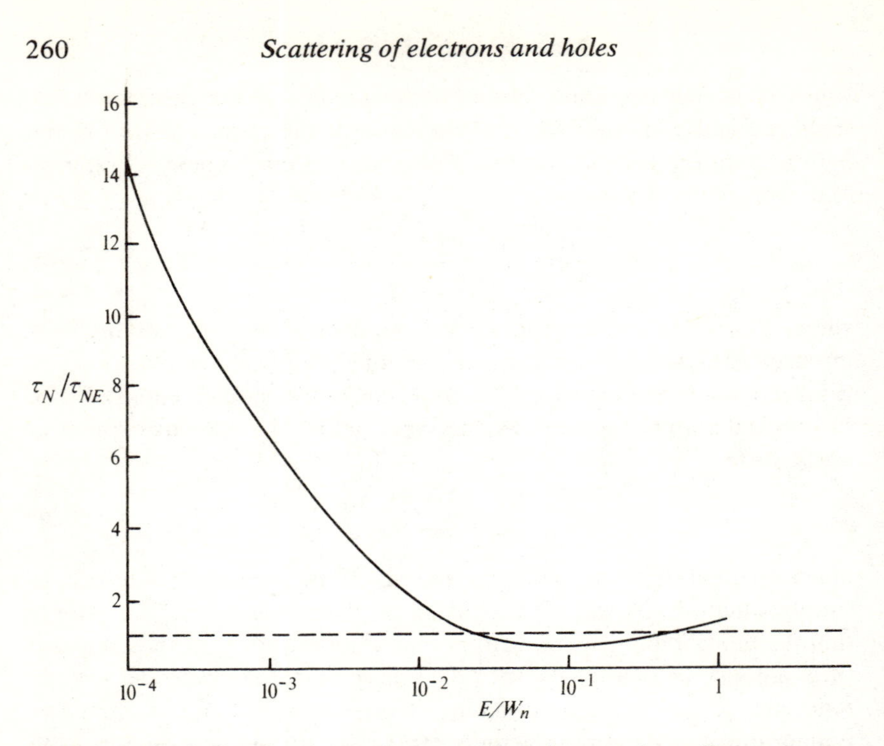

Fig. 8.4. Ratio of relaxation time τ_N for neutral impurity scattering to Erginsoy's value τ_{NE} as a function of the ratio of the energy E to the binding energy W_n of the impurity. (After T. C. McGill and R. Baron, *loc. cit.*)

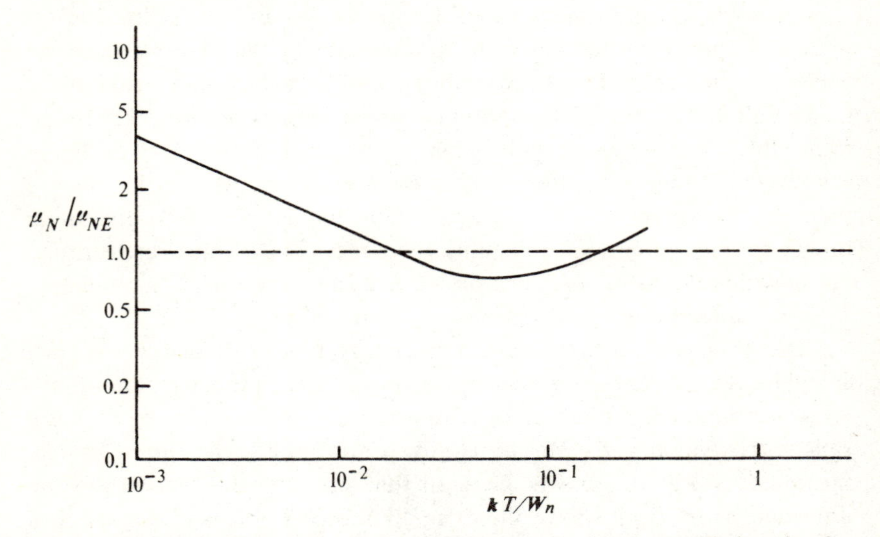

Fig. 8.5. Ratio of mobility μ_N due to neutral impurity scattering alone to Erginsoy's value μ_{NE} as a function of temperature. (After T. C. McGill and R. Baron, *loc. cit.*)

0.001 the mobility is about four times greater than Erginsoy's value. For $W_n = 0.1$ eV Erginsoy's value would hold for $T > 12$ °K, and for $W_n = 0.01$ eV for $T > 1.2$ °K. It is for the deeper lying impurities that the more elaborate calculation is required. This has been verified experimentally by R. Baron, M. H. Young and T. C. McGill[40] for various impurities Al, As, Ga and In in Si. Even when the 'hydrogenic' model does not give a good value for the binding energy the calculated value of τ_N still seems to be in quite good agreement with the experimental value.

The quantity $r = \langle\tau_N^2\rangle/\langle\tau_N\rangle^2$ which occurs in the expression for the Hall constant (see § 5.3.3) has also been calculated by Baron, Young and McGill. For values of $\boldsymbol{k}T/W_n$ above 0.01 r is only slightly in excess of unity (as would be expected if τ_N does not vary much with E) but rises to a value of 1.17 when $\boldsymbol{k}T/W_n = 0.001$.

Neutral impurity scattering has also been discussed by L. E. Blagosklonskaya, E. M. Gershenzon, Y. P. Ladyzhinskii and A. B. Popova,[41] who have shown that it can dominate the scattering at low temperatures. It is well known that scattering of electrons by atomic hydrogen is spin-dependent (Massey and Moiseiwitsch, *loc. cit.*). At very low temperatures spins of conduction electrons can be almost completely polarized in a strong magnetic field (see § 12.6) so that marked spin dependence of the mobility under such conditions is to be expected. This is discussed in some detail by I. Solomon[42] and has been demonstrated experimentally by R. Maxwell and A. Honig.[43]

8.11 Scattering by dislocations

Distortions of the crystalline lattice due to dislocations[44] will cause scattering of electrons and holes. This form of scattering has been discussed theoretically by D. L. Dexter and F. Seitz[45] and shown to make a negligible contribution unless the dislocation density is in excess of 10^8 cm^{-2}. Another mechanism whereby dislocations may cause scattering has been suggested and treated theoretically by W. T. Read,[46] in which dislocations may act as acceptor centres (see § 3.1.3). A dislocation in an n-type crystal would behave like a negative line charge

[40] *Proc. XIIIth Conf. on Phys. of Semiconductors* (Tipographia Marves, 1976), p. 1158.
[41] *Soviet Phys. Solid State* (1969) **11**, 2402.
[42] *Proc. XIIth Int. Conf. on Phys. of Semiconductors* (Polish Sci. Publ., 1972), p. 1.
[43] *Phys. Rev. Lett.* (1966) **17**, 188.
[44] See § 3.1.3. [45] *Phys. Rev.* (1952) **86**, 964.
[46] *Phil. Mag.* (1954) **45**, 775, 1119; *ibid.* (1945) **46**, 111.

and would have round it a positive space-charge (see § 7.9); this would repel incident electrons and so would cause scattering. A dislocation may thus be regarded as a cylinder of radius R, and Read has shown that, for a concentration of electrons of the order of 10^{15} cm^{-3}, R has a value of about 3×10^{-5} cm for Ge. The scattering by such cylinders should be highly anisotropic, being maximum when electrons move at right angles to the cylinders and zero for parallel motion. Dislocations may be arranged in parallel arrays by plastic bending (see § 3.1.3) and so it might be expected that anisotropic scattering should be observed in such a case.

On averaging over various directions Read shows that the scattering by dislocations may be described by means of a relaxation time τ_d given, for an electron of velocity v, by

$$\tau_d = 3/(8RNv), \quad (51)$$

where N is the number of dislocations per unit area. Taking $v = 10^7$ cm s^{-1}, $R = 3\times10^{-5}$ cm, $N = 10^6$ cm^{-2} we have $\tau_d = 1.25\times10^{-9}$ s. It will be seen that τ_d has the same velocity dependence as τ_l, the relaxation time for lattice scattering, and Read shows that, in spite of its anisotropic character, dislocation scattering may be combined in the usual way with lattice scattering, i.e. by adding the reciprocals of the relaxation times. The value of τ_d calculated above is greater than that for lattice scattering at room temperature by about 10^3 so that dislocation scattering is unlikely to be important at ordinary temperatures unless N is as high as 10^9 cm^{-2}, which would correspond to very heavily strained material; at very low temperatures scattering by dislocations might, however, become significant. There is very little experimental data on scattering by dislocations, though some experiments by G. L. Pearson, W. T. Read and F. J. Morin[47] seemed to give some evidence for it; there is now, however, some doubt as to the interpretation of the experimental data.

When the impurity concentration is high (above about 10^{19} cm^{-3} for Si and Ge) strain may be introduced into the crystal lattice resulting in dislocations. Departures from the mobility given by the equation (47*a*) at such concentrations have been attributed to dislocations by P. O. Daga and W. S. Khokle.[48]

[47] *Phys. Rev.* (1954) **93**, 666.
[48] *J. Phys. C: Solid State Phys.* (1971) **4**, 190; *ibid.* (1972) **5**, 3473.

8.12 Scattering contributions to mobility

Determining the contributions to the mobility of the different types of scattering for various temperature ranges for any particular semiconductor is a very complex process. The relative contributions vary a great deal from one material to another, so it is impossible to give any very general result. Broadly speaking, however, one may say that in the higher-temperature range lattice-vibration scattering of various kinds makes the major contribution and the purer the material the lower the temperature at which this will apply. For polar semiconductors optical-mode scattering will be very important. Impurity scattering varies naturally a great deal with impurity content, and dislocation scattering with state of strain. At very low temperatures scattering due to imperfections clearly predominates and neutral impurity scattering will be very important. We defer further detailed discussion of particular materials to Chapter 13.

9

Recombination of electrons and holes

9.1 Recombination mechanisms

Having discussed in Chapter 8 the various mechanisms which determine the relaxation times τ_e, τ_h we must now turn to a similar discussion of the processes by means of which electrons and holes recombine so as to re-establish equilibrium when this has been disturbed. These may be divided into three main categories: (*a*) direct recombination, (*b*) recombination through traps, (*c*) surface recombination. In the first of these, the electrons and holes combine through an electron dropping from a state in the conduction band to an empty state in the valence band in a single transition. An amount of energy in excess of ΔE must be disposed of by some means, and the simplest process by which this can be done is by radiation. This radiation may or may not also involve the emission or absorption of phonons.

Instead of making a direct recombination with a hole an electron may first of all join up with the hole to form an exciton (see § 3.5). The exciton may then decay by radiation, the frequency of this corresponding to an energy slightly *less* than ΔE. For free excitons this contribution will be taken account of, as we shall see, by including the exciton decay radiation along with the band-to-band radiation when calculating the probability of a radiative transition taking place. If, however, the exciton is trapped it may stay quite a long time immobile and will not contribute to the radiative lifetime (except in so far as the trapping mechanism removes electrons and holes in pairs). In this case the emitted radiation is frequently in the form of a series of very sharp lines and may be clearly separated from the continuous spectrum of the band-to-band radiation (see § 10.14).

Another means of disposing of the energy E in a single event is to transfer it to another electron in the conduction band or hole in the valence band thus creating a highly energetic charge carrier which will soon, however, share its energy with other carriers in the band. This process is known as Auger recombination since it is similar in principle to the well-known Auger effect whereby an atom may make a transition to a lower state, the energy being emitted in the form of an electron rather than as radiation.[1] This type of recombination was once thought to be too unlikely to be of importance in semiconductors but it is now known that this is not so and we shall consider it after radiative recombination.

A third process whereby the energy ΔE may be disposed of is for it to be transferred to a number of phonons. Even the most energetic phonons – the optical phonons – have energies of the order of 0.1 eV so that about ten would be required for $\Delta E = 1$ eV. The emission of so many phonons in a band-to-band transition is thought to be very unlikely, but if the process is carried out via electron or hole traps multi-phonon processes may be important.

A second type of recombination mechanism is one in which an electron makes the jump from the conduction band to the valence band in two steps, which may be separated in time by a relatively large amount. (Such a process should be clearly distinguished from the double transitions considered later in connection with indirect transitions; these follow each other so rapidly that the energy of the intermediate state is not clearly defined.) For example, an electron may be captured into a deep trap whose energy lies near the middle of the forbidden energy gap; subsequently, the occupied trap may capture a hole and the recombination process will then be completed. As we shall see, the probability of a radiative transition decreases rapidly as the energy radiated is increased, so that, by making the transition in two steps, the energy to be disposed of at each step is considerably reduced and this type of process may be much more probable than the single-step process. The third process listed above is very similar to the second, the traps being now at the surface of the semiconductor. In this situation we also see that fewer phonons would be required for a non-radiative process in which the energy is transferred to the crystal lattice vibrations, and so would be more probable than a direct multi-phonon process.

[1] See for example, D. Chattarji, *The Theory of Auger Transitions* (Academic Press, 1976).

9.2 Radiative recombination

We shall first consider radiative recombination. We shall not require to separate direct and indirect transitions (see §§ 10.5.1 and 10.5.3) in order to determine the rate of recombination, as this depends only on the total absorption coefficient due to band-to-band transitions. Let $\alpha(\nu)$ be the absorption coefficient at frequency ν; then the mean free path of a photon of frequency ν before producing a band-to-band transition is simply α^{-1}. The mean lifetime $\tau(\nu)$ of such a photon is then equal to $(\alpha V_g)^{-1}$, where V_g is the velocity of travel of the photon through the medium; this is clearly the group velocity, i.e. the velocity with which a group of photons travels through the medium. V_g is equal to $d\nu/d(1/\lambda) = c\, d\nu/d(\mathrm{n}\nu)$, where n is the refractive index, and we therefore have

$$\frac{1}{\tau(\nu)} = \frac{\alpha c\, d\nu}{d(\mathrm{n}\nu)}. \tag{1}$$

In terms of the absorption index k, we have $\alpha = 4\pi \mathrm{nk}/c$ (see § 10.1) so that

$$\frac{1}{\tau(\nu)} = \frac{4\pi\nu \mathrm{nk}\, d\nu}{d(\mathrm{n}\nu)}. \tag{2}$$

In order to obtain the generation rate in thermal equilibrium, for the frequency range ν to $\nu + d\nu$, we must multiply $1/\tau(\nu)$ by the number of photons having frequencies in this range; this number is given by the well-known Planck formula, which is usually expressed in terms of wavelength, for in this case it does not depend on the refractive index. The number of photons $N'(\lambda)\, d\lambda$ with wavelengths in the range λ to $\lambda + d\lambda$ is given by the equation

$$N'(\lambda)\, d\lambda = \frac{8\pi}{\lambda^4} \frac{d\lambda}{\exp(h\nu/kT) - 1} \tag{3}$$

from which we may readily obtain $N(\nu)\, d\nu$, the number having frequencies in the interval ν to $\nu + d\nu$. We have $\lambda = c/\mathrm{n}\nu$, so that

$$-d\lambda = \frac{c}{\mathrm{n}^2\nu^2} \frac{d(\mathrm{n}\nu)}{d\nu}\, d\nu \tag{4}$$

and on inserting this value in equation (3) we have

$$N(\nu)\, d\nu = \frac{8\pi\nu^2 \left[\mathrm{n}^2 \dfrac{d(\mathrm{n}\nu)}{d\nu}\right] d\nu}{c^3 [\exp(h\nu/kT) - 1]}. \tag{5}$$

The recombination rate in thermal equilibrium is equal to the generation rate for each interval $d\nu$, and if we denote this by $\mathscr{R}(\nu)\,d\nu$, we have

$$\mathscr{R}(\nu)\,d\nu = \frac{N(\nu)}{\tau(\nu)}\,d\nu$$

$$= \frac{32\pi^2 \mathrm{n}^3 \nu^3 \mathrm{k}\,d\nu}{\boldsymbol{c}^3[\exp(\boldsymbol{h}\nu/\boldsymbol{k}T)-1]}. \tag{6}$$

This expression gives the spectral density of radiation which should be observed in emission.

To obtain the total recombination rate $\mathscr{R}$ we must integrate over all values of ν. In this way we obtain the value

$$\mathscr{R} = \frac{32\pi^2 \boldsymbol{c}^3 \boldsymbol{k}^4 T^4}{\boldsymbol{h}^4} \int_0^\infty \frac{\mathrm{k}\mathrm{n}^3 x^3\,dx}{e^x - 1}, \tag{7}$$

where $x = \boldsymbol{h}\nu/\boldsymbol{k}T$; the factor outside the integral may be expressed numerically in the form $1.785\times 10^{28}(T/300)^4\ \mathrm{m}^{-3}\ \mathrm{s}^{-1}$. This formula was obtained by W. van Roosbroeck and W. Shockley[2] who have used it to calculate the radiative recombination rate for Ge. The integral in equation (7) is appreciable only over a fairly narrow range of frequencies near the fundamental absorption edge because of the rapid increase in e^x as ν increases, and rapid decrease in k for values of ν smaller than that corresponding to the absorption edge (see § 10.5). Thus we see that the emission corresponding to direct recombination will be in a fairly narrow band of wavelengths near that corresponding to the forbidden energy gap.

We may now calculate the minority-carrier lifetime. Since the non-equilibrium recombination rate $\mathscr{R}'$ is proportional to p and to n we may write $\mathscr{R}' = Apn$; for thermal equilibrium under non-degenerate conditions we also have $pn = n_i^2$ so that $A = \mathscr{R}/n_i^2$. If we write $p = p_0 + \Delta p$, $n = n_0 + \Delta p$ we have when $p_0 \ll n_0$ and $\Delta p \ll n_0$, by definition of τ_p,

$$\mathscr{R}' - \mathscr{R} = \frac{\Delta p}{\tau_p} = \frac{\mathscr{R}}{n_i^2}(pn - n_i^2)$$

$$= \frac{\mathscr{R} n_0 \Delta p}{n_i^2}, \tag{8}$$

so that we have

$$\frac{1}{\tau_p} = \frac{\mathscr{R}}{p_0}. \tag{9}$$

[2] *Phys. Rev.* (1954) **94**, 1558.

For intrinsic material, when $\Delta p = \Delta n \ll n_i$, we have

$$\frac{1}{\tau_i} = \frac{2\mathcal{R}}{n_i}, \tag{10}$$

so that
$$\tau_p/\tau_i = 2p_0/n_i. \tag{11}$$

The value found by van Roosbroeck and Shockley for the recombination rate in Ge is $1.57 \times 10^{19}\ \mathrm{m^{-3}\ s^{-1}}$, for $T = 300\ °\mathrm{K}$.

The value of τ_n for near-intrinsic Te (so that $\tau_n \simeq \tau_i$) as a function of temperature is shown in Fig. 9.1 as calculated by J. S. Blakemore[3] and

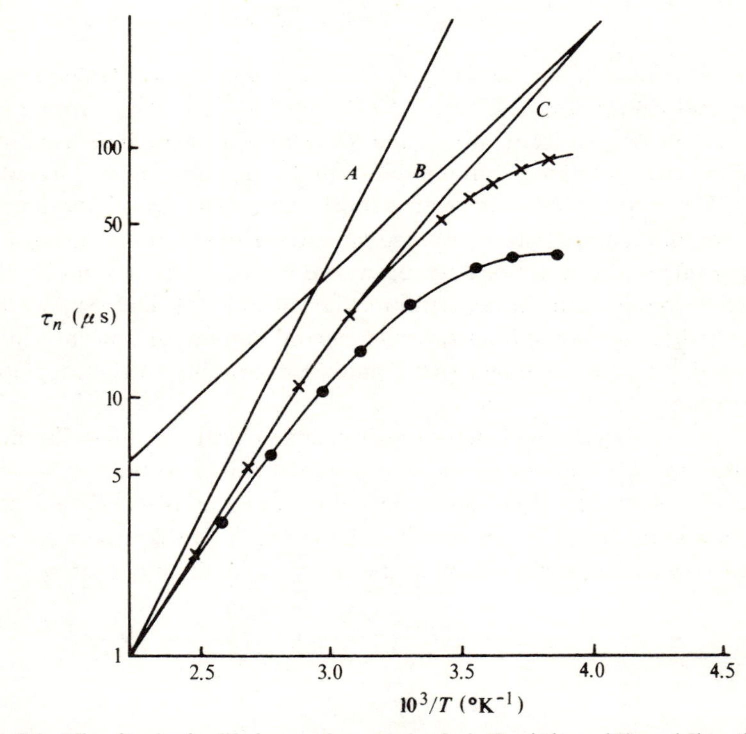

Fig. 9.1. Minority-carrier lifetime τ_n for near-intrinsic Te. (*A*) τ_R, (*B*) τ_a, (*C*) τ_n ($\tau_n^{-1} = \tau_R^{-1} + \tau_a^{-1}$). × experimental curve for near-intrinsic sample, ● experimental curve for less pure sample. (After J. S. Blakemore, *loc. cit.*)

[3] *Proc. Vth Int. Conf. on Phys. of Semiconductors* (Czech. Sci. Publ., 1961), p. 981.

compared with experimental observations. It will be seen that τ_i decreases as the temperature is lowered (mainly because of the factor n_i). At temperatures of the order of 400 °K the radiative process accounts for most of the recombination, τ_n having a value of about 2 μs at 400 °K.

The value of the recombination rate $\mathcal{R}$ increases rapidly as ΔE decreases and also increases with temperature. It would appear that this kind of recombination is of most importance for semiconductors with small energy gaps, at room temperature and above. At lower temperatures much higher rates will be produced by other processes such as Auger recombination (see § 9.3) and recombination through trapping of electrons and holes at imperfections.

The room-temperature value of $\mathcal{R}$ for Ge has been measured by Patricia H. Brill and Ruth F. Schwarz[4] and found to be $2.5 \times 10^{19}\ \mathrm{m}^{-3}\ \mathrm{s}^{-1}$. These values lead for intrinsic Ge to values of τ_i of the order of 1 s, a value much greater than τ_n or τ_p observed. It is however, important to know the value of $\mathcal{R}$ for a semiconductor in order to assess the probability of emission of radiation from it even though other processes predominate in determining the minority-carrier lifetime.

Even though the light emission process is relatively inefficient as a means of causing electrons and holes to recombine it can be of very great importance technologically in the provision of light sources. For this reason the radiative recombination for materials such as GaAs has been extensively studied (see § 10.14). An extensive review of radiative recombination in a number of element and compound semiconductors, including several of the group III–V compounds has been given by V. P. Varshni.[5]

Recombination of electrons with donor centres whose energy levels lie near the conduction band will be of importance at very low temperatures, and similarly for holes we may have recombination with acceptor centres whose energy levels lie near the valence band.

From the above analysis it will be seen that the radiative recombination rate depends on the *total* contribution of band-to-band transitions at any given frequency ν. The relative magnitudes of the contributions due to direct and indirect transitions can be obtained only from a knowledge of the transition probabilities.

[4] *Proc. IVth Int. Conf. on Phys. of Semiconductors, J. Phys. Chem. Solids* (1959) **8**, 75.
[5] *Phys. Stat. Sol.* (1967) **19**, 459; *ibid.* (1967) **20**, 9.

9.3 Auger recombination

The calculations of the Auger recombination rate involves a detailed quantum-mechanical treatment of the collision process described in § 9.1. This has been carried out by P. T. Landsberg and A. R. Beattie[6] who find for intrinsic material the expression

$$\tau_{ai} = \frac{A\epsilon^2 \boldsymbol{m}(m_v + 2m_c)(m_c + m_v)^{\frac{1}{2}}}{\epsilon_0^2 m_c m_v^{\frac{3}{2}}} \left(\frac{\Delta E}{kT}\right)^{\frac{3}{2}} \exp\left[\frac{\Delta E}{kT}\left(\frac{2m_c + m_v}{m_c + m_v}\right)\right] \tag{12}$$

where as usual ϵ/ϵ_0 is the dielectric constant and A is a constant approximately equal to $4 \times 10^{-16}\,\mathrm{s}^{-1}$. It will be seen that τ_{ai} increases rapidly as the temperature T becomes small and, like radiative recombination, Auger recombination is most important at high temperatures. It increases less rapidly than does radiative recombination. This has been illustrated well by J. S. Blakemore (*loc. cit.*) with calculations and experiments on near-intrinsic Te as shown in Fig. 9.1. It will be seen that at the lower temperatures the experimental value of τ_a falls well below that given by radiative recombination τ_R but agrees well with that obtained by combining Auger and radiative recombination. The combined lifetime τ_n is given by

$$\frac{1}{\tau_n} = \frac{1}{\tau_R} + \frac{1}{\tau_a} \tag{13}$$

and except at very low temperatures the experimental values for near-intrinsic material agree well with this combination. For a less pure sample the value of τ_n falls off much more rapidly at lower temperature and reaches a nearly constant value dependent on the impurity concentration.

The Auger recombination will, in thermal equilibrium, be exactly balanced by the converse processes of generation of electron–hole pairs by fast electrons and fast holes. Let G_e and G_h be the generation rates for these two processes; then we must have in equilibrium

$$G_e = R_{ae}, \quad G_h = R_{ah} \tag{14}$$

where R_{ae} and R_{ah} are the recombination rates for the two processes in which the energy is given to a fast electron and alternatively to a fast

[6] *J. Phys. Chem. Solids* (1959) **8**, 73; *Proc. Roy. Soc.* A (1959) **249**, 16; *ibid.* (1960) **258**, 486.

hole. We have then

$$R_a = R_{ae} + R_{ah} = G_{ae} + G_{ah}. \tag{14a}$$

For non-equilibrium conditions let the recombination rates be R'_{ae} and R'_{ah}. For non-degenerate conditions we have R'_{ae} proportional to n^2, since two electrons are involved, and to p so that if n_0 and p_0 are the electron and hole equilibrium concentrations

$$R'_{ae} = n^2 p R_{ae}/n_0^2 p_0. \tag{15}$$

The generation rate G'_{ae} on the other hand is proportional to n since it depends only on the number of fast electrons present so that

$$G'_{ae} = G_{ae} n/n_0 = R_{ae} n/n_0. \tag{16}$$

In non-equilibrium conditions the net recombination rate is given by

$$R'_{ae} - G'_{ae} = G_{ae}\left(\frac{np}{n_0 p_0} - 1\right)\frac{n}{n_0} \tag{17}$$

with a similar expression for holes. Thus we have for the relaxation time τ_a

$$\frac{1}{\tau_a} = \frac{n(n+p_0)G_{ae}}{n_0^2 p_0} + \frac{p(p+n_0)G_{ah}}{p_0^2 n_0}. \tag{18}$$

It turns out that one or other of the terms in (18) generally predominates.

In fact if $m_e \ll m_h$ the first term is much larger. In this case we may write

$$\frac{1}{\tau_a} = \frac{n(n+p_0)G_{ae}}{n_0 n_i^2}. \tag{19}$$

If $\Delta n = \Delta p$ is small

$$\frac{1}{\tau_a} = \frac{(n_0+p_0)}{n_i^2} G_{ae}. \tag{20}$$

Since clearly G_{ae} is proportional to n_0 we may write

$$G_{ae} = n_0 g_{ae} \tag{21}$$

and

$$\frac{1}{\tau_a} = \frac{n_0(n_0+p_0)}{n_i^2} g_{ae}. \tag{22}$$

For intrinsic material under these circumstances we have

$$\frac{1}{\tau_i} = 2g_{ae} \tag{23}$$

and we may write
$$\tau_i/\tau_a = \frac{n_0(n_0+p_0)}{2n_i^2}. \tag{24}$$

This gives the variation of τ_a with 'doping' of the semiconductor.

More generally we may write

$$\frac{\tau_i}{\tau_a} = \frac{(n+p_0)(n+cp)}{2n_i^2} \tag{25}$$

where
$$c = g_{ah}/g_{ae} \tag{26}$$

and
$$G_{ah} = p_0 g_{ah}. \tag{27}$$

The expression (25) clearly reduces to (24) when $cp \ll n$. If $c \ll 1$, as is usual, the simpler equation holds unless $p \gg n$, i.e. for strongly p-type material.

When $\Delta n = \Delta p$ is much smaller than n or p equation (25) may be written in the simpler form

$$\frac{\tau_i}{\tau_a} = \frac{n_0^2}{2n_i^2} + \tfrac{1}{2}(1+c) + \frac{cn_i^2}{2n_0^2} \tag{28}$$

$$= \frac{n_i^2}{2p_0^2} + \tfrac{1}{2}(1+c) + \frac{cp_0^2}{2n_i^2}. \tag{29}$$

It will be seen that τ_a passes through a maximum value $(1+c)\tau_i$ when $n_0 = n_i$ and falls off as n_0 increases beyond this value and also as p_0 increases beyond the value n_i. The rate of fall-off is however much less rapid with p-type material if $c \ll 1$.

A detailed discussion of the variation of τ_a with carrier concentration and with temperature has been given by J. S. Blakemore.[7]

When there exists a split-off valence band due to coupling of the spin with the hole motion (see § 2.3) and the 'spin-orbit' splitting ΔE_s is nearly equal to the forbidden-energy gap ΔE, the energetic hole created in an Auger recombination may be transferred to the split-off band with very little change in crystal momentum. In this circumstance the probability of an Auger recombination is very high. This situation exists for example in GaSb for which G. Benz and R. Conradt[8] have studied this process. A considerable number of the Auger holes in the split-off band appear, in this case, to return to combine with electrons in shallow neutral donors at low temperatures by a radiative process. The recombination in a sense therefore might be regarded as a combined Auger–recombination process.

[7] *Semiconductor Statistics* (Pergamon Press, 1962).
[8] *Proc. XIIth Conf. on Phys. of Semiconductors* (Teubner, 1974), p. 1262.

If we have a heavily doped semiconductor in which the electron or hole concentration is degenerate the relationship given by equation (15) in which the recombination rate is proportional to n^2p no longer applies. This situation has been considered by A. Haug[9] who has shown that under certain conditions a variation of the form $\mathscr{R}'_{ae} \propto n^{\frac{5}{3}}p$ should be expected.

The subject of band-to-band Auger recombination has been fully discussed by P. T. Landsberg[10] as part of a review of all forms of non-radiative recombination processes.

9.4 Recombination through traps

The theory of recombination through traps has been given by W. Shockley and W. T. Read;[11] the following is a simplified version of their treatment. We first of all calculate the rate of trapping of electrons from the conduction band; this will be proportional to n and to $(1-f_t)$ where f_t is the fraction of the traps occupied by electrons, and we may write the capture rate per unit volume $\mathscr{R}_c$ in the form

$$\mathscr{R}_c = C_n n(1-f_t), \tag{30}$$

where C_n gives the capture rate per electron when all the traps are empty. C_n will be proportional to the trap concentration N_t. The rate of emission $\mathscr{R}_e$ from the traps will be proportional to f_t and may be written in the form

$$\mathscr{R}_e = C'_n f_t. \tag{31}$$

In equilibrium we have $\mathscr{R}_c = \mathscr{R}_e$ so that

$$C'_n = C_n n_0(1-f_{t0})/f_{t0},$$

where n_0 and f_{t0} are the equilibrium values of n and f_t; f_{t0} is obtained from the Fermi function and

$$(1-f_{t0})/f_{t0} = \exp\left[(E_t - E_F)/\boldsymbol{k}T\right], \tag{32}$$

where E_t is the energy of the trapping level. (We shall consider a single level, for the moment.) Also we have for a non-degenerate condition[12]

$$n_0 = N_c \exp(E_F/\boldsymbol{k}T) \tag{33}$$

[9] *Proc. XIIIth Int. Conf. on Phys. of Semiconductors* (Tipographia Marves, 1976), p. 1106.

[10] *Phys. Stat. Sol.* (1970) **41**, 457.

[11] *Phys. Rev.* (1952) **87**, 835.

[12] The effect of degeneracy has been discussed by P. T. Landsberg, *Proc. Phys. Soc.* B (1957) **70**, 282.

(§ 4.3, equation (29)), so that

$$C'_n = C_n N_c \exp(E_t/\boldsymbol{k}T) = n_1 C_n, \tag{34}$$

where n_1 is the number of electrons in the conduction band when the Fermi level coincides with the trapping level. We may therefore write the net trapping rate $\mathscr{R}_n$ for electrons in the form

$$\mathscr{R}_n = C_n[(1-f_t)n - n_1 f_t]. \tag{35}$$

In exactly the same way, we may obtain the net trapping rate $\mathscr{R}_p$ for holes, which may be written in the form

$$\mathscr{R}_p = C_p[f_t p - p_1(1-f_t)]. \tag{36}$$

If we now have a disturbing influence creating electron–hole pairs at a constant rate $\mathscr{R}$, when steady conditions have been established, electrons and holes will be trapped at the same rate, since they recombine in pairs, and we have $\mathscr{R}_n = \mathscr{R}_p = \mathscr{R}$. This gives us an equation for the fraction f_t of traps which are occupied. (This will, in general, differ from the equilibrium value f_{t_0}.) We have, on equating $\mathscr{R}_n$ and $\mathscr{R}_p$,

$$f_t = \frac{nC_n + p_1 C_p}{C_n(n+n_1) + C_p(p+p_1)}. \tag{37}$$

The recombination rate $\mathscr{R}$ is therefore given by the equation

$$\mathscr{R} = \frac{C_n C_p(pn - n_i^2)}{C_n(n+n_1) + C_p(p+p_1)}, \tag{38}$$

since we have $n_1 p_1 = n_i^2$. If we write $n = n_0 + \Delta p$, $p = p_0 + \Delta p$ we have

$$\mathscr{R} = \frac{n_0 \Delta p + p_0 \Delta p + \Delta p^2}{C_p^{-1}(n_0 + n_1 + \Delta p) + C_n^{-1}(p_0 + p_1 + \Delta p)} \tag{39}$$

assuming that $\Delta n = \Delta p$; also, if we define τ by means of the equation $\mathscr{R} = \Delta p/\tau$, we have

$$\frac{1}{\tau} = \frac{n_0 + p_0 + \Delta p}{C_p^{-1}(n_0 + n_1 + \Delta p) + C_n^{-1}(p_0 + p_1 + \Delta p)}. \tag{40}$$

When Δp is small, $n_0 \gg p_0$, and also $n_0 \gg n_1$, we have

$$\frac{1}{\tau} = \frac{1}{\tau_{p_0}} = C_p. \tag{41}$$

Similarly, if $p_0 \gg n_0$ and $p_0 \gg p_1$ we have

$$\frac{1}{\tau} = \frac{1}{\tau_{n_0}} = C_n. \tag{42}$$

We may therefore write equation (40) in the form

$$\frac{1}{\tau} = \frac{n_0 + p_0 + \Delta p}{\tau_{p_0}(n_0 + n_1 + \Delta p) + \tau_{n_0}(p_0 + p_1 + \Delta p)}, \tag{43}$$

and when Δp is small we have

$$\tau = \tau_{p_0}\frac{n_0 + n_1}{n_0 + p_0} + \tau_{n_0}\frac{p_0 + p_1}{n_0 + p_0}. \tag{44}$$

For highly extrinsic n-type and p-type material τ_p and τ_n are constant, i.e. they do not vary with carrier concentration; this is in contrast to radiative recombination for which τ_p is proportional to p_0^{-1}.

The variation of lifetime with carrier concentration as given by equation (44) is shown in Fig. 9.2(a) for small values of Δp. The value of τ passes through a maximum at a value of n_0 which depends on n_1. If $\tau_{n_0} = \tau_{p_0}$ the maximum occurs when $n_0 = p_0 = n_i$, and we have

$$\tau = \tau_{p_0}\left(1 + \frac{n_1 + p_1}{n_0 + p_0}\right). \tag{45}$$

When $n = n_i$

$$\tau_i = \tau_{p_0}\left(1 + \frac{n_1 + p_1}{2n_i}\right), \tag{46}$$

$$= \tau_{p_0}\{1 + \cosh\left[(E_t - E_i)/\boldsymbol{k}T\right]\}, \tag{47}$$

where E_i is the Fermi level for intrinsic material; thus we see that if $(E_t - E_i) \gg \boldsymbol{k}T$, $\tau_i \gg \tau_{p_0}$, and when $E_t = E_i$ we have $\tau_i = 2\tau_{p_0}$.

The variation of τ with carrier concentration for both n-type and p-type samples of Ge has been investigated by J. A. Burton, G. W. Hull, F. J. Morin and J. C. Severiens,[13] and excellent agreement found with equation (44). Variation of n and p were obtained by doping the Ge crystals with Sb and In, while trap concentrations were varied by doping with Cu and Ni, which produce deep-lying recombination centres with energy levels near the middle of the forbidden band; in this case it was found that $\tau_{p_0}/\tau_{n_0} = 0.1$.

[13] *J. Phys. Chem.* (1953) **57**, 853.

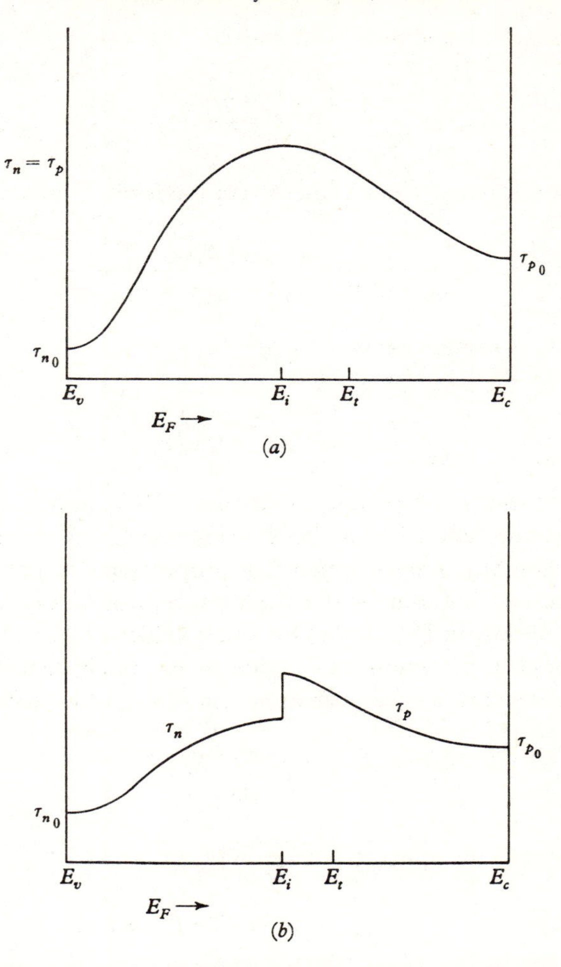

Fig. 9.2. Variation with carrier concentration of lifetime due to recombination through traps.

When N_t is not small compared with n_0 or p_0 we are not justified in assuming that $\Delta n = \Delta p$; we have actually

$$\Delta n = \Delta p - N_t \Delta f_t. \tag{48}$$

Moreover, we do not then have $\tau_n = \tau_p$, we have

$$\mathscr{R} = \frac{\Delta p}{\tau_p} = \frac{\Delta n}{\tau_n}. \tag{49}$$

When $n > p$ we are generally concerned with τ_p and when $n < p$ with τ_n. We shall now consider only small values of Δn and Δp, but shall not neglect Δf_t. From the equality of $\mathscr{R}_n$ and $\mathscr{R}_p$ we have (equations (35), (36))

$$C_n[(1-f_t)\Delta n - (n_0+n_1)\Delta f_t] = C_p[f_t \Delta p + (p_0+p_1)\Delta f_t],$$

so that

$$\Delta f = \frac{\tau_{p_0}(1-f_t)\Delta n + \tau_{n_0} f_t \Delta p}{\tau_{n_0}(p_0+p_1) + \tau_{p_0}(n_0+n_1)}. \tag{50}$$

We may use to this order of approximation the equilibrium value for f_t, namely, $f_t = n_0/(n_0+n_1) = 1 - p_0/(p_0+p_1)$ giving

$$\Delta f_t = \frac{\tau_{p_0}(n_0+n_1)p_0\Delta n + \tau_{n_0}(n_0+n_1)n_0\Delta p}{(n_0+n_1)(p_0+p_1)[\tau_{n_0}(p_0+p_1) + \tau_{p_0}(n_0+n_1)]}. \tag{51}$$

We therefore see that $\Delta f_t \ll \Delta n$ or Δp if $N_t \ll (n_0+n_1)$ or (p_0+p_1). Substituting for Δn and Δf_t in terms of Δp in equation (38) we obtain

$$\tau_p = \frac{\tau_{n_0}(p_0+p_1) + \tau_{p_0}[(n_0+n_1) + N_t n_1/(n_1+n_0)]}{n_0 + p_0 + N_t n_1 n_0/(n_1+n_0)^2}, \tag{52}$$

$$\tau_n = \frac{\tau_{p_0}(n_0+n_1) + \tau_{n_0}[(p_0+p_1) + N_t p_1/(p_1+p_0)]}{n_0 + p_0 + N_t p_1 p_0/(p_1+p_0)^2}. \tag{53}$$

When n_0 or $p_0 \to n_i$ we no longer have $\tau_p \to \tau_n$; if we plot τ_p for $n > n_i$ and τ_n for $n < n_i$ we obtain a curve of the form shown in Fig. 9.2(*b*). Again the maximum values of τ_n and τ_p occur near, but not necessarily for, the intrinsic condition.

The lifetimes given by equations (52) and (53) are derived from steady-state conditions. When the terms containing N_t are of importance it does not follow that the same lifetime will be obtained for transient conditions. When the number of electrons in the traps is changing, the rate of change of Δn is not necessarily equal to the rate of change of Δp. This effect may be of importance in determining the rate of decay of photo-conductivity induced by light, after the light has been switched off, and will be discussed in § 10.9.2. In the above calculation the quantities C_n, C_p and their related lifetimes τ_{n_0}, τ_{p_0} have been introduced as unknown parameters. Various attempts to calculate them have been made with limited success. We shall discuss methods used for making *ab initio* estimates of their magnitude. However, the various formulae which we have given enable them to be estimated from experimental studies of the variation of τ_n and τ_p with initial carrier concentration and with strength of external excitation. A number of methods for obtaining such estimates have been described by A. Many and R. Bray.[14]

Various processes have been suggested whereby an electron can lose a substantial amount of energy in falling from the conduction band into a trapping centre. Similarly, energy can be lost through a hole rising from the valence band to a trapping centre. If phonons are involved, as we have already seen, a considerable number are required. A proposal by M. Lax[15] gets over this difficulty by assuming that the electron (or hole) is captured into a highly excited state of the impurity centre forming the trap. This can be done with only a small amount of energy to be disposed of and so need only involve a single acoustic phonon. After being captured, the electron cascades down through the excited states of the centre, each step down requiring the emission of only a single phonon. Lax's calculation was primarily 'classical' and has been elaborated by R. A. Brown and S. Rodriguez[16] using a quantum-theoretical treatment and taking account of band-structure effects. They compare the calculated values with experimental results for 'hydrogenic' traps in Si and Ge at low temperatures and while some agreement is found it is not very good.

Some other mechanism would appear also to be operating, and it has been suggested by I. V. Karpova and S. G. Kalashnikov[17] that this might be an Auger recombination in which an electron is trapped rather than falling into the valence band. This process has been treated theoretically by V. L. Bonch-Bruevich and Y. V. Gulyaev[18] and by P. T. Landsberg, D. A. Evans and C. Rhys-Roberts.[19] It explains certain features of the variation of τ_n and τ_p for Ge containing recombination centres found by Karpova and Kalashnikov (*loc. cit.*), in particular the rapid fall-off of the value of both of these for carrier concentrations in excess of 10^{17} cm^{-3}.

Various more complex Auger processes involving traps have been suggested and examined theoretically by a number of authors including M. K. Sheinkmann[20] and have been discussed by P. T. Landsberg[21] in a review article on recombination processes.

Another proposal as to how the excess energy might be disposed of to the crystal lattice involves consideration of the effect of a displacement

[14] *Progress in Semiconductors* (Heywood, 1958) **3**, 117.

[15] *Proc. IVth Int. Conf. on Phys. of Semiconductors, J. Phys. Chem. Solids* (1959) **8**, 66.

[16] *Phys. Rev.* (1967) **153**, 890.

[17] *Proc. VIth Int. Conf. on Phys. of Semiconductors* (Inst. of Phys. & Phys. Soc., 1962), p. 880.

[18] *Soviet Phys. Solid State* (1960) **2**, 431.

[19] *Proc. Phys. Soc.* (1964) **83**, 325.

[20] *Soviet Phys. Solid State* (1964) **5**, 2035; *ibid.* (1965) **7**, 18.

[21] *Phys. Stat. Sol.* (1970) **41**, 457.

of a recombination centre from its position of equilibrium relative to its nearest neighbours. This will, in general, be different when it is or is not occupied by a hole or an electron. A situation may then arise in which the electron energy plotted as a function of displacement may, as in a molecule, have two energy curves which try to overlap. At the point where they cross (or approach each other closely) a transition without radiation may take place from one to the other. If the crossing position differs from the equilibrium position when occupied, the transition will set up a strong vibration as the lattice reverts to its equilibrium position when the centre is unoccupied. This would be equivalent to the emission of a number of phonons.

This idea has been expressed in quantum-mechanical terms by R. Englman and J. Jortner[22] and has been further developed by N. F. Mott, A. E. Davis and R. Street[23] to calculate the non-radiative recombination from centres, particularly for amorphous semiconductors. A similar method of calculation has also been applied by M. Jaros and S. Brand[24] who have used it to account for the recombination at deep levels in GaP and GaAs.

The variation of τ_n and τ_p with carrier concentration exhibits a great variety of forms but generally there is a 'plateau' region over which the value is constant, rising to larger values for low concentrations and falling off rapidly at concentrations above a certain critical value (10^{17} cm^{-3} for Ge). For example, the variation τ_n for B-doped samples of Ge containing Cu recombination centres is shown in Fig. 9.3 as obtained by S. G. Kalashnikov[25] who has also discussed in some detail the values of the various rate constants for Si and Ge containing a great variety of trapping centres.

9.5 Recombination through excitons

In studies of the radiation in the near infra-red from semiconductors it became clear that in addition to band-to-band transitions those involving excitons made up a large part of the radiation (see § 10.14). This indicates that *before recombining* with a hole an electron frequently becomes bound to the hole and the radiation is emitted in the course of exciton decay. Not only is radiation emitted from free excitons but also

[22] *Molec. Phys.* (1970) **18**, 145. [23] *Phil. Mag.* (1975) **32**, 961.
[24] *Proc. XIIIth Int. Conf. on Phys. of Semiconductors* (Tipographia Marves, 1976), p. 1090.
[25] *Proc. IVth Int. Conf. on Phys. of Semiconductors, J. Phys. Chem. Solids* (1959) **8**, 52; *Proc. VIth Int. Conf. on Phys. of Semiconductors* (Czech. Acad. Sci., 1961), p. 241.

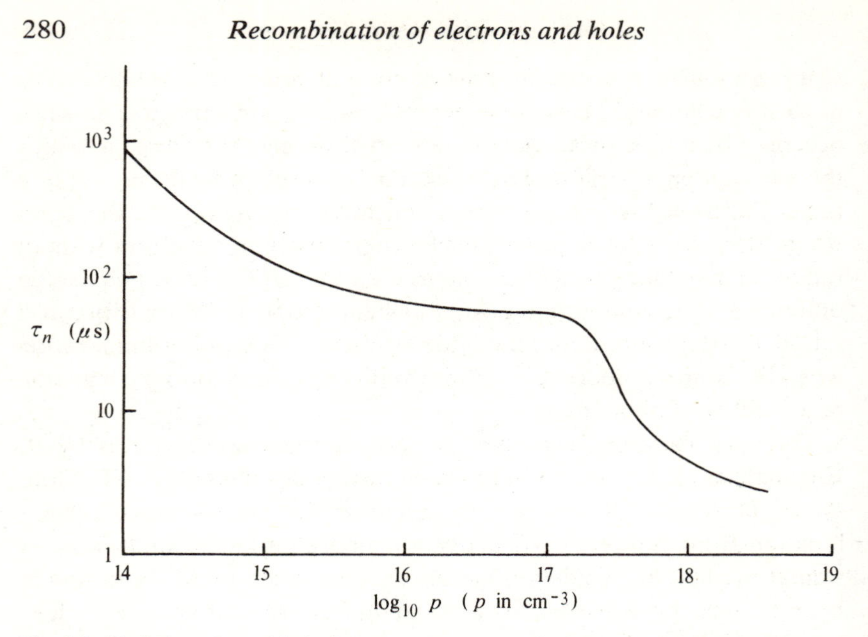

Fig. 9.3. Variation of τ_n with the hole concentration p. (After S. G. Kalashnikov, *loc. cit.*)

from excitons bound to donors and acceptors, especially in semiconductors (such as GaP) with somewhat larger energy gaps (see § 10.14).

For the free exciton the emitted spectrum merges with that due to band-to-band transitions. In the method of calculation of the radiative lifetime in § 9.2 we have integrated over the absorption spectrum and so the exciton recombination is included. If, however, we can separate the exciton contribution, as seems possible for the highly purified semiconductors, the free exciton lifetime may be determined from the absorption. Excitons bound to some neutral donors and acceptors continue to exist because of their lowered energy down to very low temperatures. When the electron and hole forming the exciton recombine the energy is emitted as a very sharp line which can be readily identified (see § 10.14). The line may also be observed in absorption and calculation of the radiative lifetime as in § 9.2 by integration of the absorption over the line gives the radiative lifetime. The lifetime may also be observed by studying with fast pulses the rate of decay of the emitted radiation. This turns out to be several orders of magnitude less than the calculated radiative lifetime. For example D. F. Nelson[26] finds that for excitons trapped at S atoms in GaP at 4.2 °K about one quantum is emitted

[26] *Electronic Structure in Solids*, ed. E. D. Hademenakis (Plenum Press, 1969), p. 122.

in 700 recombinations, while for As in Si at 1.6 °K the number is about one in 8000. There must therefore be a very much more efficient process in operation and Nelson has identified this with Auger recombination with ejection of a fast electron from the exciton complex.

This process is much more like the original Auger effect than that involving free electrons and holes and, although detailed quantum-mechanical calculations have not yet been carried out, the order of magnitude may be estimated from the well-known internal conversion coefficient for atoms, adapted to take account of dielectric constant and electron and hole effective masses. The order of magnitude so estimated agrees well with the experimental values and leaves little doubt that Auger recombination is the dominant process. This adds strength to the view that in recombination of electrons and holes through traps it also plays an important part.

The subject of exciton decay has been discussed in some detail by P. T. Landsberg[27] in his review of recombination processes already referred to.

9.6 Recombination at dislocations

One form of trap of particular interest is that associated with dislocations; as we have seen in § 3.1.3 these provide lines of traps in the crystal through the unsaturated bonds which exist at the dislocation. The effect of the space-charge associated with the dislocations has been examined by S. R. Morrison[28] who has shown that this may have a marked effect on the recombination constants C_n, C_p. Experiments with plastically deformed Ge crystals by J. P. McKelvey,[29] have shown that in n-type material the lifetime is affected by the presence of edge dislocations when their density exceeds about 10^4 cm^{-2}; this has also been found by G. K. Wertheim and G. L. Pearson;[30] from these measurements the contribution to the minority-carrier lifetimes in Ge due to dislocations τ_{pd}, τ_{nd} may be expressed in the form

$$\left.\begin{aligned} \tau_{pd} &= 2.5/N \text{ s} \quad (n\text{-type}), \\ \tau_{nd} &= 0.7/N \text{ s} \quad (p\text{-type}), \end{aligned}\right\} \qquad (54)$$

where N is the dislocation density in cm^{-2}. Experiments by J. S. Blakemore, J. Y. Schultz and K. C. Nomura[31] on Te showed a nearly

[27] *Phys. Stat. Sol.* (1970) **41**, 457.
[28] *Phys. Rev.* (1956) **104**, 619.
[29] *Phys. Rev.* (1957) **106**, 910.
[30] *Phys. Rev.* (1957) **107**, 694.
[31] *J. Appl. Phys.* (1960) **31**, 2226.

linear increase in recombination rate with dislocation density for densities below $5\times10^5\ \text{cm}^{-2}$ but a less rapid increase for higher densities. The minority-carrier lifetime which was of the order of 10^{-5} s at room temperature in an unstressed sample fell to about 10^{-6} s when a dislocation density of the order of $10^6\ \text{cm}^{-2}$ was introduced.

9.7 Recombination with donors or acceptors at low temperatures

At low temperatures the number of holes in n-type material is so small as to be negligible and we are concerned only with the rate of decay to the donors of excited electrons, i.e. of *majority* carriers. We may assume that the trapping centres are the donors themselves, that p_0 and p_1 are negligible, and also that we have $n=(1-f_t)N_d$. The rate of recombination of electrons with the donors is then given by $\mathcal{R}_n$ (equation (35)). If we assume that $n \ll N_d$ we have, since C_n is proportional to $N_t(=N_d)$,

$$\mathcal{R}_n = B_d(n^2-n_0^2), \tag{55}$$

where n_0 is the equilibrium concentration of electrons and B_d is a constant; we have therefore

$$\frac{1}{\tau_n}=B_d(2n_0+\Delta n). \tag{56}$$

At low temperatures n_0 is very small so that for an appreciable increase in carrier density $\tau_n^{-1}\propto\Delta n$ and the recombination rate is proportional to Δn^2 and not to Δn; we have then

$$\frac{d\Delta n}{dt}=-B_d\Delta n^2. \tag{57}$$

The solution of this equation corresponding to $\Delta n=\Delta n_0$ at $t=0$ is

$$\Delta n=\frac{\Delta n_0}{1+\Delta n_0 B_d t}, \tag{58}$$

so that the excess electrons decay according to a hyperbolic law, and not exponentially, as is more usual.

Radiation from shallow donor impurities in InSb has been observed by E. Gornik[32] in the millimetre-wave region of the spectrum, so some of the centres must recombine by radiating the small amount of excess energy. This is likely to be an inefficient process compared with recom-

[32] *Phys. Rev. Lett.* (1972) **29**, 595.

bination through traps. The Auger trap mechanism is probably responsible and has been applied to this problem also by R. A. Brown and S. Rodriguez[33] for n-type donors in Ge and Si.

9.8 Surface recombination

In addition to recombining in the body of a semiconductor, electrons and holes may diffuse to the surface and recombine there, and, under some conditions, the rate of recombination at the surface may even be much greater than the body recombination rate. Indeed, for many technological applications, special precautions must be taken to ensure that the surface recombination rate is not intolerably high; for example, if the surface recombination rate near a point contact is unduly high it may affect adversely the properties of the contact. Also in thin films such as are sometimes used for photo-conductive infra-red detectors the surface recombination, unless greatly reduced by suitable treatment, may dominate the decay process and negate the advantages of a long bulk lifetime (see § 10.9.3).

In equilibrium, the number of holes flowing to the surface of an n-type crystal must be equal to the number flowing from it, the latter number including holes reflected and holes emitted; when equilibrium is disturbed these will no longer be equal. For small deviations from equilibrium, we should expect the number emitted to be independent of the deviation. Let r be the average probability of reflection of a hole striking the surface, then the rate at which holes flow to the surface is equal to $\frac{1}{4}v_t p$ for unit area, where v_t is the average thermal velocity; the rate at which holes flow *from* the surface is $\frac{1}{4}rv_t p + S$ where S is the rate of emission per unit area. In equilibrium we have

$$\tfrac{1}{4}v_t p_0 = \tfrac{1}{4}rv_t p_0 + S_0, \tag{59}$$

so that

$$S_0 = \tfrac{1}{4}(1-r)v_t p_0. \tag{60}$$

The net rate S_a at which holes are absorbed by the surface is then given by the equation

$$S_a = \tfrac{1}{4}(1-r)v_t(p-p_0), \tag{61}$$

and this may be written in the form

$$S_a = s\Delta p, \tag{62}$$

[33] *Phys. Rev.* (1967) **153**, 890.

where s has the dimensions of a velocity and is called the surface recombination velocity. It will be clear that, in the steady state, the rate at which holes are absorbed by the surface is equal to the rate at which they recombine with electrons, otherwise the surface would continue to receive a charge.

We have already discussed the two kinds of traps which are known to exist at the surface of a semiconductor, the 'slow' traps, and the 'fast' traps (see § 7.11.1). The latter, being nearest to the body of the semiconductor, are more readily available to act as recombination centres. Their energy levels lie below the conduction band and above the valence band and they should therefore act like the traps in the Shockley–Read recombination process. These traps have been discussed by a number of authors (see § 7.11.1), but have been particularly studied as recombination centres by W. H. Brattain and J. Bardeen,[34] who have shown that both electron traps and hole traps are necessary to explain the observations on Ge, and that the electron traps do not lie very far below the conduction band, or the hole traps very far above the valence band. If either $n_1 \gg p_1$ or $n_1 \ll p_1$ in equation (43) we have

$$\mathscr{R} \propto (n_0 + p_0 + \Delta p),$$

and this will be approximately true even when two types of trap are in operation, each of which satisfies this condition. Applying this argument to the surface traps we find that

$$s = B_s(n_0 + p_0 + \Delta p), \tag{63}$$

where B_s is a constant; an equation of this form has been derived by Brattain and Bardeen from a more detailed analysis. The variation of s with n_0 and p_0 for Ge has been studied by B. H. Schultz[35] who has found fairly good agreement with equation (63).

The determination of the quantity B_s for a surface is difficult and has been discussed by a number of authors. Various ambiguities have to be resolved and some doubt must remain regarding some of the determinations. A method which determines the recombination parameters uniquely and also the quantity s has been described by Y. Margoninski and Y. Walzer.[36] This combines recombination measurements with the pulsed field effect. It has been applied mainly to Ge and Si.

[34] *Bell Syst. Tech. J.* (1953) **32**, 1.
[35] *Physica* (1954) **20**, 1031.
[36] *Phys. Rev.* (1967) **156**, 903.

The chemical nature of the trapping and recombination centres has been discussed by S. R. Morrison[37] together with their role in providing hole and electron traps, both of which must be present if recombination is to take place through them without radiation. Surface recombination velocities found by Brattain and Bardeen for Ge varied from about 10^6 to 10^2 cm s^{-1}; if all holes reaching the surface are absorbed $s \simeq \frac{1}{4}v_t \simeq 2\times10^6$ cm s^{-1} for $T = 300$ °K. Large surface recombination velocities are found for ground or sand-blasted surfaces and low values for surfaces polished in a suitable etching solution.

The influence of surface recombination on the field effect (see § 7.11.1) has also been studied. A number of papers are included in books[38] dealing with this and other problems concerning surface recombination. More recent treatments have been given by A. Many, Y. Goldstein and N. B. Grover[39] and also by D. R. Frankl.[40]

Let us now see how surface recombination affects the recovery of equilibrium; at a distance from a surface greater than a diffusion length the effect will clearly be small. Let us consider a plane surface, and let x be the distance of a point in the semiconductor from the surface; we shall suppose that we have a uniform generation of *excess* of carriers throughout the bulk of the material at rate $\mathcal{R}_e$; we then have to solve the equation

$$D_h\frac{d^2\Delta p}{dx^2}=\frac{\Delta p}{\tau_p}-\mathcal{R}_e. \tag{64}$$

The surface recombination enters into the problem as a boundary condition at $x = 0$; there we have

$$D_h\frac{\partial\Delta p}{\partial x}=s\,\Delta p. \tag{65}$$

As a second boundary condition we take $\Delta p \to \mathcal{R}_e\tau_p$ as $x\to\infty$, and the appropriate solution of equation (64) is of the form

$$\Delta p=\Delta p_1\,e^{-x/L_p}+\mathcal{R}_e\tau_p. \tag{66}$$

[37] *Treatise on Solid State Chemistry*, ed. N. B. Hannay (Plenum Press, 1976) vol. 6B, p. 203.

[38] *Semiconductor Surface Physics*, ed. R. H. Kingston (Penn. Univ. Press, 1957); *Surface Properties of Semiconductors*, ed. A. N. Frumkin (Consultants Bureau, N.Y., 1964).

[39] *Semiconductor Surfaces* (North Holland, 1965).

[40] *Electronic Properties of Semiconductor Surfaces* (Pergamon Press, 1967).

Applying the boundary condition at $x=0$ we obtain

$$-\frac{D_h\Delta p_1}{L_p}=s(\Delta p_1+\mathcal{R}_e\tau_p), \tag{67}$$

giving

$$\Delta p_1=-s\mathcal{R}_e\tau_p^2/(L_p+s\tau_p) \tag{68}$$

on writing $D_h/L_p=L_p/\tau_p$. We have therefore

$$\Delta p=\mathcal{R}_e\tau_p[1-s\tau_p\,e^{-x/L_p}/(L_p+s\tau_p)] \tag{69}$$

and the value Δp_0 of Δp at $x=0$ is given by

$$\Delta p_0=\mathcal{R}_e\tau_p[1-s\tau_p/(L_p+s\tau_p)]. \tag{70}$$

The decrease in the value of Δp near the surface, due to surface recombination, extends only to a distance of the order of a few times L_p from the surface as we should expect. The magnitude of the decrease depends on the ratio s/v_p where v_p is a velocity equal to L_p/τ_p; for $L_p=0.05$ cm, $\tau_p=100\ \mu$s, which are fairly typical values for Ge, we have $v_p=500$ cm s^{-1}. If $s\gg v_p$, Δp_0 is very small, and we have a large decrease in the excess hole concentration near the surface.

9.9 Mean lifetime in filaments and thin strips

The general solution of the problem of the motion of electrons and holes in a uniform crystal of arbitrary shape under an electric field, taking account of surface recombination, cannot be solved by elementary means. Some of the mathematical difficulties associated with this type of problem have been discussed by W. van Roosbroeck[41] who has obtained some fairly general solutions. For certain simple geometrical shapes, however, a simple solution may be given and from such solutions the general effect of surface recombination may be seen. Such an example is a long bar or filament of rectangular cross-section; this problem has also been discussed by W. Shockley[42] who has shown that simple analytical solutions may be obtained for a number of problems concerning motion of electrons in filaments.

We shall treat the simpler problem of a long, thin strip whose length and width are much greater than its thickness. Let us take the z-axis along the length of the strip, parallel to its edges, the y-axis in the plane of the strip and the x-axis at right angles to its plane, the faces being the

[41] *J. Appl. Phys.* (1955) **26**, 380.
[42] *Electrons and Holes in Semiconductors* (Van Nostrand, 1950), p. 318.

planes $x = \pm a$ so that the thickness of the strip is $2a$; for simplicity we shall assume that the surface recombination velocity is the same for both faces.

The first problem we shall discuss is that of finding the average excess carrier density created by a source which produces electron–hole pairs at a uniform rate $\mathscr{R}_e$ throughout the material; for example, this could be due to illumination of the strip by light whose absorption coefficient α is much less than a^{-1}. We must now solve equation (64) with the boundary condition (65) applied at $x = a$ and at $x = -a$; the appropriate solution has the form (by symmetry considerations)

$$\Delta p = A \cosh (x/L_p) + \mathscr{R}_e\tau_p. \tag{71}$$

Applying the boundary conditions we obtain

$$\Delta p = \mathscr{R}_e\tau_p\left\{1 - \frac{s\tau_p \cosh (x/L_p)}{s\tau_p \cosh (a/L_p) + L_p \sinh (a/L_p)}\right\}. \tag{72}$$

The average value of Δp, $\overline{\Delta p}$, is given by the equation

$$\overline{\Delta p} = \mathscr{R}_e\tau_p\left\{1 - \frac{(L_p/a)s\tau_p \sinh (a/L_p)}{s\tau_p \cosh (a/L_p) + L_p \sinh (a/L_p)}\right\} \tag{73}$$

and we may define in this way a mean value τ_p for the strip through the equation $\overline{\Delta p} = \mathscr{R}_e\overline{\tau_p}$. When $a \ll L_p$, Δp is approximately constant and is given by

$$\Delta p = \mathscr{R}_e\tau_p\left\{\frac{a}{s\tau_p + a}\right\}, \tag{74}$$

so that we have

$$\frac{1}{\overline{\tau_p}} = \frac{1}{\tau_p} + \frac{s}{a}. \tag{75}$$

The solution of the corresponding problem for a bar with rectangular cross-section is not easy, because of the difficulty of applying the non-homogeneous boundary conditions; if the sides of the rectangular cross-section are $2a$, $2b$ and $a \ll L_p$, $b \ll L_p$ it may be shown that we have approximately

$$\frac{1}{\overline{\tau_p}} = \frac{1}{\tau_p} + \frac{s}{a} + \frac{s}{b}. \tag{76}$$

Let us now consider how an initial uniform excess concentration of carriers will decay with time after the generating source is removed. We have now to solve the equation

$$D_h \frac{\partial^2 \Delta p}{\partial x^2} = \frac{\Delta p}{\tau_p} + \frac{\partial \Delta p}{\partial t} \tag{77}$$

and our solution must now contain both t and x. The solutions of this equation previously discussed in § 7.6 are not appropriate to this problem as they cannot represent the initial uniform distribution. We may seek solutions of the form $\Delta p = \exp(\alpha' x - \beta t)$, and we find that such solutions do exist; the relationship between α' and β is found by substitution in equation (77) to be

$$D_h \alpha'^2 = \frac{1}{\tau_p} - \beta. \tag{78}$$

Physical considerations show that the rate of decay will be faster than $1/\tau_p$ so that $1/\tau_p - \beta$ will be negative; we are therefore led to a trigonometric type of solution for the x-variation, and by symmetry choose solutions of the type

$$\Delta p = A \cos \alpha x \, e^{-\beta t}. \tag{79}$$

The relationship between α and β is now

$$D_h \alpha^2 = \beta - 1/\tau_p \tag{80}$$

and α is found by applying the boundary condition (equation (65)), which gives

$$D_h a \alpha \sin \alpha a = as \cos \alpha a. \tag{81}$$

The equation

$$\eta \tan \eta = as/D_h \tag{82}$$

is a transcendental one having an infinite number of solutions η_1, η_2, $\eta_3, \ldots$, etc.; corresponding to the solution η_r we have $\alpha_r = \eta_r/a$, and β given by

$$\beta_r = 1/\tau_p + D_h(\eta_r/a)^2. \tag{83}$$

The original value Δp_0 of Δp at $x = 0$ may be built up from a Fourier series of the form

$$\Delta p_0 = \Sigma A_r \cos \eta_r x \quad (-a \leqslant x \leqslant a), \tag{84}$$

but we shall not trouble to find the values of the constants A_r, since we note that $\eta_1 < \eta_2 < \eta_3, \ldots$, and from equation (79) we see that the higher-order solutions decay more rapidly with time than the first, and after an initial 'transient' condition may be neglected; the complete solution has been studied by S. Visvanathan and J. F. Battey.[43] If we write $\beta_1 = 1/\tau_1$, so that τ_1 is the effective lifetime for the strip, we have

$$\frac{1}{\tau_1} = \frac{1}{\tau_p} + \frac{\eta_1^2 D_h}{a^2}. \tag{85}$$

For large values of s, the smallest solution of equation (82) tends to the value $\eta_1 = \frac{1}{2}\pi$, and we have

$$\frac{1}{\tau_1} \to \frac{1}{\tau_p} + \frac{\pi^2 D_h}{4a^2}. \tag{86}$$

For small values of s we have $\eta_1^2 = as/D_h$, so that

$$\frac{1}{\tau_1} = \frac{1}{\tau_p} + \frac{s}{a}; \tag{87}$$

this is the same value as we obtained for $\overline{\tau_p^{-1}}$ for small values of s. In general, however, the lifetimes τ_1 and $\overline{\tau_p}$ are not the same; τ_1 is generally called the strip lifetime.

The problem of a filament with rectangular cross-section may be treated in exactly the same way.[44] We have in this case

$$\frac{1}{\tau_1} = \frac{1}{\tau_p} + D_h\left(\frac{\eta_1^2}{a^2} + \frac{\zeta_1^2}{b^2}\right), \tag{88}$$

where ζ_1 is the smallest solution of the equation

$$\zeta \tan \zeta = bs/D_h. \tag{89}$$

For small values of s equation (88) reduces to the same form as equation (76) and $\tau_p = \overline{\tau_p}$. Exactly similar equations hold when we have electrons as minority carriers.

We next consider the motion of minority carriers along the strip under the influence of an electric field $\mathscr{E}_z$, and we have now to solve, for the steady state, the equation

$$D_p\left(\frac{\partial^2 \Delta p}{\partial x^2} + \frac{\partial^2 \Delta p}{\partial z^2}\right) - \mu_h \mathscr{E}_z \frac{\partial \Delta p}{\partial z} - \frac{\Delta p}{\tau_p} = 0. \tag{90}$$

From the previous considerations it is clear that the appropriate solution which tends to zero as $z \to \infty$ has the form

$$\Delta p = e^{-\gamma z} \cos \alpha x, \tag{91}$$

[43] *J. Appl. Phys.* (1954) **25**, 99.
[44] W. Shockley, *Electrons and Holes in Semiconductors* (Van Nostrand, 1950), p. 322.

and for the solution that decays least rapidly, α is equal to η_1/a, so that on substituting in equation (90) we obtain

$$D_p(\gamma^2-\alpha^2)+\mu_h\gamma\mathscr{E}_z-1/\tau_p=0.$$

If we write $\gamma=1/L_{d1}$ or $1/L_{u1}$ according as $\mathscr{E}_z$ is positive or negative, we get the same equations as in § 7.5.6 (equations (75) and (76)), provided we replace τ_p by τ_1. In particular, the high-field diffusion length L_1 for holes is given by

$$L_1=\mathscr{E}_z\mu_h\tau_1, \tag{92}$$

and we have a similar equation for electrons. This equation is also valid for a filament of rectangular cross-section. By measuring the minority concentration along the strip using a collector contact (see § 7.9.2) L_1 may readily be found; by comparing this with L_d, the value found for a thick strip and corresponding to bulk material, the surface recombination velocity s may be determined. By measuring τ_1 for a filament of rectangular cross-section with a high surface recombination velocity, and using equation (88) with $\eta_1=\zeta_1=\frac{1}{2}\pi$, A. Many[45] has been able to make a *direct* measurement of the diffusion constants D_h and D_e for electrons and holes.

[45] *Physica* (1954) **20**, 989.

10

Optical and high-frequency effects in semiconductors

10.1 Optical constants of semiconductors

Many semiconductors are almost indistinguishable in appearance from metals, having the characteristic lustre of polished metals; the reason for this is their high reflectivity. In the visible region of the spectrum, like metals, they generally absorb strongly, having coefficients of absorption of the order of $10^5\ \mathrm{cm}^{-1}$. The optical constants of a number of semiconductors have been measured over a wide range of wavelengths; the characteristic feature of all semiconductors, in the pure state, is that at a certain wavelength, generally in the near or intermediate infra-red, the absorption coefficient drops rapidly and the material becomes fairly transparent at longer wavelengths. This marked drop in absorption is called the fundamental absorption edge; it is sometimes referred to in the literature as the 'lattice absorption edge'. The transparency of semiconductors at wavelengths beyond the absorption edge is frequently only apparent when they have been purified to such an extent that the absorption due to free carriers is small enough to show up the fundamental absorption; in an impure state they are generally opaque from the ultra-violet right up to radio wavelengths.

The complex refractive index n*, which may be written in real and imaginary parts in the form

$$\mathrm{n}^* = \mathrm{n}(1 - i\mathrm{k}), \tag{1}$$

is frequently used to describe the optical properties of a material.[1] The significance of n* is that a plane electromagnetic wave of frequency ν and wavelength λ propagated through the material in the x-direction has the form

$$\psi = A \exp\{2\pi i\nu[t - \mathrm{n}^* x/c]\}, \tag{2}$$

[1] k as used in equation (1) should not be confused with $k = 2\pi/\lambda$.

c being the velocity of light in free space. Writing n^* in terms of its real and imaginary parts we have

$$\psi = A \exp\{-2\pi\nu nkx/c\} \exp\{2\pi i\nu[t - nx/c]\}. \tag{3}$$

The phase velocity V is equal to c/n and the wavelength λ, when the attenuation is not too great, is equal to $c/\nu n$. In terms of λ we have

$$\psi = A \exp[-\alpha x/2][\exp 2\pi i\nu(t - x/V)], \tag{4}$$

where

$$\alpha = 4\pi k/\lambda = 4\pi nk/\lambda_0 \tag{5}$$

and λ_0 is the free-space wavelength at frequency ν; k is generally called the absorption index and α the absorption coefficient.[2] By means of classical electromagnetic theory we may relate n^* to the complex permittivity ϵ^* defined by means of the equation

$$\epsilon^* = \epsilon - i\sigma/2\pi\nu, \tag{6}$$

where ϵ is the ordinary permittivity and σ the conductivity. The complex refractive index n^* and ϵ^* are related by means of the equation

$$n^* = (\epsilon^*/\epsilon_0)^{\frac{1}{2}}, \tag{7}$$

ϵ_0 being the permittivity of free space; ϵ/ϵ_0 is equal to the dielectric constant K. Equation (7) follows at once from the equations for the propagation of electromagnetic waves in a conducting medium. (N.B. equation (6) is in M.K.S. units, involving a factor $1/4\pi$ in the second term as compared with C.G.S. units.) Taking the real and imaginary parts of equation (7) we have

$$\epsilon = \epsilon_0 n^2(1 - k^2), \tag{8}$$

$$\sigma = 4\pi n^2 k\nu\epsilon_0, \tag{9}$$

and from equations (5) and (9) we have, if $\epsilon > 0$,

$$\alpha = \sigma/n^2\lambda\nu\epsilon_0 = \sigma/nc\epsilon_0. \tag{10}$$

The quantities $n^2(1 - k^2)$ and $2n^2k$ are most frequently plotted in order to describe the optical properties of a material. They may be obtained from measurements of the reflection coefficient. For normal incidence,

[2] Some writers use k where we have used nk.

the reflection coefficient R is given by

$$R = \left|\frac{1-n^*}{1+n^*}\right|^2$$

$$= \frac{(1-n)^2+n^2k^2}{(1+n)^2+n^2k^2}. \tag{11}$$

It will be seen that if both n and k are large, R will not be very different from unity. For angles of incidence other than normal, the reflection coefficient depends on the polarization; by making observations for different angles of incidence both n and k may be determined when k is not too small.

When the incident radiation cannot excite electrons from the valence band to the conduction band, we may expect the optical effects to be due to transitions within a band. In this case, as we shall see, the interaction of radiation with the free charge carriers may be approximately treated by means of classical electromagnetic theory, taking account of the effective masses of the electrons and holes. When, however, the radiation is able to excite electrons from the valence band to the conduction band, marked quantum effects will take place; in this case, a large increase in absorption may be expected, and this is observed. The form of the absorption usually observed is shown as a function of wavelength in Fig. 10.1; it will be seen that the absorption coefficient passes through a minimum whose value depends on the impurity content, being smallest for the purest samples. The wavelength λ_e and frequency ν_e corresponding to the fundamental absorption edge are given approximately by the equation

$$\boldsymbol{hc}/\lambda_e = \boldsymbol{h}\nu_e = \Delta E, \tag{12}$$

where ΔE is the forbidden energy gap, since ΔE is the energy required to excite an electron from the top of the valence band to the bottom of the conduction band. As we shall see, this transition is not always possible without involving other energy changes, and equation (12) will have to be modified; unless very precise measurements are available, it is, however, a good approximation. As will be seen from Fig. 10.1, the precise position of the absorption edge is not very well defined; its approximate position may generally be determined to an accuracy of about 0.1 eV without much difficulty, but for higher accuracy much more detailed consideration must be given to the types of electronic transition involved. It will also be clear from the form of the curves of Fig. 10.1 that, in order to obtain an accurate value of the position of the

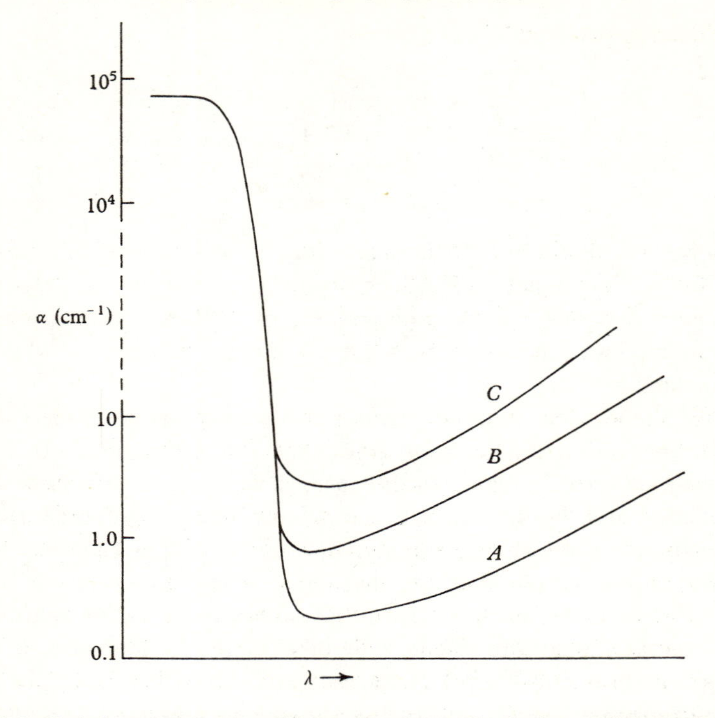

Fig. 10.1. Variation of absorption with wavelength near the fundamental absorption edge for specimens of different purity. (Curve *A* represents the purest sample.)

absorption edge, specimens of the highest purity should be used. They should also be high-quality single crystals, if available, since strains tend to blur the edge.

On the short-wave side of the absorption edge the value of the absorption coefficient rises rapidly and may reach a value of the order of a few times 10^4 cm^{-1}. It may then remain fairly constant over a considerable range of wavelengths, but sometimes exhibits a further rise to a value in excess of 10^5 cm^{-1}. The edge itself may also show some fine structure, but we must consider the theory of the fundamental absorption before examining the experimental results (see § 10.5.2).

10.2 Free-carrier absorption

When the frequency of the incident radiation is insufficiently high to cause band-to-band transitions or formation of excitons, absorption is due to excitation of lattice vibrations or to transitions between states in a single band; we shall first of all deal with the latter. Clearly, for a completely filled band, no such transitions are possible. For the conduction band, the absorption will be proportional to the number of free electrons, in the non-degenerate condition, and, as we shall see, there is an added absorption proportional to the number of free holes in the valence band. The free-carrier absorption may be treated using the quantum theory of transition probabilities between states of different energy, but not only electronic transitions but also transitions between different states of the lattice must be considered. Such a treatment has been given by a number of authors including H. Y. Fan[3] and J. Bardeen.[4] Except when the band is degenerate (as for *p*-type Ge) a semi-classical treatment gives essentially the same results as the quantum theory.[5] Since the former is much simpler, we shall follow this method, in which the absorption coefficient is deduced from the high-frequency conductivity.

We must now consider the modification to the treatment of electrical conductivity given in § 5.2.1 when we have an alternating electric field applied to the semiconductor. The equation of motion of an electron in an electric field of frequency $\omega/2\pi$ in the x-direction is

$$m_e\dot{v}_x = -e\mathscr{E}_x = -e\mathscr{E}\,e^{-i\omega t}. \tag{13}$$

The appropriate solution is, when v_{x0} is the velocity at $t = t_0$ $(t_0 < t)$,

$$v_x = v_{x0} + \frac{e\mathscr{E}}{i\omega m_e}(e^{-i\omega t_0} - e^{-i\omega t}). \tag{14}$$

Multiplying the right-hand side of equation (14) by $\tau^{-1}\,e^{-(t-t_0)/\tau}$ and averaging as in § 5.2.1, taking t_0 from $-\infty$ to t, we obtain, on putting $\overline{v_{x0}} = 0$

$$\overline{v_x} = -\left\{\frac{e\tau}{m_e(1+\omega^2\tau^2)} - \frac{ie\omega\tau^2}{m_e(1+\omega^2\tau^2)}\right\}\mathscr{E}_x, \tag{15}$$

where τ is the relaxation time for electrons of energy E. The current density J_x is obtained by averaging over the allowed values of E and we obtain

$$J_x = \frac{ne^2}{m_e}\left\{\left\langle\frac{\tau}{1+\omega^2\tau^2}\right\rangle - i\omega\left\langle\frac{\tau^2}{1+\omega^2\tau^2}\right\rangle\right\}\mathscr{E}_x. \tag{16}$$

[3] *Semiconductors and Semimetals* (Academic Press, 1967) **3**, 406.
[4] *Phys. Rev.* (1950) **79**, 216. See also H. Y. Fan, *Rep. Progr. Phys.* (1956) **19**, 107.
[5] See, for example, F. Stern, *Solid State Physics* (Academic Press, 1963) **15**, 300.

Instead of the complex permittivity ϵ^* defined by means of equation (6) we may define a complex conductivity σ^* by means of the equation $J_x = \sigma^* \mathscr{E}_x$. We have on comparing with equation (6)

$$\sigma^* = i\omega\epsilon_e^* = \sigma + i\omega\epsilon_e, \tag{17}$$

where we have written ϵ_e for the electronic contribution to the permittivity. There will also be a contribution due to the polarization of the atoms of the lattice, and in polar crystals to their movement from their equilibrium positions, which we shall write as ϵ_L. We then have for the total permittivity ϵ

$$\epsilon = \epsilon_L + \epsilon_e. \tag{18}$$

From equation (16) we have

$$\sigma = \frac{ne^2}{m_e}\left\langle\frac{\tau}{1+\omega^2\tau^2}\right\rangle \tag{19}$$

$$= \sigma_0\left\langle\frac{\tau}{1+\omega^2\tau^2}\right\rangle \Big/ \langle\tau\rangle, \tag{20}$$

where σ_0 is the low-frequency conductivity; also we have

$$\epsilon = \epsilon_L - \frac{ne^2}{m_e}\left\langle\frac{\tau^2}{1+\omega^2\tau^2}\right\rangle \tag{21}$$

$$= \epsilon_L - \sigma_0\langle\tau\rangle\{\langle\tau^2/(1+\omega^2\tau^2)\rangle/\langle\tau\rangle^2\}. \tag{22}$$

When $\omega\langle\tau\rangle \ll 1$ we have to the first order in ω, $\sigma = \sigma_0$ and

$$\epsilon = \epsilon_1 = \epsilon_L - \sigma_0\langle\tau^2\rangle/\langle\tau\rangle, \tag{23}$$

where σ_0 and ϵ_1 are respectively the low-frequency values of the conductivity and permittivity. We have already used the quantity $\langle\tau^2\rangle/\langle\tau\rangle^2$ in connection with the Hall effect (see § 5.3.3) and have written it as r. We have then

$$\epsilon_1 = \epsilon_L - r\sigma_0\langle\tau\rangle. \tag{24}$$

We note that the first-order contribution of the electrons to the permittivity is negative and that it is independent of frequency. To the next order of approximation we have

$$\sigma = \sigma_0(1 - a\omega^2\langle\tau\rangle^2), \tag{25}$$

where $$a = \langle\tau^3\rangle/\langle\tau\rangle^3; \tag{26}$$

we have also $$\epsilon = \epsilon_1 + b\sigma_0\omega^2\langle\tau\rangle^3, \tag{27}$$

where $$b = \langle\tau^4\rangle/\langle\tau\rangle^4. \tag{28}$$

One of the first methods of measuring $\langle\tau\rangle$ directly, and hence of deducing the value of the effective mass, was that proposed by T. S. Benedict and W. Shockley.[6] This method depends on the comparison of the dielectric constant ϵ/ϵ_0 of very pure material with that of a sample containing a known number of charge carriers. The original experiments were carried out with Ge, using a frequency in the microwave range.

Another extreme situation arises when $\omega\langle\tau\rangle \gg 1$. For a wavelength of 10 μm, $\omega \simeq 2\times10^{14}$ radians s^{-1}, and, if we take $T = 300$ °C, the value of $\omega\tau_e$ is about 260, so that this condition clearly holds in the near infra-red, except possibly at very low temperatures. With this assumption we have, replacing m_e by the conductivity effective mass m_c,

$$\sigma = \frac{ne^2}{m_c\omega^2}\left\langle\frac{1}{\tau}\right\rangle = \frac{g}{\omega^2\langle\tau\rangle^2}\frac{ne^2\langle\tau\rangle}{m_c}, \tag{29}$$

$$= \frac{g\sigma_0}{\omega^2\langle\tau\rangle^2}, \tag{30}$$

where $$g = \langle\tau\rangle\left\langle\frac{1}{\tau}\right\rangle, \tag{31}$$

and if τ has the form aE^{-s} we have

$$g = \Gamma(\tfrac{5}{2}-s)\Gamma(\tfrac{5}{2}+s)/[\Gamma(\tfrac{5}{2})]^2 \tag{32}$$

(cf. § 5.3.5, equation (162)). We may also write equation (29) in the form

$$\sigma/\sigma_0 = g(e^2/\omega^2\mu_c^2)/m_c^2, \tag{33}$$

where μ_c is the mobility deduced from the d.c. conductivity. The absorption coefficient α may be deduced, using equation (10), and we have

$$\alpha = \sigma_0 g(e^2/\omega^2\mu_c^2)/m_c^2 c\epsilon_0 \mathrm{n}, \tag{34}$$

where n is the refractive index; this equation may also be written in the form

$$\alpha = n\lambda_0^2 ge^3/4\pi^2\epsilon_0 m_c^2 c^3 \mathrm{n}\mu_c, \tag{35}$$

where n is the electron concentration in the conduction band. We see that the absorption coefficient α is proportional to n, the number of free carriers, to the square of the free-space wavelength λ_0, and is inversely

[6] *Phys. Rev.* (1953) **89**, 1152.

proportional to the conductivity mobility. Equation (35) has been verified for many semiconductors by measuring the free-carrier absorption at wavelengths in the infra-red greater than that corresponding to the fundamental absorption edge. The minimum in the absorption curves in Fig. 10.1 is due to the combination of fundamental absorption, which decreases with wavelength, and free-carrier absorption which increases as λ^2; the value of the absorption at the minimum increases as the number of free carriers is increased. A notable exception is *p*-type Ge, which we shall discuss later. For *n*-type Ge, equation (35) was thought at one time to give too small an absorption by a factor of about 100;[7] this was before the band structure of Ge had been elucidated or the cyclotron resonance experiments carried out to determine the effective masses: the factor $g(\boldsymbol{m}/m_c)^2$ for *n*-type Ge is about 70 and brings equation (35) into line with experiment. By increasing the carrier density artificially either by injection of carriers from electrodes (see § 7.5.3) or by optical injection (creation of electron–hole pairs by light) it has been shown that the absorption due to electrons and holes is additive. The absorption coefficient for holes is given also by equation (35) with the appropriate values of g, m_c and μ_c.

While, as we have seen, the absorption due to free carriers in the intermediate infra-red varies as λ^2, this is no longer true at the somewhat shorter wavelengths of the near infra-red. M. Becker and H. Y. Fan (*loc. cit.*) have shown that α varies as $\lambda^{1.5}$ for acoustic mode scattering of the carriers and as $\lambda^{2.5}$ for optical-mode scattering, while for impurity scattering α varies between λ^3 and $\lambda^{3.5}$. Since in general all these forms of scattering will be present to some extent, the variation will not be given accurately by a simple power law. The absorption does, however, increase fairly rapidly with wavelength.

Since an electron in the conduction band must change its wave vector **k** in making a transition to a state with different energy within the band, the transition must involve a change in crystal momentum **P**. As we shall see later such a change is too big to be taken up by the momentum of the absorbed photon and so must be taken up by the crystal lattice or by an impurity. In other words the electron must be scattered in the absorption process. It is not surprising then that the absorption coefficient should depend on the kind of scattering involved. At the longer wavelengths this dependence is small and we have the λ^2 variation of absorption coefficient.

[7] See, for example, M. Becker and H. Y. Fan, *Semiconducting Materials*, ed. H. K. Henisch (Butterworth, 1951), p. 132; A. F. Gibson, *Proc. Phys. Soc.* B (1953) **66**, 588.

If $\omega\langle\tau\rangle \ll 1$ we have the absorption coefficient

$$\alpha = \sigma_0\left(\frac{\epsilon_0}{\epsilon_1}\right)^{\frac{1}{2}} \Big/ \epsilon_0 c, \tag{36}$$

$$= 375(\epsilon_0/\epsilon_1)^{\frac{1}{2}}\sigma_0\ \mathrm{m}^{-1} \tag{36a}$$

if σ_0 is expressed in $\Omega^{-1}\ \mathrm{m}^{-1}$ (or α is in cm^{-1} if σ is in $\Omega^{-1}\ \mathrm{cm}^{-1}$). This equation has been verified by A. F. Gibson[8] for frequencies up to 3.9×10^{10} Hz ($\lambda_0 = 8.3$ mm).

For material having an electrical conductivity of the order of $0.1\ \Omega^{-1}\ \mathrm{cm}^{-1}$ and high dielectric constant of the order of 10, the electronic contribution to the dielectric constant is fairly small. For example, if $\langle\tau\rangle = 10^{-12}\ \mathrm{s}^{-1}$ we have $\epsilon_e/\epsilon_0 \simeq 1$; however, for higher conductivity material, ϵ may become negative. If we have high conductivity material so that $|\epsilon_e| \gg \epsilon_L$ we have approximately

$$\epsilon = -r\sigma_0\langle\tau\rangle. \tag{37}$$

The complex permittivity ϵ^* is then given by the equation

$$\epsilon^* = -\sigma_0 r\langle\tau\rangle - i\sigma_0/\omega. \tag{38}$$

In terms of n and k we have then

$$\mathrm{n}^2(1-\mathrm{k}^2) = -\sigma_0 r\langle\tau\rangle/\epsilon_0, \tag{39}$$

$$2\mathrm{n}^2\mathrm{k} = \sigma_0/\omega\epsilon_0, \tag{40}$$

and from these equations we obtain an equation for k, namely

$$\frac{2\mathrm{k}}{\mathrm{k}^2-1} = \frac{1}{r\langle\tau\rangle\omega}. \tag{41}$$

We shall suppose that $\omega\langle\tau\rangle \ll 1$ so that $k \simeq 1$. This gives for n the value

$$\mathrm{n} = (\sigma_0/2\omega\epsilon_0)^{\frac{1}{2}} \tag{42}$$

and for the absorption coefficient

$$\alpha = (2\sigma_0\omega/\epsilon_0 c^2)^{\frac{1}{2}}. \tag{43}$$

The material in this condition is metallic in behaviour and we have the well-known skin effect in which the radiation penetrates to a depth inversely proportional to the $\frac{1}{2}$ power of the frequency and also to the $\frac{1}{2}$ power of the conductivity.

[8] *Proc. Phys. Soc.* B (1956) **69**, 488.

For *p*-type Ge, the λ^2 variation of the free-carrier absorption is not obeyed, and a well-marked structure between 2 μm and 10 μm has been observed by H. B. Briggs and R. C. Fletcher[9] and also by W. Kaiser, R. J. Collins and H. Y. Fan.[10] In this case the absorption coefficient is about ten times greater than for *n*-type Ge of comparable purity. By considering transitions between the various overlapping bands, which together make up the valence band of Ge, A. H. Kahn[11] has been able to account in considerable detail for the observed structure. The types of transition involved are shown in Fig. 10.2; this also shows the structure of the

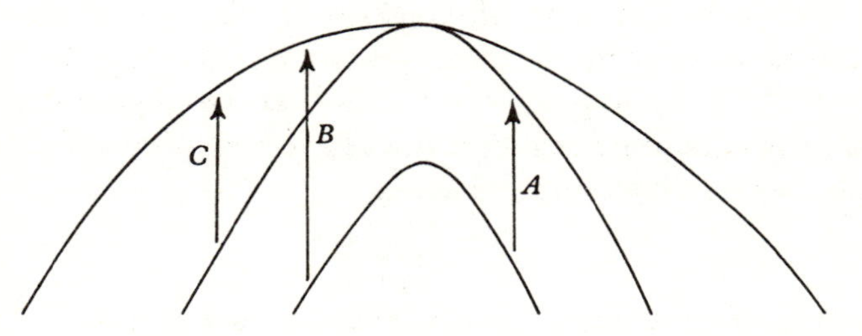

Fig. 10.2. Transitions between overlapping bands.

valence band of Ge, which is believed to consist of two nearly 'spherical' bands degenerate at $\mathbf{k}=0$, with another band split off from these. Agreement with the absorption measurements is obtained if the splitting is taken as equal to 0.28 eV. No similar fine structure has been observed for *p*-type Si which obeys the λ^2 law quite well.[12] In this case the band splitting is thought to be much less (~0.05 eV).

10.2.1 Absorption in the far infra-red at low temperatures

We have seen in § 5.4 that some electrical conductivity remains at very low temperatures in some semiconductors even although all the current carriers should have been frozen into donor or acceptor centres. This has been attributed to a hopping process which can take place when some of the donors or acceptors are empty due to compensation. Under these conditions according to equation (30) we should expect some residual absorption of far infra-red radiation, although because of the

[9] *Phys. Rev.* (1952) **87**, 1130; *ibid.* (1953) **91**, 1342.
[10] *Phys. Rev.* (1953) **91**, 230; *ibid.* (1953) **91**, 1380.
[11] *Phys. Rev.* (1955) **97**, 1647.
[12] H. Y. Fan, M. L. Shepherd and W. G. Spitzer, *Proceedings of Atlantic City Photoconductivity Conference* (John Wiley and Sons, and Chapman and Hall, 1956), p. 184.

high frequency of such radiation the absorption would not be expected to be simply related to the d.c. conductivity.

Such absorption has been found in Si by L. J. Neuringer, R. C. Milward and R. L. Aggarwal[13] in the wavelength range 200–1000 μm, as shown in Fig. 10.3. Based on the hopping mechanism, a theoretical

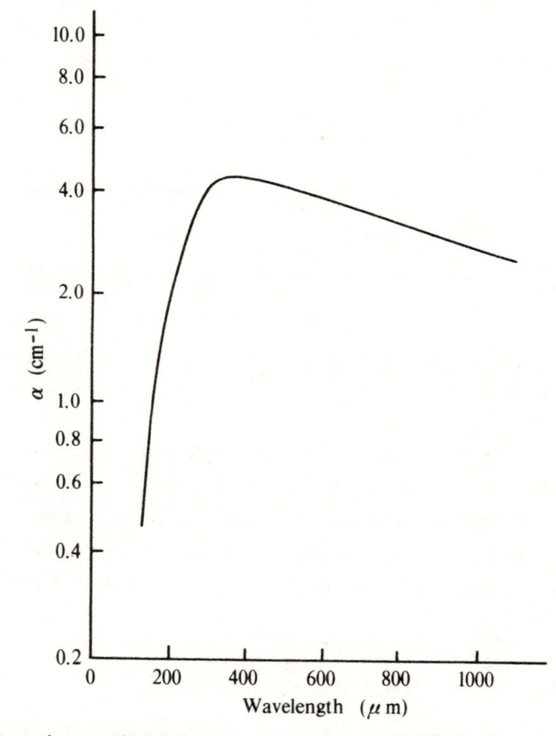

Fig. 10.3. Absorption coefficient α for n-type Si at 2.5 °K. $N_d = 1.4 \times 10^{17}$ cm^{-3}, $K = 0.13$. (After L. J. Neuringer *et al., loc. cit.*)

treatment by J. Blinowski and J. Myceilski[14] gives very good agreement with the experimentally determined absorption. Similar absorption at low temperature in the far infra-red has been found for compensated p-type Ge by R. A. Smith, S. Zwerdling, S. N. Dermatis and J. P. Theriault[15] as shown in Fig. 10.4. The absorption was found to vary with compensation ratio $K = N_d/N_a$ but to be considerably greater than

[13] *Phys. Rev. Lett.* (1965) **15**, 664; *Proc. VIIIth Int. Conf. on Phys. of Semiconductors, J. Phys. Soc. Japan* (1966) **21**, 582.

[14] *Phys. Rev.* (1964) **136**, A266; (1965) **140**, A1024.

[15] *Proc. IXth Int. Conf. on Phys. of Semiconductors* (Nauka, 1968), p. 149.

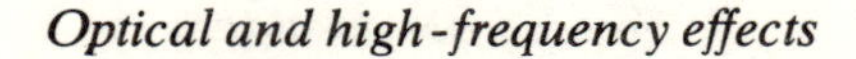

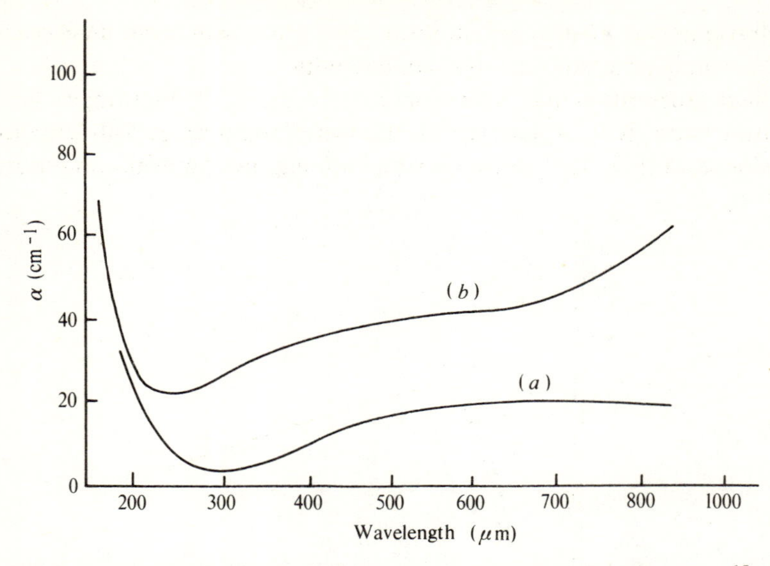

Fig. 10.4. Absorption coefficient α for p-type Ge at 4.3 °K. (a) $N_a = 9\times10^{15}$ cm^{-3}, $K = 0.11$. (b) $N_a = 1.6\times10^{16}$ cm^{-3}, $K = 0.50$. (After R. A. Smith *et al.*, *loc. cit.*)

would be given by the theory of Blinowski and Myceilski. A modified version of the theory by E. Kaczmarck and Z. W. Gortel,[16] taking account of the degeneracy of the valence band, gives much better agreement though not in such detail as for Si. The rise of absorption at wavelengths less than 200 μm is due to ionization of shallow impurity centres.

10.3 Plasma resonance

When we have the condition $\omega\langle\tau\rangle \gg 1$, the permittivity ϵ is given by the equation

$$\epsilon = \epsilon_L - \frac{ne^2}{m_c\omega^2}, \tag{44}$$

the conductivity σ being given by equation (33). The condition $\epsilon = 0$ is of special interest. The frequency ω_p for which this takes place is given by

$$\omega_p = e(n/m_c\epsilon_L)^{\frac{1}{2}} \tag{45}$$

and is known as the plasma frequency; it is the frequency at which an undamped plasma of electrons and positive ions can oscillate as a whole.

[16] *Proc. XIIth Int. Conf. on Phys. of Semiconductors* (Teubner, 1974), p. 1207.

When $\omega\langle\tau\rangle$ is not very large the resonance will not be sharp, neither will the absorption coefficient be infinite. It is fairly easy to show from equations (21), (23) that the resonance will not be at all sharp, unless $\omega\langle\tau\rangle$ is quite appreciably greater than unity – say $\omega\langle\tau\rangle \sim 10$ at the resonant frequency.

10.4 High-frequency effects in a magnetic field

When we have a magnetic field present we must see what modifications to the conductivity equations are required when the frequency is high. When $\omega\langle\tau\rangle \ll 1$, the formula for the low-field Hall constant is unchanged. H. E. M. Barlow[17] has measured Hall constants for Ge at frequencies up to 10^{10} Hz and found substantially the same values as for d.c. measurements.

To obtain the equations of motion of electrons in a magnetic field it is simpler to proceed in two stages. In the absence of a magnetic field the equation of motion of an electron after averaging over various collision times may be written in the form

$$m_e \frac{d\overline{v_x}}{dt} + m_e \frac{\overline{v_x}}{\tau} = -e\mathscr{E}_x. \tag{46}$$

This is clearly valid for a steady field as we obtain the steady-state solution

$$\overline{v_x} = -e\tau\mathscr{E}_x/m_e. \tag{47}$$

For an alternating field of frequency $\omega/2\pi$ we have

$$\overline{v_x} = -\frac{e\tau\mathscr{E}_x}{m_e(1+i\omega\tau)}. \tag{48}$$

This is exactly equivalent to equation (15) and we may then proceed to average over the values of τ as a function of the energy E. When we have a magnetic field in the z-direction, the equations for motion in the (x, y)-plane corresponding to equation (46) are

$$\begin{aligned} \frac{d\overline{v_x}}{dt} + \frac{\overline{v_x}}{\tau} &= -\frac{e}{m_e}\mathscr{E}_x - \omega_c\overline{v_y}, \\ \frac{d\overline{v_y}}{dt} + \frac{\overline{v_y}}{\tau} &= -\frac{e}{m_e}\mathscr{E}_y + \omega_c\overline{v_x}, \end{aligned} \tag{49}$$

[17] *Proc. Instn Elect. Engrs.* (1954) **102**B, 179, 186; *ibid.* (1957) **104**C, 35.

where $\omega_c = eB/m_e$, and we may now use these equations to find $\overline{v_x}$ and $\overline{v_y}$.

10.4.1 Cyclotron resonance

We shall use equations (49) to treat cyclotron resonance, which we have already discussed in § 2.5 without taking account of relaxation effects. It turns out to be simpler if we consider a circularly polarized electric field; for right-hand circular polarization we have

$$\left.\begin{aligned}\mathscr{E}_x &= \mathscr{E}\cos\omega t,\\ \mathscr{E}_y &= \mathscr{E}\sin\omega t.\end{aligned}\right\} \tag{50}$$

Multiplying the second of equations (49) by i and adding to the first we have

$$\frac{d}{dt}(\overline{v_x}+i\overline{v_y})+\left(\frac{1}{\tau}-i\omega_c\right)(\overline{v_x}+i\overline{v_y}) = -\frac{e\mathscr{E}}{m_e}e^{i\omega t}. \tag{51}$$

The solution of equation (51) corresponding to a forced oscillation is given by

$$(\overline{v_x}+i\overline{v_y}) = -\frac{\tau e(\mathscr{E}_x+i\mathscr{E}_y)/m_e}{1+i(\omega-\omega_c)\tau}. \tag{52}$$

The average velocity is therefore out of phase with the field unless $\omega = \omega_c$. Comparing with equations (15), (17), (18), we see that

$$\sigma_r = \frac{ne^2}{m_e}\left\langle\frac{\tau}{1+(\omega-\omega_c)^2\tau^2}\right\rangle, \tag{53}$$

$$\epsilon_r = \epsilon_L - \frac{ne^2}{m_e}\left\langle\frac{\tau^2}{1+(\omega-\omega_c)^2\tau^2}\right\rangle\left(\frac{\omega-\omega_c}{\omega}\right); \tag{54}$$

for left-hand circular polarization we have similarly

$$\sigma_l = \frac{ne^2}{m_e}\left\langle\frac{\tau}{1+(\omega+\omega_c)^2\tau^2}\right\rangle, \tag{55}$$

$$\epsilon_l = \epsilon_L - \frac{ne^2}{m_e}\left\langle\frac{\tau^2}{1+(\omega+\omega_c)^2\tau^2}\right\rangle\left(\frac{\omega+\omega_c}{\omega}\right). \tag{56}$$

If $\omega_c\langle\tau\rangle = B\mu_c \gg 1$ we clearly have a sharp resonance for right-hand circular polarization when $\omega = \omega_c = eB/m_e$. This is the cyclotron resonance already discussed in § 2.5. The value of σ at resonance is equal to the d.c. value and falls on either side to a much lower value if $\omega_c\langle\tau\rangle \gg 1$, so that there will therefore be a marked increase in the absorption coefficient near resonance; the width of the resonance will be of the order of $1/\langle\tau\rangle$ radians s^{-1}, and unless $\omega_c\langle\tau\rangle \gg 1$ the resonance will be very flat. Since for fields of the order of 0.1 T (10^3 G) ω_c comes in the microwave range ($\omega/2\pi = 2.8\times10^9$ Hz for $m_e = \boldsymbol{m}$), $\langle\tau\rangle$ must be somewhat greater than 10^{-10} s; this can only be achieved at low temperatures for pure single crystals. The condition $\omega_c\langle\tau\rangle \gg 1$ for a sharp resonance is readily understood; it is simply the condition that a particle may describe its periodic orbit a number of times before making a collision which throws it out of phase with the exciting field. While the particle remains in phase it may extract energy from the alternating field. The resonant frequency depends only on the effective mass m_e, the electronic charge and the magnetic induction B. We have, therefore, a direct measure of m_e. We have dealt with the simple case in which the effective mass is scalar. We shall deal later with the extension to a more general tensor form of effective mass.

We may note that only the right-hand circular polarization gives a resonance, left-hand circular polarization does not. The difference in permittivity for the two polarizations leads to two refractive indices n^+, n^- which differ slightly. This leads to the Faraday rotation of the plane of polarization of plane polarized radiation incident on the crystal. The above calculation has been made for electrons; for positive holes we must replace ω_c by $-\omega_c$ and so we get a resonance with left-hand, circular polarization. This has been used to identify the sign of the carriers associated with a particular resonance. Another condition for observation of a resonance is that the a.c. field should be able to penetrate the sample, i.e. that the absorption coefficient α should not be too high at the operating frequency. The condition for this has already been discussed. Faraday rotation has been observed in a number of semiconductors[18] but is especially marked in those with small effective electron mass such as InSb. It has been discussed in some detail by J. Houghton and S. D. Smith[19] and also by H. Piller.[20]

When we have a number of closed constant-energy surfaces, as for n-type Ge, the equations of motion for the electrons become extremely complex. Fortunately, each surface can be treated independently since the resonance, except in certain directions of the magnetic field, will occur at different frequencies for each. To obtain the resonance frequency we may assume $\langle\tau\rangle$ to be infinite and simply find the condition

[18] S. D. Smith, T. S. Moss and K. W. Taylor, *Phys. Chem. Solids* (1959) **11**, 131.
[19] *Infra-red Physics* (Oxford University Press, 1966), Ch. 4.
[20] *Semiconductors and Semimetals* (Academic Press, 1972) **8**, 103.

for periodic motion of electrons or holes in the static magnetic field; this frequency, being the cyclotron frequency, will also be the resonant frequency. Let us take axes x, y, z along the principal axes of one of the constant-energy ellipsoids, the principal values of the effective masses being m_1, m_2, m_3. The equations for motion for electrons in a constant magnetic field, the components of magnetic induction along the axes being B_x, B_y, B_z, are

$$\left.\begin{aligned} m_1\frac{dv_x}{dt} &= -eB_zv_y + eB_yv_z, \\ m_2\frac{dv_y}{dt} &= -eB_xv_z + eB_zv_x, \\ m_3\frac{dv_z}{dt} &= -eB_yv_x + eB_xv_y. \end{aligned}\right\} \quad (57)$$

For periodic motion with frequency $\omega/2\pi$ in a closed orbit we may take v_x, v_y, v_z each proportional to $e^{i\omega t}$. On substituting in equation (57) we obtain three homogeneous equations for v_x, v_y, v_z. These have a non-zero solution only if the determinant of the coefficients of v_x, v_y, v_z vanishes. There are two solutions

$$\omega = 0, \quad (58)$$

or

$$\omega^2 = \frac{e^2}{m_1m_2m_3}\{m_1B_x^2 + m_2B_y^2 + m_3B_z^2\}. \quad (59)$$

For the first solution we have $v_x : v_y : v_z = B_x : B_y : B_z$ and this represents uniform motion parallel to the direction of the magnetic field. When the direction of the magnetic field lies along one of the principal axes of the constant-energy ellipsoid we have

$$\left.\begin{aligned} \omega = \omega_1 &= \frac{eB}{(m_2m_3)^{\frac{1}{2}}}, \\ \text{or} \quad \omega = \omega_2 &= \frac{eB}{(m_1m_3)^{\frac{1}{2}}}, \\ \text{or} \quad \omega = \omega_3 &= \frac{eB}{(m_2m_1)^{\frac{1}{2}}}. \end{aligned}\right\} \quad (60)$$

If the field has direction cosines (l, m, n) with respect to the principal axes we have

$$\omega = \{l^2\omega_1^2 + m^2\omega_2^2 + n^2\omega_3^2\}^{\frac{1}{2}}. \quad (61)$$

The semi-classical treatment which we have given fails when $\omega_c\langle\tau\rangle$ is much greater than one, and quantum theory must be used. We defer such a treatment till we have considered further the effective mass approximation for high magnetic fields (see § 12.5.2). We may say, however, that the resonant condition obtained by quantum theory is the same as that given by equation (61).

In order to be able to calculate the shape of the resonance lines a more elaborate approach is required;[21] L. Gold, W. M. Bullis and R. A. Campbell[22] have carried through the analysis with an energy-dependent relaxation time for the 'many-valley' condition, and H. J. Zeiger, B. Lax and R. N. Dexter[23] have dealt with warped constant-energy surfaces as found in p-type Si and Ge.

Cyclotron resonance has now been observed in a considerable number of semiconductors, the number increasing as purification techniques enable the condition $\omega_c\langle\tau\rangle$ somewhat greater than one to be obtained for more materials. It provides the most direct method of obtaining not only effective masses at maxima and minima in the energy bands but also the general form of the bands and without this vital technique our knowledge of semiconductors would be meagre indeed. An extensive review of the subject has been given by B. Lax and J. G. Mavroides.[24]

10.4.2 Magnetic quantization

We have pointed out previously that the treatment which was given of the motion of electrons in a strong magnetic field would be expected to break down if $\omega_c\langle\tau\rangle \gg 1$, a condition which may also be expressed in the form $B\mu \gg 1$. In this condition we should expect the motion of an electron to be in a periodic orbit and so to be quantized; this also is just the condition under which cyclotron resonance should be observed. It is fortunate that quantization of the motion leads to the same values for the resonant frequencies as the semi-classical treatment given in § 10.4.1; this may be shown in an elementary way as follows. An electron moving in a circular orbit of radius r with period $2\pi/\omega_c$ will have angular momentum $m_e r^2 \omega_c$, where $\omega_c = eB/m_e$, and the angular momentum according to elementary quantum theory should be equal to

[21] See, for example, J. M. Luttinger and R. R. Goodman, *Phys. Rev.* (1955) **100**, 673; J. M. Luttinger, *Phys. Rev.* (1956) **102**, 1030.

[22] *Phys. Rev.* (1956) **103**, 1250.

[23] *Phys. Rev.* (1957) **105**, 495.

[24] *Solid State Physics* (Academic Press, 1960) **11**, 261.

$s\hbar$, where s is an integer greater than or equal to one; thus we have

$$r^2 = s\hbar/m_e\omega_c. \tag{62}$$

The kinetic energy E of the electron is given by the equation

$$E = \tfrac{1}{2}m_e r^2\omega_c^2 = \tfrac{1}{2}s\hbar\omega_c \tag{63}$$

so that the lowest energy in the conduction band is now no longer equal to zero but to $\frac{1}{2}\hbar\omega_c$; the bottom of the conduction band is therefore raised by an amount proportional to the magnetic field. A proper quantum-mechanical treatment by L. Landau[25] shows that for free electrons only *odd* values of s are permitted; the allowed energy levels are then given by

$$E_s = (s + \tfrac{1}{2})\hbar\omega_c, \tag{64}$$

where now s is *any* positive integer. If the magnetic field is directed along the z-axis the quantization will only refer to motion in the x- and y-directions, the particle being free to move parallel to the z-axis, the energy of an electron in the conduction band is then given by the equation (see § 12.5)

$$E = (s + \tfrac{1}{2})\hbar\omega_c + (\hbar^2/2m_e)k_z^2. \tag{64a}$$

Transitions between two levels having quantum numbers s and $(s+1)$ may be excited with radiation having quantum energy $\hbar\omega_c$, and hence frequency $\omega_c/2\pi$; this is just the cyclotron frequency.

We shall later consider in greater detail the effect of a strong magnetic field on the electron energy levels of a semiconductor. A treatment using quantum theory also gives equation (61) for the resonance condition but in addition introduces some new features (see § 12.5.2).

10.4.3 The magnetic band-shift

The shift in the position of the bottom of the conduction band will be large when m_e is small, making ω_c large. For InSb for which $m_e = 0.013\boldsymbol{m}$, the shift corresponds to 3.85×10^{-3} eV for $B = 1$ T. For the valence band a similar shift due to holes will be much smaller, owing to the much larger value of m_h. This effect has been observed in InSb by E. Burstein, G. S. Picus, H. A. Gebbie and F. Blatt[26] as a shift in the fundamental absorption edge, on application of a magnetic induction up to nearly 6 T. The effect has also been observed in InSb and in InAs by

[25] *Z. Phys.* (1930) **64**, 629. [26] *Phys. Rev.* (1956) **103**, 826.

S. Zwerdling, R. J. Keyes, S. Foner, H. H. Kolm and B. Lax[27] using pulsed magnetic fields giving values of B up to 25 T. The value of the shift found is somewhat less at high fields than that predicted from equation (64); this shows that the effective mass for electrons increases as the energy is increased.

10.5 The fundamental absorption

In the previous sections we have considered optically induced transitions between states of different energy *within* either the conduction or valence band with the absorption of a photon. We now consider transitions in which an electron is excited from the valence band to the conduction band with absorption of a photon of energy approximately equal to ΔE, the energy of the forbidden gap. This we have called (see § 10.1) the fundamental absorption. Here we must distinguish two types of transition, those involving only one (or more) photons and those involving also lattice energy in the form of phonons. For the present we shall assume that only *one* photon is involved in a transition, leaving for later consideration multi-photon transitions which have much smaller probability (see § 10.5.5). Transitions involving no phonons we call *direct* transitions and those involving phonons *indirect* transitions.

We must also distinguish between two types of semiconductor since their behaviour is rather different; the first type consists of semiconductors which have the same value $\mathbf{k}_{\text{min.}}$ of the wave-vector $\mathbf{k}$ (usually $\mathbf{k}=0$) for the lowest energy state in the conduction band, as for the highest energy state in the valence band ($\mathbf{k}_{\text{max.}}$); the second type consists of those semiconductors for which this is not so. Si and Ge are typical of this second class. Examples of the first type are the intermetallic compounds InSb and GaAs and the lead chalcogenides PbS etc. The transitions leading to the fundamental absorption in the first class are somewhat simpler than those for the second and we shall first of all consider semiconductors of this type. To begin with we shall assume that we have non-degenerate bands as shown in Fig. 10.5 in which we take $\mathbf{k}_{\text{min.}} = \mathbf{k}_{\text{max.}} = 0$.

10.5.1 Direct transitions, $\mathbf{k}_{\text{min.}} = \mathbf{k}_{\text{max.}}$

In an absorption process, crystal momentum must be conserved (see below). The momentum $\mathbf{P}_p$ given up by the incident photon, is equal to

[27] *Phys. Rev.* (1956) **104**, 1805.

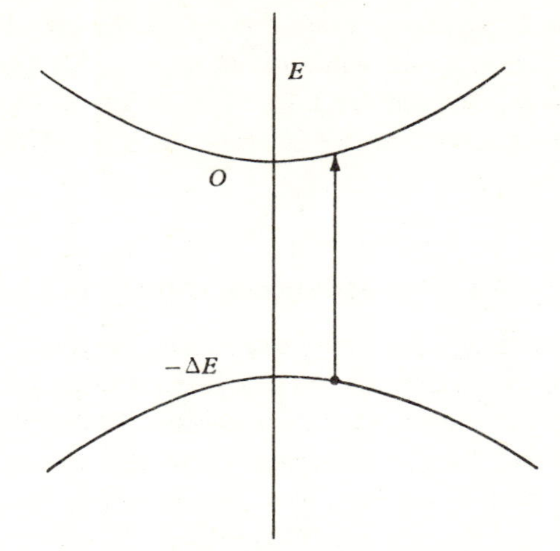

Fig. 10.5. Direct transitions, $\mathbf{k}_{\min} = \mathbf{k}_{\max}$.

$(\boldsymbol{h}/\lambda)\mathbf{i}$, where $\mathbf{i}$ is a unit vector along the direction in which the photon is travelling before it is absorbed, and λ is the wavelength of the incident radiation. The equation expressing conservation of momentum is then

$$\mathbf{P}_2 - \mathbf{P}_1 = (\boldsymbol{h}/\lambda)\mathbf{i}, \tag{65}$$

where $\mathbf{P}_1$ and $\mathbf{P}_2$ are the crystal momenta associated with the states occupied by the electron before and after absorption. If λ_1 is the wavelength associated with an electron whose crystal momentum is $\mathbf{P}_1$ ($|P_1| = \boldsymbol{h}/\lambda_1$), then for an electron of energy of the order of $\boldsymbol{k}T$, with $T = 300\,°\text{K}$, λ_1 is of the order of 5×10^{-7} cm; if λ corresponds to a wavelength in the near infra-red, say $\lambda = 10^{-4}$ cm, then $\lambda \gg \lambda_1$. Thus $|\mathbf{P}_1 - \mathbf{P}_2|$ is very small compared with $|\mathbf{P}_1|$ and we may take $\mathbf{P}_1 = \mathbf{P}_2$.

Conservation of crystal momentum is a consequence of quantum mechanics. The probability of a transition taking place under a perturbing potential F is proportional to the matrix element $|M_{if}|^2$, where

$$M_{if} = \int \psi_i F \psi_f^* \, d\mathbf{r} \tag{66}$$

and ψ_i and ψ_f are the initial and final wave-functions. In this case F has the form

$$F = A \exp[2\pi i(\mathbf{r} \cdot \mathbf{i})/\lambda],$$

and

$$\psi_i = U_{k_1}(\mathbf{r}) \exp[i(\mathbf{k}_1 \cdot \mathbf{r})], \tag{67}$$

$$\psi_f = U_{k_2}(\mathbf{r}) \exp[i(\mathbf{k}_2 \cdot \mathbf{r})] \tag{68}$$

(see § 2.2). The integral therefore contains the factor

$$\exp\,[i(\mathbf{k}_1-\mathbf{k}_2+2\pi\mathbf{i}/\lambda)\cdot\mathbf{r}],$$

which varies rapidly and periodically unless

$$\mathbf{k}_1-\mathbf{k}_2+2\pi\mathbf{i}/\lambda=0. \tag{69}$$

Unless equation (69) is satisfied the matrix element M_{if} will be very small. On multiplying the equation (69) on both sides by $\hbar$ we obtain equation (65), since $\mathbf{k}=\mathbf{P}/\hbar$.

In transitions from one band to another we thus see that only 'vertical' transitions are 'allowed', and that we have a selection rule which may be expressed in the form $\Delta\mathbf{k}=0$. This is illustrated in Fig. 10.5. It is interesting to note that if an electron is in a state with crystal momentum $\mathbf{P}_1$ in the full band, it goes into a state with crystal momentum $\mathbf{P}_1$ in the conduction band leaving behind a hole in the full band. The velocity of the hole is equal to $-P_1/m_h$ (see § 2.5) so that the electron and hole move off in opposite directions. In this case the *minimum* frequency for which such a direct transition will take place is given exactly by equation (12), so that the absorption edge begins at this frequency if no other types of transition are of importance in this frequency range, and it is of considerable interest to determine the form of the absorption curve as a function of wavelength immediately on the short-wave side of the edge. The form depends on whether the transition is or is not an allowed one, apart from the momentum condition, i.e. on the form of the functions $U_{k_1}(\mathbf{r})$, $U_{k_2}(\mathbf{r})$ in equations (67) and (68). If these are such as to give an allowed transition then the probability P_k of a transition between two such states is very nearly independent of $\mathbf{k}$. Let us consider transitions from the valence band originating from states having values of $\mathbf{k}$ in the interval $\mathbf{k}$ to $\mathbf{k}+d\mathbf{k}$. The energy of a photon required to induce a transition is then given by

$$\boldsymbol{h}\nu=E_c(\mathbf{k})-E_v(\mathbf{k}). \tag{70}$$

The probability of absorption of a quantum in the frequency interval $d\nu$ is then proportional to the number of states in the *valence* band with energies lying between $-\Delta E-E'$ and $-\Delta E-E'-dE'$. Let us assume for the moment that the conduction band is flat near $\mathbf{k}=0$, i.e. that $E_c(\mathbf{k})$ is constant. Then if we write $E_v(\mathbf{k})=-\Delta E-E'$ we have

$$\boldsymbol{h}\,d\nu=dE'. \tag{71}$$

The number of transitions $N_t\,d\nu$ in the interval $d\nu$ is then given by

$$N_t d\nu = P_k N(E')\,dE',$$

where $N(E')\,dE'$ is the number of states with E' in the interval E' to $E'+dE'$; as we have seen

$$N(E') = AE'^{\frac{1}{2}}, \tag{72}$$

where A is a constant (§ 2.3). The absorption coefficient α_d may therefore be written in the form

$$\left.\begin{aligned} \alpha_d &= A'P_k(\boldsymbol{h}\nu - \Delta E)^{\frac{1}{2}}, & \boldsymbol{h}\nu &> \Delta E, \\ &= 0, & \boldsymbol{h}\nu &\leqslant \Delta E, \end{aligned}\right\} \tag{73}$$

where A' is a constant, since $E' = \boldsymbol{h}\nu - \Delta E$ from equation (70). If the conduction band is not flat we have

$$\boldsymbol{h}\,d\nu = \frac{\partial E_c}{\partial k}\,dk - \frac{\partial E_v}{\partial k}\,dk$$

and so the effect of this is to multiply the absorption coefficient by a constant; for allowed transitions we may therefore write, since P_k is constant,

$$\left.\begin{aligned} \alpha_d &= B(\boldsymbol{h}\nu - \Delta E)^{\frac{1}{2}}, & \boldsymbol{h}\nu &> \Delta E, \\ &= 0, & \boldsymbol{h}\nu &\leqslant \Delta E, \end{aligned}\right\} \tag{74}$$

where B is a constant. This holds only for a limited range of values of $(\boldsymbol{h}\nu - \Delta E)$. For larger values the absorption rises to a value which varies only slowly with ν, the exact slope of the curve depending on the band shape.

For values of $(\boldsymbol{h}\nu - \Delta E)$ which are comparable with the exciton binding energy (see § 3.5), it has been shown by G. G. Macfarlane, T. P. McLean, J. E. Quarrington and V. Roberts[28] that equation (74) must be modified when account is taken of the Coulomb interaction between the free hole and electron. As $\boldsymbol{h}\nu \to \Delta E$ the absorption does not tend to zero but to a steady value. For $\boldsymbol{h}\nu < \Delta E$ the absorption merges continuously into that due to the higher excited states of the exciton. This is shown in Fig. 10.6 (see also Fig. 10.9).

Allowed transitions take place, for example, if the valence band wave-functions are derived from *s*-states of the individual atoms, and the conduction band wave-functions from *p*-states; if the latter were

[28] *Proc. Phys. Soc.* (1958) **71**, 863.

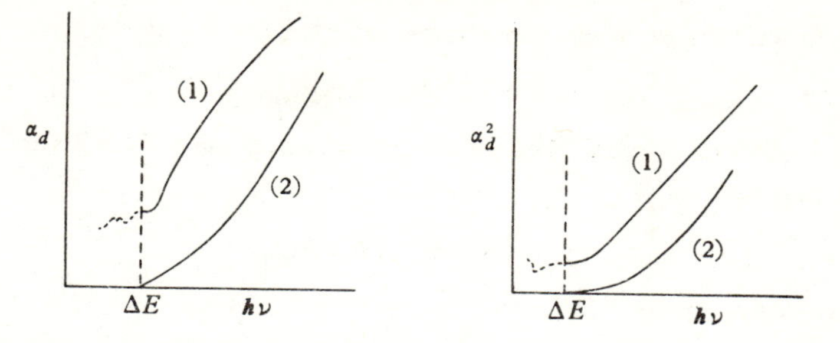

Fig. 10.6. Variation of absorption coefficient with frequency for direct transitions. (Curves marked (1) refer to 'allowed' transitions and curves marked (2) to 'forbidden' transitions.)

derived, for example, from *d*-states the transition would not be allowed. In this case it may readily be shown that when $k=0$, $P_k=0$, but as k departs from zero the transition probability is proportional to k^2, i.e. to $(h\nu-\Delta E)$.[29] Thus we may write

$$P_k = \text{const.} \times (h\nu - \Delta E), \tag{75}$$

so that the absorption coefficient α'_d is given by the equation

$$\alpha'_d = C(h\nu - \Delta E)^{\frac{3}{2}}, \tag{76}$$

where C is a constant. The values of B and C may be found by means of quantum mechanics; they have been given by J. Bardeen, F. J. Blatt and L. H. Hall.[30] B may be expressed in the form[31]

$$B = \frac{\pi e^2 (2m_r)^{\frac{3}{2}}}{\text{n}ch^2 m\epsilon_0} f_{if}, \tag{77}$$

where n is the real part of the refractive index and m_r a reduced mass given by the equation

$$\frac{1}{m_r} = \frac{1}{m_e} + \frac{1}{m_h}; \tag{78}$$

f_{if} is a factor of the order of unity known as the oscillator strength of the transition. Equation (77) differs from that given by Bardeen, Blatt and Hall in that m_h in their formula is replaced by m_r. This takes account of the variation of E with **k** in the conduction band; in addition, an extra factor $4\pi\epsilon_0$ is included in the denominator through the use of M.K.S.

[29] *W.M.C.S.*, § 13.3.
[30] *Proceedings of Atlantic City Photo-conductivity Conference* (1954) (John Wiley and Sons and Chapman and Hall, 1956), p. 146. [31] *W.M.C.S.*, § 13.3.

units. If we take $m_e = m_h = m$ we find that for n $= 4$

$$\alpha_d = 6.7 \times 10^4 (h\nu - \Delta E)^{\frac{1}{2}} f_{if} \text{ cm}^{-1} \tag{79}$$

if $h\nu$ is expressed in eV. The value of C in equation (77) is given in the same notation by[31]

$$C = \tfrac{2}{3} B (2m_r/m) f'_{if}/h\nu f_{if}. \tag{80}$$

In this case we have

$$\alpha'_d = 4.5 \times 10^4 f'_{if} \frac{(h\nu - \Delta E)^{\frac{3}{2}}}{h\nu} \text{ cm}^{-1}. \tag{81}$$

For $h\nu = 1$ eV and $(h\nu - \Delta E) = 0.01$ eV, if we take $f_{if} = f'_{if} = 1$, we have $\alpha_d = 6.7 \times 10^3 \text{ cm}^{-1}$ and $\alpha'_d = 45 \text{ cm}^{-1}$.

If we plot α_d^2 against $h\nu$ we should obtain for allowed direct transitions a straight line which we could extrapolate to $\alpha_d = 0$ to obtain ΔE; for other transitions this will not be so. It is more usual to plot α_d against $h\nu$. The form of curves obtained in this way for allowed and forbidden direct transitions are shown in Fig. 10.6. In most cases for which direct transitions have been observed the variation follows more nearly the $\frac{3}{2}$ power law than the $\frac{1}{2}$ power law for allowed transitions.

InSb shows a very steep rise in absorption when pure and is thought from other evidence (see § 13.5) to be a semiconductor of the kind under discussion, although there is a complication due to degeneracy of the valence band. However, the absorption curve found by V. Roberts and J. E. Quarrington[32] shows a long-wave 'tail' (see Fig. 10.7). There are various causes to which this might be attributed – for example, stresses causing distortion of the crystal lattice; these experiments were, however, carried out with high-quality single crystals and this cause would appear to be unlikely. It has also been suggested that transitions involving phonons may be the cause of this tail; we postpone the discussion of this point till we have discussed semiconductors for which $\mathbf{k}_{\text{min.}} \neq \mathbf{k}_{\text{max.}}$.

10.5.2 Direct transitions, $\mathbf{k}_{\text{min.}} \neq \mathbf{k}_{\text{max.}}$

When the lowest energy state in the conduction band does not have the same value of $\mathbf{k}$ as the highest energy state in the valence band, direct transitions may still take place, but these are not the transitions corresponding to the lowest value of $h\nu$ for which transitions between the

[32] *J. Electron.* (1955) **1**, 152.

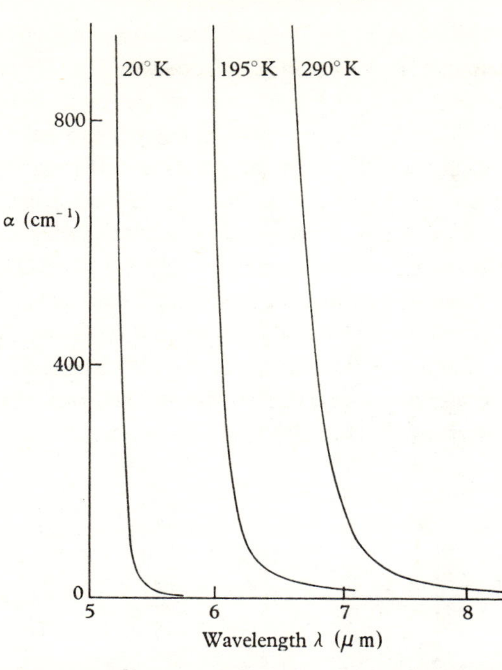

Fig. 10.7. Fundamental absorption of InSb. (After V. Roberts and J. E. Quarrington, *loc. cit.*)

bands are possible. This is illustrated in Fig. 10.8 in which $\mathbf{k}_{max.}$ is taken at $\mathbf{k} = 0$ and $\mathbf{k}_{min.}$ near the edge of the first Brillouin zone in a particular direction in the crystal; $\mathbf{k}_{min.}$ may actually be equal to $\mathbf{k}_e$, a value of $\mathbf{k}$ corresponding to the edge of the zone, as for Ge (see § 13.3), but this need not be so. If the value of E_c, the energy corresponding to the bottom of the conduction band, is considerably less than E_0, the value of the energy at $\mathbf{k} = 0$, the energy involved in making the 'vertical' or direct

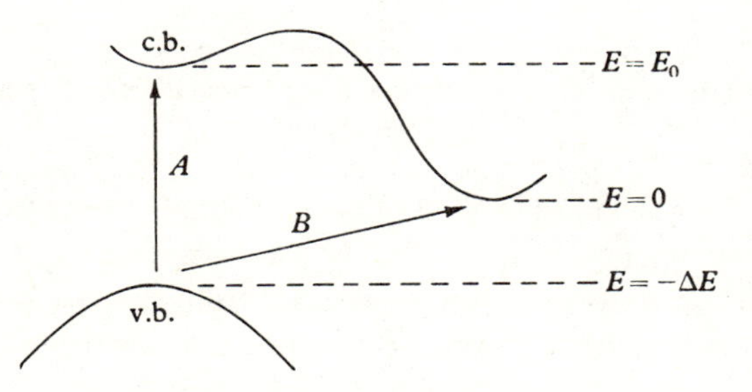

Fig. 10.8. Direct and indirect transitions, $\mathbf{k}_{max} \neq \mathbf{k}_{min}$.

transition (marked A in Fig. 10.8) will be somewhat greater than that involved in the indirect transition (marked B in Fig. 10.8). The absorption of pure Ge obtained with very thin single crystals by W. C. Dash and R. Newman[33] shows a steep rise in absorption near a value of $h\nu$ corresponding to 0.80 eV and is interpreted as being due to a direct transition from the valence band to the conduction band at $\mathbf{k}=0$. This region of the spectrum has been examined with higher spectral resolution over a wide range of temperatures by G. G. Macfarlane, T. P. McLean, J. E. Quarrington and V. Roberts,[34] who have shown that the steep rise in absorption is not only associated with the direct band-to-band transition, but also with an exciton line which they have been able to separate from the continuous absorption. The form of the absorption curve at 77 °K is shown in Fig. 10.9. It will be seen that the continuous

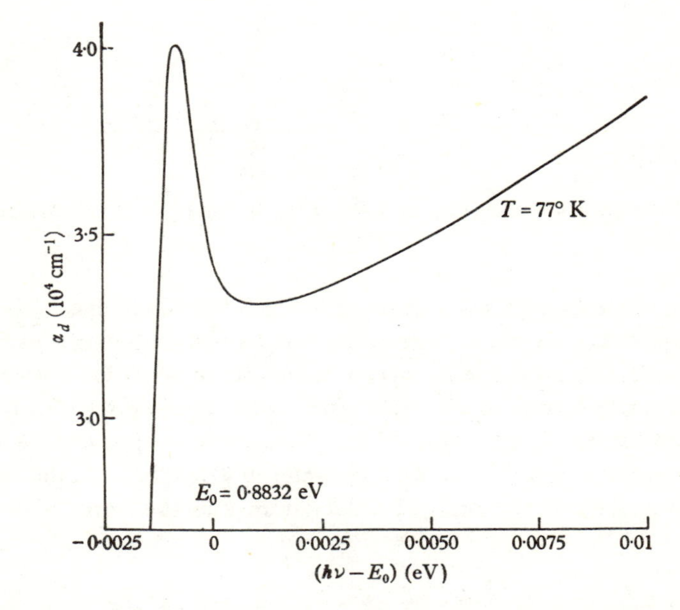

Fig. 10.9. Absorption in Ge due to 'vertical' transitions at $\mathbf{k}=0$. (After G. G. Macfarlane *et al., loc. cit.*)

absorption does not tend to zero as $h\nu \rightarrow \Delta E_0$, ΔE_0 being the value of the forbidden energy gap at $\mathbf{k}=0$, but to a constant value. Moreover, the form of the curve departs quite markedly from the law $\alpha \propto (h\nu - \Delta E)^{\frac{1}{2}}$. Good agreement with the observed variation of absorption with $h\nu$ has

[33] *Phys. Rev.* (1955) **99**, 1151. [34] *Proc. Phys. Soc.* B (1958) **71**, 863.

been obtained by taking account of the Coulomb interaction of the electron and free hole created by the transition. The exciton line has also been observed by S. Zwerdling, Laura M. Roth and B. Lax who have studied its variation in a strong magnetic field.[35]

The absorption at longer wavelengths is due to indirect transitions of the type B. The long-wave limit of the latter corresponds to the true absorption edge, but the energy relationships are now complicated by the fact that the transition B cannot take place as a direct transition, since momentum is very far from being conserved. In order to conserve momentum it must be taken from or given to the crystal lattice by emission or absorption of one or more phonons. It may readily be shown that multi-phonon processes are much less probable than those involving a single phonon; we shall therefore consider only the latter.

10.5.3 Indirect transitions, $\mathbf{k}_{min.} \neq \mathbf{k}_{max.}$

As before, let $\mathbf{P}_1$ be the crystal momentum of an electron before the transition and $\mathbf{P}_2$ the momentum after the transition; then, neglecting the momentum of the absorbed photon, momentum of amount $(\mathbf{P}_1 - \mathbf{P}_2)$ must be supplied by the lattice. This may be achieved in two ways, either by absorption of a phonon of momentum $-(\mathbf{P}_1 - \mathbf{P}_2)$ or by emission of a phonon of momentum $(\mathbf{P}_1 - \mathbf{P}_2)$. For simplicity let us take $\mathbf{k}_1 = 0$, $\mathbf{k}_2 = \mathbf{P}_2/\hbar = \mathbf{k}_{min.}$; then a phonon whose wave-number is $\mathbf{k}_{min.}$ must be absorbed or emitted. From the energy–wave-number relationship for the phonon, obtained from the lattice vibration spectrum (see §§ 8.3 and 10.8), one may obtain E_p the energy of the phonon corresponding to $\mathbf{k}_{min.}$. The minimum frequency for which a transition of type B may take place is now given by

$$h\nu = \Delta E - E_p \tag{82}$$

for absorption of a phonon, and by

$$h\nu = \Delta E + E_p \tag{83}$$

for emission of a phonon, instead of by equation (12). The lowest value of $h\nu$ for this type of transition is given by equation (82), which defines the edge of the fundamental absorption band (apart from very improbable transitions involving absorption of more than one phonon). Since E_p is generally quite small compared with ΔE, both types of transition affect the form of the absorption curve for values of $h\nu$ near ΔE.

[35] *Phys. Rev.* (1958) **109**, 2207.

The values for the absorption coefficient for transitions involving emission and absorption of a single phonon were first treated theoretically by J. Bardeen, F. J. Blatt and L. H. Hall[36] who give a formula for the absorption coefficient involving an electron–phonon interaction constant, which may be evaluated in terms of the relaxation time associated with lattice scattering. This formula does not contain explicitly terms which separate the different phonon contributions. Explicit formulae for these have been given in *W.M.C.S.*, § 13.5. The transitions are considered as taking place through a virtual state which may be regarded as very short lived. Energy is not conserved in the transition to the virtual state but momentum is conserved; energy is conserved only in the overall transition. For either emission or absorption of a phonon the double transition may take place in two ways (see Fig. 10.10). First,

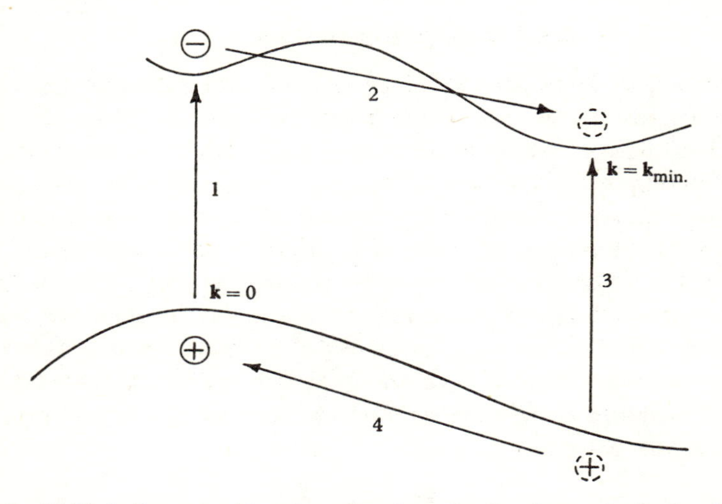

Fig. 10.10. Indirect transitions from the valence band to the conduction band.

an electron may be excited without appreciable change of wave-vector **k** from near the top of the valence band ($\mathbf{k}=0$) and leave a hole with $\mathbf{k}=0$; the electron goes into a state with the *same* value of **k** in the conduction band (transition 1 in Fig. 10.10). Since this state has somewhat higher energy than the minimum of the band, the electron quickly makes a transition to a state with energy near the minimum energy of the band with emission or absorption of a phonon of momentum almost

[36] *Proceedings of Atlantic City Photo-conductivity Conference* (John Wiley and Sons and Chapman and Hall, 1956), p. 146.

equal to $\mathbf{k}_{\text{min.}}$ (transition 2 in Fig. 10.10). The life of the electron in the intermediate state is considered to be so short that energy is not well defined, and is only conserved in the double transition. Alternatively, an electron may be excited 'vertically' from deep in the valence band to a state having a value of $\mathbf{k}$ nearly equal to $\mathbf{k}_{\text{min.}}$ leaving a positive hole deep in the valence band (transition 3 in Fig. 7.7); this hole then makes a transition to a state with energy near the maximum energy of the band, i.e. with $k \simeq 0$, with emission or absorption of a phonon. The formula given by J. Bardeen, F. J. Blatt and L. H. Hall (*loc. cit.*) for the absorption coefficient includes both types of transition but does not separate the contributions from emission and absorption of the phonons. In order to explain the form of the absorption curves for Ge and Si it is necessary to do this, as shown by G. G. Macfarlane and V. Roberts.[37] The two contributions have in fact different long-wave limits as will be seen from equations (82) and (83). If we write α_i, the absorption coefficient due to indirect transitions, in the form

$$\alpha_i = \alpha_e + \alpha_a, \tag{84}$$

α_e and α_a being respectively the contributions due to emission and absorption of phonons, we have

$$\alpha_e = 0, \quad h\nu < \Delta E + E_p, \tag{85}$$

$$\alpha_a = 0, \quad h\nu < \Delta E - E_p. \tag{86}$$

We may determine the shape of the absorption curve (but not the magnitude) from fairly simple considerations. If the wave-vector $\mathbf{k}_{\text{min.}}$ is well separated from $\mathbf{k} = 0$, the transition probability between a state with a small value of $\mathbf{k}$ situated near the top of the valence band and a state with $\mathbf{k} = \mathbf{k}_{\text{min.}} - \mathbf{k}'$ situated near the bottom of the conduction band ($k \ll k_{\text{min.}}$ and $k' \ll k_{\text{min.}}$) will not vary very rapidly with $\mathbf{k}$ and $\mathbf{k}'$, and may as a first approximation be taken as constant. The absorption coefficient then depends on the density of states from which transitions may take place and on the relative probability of emission and absorption of phonons. Clearly, the energy E_p may also be taken as constant and equal to that corresponding to $\mathbf{k}_{\text{min.}}$. For direct transitions, if the value of $\mathbf{k}$ for the initial state is fixed, then the value of $\mathbf{k}$ for the final state is fixed. This is no longer so for indirect transitions and $\mathbf{k}'$ may vary, the momentum difference being taken up by having a phonon of slightly different momentum from that corresponding to $\mathbf{k}' = 0$. We have for the

[37] *Phys. Rev.* (1955) **97**, 1714; *ibid.* (1955) **98**, 1865.

condition of conservation of energy

$$\boldsymbol{h}\nu = \Delta E \pm E_p + E + E', \tag{87}$$

where E is the energy of the electron in the final state in the conduction band and $-\Delta E - E'$ its energy in the initial state in the full band, the zero of energy being taken, as before, at the bottom of the conduction band. Now suppose we fix the value of E'. Then if the frequency ν lies in the interval ν to $\nu + d\nu$ we have $dE = \boldsymbol{h}\, d\nu$. The number of states in this interval is equal to

$$N_c(E)\, dE = aE^{\frac{1}{2}}\, dE = a(\boldsymbol{h}\nu - \Delta E \mp E_p - E')^{\frac{1}{2}}\, dE,$$

where a is a constant. To find the total number of pairs of states between which transitions corresponding to a frequency between ν and $\nu + d\nu$ may take place, we must integrate with respect to E' over the portion of the valence band for which equation (87) can be satisfied. The largest possible value E'_m of E' is given by

$$E'_m = \boldsymbol{h}\nu - \Delta E \pm E_p, \tag{88}$$

the negative sign being for phonon emission and the positive sign for phonon absorption. The density of states in the valence band $N_v(E')\, dE'$ in the interval dE' we may write as $a'E'^{\frac{1}{2}}\, dE'$. The total number of pairs of states $N\, d\nu$ between which transitions may take place when the frequency lies in the interval ν to $\nu + d\nu$ is given by the equation

$$\begin{aligned} N(\nu)\, d\nu &= aa'\boldsymbol{h}\, d\nu \int_0^{E'_m} (E'_m - E')^{\frac{1}{2}} E'^{\frac{1}{2}}\, dE' \\ &= DE'^2_m\, d\nu, \end{aligned} \tag{89}$$

where D is a constant. The absorption coefficient for transitions with absorption of a phonon is therefore proportional to $(\boldsymbol{h}\nu - \Delta E + E_p)^2$. It is also proportional to the number N_p of phonons of energy E_p present, which is given by[38]

$$N_p = 1/(\exp[E_p/\boldsymbol{k}T] - 1). \tag{90}$$

The contribution to the absorption coefficient α_a is therefore given by the equation

$$\left.\begin{aligned} \alpha_a &= \frac{A(\boldsymbol{h}\nu - \Delta E + E_p)^2}{[\exp(E_p/\boldsymbol{k}T) - 1]} \quad (\boldsymbol{h}\nu > \Delta E - E_p), \\ &= 0 \qquad\qquad\qquad\qquad\quad (\boldsymbol{h}\nu \leqslant \Delta E - E_p), \end{aligned}\right\} \tag{91}$$

[38] *W.M.C.S.*, § 8.16.

where A is a slowly varying function of ν.[39] The ratio of the probabilities for emission and absorption of a phonon is equal to $(N_p+1)/N_p$ (see § 14.6). The contribution α_e to the absorption coefficient is therefore given by the equation

$$\left.\begin{aligned} \alpha_e &= \frac{A(h\nu-\Delta E-E_p)^2}{[1-\exp(-E_p/kT)]} \quad (h\nu > E+E_p), \\ &= 0 \qquad\qquad\qquad\qquad\quad (h\nu \leqslant E+E_p). \end{aligned}\right\} \tag{92}$$

The absorption coefficient α_i arising from indirect transitions has therefore the following values, if we write $A' = A\Delta E^2$,

$$\left.\begin{aligned} \alpha_i &= \frac{A'}{\Delta E^2}\left\{\frac{(h\nu-\Delta E-E_p)^2}{1-\exp(-E_p/kT)}+\frac{(h\nu-\Delta E+E_p)^2}{\exp(E_p/kT)-1}\right\} \quad (h\nu > \Delta E+E_p), \\ &= \frac{A'(h\nu-\Delta E+E_p)^2}{\Delta E^2[\exp(E_p/kT)-1]} \quad (\Delta E-E_p < h\nu \leqslant \Delta E+E_p), \\ &= 0 \qquad\qquad (h\nu \leqslant \Delta E-E_p). \end{aligned}\right\} \tag{93}$$

We have so far considered only one type of phonon, but should really have taken account of both longitudinal and transverse acoustic modes and optical modes as well. When the two acoustic modes are taken into account the absorption coefficient α_i may be written in the form

$$\alpha_i = \alpha_{el}+\alpha_{al}+\alpha_{et}+\alpha_{at}; \tag{94}$$

α_{el} and α_{et} are given by equations like (92) with A replaced by A_l, A_t and E_p by E_{pl}, E_{pt}; α_{al} and α_{at} are given by equations like (91) with A replaced by A_l, A_t and E_p by E_{pl}, E_{pt}. In this case the absorption curve breaks into four distinct sections separated by the values of $h\nu$ given by $(\Delta E-E_{pl})$, $(\Delta E-E_{pt})$, $(\Delta E+E_{pt})$, $(\Delta E+E_{pl})$.

For simplicity, let us consider first the case where only a single phonon is of importance. If we plot $\alpha_a^{\frac{1}{2}}$ against $h\nu$ we obtain a straight line intersecting the $h\nu$-axis at $h\nu = \Delta E-E_p$, and whose slope is $A^{\frac{1}{2}}[\exp(E_p/kT)-1]^{-\frac{1}{2}}$. For very low temperatures $E_p/kT \gg 1$ and the slope tends to zero; this is shown in Fig. 10.11. The contribution from α_a therefore tends to zero as T tends to zero, the reason for this being simply that at the low temperature few phonons are excited and so phonon absorption is very improbable. We may note that ΔE is a function of T and so the intercepts with $h\nu$-axis will vary with tempera-

[39] *W.M.C.S.*, § 13.5.2.

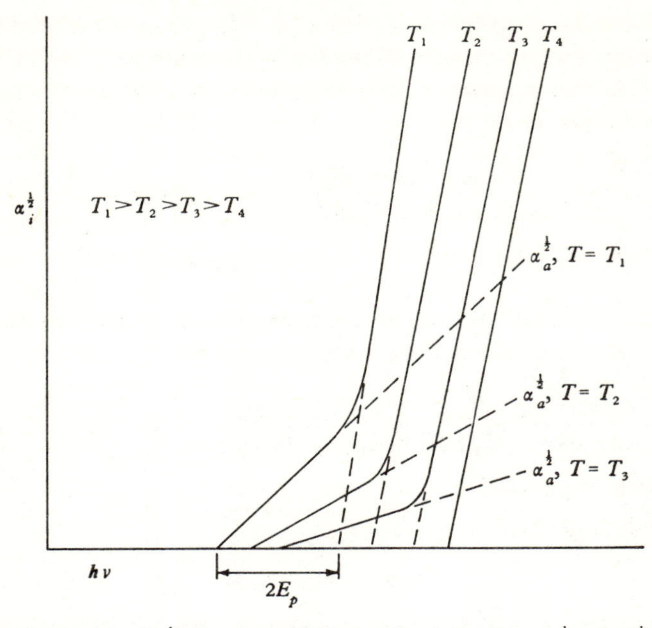

Fig. 10.11. Variation of $\alpha_i^{\frac{1}{2}}$ with temperature. Broken lines show $\alpha_a^{\frac{1}{2}}$ and $\alpha_e^{\frac{1}{2}}$, full lines $\alpha_i^{\frac{1}{2}} = (\alpha_a + \alpha_e)^{\frac{1}{2}}$.

ture. If we plot $\alpha_e^{\frac{1}{2}}$ against $\boldsymbol{h}\nu$ we shall obtain a straight line of slope

$$A^{\frac{1}{2}}[1 - \exp(-E_p/\boldsymbol{k}T)]^{-\frac{1}{2}};$$

this slope tends to the constant value $A^{\frac{1}{2}}$ at low temperatures and increases slightly as the temperature increases. The straight lines corresponding to $\alpha_e^{\frac{1}{2}}$ are also shown broken in Fig. 10.11 (the straight lines of steeper slope). We note that the difference between the intercepts of the lines representing $\alpha_e^{\frac{1}{2}}$ and $\alpha_a^{\frac{1}{2}}$ for a given temperature is equal to $2E_p$ and the value for $\boldsymbol{h}\nu$ corresponding to ΔE is midway between the intercepts. The full-line curves in Fig. 10.11 represent $\alpha_i^{\frac{1}{2}}$.

If the transition between highest state in the valence band is 'forbidden' in the sense already discussed, an extra factor $(\boldsymbol{h}\nu - \Delta E \pm E_p)$ is introduced into the expressions for α_e and α_a as for direct transitions (with $E_p = 0$).[40]

Measurements on pure single crystals of Ge and Si of good quality with high spectral resolution have been reported by G. G. Macfarlane, T. P. McLean, J. E. Quarrington and V. Roberts.[41] The analysis of these

[40] *W.M.C.S.*, p. 513. [41] *Phys. Rev.* (1957) **108**, 1137; *ibid.* (1958) **111**, 1245.

measurements requires an expression for α_i of the form given by equation (94) and reveals contributions to the absorption by phonons associated with both the longitudinal and transverse acoustic modes of the lattice vibration spectrum. For Ge the lowest states in the conduction band are in the $\langle 111 \rangle$ directions and occur at the edge of the first Brillouin zone. The values of E_{pt} and E_{pl} corresponding to the band-edge energies of the transverse and longitudinal acoustic branches of the lattice vibration spectrum for Ge found to fit the absorption measurements are 0.008 eV ($\theta_t = 90$ °K) and 0.027 eV ($\theta_l = 320$ °K). These values of θ_l and θ_t do not agree very well with values calculated theoretically from the form of the diamond-type lattice vibrational spectrum by Y. C. Hsieh[42] who estimates that $\theta_t = 150$ °K and $\theta_l = 350$ °K. In particular, the value deduced for θ_t is much lower than Hsieh's value. This disagreement is not altogether surprising in view of the approximations made in deriving the form of the lattice vibration spectrum. Recent measurements of the lattice vibration spectrum by means of neutron scattering by B. N. Brockhouse and P. K. Iyengar[43] have, however, confirmed the values of θ_t and θ_l deduced from the infra-red absorption measurements. It might also be expected that, since the band edge limits of the spectrum corresponding to the optical modes do not lie very far above that due to the longitudinal acoustic mode, these would also contribute to the absorption. In Ge absorption due to the optical phonons is very weak, since certain selection rules due to symmetry conditions at the band edge in the $\langle 111 \rangle$ directions tend to make transitions involving the optical phonons rather improbable.[44]

For Si both acoustical and optical phonons give an appreciable contribution to the absorption and for each both the longitudinal and transverse types are required to fit the experimental curves obtained with high spectral resolution. The phonon energies and the corresponding values of $\theta(E_p = k\theta)$ corresponding to $\mathbf{k} = \mathbf{k}_{\text{max.}}$ are as follows:

Acoustic phonons	$E_{pt} = 0.0185$ eV	$E_{pl} = 0.0575$ eV,
	$\theta_t \doteq 212$ °K	$\theta_l = 670$ °K.
Optical phonons	$E_{pt} = 0.120$ eV	$E_{pl} = 0.091$ eV,
	$\theta_t = 1420$ °K	$\theta_l = 1050$ °K.

There is now some doubt as to the validity of the interpretation given to

[42] *Chem. Phys.* (1954) **22**, 306.
[43] *Phys. Rev.* (1957) **108**, 894.
[44] R. J. Elliott, *Phys. Rev.* (1957) **108**, 1384.

these phonons. It is thought that the phonon with $\theta = 212$ °K is indeed the TA phonon but that $\theta = 670$ °K corresponds to an optical phonon; $\theta = 1050$ °K and $\theta = 1420$ °K are thought to correspond to combinations TA+O and O+O. If this is so it means that the LA phonon in Si is not observed.[45]

The high resolution used in these measurements has shown up a number of sharp rises in absorption both in Ge and in Si which cannot be interpreted in terms of the theory of indirect transitions between the bands. Some of the experimental curves for Ge are shown in Fig. 10.12

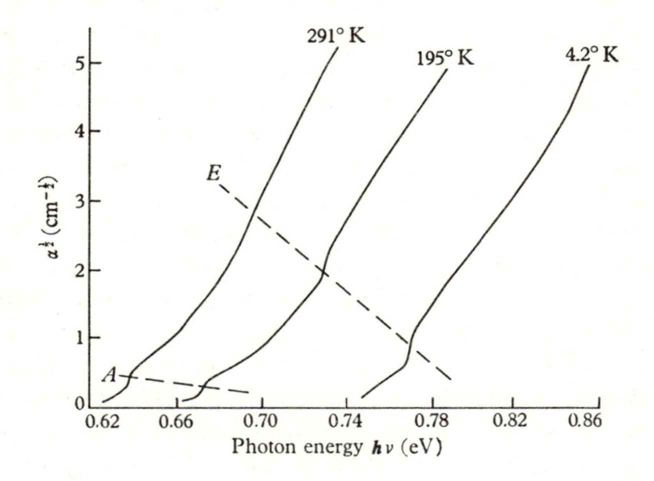

Fig. 10.12. High-resolution absorption curves for Ge. (After G. G. Macfarlane *et al.*, *Phys. Rev.* (1957) **108**, 1377.)

and for Si in Fig. 10.13. The sharp rises in absorption have been interpreted by G. G. Macfarlane, T. P. McLean, V. Roberts and J. E. Quarrington[46] due to the formation of excitons. We shall discuss absorption due to exciton formation in § 10.6.

A great deal of experimental work has developed from these considerations. This is not surprising since the accurate measurement of the absorption near the fundamental edge gives not only the most accurate information on the value of the forbidden energy gap ΔE but also information on the phonon values corresponding to extrema in the lattice vibration spectra, thus supplementing values obtained from neutron scattering. These matters have been dealt with extensively in a

[45] B. N. Brockhouse, *J. Phys. Chem. Solids* (1959) **8**, 400; F. A. Johnson, *Proc. Phys. Soc.* (1959) **73**, 265.
[46] *Phys. Rev.* (1957) **108**, 1377; *ibid.* (1958) **111**, 1245.

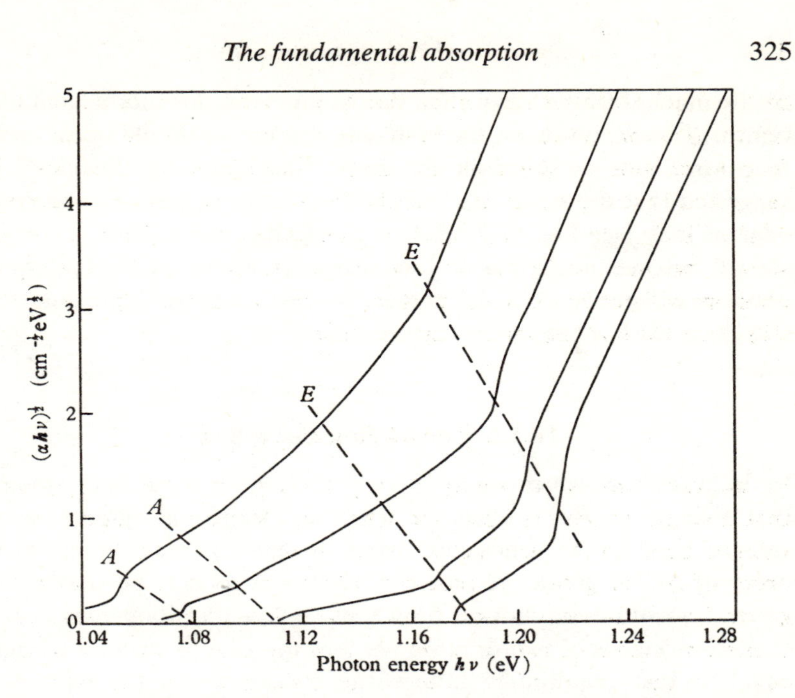

Fig. 10.13. High-resolution absorption curves for Si. (After G. G. Macfarlane *et al.*, *Phys. Rev.* (1958), **111**, 1245.)

number of books and reviews.[47] The last two references give extensive data on the absorption spectra of numerous semiconductors.

10.5.4 Indirect transitions, $\mathbf{k}_{\text{min.}} = \mathbf{k}_{\text{max.}}$

Indirect transitions involving emission and absorption of phonons may also take place as well as the direct 'vertical' transitions when $\mathbf{k}_{\text{min.}} = \mathbf{k}_{\text{max.}}$. Their probability will, however, be much less than for the direct transitions. If we take $\mathbf{k}_{\text{min.}} = \mathbf{k}_{\text{max.}} = 0$, then only phonons with very small values of **k** will be of importance. The acoustic phonons with $k \simeq 0$ have very small energy and will be of no account. It is, however, possible for an optical phonon with $k \simeq 0$ to be emitted or absorbed, the lattice vibration frequency being the *reststrahl* frequency ν_0. When such a phonon is emitted, absorption will take place on the *short-wave* side of the limit due to direct transitions, and will almost certainly be obscured

[47] See, for example, T. P. McLean, *Progress in Semiconductors* (Heywood, 1960) **5**, 53; F. Bassani, *Optical Properties of Solids*, ed. E. D. Haidemenakis (Gordon and Breach, 1970); D. L. Greenaway and G. Harbeke, *Optical Properties and Band Structure of Semiconductors* (Pergamon Press, 1968); T. S. Moss, G. J. Burrell and B. Ellis, *Semiconductor Opto-electronics* (Butterworths, 1973).

by the much stronger absorption due to the latter. For absorption of an optical phonon, however, the resultant absorption should occur on the *long-wave* side of the limit for direct transitions; W. Dumke[48] has suggested that the small 'tail' on the long-wave side of the absorption edge of InSb (see Fig. 10.7) may be due to this cause. Such absorption should become negligible at low temperatures at which the optical phonons will not be excited, but there is some evidence that a small 'tail' still exists even at the lowest temperatures.

10.5.5 Two-photon absorption

In the transitions which we have considered so far it has been assumed that a single photon is absorbed when an electron is raised from the valence band to the conduction band so that a photon energy of the order of ΔE or greater is required. However, as may be shown from general quantum-mechanical principles such a transition may also be excited by means of two or more photons the sum of whose energies is equal to that required to induce the transition. As the number of photons increases the transition probability decreases rapidly and we shall only consider transitions involving two photons. We must also consider transitions with and without the aid of phonons.

When the two photons are derived from the same nearly monochromatic source they will each have half the energy, and therefore twice the wavelength of a single photon required to excite the transition, in the case of a direct transition. For example two 14 μm photons would be required to excite a transition at room temperature from the valence to conduction band of InSb which has its fundamental absorption edge at approximately 7 μm. Since the momentum of the two photons even if moving in the same direction is still very small, the transition will take place with very little change in **k**, i.e. it will be a direct or 'vertical' transition. The probability for the two-photon transition, as for the one-photon transition will depend on whether or not it is 'allowed'.

The probability of an electron being excited by absorption of a single photon is proportional to the number of photons per unit volume in the exciting beam, i.e. to the intensity I. For a two-photon process, however, we should expect the probability to be proportional to I^2. We cannot therefore define a unique absorption coefficient α_2 but have to define it per unit power since α_2 will be proportional to the power; α_2 will

[48] *Phys. Rev.* (1957) **108**, 1419.

therefore be measured in cm^{-1} per watt per cm^2 or cm W^{-1}. (Since high powers are needed the units used are normally cm MW^{-1}.) Since the probability of a one-photon transition increases linearly with I while that of a two-photon transition increases quadratically it is clear that at some power the latter will overtake the former. This only happens at very large power of the order of GW cm^{-2}, but nevertheless quite appreciable two-photon absorption is observable with powers of the order of 1 MW cm^{-2} such as are available from quite modest lasers (see § 14.6).

A two-photon transition may also be excited using two sources of different frequencies ν_1 and ν_2. If E_t is the energy required for the transition it is simply necessary to have $h\nu_1 + h\nu_2 = E_t$. The theory of such transitions follows very closely that of phonon-assisted transitions, one photon replacing the phonon. The electron is excited by the frequency ν_1 to a virtual state at energy E_1 above the initial state where $h\nu_1 = E_1$ and then from there to the final state at energy $E - E_1$ above the virtual state with $h\nu_2 = E - E_1$. The virtual state is very short-lived so that energy is only conserved overall. Such transitions were discussed for simple bands by R. Braunstein[49] and the effect of excitons, which must play a part in a direct transition (see § 10.6) were included by R. Loudon.[50] The theory was also considered by N. G. Basov *et al.*[51] The effect of phonon assistance in the case of indirect transitions was considered by J. H. Yee,[52] who showed that, as might be expected, the probability of such transitions is smaller than for direct ones but by no means negligible.

Two-photon transitions will create photo-conductivity if the sum of the photon energies is large enough to produce a free electron (rather than an exciton). This has been used to deduce the magnitude of the absorption but requires an accurate knowledge of the recombination and trapping processes (see § 10.9). We are here concerned with direct measurement of the absorption. Such absorption was observed by K. J. Button *et al.*[53] for direct transitions in InSb and PbTe using a CO_2 laser. Two-photon transitions have been observed in a number of semiconductors including ZnTe, GaAs, InP and InSb by C. C. Lee and H. Y. Fan[54] who also have considered the effect of excitons.

While most of these measurements gave order of magnitude values for the absorption coefficients α_2 it has been shown by A. F. Gibson *et*

[49] *Phys. Rev.* (1962) **125**, 475.
[50] *Proc. Phys. Soc.* (1962) **80**, 952.
[51] *Soviet Phys. J.E.T.P.* (1966) **23**, 366.
[52] *J. Phys. Chem. Solids* (1972) **33**, 693.
[53] *Phys. Rev. Lett.* (1966) **17**, 1005.
[54] *Phys. Rev.* (1974) **B9**, 3502.

al.[55] that great care has to be taken with their interpretation if accurate values are to be obtained. This is mainly due to absorption caused by the free carriers created by the two-photon process and by the need to consider carefully their recombination. Their values of α_2 for InSb and Ge are probably the most accurate available and they have shown how the measurements of Lee and Fan (*loc. cit.*) can be reinterpreted. For InSb they found a value of 0.2 cm MW^{-1} at room temperature at a wavelength of 10.6 μm, nearly two orders of magnitude less than previous values where account had not been taken of free-carrier absorption. For Ge the value found was 0.75 cm MW^{-1} at 2.6 μm decreasing to zero at about 3 μm. This is for the direct transition at $\mathbf{k} \simeq 0$ and not that corresponding to the fundamental absorption edge for which the absorption is much smaller, since the transition there is indirect.

10.6 Exciton absorption

We have already discussed the formation of excitons in § 3.5, and have seen in § 10.5.2 that when these are associated with a direct transition they give rise to a line absorption spectrum on the long-wave side of the fundamental absorption edge. It is of interest to consider the momentum conservation laws as applied to exciton formation and to compare the absorption processes involved with those resulting from band-to-band transitions. The momentum of the exciton comes from the motion of its centre of mass. If the latter is described by means of a wave-vector **K** then the energy E_{ex} of the exciton will be given by the equation

$$E_{ex} = \frac{\hbar^2}{2M} K^2 - W_{ex}^n, \tag{95}$$

where W_{ex}^n is the binding energy of the exciton in its nth state given by equation (10) of § 3.5, and $M = m_e + m_h$.

In formation of excitons by direct transitions it is easy to show that we must have $K \simeq 0$. When a separated hole–electron pair is formed we have seen that the electron and hole move in *opposite* directions with velocity proportional to the wave-number k of the initial state, in order to conserve momentum; for an exciton, the hole and electron are bound and must move in the *same* direction through the crystal. These conditions can both be satisfied only when $k \simeq 0$ implying also $K \simeq 0$. The exciton energy is therefore sharply determined and we have a *line*

[55] *Proc. Phys. Soc.* C (1976) **9**, 3259.

absorption spectrum. There will be a finite probability of transitions taking place with *very* small values of k leading to a broadening of the line.

Equation (95) has been obtained by an intuitive application of the effective-mass concept. The more sophisticated effective-mass approximation discussed in § 11.5 applies only to either single electrons or single holes. It has been shown, however, by G. Dresselhaus[56] and by R. J. Elliott[57] how this approximation may be extended to deal with two quasi-particles, namely a hole and an electron. In this way equation (95) for the exciton energy is given a more formal theoretical basis. Moreover the intensity of the absorption lines as a function of the quantum number n has been shown by Elliott to be proportional to n^{-3} for transitions allowed in the sense of § 10.5 and as $(n^2-1)/n^5$ for forbidden transitions. For the latter the $n = 1$ line is missing. This theory has also been discussed in some detail by J. O. Dimmock[58] and a rather more sophisticated treatment of excitons in which the electronic excitation is closely linked with the lattice has been given by R. S. Knox.[59]

These treatments have been based on the use of scalar effective masses. The difficulties introduced by the complexities in the structures of the conduction bands of Ge, Si and various group III–IV compounds have been discussed by M. Altarelli and N. O. Lipari[60] who have shown that many of the finer details of the exciton spectra can be accounted for in this way. The intensity of the absorption is less than for the corresponding excited state of a collection of free atoms in the ratio $(d/r_{ex})^3$, where d is the lattice spacing and r_{ex} the radius of the exciton orbit (see § 3.5). This may amount to about 10^{-5}. We should perhaps point out that observation of a hydrogen-like absorption line spectrum is not always evidence for exciton formation, since absorption by certain types of impurity can also give rise to a spectrum of this kind (see § 3.4.2).

Absorption spectra have been found, showing a considerable number of sharp lines, for some compound semiconductors with fairly large forbidden-energy gaps, such as CdS, HgI, PbI, CdI.[61] Line spectra from Cu_2O have been reported by a number of authors including E. F. Gross and B. P. Zakharchenya[62] and S. Nikitine.[63] A puzzling feature of the

[56] *Phys. Rev.* (1957) **106**, 76. [57] *Phys. Rev.* (1957) **108**, 1384.
[58] *Semiconductors and Semimetals* (Academic Press, 1967) **3**, 259.
[59] *Theory of Excitons, Solid State Physics* (Academic Press, 1963), suppl. 5.
[60] *Proc. XIIIth Int. Conf. on Phys. of Semiconductors* (Tipographia Marves, 1976), p. 811. [61] E. F. Gross, *Nuovo Cim.* (1956) **3**, 672.
[62] *C.R. Acad. Sci., URSS* (1953) **90**, 715. [63] *J. Phys. Radium* (1956) **17**, 817.

Cu_2O spectra was that there were two series and that the $n = 1$ lines were missing. It is now known that the two series arise from a split-off valence band and the missing lines from the fact that the transition is 'forbidden'.

For some time it was puzzling that line exciton spectra had not been found for the semiconductors Ge and Si, associated with the fundamental absorption edge. Here very pure and nearly perfect crystals were available so that no confusion with line spectra due to impurities should take place. It was not until it was known that the fundamental absorption edge in these materials was due to an indirect transition that the reason for not having a line spectrum became clear (see below). The exciton associated with the direct transition from the top of the valence band to the minimum in the conduction band at $\mathbf{k} = 0$ (which is about 0.15 eV above the *lowest* minimum) was found by G. G. Macfarlane, T. P. McLean, J. E. Quarrington and V. Roberts[64] as shown in Fig. 10.9 (see § 10.5.2). The variation in the shape of this line with doping (Sb) has been studied by A. A. Rogachev[65] who has shown that the line may be still obtained for concentrations of Sb up to about 10^{16} cm^{-3}.

GaAs having its fundamental absorption due to a direct transition should provide a clear case for exciton lines appearing in the absorption spectrum. The $n = 1$ line was clearly observed by M. D. Sturge[66] who was also able to observe the increased line broadening as the temperature is raised. This is shown in Fig. 10.14.

Excitons may also be formed as a result of indirect transitions with emission or absorption of a phonon. Only in the case of emission of the most energetic phonon would the exciton absorption lie to the long-wave side of the absorption due to indirect band-to-band transitions. As we shall see, the spectral variation for exciton absorption is different from that for absorption in which free electron–hole pairs are created and this enables it to be distinguished from the absorption due to band-to-band transitions.

The theory of formation of excitons with emission and absorption of phonons has also been given by G. Dresselhaus and by R. J. Elliott (*loc. cit*). We have now one extra degree of freedom in choice of the initial state from which the transition may take place, since the momentum $\hbar\mathbf{K}$ of the centre of mass of the exciton may be compensated by the momentum of the phonon which is emitted or absorbed. Thus, we now

[64] *Proc. Phys. Soc.* (1958) **71**, 863.
[65] *Proc. Xth Int. Conf. on Phys. of Semiconductors* (Nauka, 1968), p. 207.
[66] *Phys. Rev.* (1962) **127**, 768.

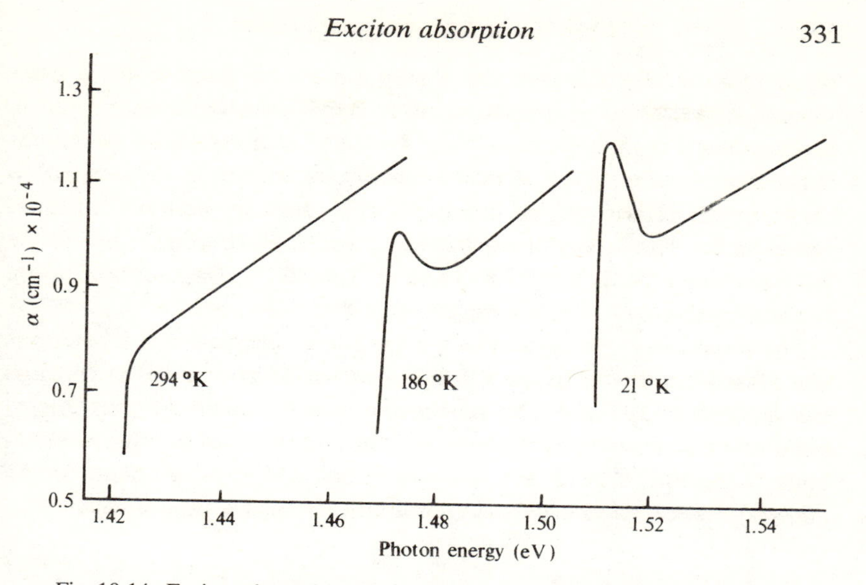

Fig. 10.14. Exciton absorption spectrum in GaAs. (After M. D. Sturge, *loc. cit.*)

have an exciton *band spectrum* with a well-defined lower limit, rather than a line spectrum. If $E_{\text{min.}}$ is the smallest energy required for the formation of an exciton and, as before E_p represents the energy of the phonon, we have for the absorption coefficient α_{ex} due to exciton formation

$$\alpha_{ex} = D(h\nu - E_{\text{min.}} \pm E_p)^{\frac{1}{2}} \tag{96}$$

for an 'allowed' transition and

$$\alpha_{ex} = G(h\nu - E_{\text{min.}} \pm E_p)^{\frac{3}{2}} \tag{97}$$

for a 'forbidden' transition, where D and G are approximately constant. The long-wave limit of the absorption spectrum for an exciton in its nth state is given by

$$h\nu = \Delta E - W_{ex}^{n} - E_p. \tag{98}$$

For absorption corresponding to excitons in their lowest state we shall drop the suffix n. If W_{ex} is the binding energy of the exciton in its lowest state, the smallest value of $h\nu$ for which a transition may be observed is given by

$$h\nu = \Delta E - E_p - W_{ex}. \tag{99}$$

This represents the true lower limit of the absorption continuum, if E_p corresponds to the most energetic phonon involved in the indirect transitions. The exciton absorption will pass steadily (but with change of slope at $h\nu = \Delta E - E_p$) into the absorption due to band-to-band transitions. At very low temperatures no free carriers should be created in the spectral range $\Delta E - E_p - W_{ex} < h\nu < \Delta E - E_p$. At higher temperatures the excitons would almost certainly be thermally dissociated so that free carriers would be formed in this range as well as for higher values of $h\nu$. The theory of the indirect transition excitons for Si and Ge has been given by T. P. McLean and R. Loudon[67] using a modification of the effective-mass approximation (see § 11.5).

The most direct method of observing the absorption due to excitons when associated with indirect transitions would appear to be through the spectral variation of this absorption as compared with that due to band-to-band transitions. If we plot $\alpha^{\frac{1}{2}}$ as a function of $h\nu$, the band-to-band transitions should give a series of straight lines, whereas for an allowed exciton band with lower limit at $h\nu = E_m$ we should have

$$\alpha_{ex}^{\frac{1}{2}} \propto (h\nu - E_m)^{\frac{1}{4}}. \tag{100}$$

This should give a very rapid rise in absorption and should be easily recognizable. Such sharp rises have been observed by G. G. Macfarlane, T. P. McLean, J. Quarrington and V. Roberts[68] in Ge (see Fig. 10.12) and have been interpreted by them as due to excitons formed with emission and absorption of phonons with E_{pt} having the value given in § 10.5.3. The rises due to emission are marked E in Fig. 10.12 and those due to absorption are marked A. Two less rapid rises are interpreted as due to exciton formation with emission and absorption of transverse acoustic phonons. The variation with $h\nu$ in this case is such as to require the transitions for these to be forbidden, i.e. $\alpha_{ex}^{\frac{1}{2}} \propto (h\nu - E_m)^{\frac{3}{4}}$, and consequently they are not so easily observed on top of the absorption due to band-to-band transitions. It will be noted that there is still a very small residual absorption (of the order of $0.04\ \text{cm}^{-1}$) which occurs at values of $h\nu$ less than the lower limit of the lowest exciton band. It is not known whether this is a small residual experimental error due to incomplete elimination of free-carrier or other absorption or is due to some higher-order transitions involving the emission of more than one phonon.

For Si, four rises are found[69] corresponding to variations in $\alpha^{\frac{1}{2}}$ given by equation (100) (see Fig. 10.13). In Ge the value of $\mathbf{k}_{\text{min.}}$ occurs at the edge of the zone in the ⟨111⟩ directions, whereas in Si it occurs at some

[67] *J. Phys. Chem. Sol.* (1960) **13**, 1. [68] *Phys. Rev.* (1957) **108**, 1377.
[69] G. G. Macfarlane, T. P. McLean, J. E. Quarrington and V. Roberts, *Phys. Rev.* (1958) **111**, 1245.

distance from the zone edge in ⟨100⟩ directions (see § 13.3). In this case the symmetry conditions will be different and transitions involving both longitudinal and transverse phonons appear to be 'allowed'; also the selection rule which makes the contribution from optical phonons small in Ge no longer holds for Si and absorption due to emission and absorption of these is also observed. Further evidence that the sudden increase in absorption is due to exciton formation is afforded by the observation that there is no corresponding sudden increase in photo-conductivity at very low temperatures, whereas at room temperature the variation in photo-conductivity follows the variation in absorption.

The binding energy of the 'indirect' excitons was estimated to be of the order of 5 meV for both Si and Ge. More recent work using differential absorption techniques has shown up not only the $n=1$ exciton but also the $n=2$ state, and much more accurate values for the exciton binding energy have been obtained. For Ge, E. F. Gross *et al.*[70] found a value of 3.6 meV while for Si K. L. Shaklee and R. E. Nabory[71] found 14.7 meV. These are in very good agreement with the theoretical calculations for Ge and Si by T. P. McLean and R. Loudon (*loc. cit.*).

Excitons associated with indirect transitions have also been observed in a number of compound semiconductors. For example GaP having a fundamental edge produced by an indirect transition has an exciton spectrum showing bands rather than lines, in marked contrast to GaAs. The spectrum as observed by M. Gershenzon, D. G. Thomas and R. E. Deitz[72] is shown in Fig. 10.15. Sharp lines in the spectrum of GaP doped with S are due to excitons bound to S donors. These have also been observed by P. J. Dean[73] for S, Se and Te donors. We shall later discuss emission from such bound excitons and excitonic molecules (see §§ 10.14.2, 14.1).

J. C. Phillips[74] who has made a very full analysis of the excitons observed in a wide range of materials has suggested that excitons may also be formed at certain other critical points in the energy band structure where $\nabla_k(E_c - E_v) = 0$. This is just the condition that electrons and holes have the same velocity, so seems not unreasonable.

It might be expected that absorption would take place due to raising of excitons from their ground state to higher excited states. This would

[70] *JETP Lett.* (1971) **13**, 235.
[71] *Phys. Rev. Lett.* (1970) **24**, 942.
[72] *Proc. VIth Int. Conf. on Phys. of Semiconductors* (Inst. of Phys., London, 1962), p. 752.
[73] *Phys. Rev.* (1967) **157**, 655.
[74] *Phys. Rev.* (1964) **136**A, 1705; *Solid State Physics* (Academic Press, 1966) **18**, 55.

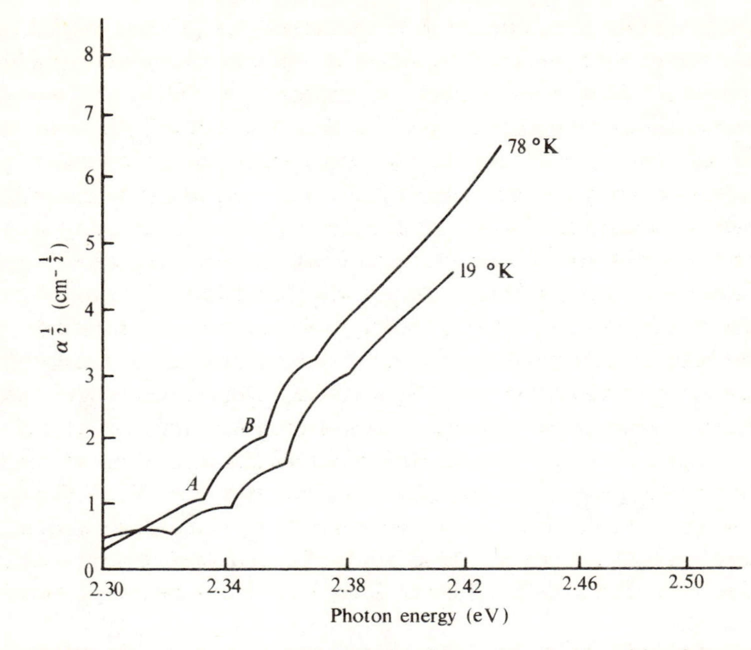

Fig. 10.15. Absorption spectrum of GaP showing exciton bands arising from various combinations of phonons. Those marked *A* and *B* are due to the emission of LA and T0, L0 phonons. (After M. Gershenzon *et al.*, *loc. cit.*)

occur in the very far infra-red. By using the high intensities available from lasers a sufficient number of excitons can now be created for this absorption to be directly observable. Absorption of this kind has been observed in pure Ge by V. L. Kononenko *et al.*[75] More recently using interferometric techniques M. Buchanan and T. Timusk[76] have found a rich variety of lines in the spectral region covering photon energies of 1–7 meV. These they have interpreted as due to excitons formed not only from the light holes in Ge but also from the heavy holes (see § 2.3).

10.7 Impurity absorption

Various optical transitions from impurity centres may be induced by incident radiation of appropriate frequency, leading to absorption spectra. If E_1 is the depth of an impurity level below the conduction band, then if $h\nu > E_1$, continuous absorption will be caused due to excitation

[75] *Proc. XIIth Int. Conf. on Phys. of Semiconductors* (Teubner, 1974), p. 152.

[76] *Proc. XIIIth Int. Conf. on Phys. of Semiconductors* (Tipographia, Marves, 1976), p. 821.

of electrons from the centre to the conduction band; if E_1 is small this will be observed only at low temperatures when a considerable fraction of the impurity centres are un-ionized. Similarly, electrons may be excited into un-ionized donor centres near the valence band. As we have seen in § 3.4.1, impurity centres have excited states, the group III and group V impurities in Si and Ge having a series of hydrogen-like states. Excitation from the ground state to the excited states of an impurity centre would lead to a line absorption spectrum. In most semiconductors the majority of donor and acceptor states lie very near the conduction and valence bands, so that the absorption corresponding to these processes would lie in the far infra-red; in Ge, for example, the ground states due to group III donors are about 0.01 eV below the conduction band and the ionization limit would be at about 100 μm. For Si, however, the ground states of group III donors are about 0.05 eV below the conduction band and so the absorption lies in a much more accessible region of the spectrum. A great deal of work has been done in studying the optical absorption due to impurities in Si and Ge and in comparing the measurements with theory. Several comprehensive reviews of this work are available.[77]

A typical absorption curve is shown in Fig. 10.16 for Al impurity centres in Si, in which three absorption lines due to excitation from the ground state are apparent before the continuous absorption due to photo-ionization of the impurity centres sets in. Measurements by H. J. Hrostowski and R. H. Kaiser[78] have shown no less than seven excited levels arising from acceptor states due to *B*. For the excited states, of donor impurities particularly, agreement with the positions of the excited levels as predicted by theory based on the simple hydrogen model is remarkably good. The agreement with the value predicted for the ground state is not so good, since no variation for different impurities is predicted and there is quite a large variation. This is not surprising, since the assumption that the Coulomb field, due to the impurity, holds right into the centre is invalid, and this is more likely to affect the ground states, with their smaller orbits, than the excited states. For example, for *B*, in Si the acceptor level lies 0.045 eV above the valence band whereas for In it is 0.16 eV above.

[77] E. Burstein and P. H. Egli, *Advances in Electronics and Electron Physics* (Academic Press, 1955) **7**, 1; E. Burstein, G. S. Picus and N. Sclar, *Proceedings of Atlantic City Photo-conductivity Conference* (John Wiley and Sons; Chapman and Hall, 1956), p. 353; W. Kohn, *Solid State Physics* (Academic Press, 1957) **5**, 257; D. H. Parkinson, in *Electronic Structure in Solids*, ed. E. D. Haidemenakis (Plenum Press, 1969), pp. 93, 102.

[78] *Bull. Amer. Phys. Soc.* (1955) **2**, 66.

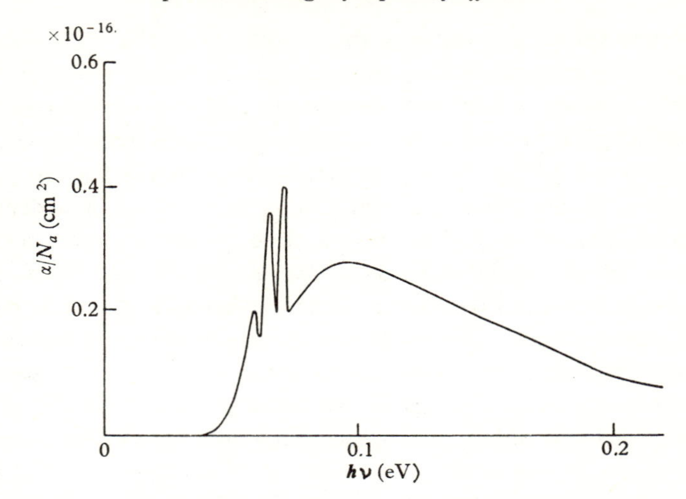

Fig. 10.16. Absorption due to Al impurity centres in Si. (After E. Burstein and P. H. Egli, *Advances in Electronics and Electronic Physics* (1955) **7**, 1.)

Impurity absorption should also take place corresponding to transitions from the valence band to empty donor levels near the conduction band; similarly, excitation from acceptor levels which are occupied by electrons should also lead to absorption, and this should be just to the long-wave side of the fundamental absorption edge. The absorption due to this cause will generally be very weak unless the impurity concentration is high, and will generally be obscured by free-carrier absorption. Experiments to determine the form of the fundamental absorption edge are generally made with crystals of the highest purity available, and this kind of absorption is not observed. For donor levels lying somewhat deeper, the impurity absorption will be well separated from the fundamental absorption, but also the chance of the donor impurities being empty will be less. It should be noted that for shallow impurities the absorption spectrum will take the form of a series of lines in approximately the same position as would be expected for the exciton spectrum; some of the spectra which have been interpreted as exciton spectra may have been due to impurities with a hydrogenic series of levels.

The simple 'hydrogen' model, as we have seen, must be modified for semiconductors such as Si and Ge to take account of the anisotropy of the conduction band and warping of the constant-energy surfaces of the valence band. A method of doing this has been described by W. Kohn (*loc. cit.*) (see also *W.M.C.S.*, § 11.2.6). Kohn has compared the experimental data for a large number of impurities in Si and Ge. For shallow impurities the agreement between theory and experiment is extremely good apart from the ground states. The situation has also been discussed in some detail by D. H. Parkinson (*loc. cit.*) and also by A. M. Stoneham.[79]

For the ground states and also for impurities whose ground states lie further from the conduction and valence bands more sophisticated theoretical treatments are required (see § 11.5.1). Experimental data on the levels of such impurities have been derived mainly from electrical measurements but the optical data obtained from the study of their absorption spectra have added greatly to our knowledge, particularly of the excited states of such impurities. D. H. Parkinson (*loc. cit.*) has also discussed in some detail the deeper impurities and has given data for a large number of such.

The value of the absorption coefficient for the hydrogenic model may readily be obtained by adapting the transition probabilities for excitation from the ground state to higher levels of the hydrogen atom. For excitation to the conduction band or valence band the situation is somewhat different to that for ionization of a hydrogen atom because of the different density of states available in the bands for free electrons or holes. Various authors have considered this problem which has been discussed in some detail by T. S. Moss, G. J. Burrell and B. Ellis.[80] For an electron excited into the conduction band from an acceptor level E_a above the valence band the absorption coefficient α is given by an expression of the form

$$\alpha = AN_a(h\nu - \Delta E + E_a)^{\frac{1}{2}} \tag{101}$$

where A is a constant and N_a is the concentration of acceptors. The order of magnitude of A is of the order of 10^{-18} if α is expressed in cm^{-1} and N_a in cm^{-3}, with the energies in eV, but varies considerably with the material.

10.7.1 Spin resonance due to impurities

If we have impurity centres occupied by single electrons it should be possible, in principle, to observe paramagnetic spin resonance due to inversion of the spin by an alternating magnetic field. The impurity levels will have their energy states split by a magnetic field, because of

[79] *Theory of Defects in Solids* (Clarendon Press, 1975).
[80] *Semiconductor Opto-electronics* (Butterworth, 1973).

the spin of the electron, into two levels separated by an amount $\beta g B$, where β is the magnetic moment of the electron (equal to $e\hbar/2\boldsymbol{m}$ in M.K.S. units), B is the magnetic induction and g is a factor very nearly equal to 2; g differs slightly from 2 owing to interaction with the crystalline field. The energy difference δE between the states may be expressed in the form

$$\delta E = ge\hbar B/2\boldsymbol{m} = (g/2)\hbar\omega_c, \qquad (102)$$

where ω_c is the cyclotron resonance frequency for a free electron ($\omega_c = 2.8 \times 10^{10}$ Hz if $B = 1$ T). The absorption due to this type of transition occurs in the microwave region of the spectrum for $B \simeq 0.1$ T; it is extremely weak but the lines are quite sharp and may be observed by having the sample in a resonant cavity.

Spin resonance has been observed for donor states in Si by R. C. Fletcher *et al.*[81] and by a number of others.[82] Its study has given valuable information about the impurity levels; indeed, the absorption lines are not single but show an interesting fine structure which gives information about the effect of the neighbouring atoms of the crystal lattice on the impurity levels.

A full discussion of the techniques involved in the application of electron spin resonance to semiconductors has been given by G. Lancaster.[83] Application to various semiconductors but especially to the III–V group has been described by G. W. Ludwig and H. H. Woodbury.[84]

10.8 Lattice absorption

In addition to the absorption due to electronic transitions within and between energy bands and those due to impurities, absorption of infrared radiation may also take place through excitation of lattice vibrations and this sometimes overlaps and obscures some of the other absorption already described. We have already discussed briefly in § 8.3 the form of the lattice vibrational spectrum. The absorption of radiation will take place when some additional phonons are created from the energy of an absorbed photon.

[81] R. C. Fletcher, W. A. Yager, G. L. Pearson, A. N. Holden, W. T. Read and F. R. Merritt, *Phys. Rev.* (1954) **94**, 1392; R. C. Fletcher, W. A. Yager, G. L. Pearson and F. R. Merritt, *ibid.* (1954) **95**, 844.

[82] See, for example, A. Honig and A. F. Kip, *Phys. Rev.* (1954) **95**, 1686; A. Honig, *ibid.* (1954) **96**, 234; G. Feher, *ibid.* (1956) **103**, 834.

[83] *Electron Spin Resonance in Semiconductors* (Hilger and Watts, 1966).

[84] *Solid State Physics* (Academic Press, 1962) **13**, 223.

First of all let us suppose that a single phonon with wave-vector $\mathbf{q}$ is created with the absorption of a photon of wave-vector $\mathbf{K}$. Then for conservation of momentum we must have $\mathbf{K}=\mathbf{q}$. As we have seen in connection with direct inter-band transitions, for an infra-red photon $\mathbf{K}$ is small, so that $\mathbf{q}$ must be also small. For an acoustic phonon the energy to $\omega(\mathbf{K})$ will be very small so that the photon created must be an optical phonon, with $\mathbf{q}\simeq 0$. As we have seen in § 8.3 this corresponds to the *reststrahl* frequency. Strong absorption at this frequency only occurs for crystals having a dipole moment arising from the displacements in the long-wave optical mode and so would not occur in semiconductors such as Si and Ge. For semiconductors with partial ionic binding, we should expect strong absorption when $\omega=\omega_{TO}$, the value of the transverse optical frequency corresponding to $\mathbf{K}=0$. This indeed is found for all the III–V compounds and also for many polar semiconductors.[85] This is the *reststrahl* absorption and occurs near the following wave-numbers for InSb, GaAs, GaP: 180 cm^{-1}, 268 cm^{-1}, 365 cm^{-1}. The longitudinal optical frequencies at $\mathbf{K}=0$ for these semiconductors are respectively 191 cm^{-1}, 291 cm^{-1}, 402 cm^{-1}.

For Si and Ge, absorption, though weaker, is still observed arising from the lattice vibration and extends over a considerable spectral range. The absorption spectrum of Si is shown in Fig. 10.17. It shows a number of marked peaks with continuous absorption between. This absorption must be due to processes involving at least two phonons. By having two (or more) phonons we can conserve momentum without having $\mathbf{q}$ small. The conservation condition is now $\mathbf{q}_1+\mathbf{q}_2=\mathbf{K}$ or, to a good approximation, $\mathbf{q}_1=-\mathbf{q}_2$. The energy condition is now, for a photon of frequency ν

$$h\nu=\hbar\omega_1(\mathbf{q}_1)+\hbar\omega_2(-\mathbf{q}_1) \tag{103}$$

where ω_1 and ω_2 may refer to different branches of the lattice vibration dispersion curves. For example, ω_2 may refer to an optical phonon and ω_1 to an acoustic phonon. This is illustrated in Fig. 10.18.

The conservation conditions have been discussed by many authors including F. A. Johnson[86] (see also *W.M.C.S.*, Ch. 8). Spectra similar to that for Si are also obtained for Ge and for group III–V semiconductors such as InSb, GaAs etc. in addition to the *reststrahl* absorption. Since

[85] See for example, M. Haaš, *Semiconductors and Semimetals* (Academic Press, 1967) **3**, 3.

[86] *Rendiconti della Scuola Internazionale di Fisica Enrico Fermi,* Corso XXII (Academic Press, 1963), p. 504; *Prog. in Semiconductors* (Heywood, 1965) **9**, 179.

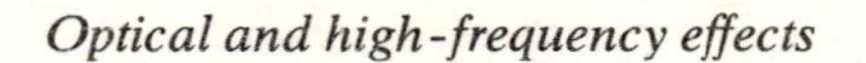

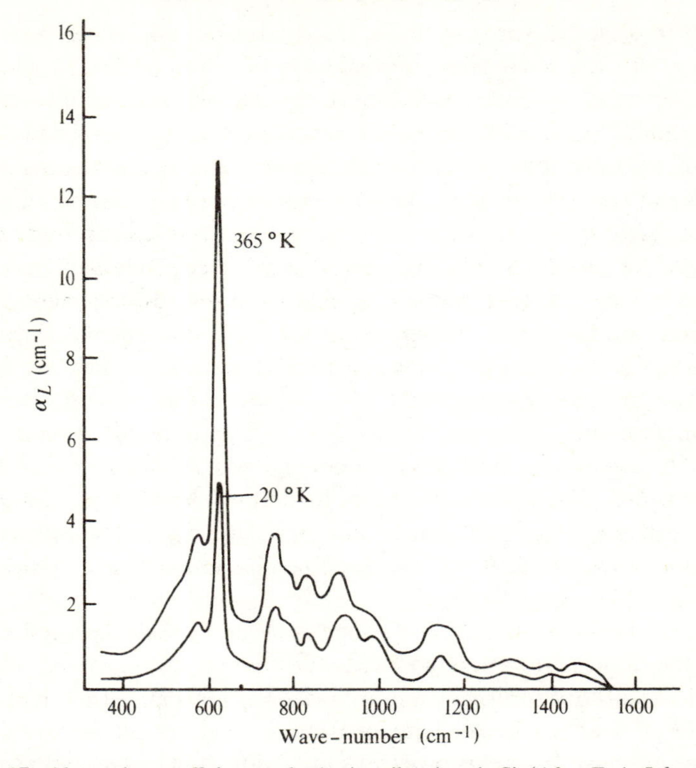

Fig. 10.17. Absorption coefficient α_L for lattice vibrations in Si. (After F. A. Johnson, *loc. cit.*)

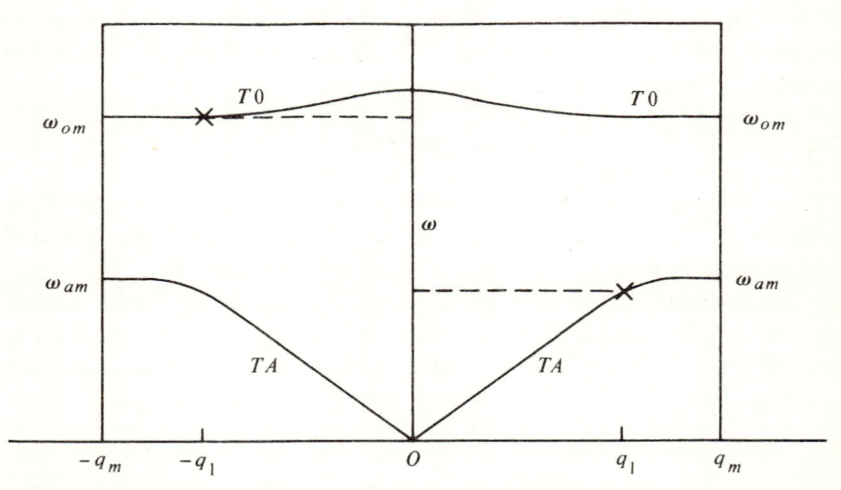

Fig. 10.18. Illustrating creation of two phonons by absorption of a photon. q_m corresponds to the zone edge, with corresponding angular frequencies ω_{om} and ω_{am}.

there is now nothing that determines the value of **q**, absorption can occur for all values of **q** in a zone. The spectra are, however, all characterized by marked peaks. These come from a high concentration of states at certain critical points within the zone. Using dispersion curves derived from theory and neutron scattering, F. A. Johnson and R. Loudon[87] have analysed the Si and Ge spectra from this point of view and have obtained not unreasonable agreement with experiment.

The theory of lattice vibrations has been greatly extended under the stimulus of both infra-red absorption studies and neutron scattering and data have now been obtained for a large number of semiconductors.[88]

Although the single-phonon process does not lead to absorption in pure semiconductors like Si and Ge with the diamond lattice, impurities may induce such absorption. This absorption has been studied and discussed in considerable detail by J. Houghton and S. D. Smith[89] who have also described the phenomenon in diamond itself. The theory of such absorption has been reviewed by R. C. Newman[90] who has extended it to deal particularly with *pairs* of impurities. He has also considered more complex aggregates. He has compared the experimental data on Ge with the theory and has also discussed absorption due to localized modes caused by impurities in GaAs.

The problem of analysing the complex spectra when more than two phonons are involved is very difficult. Three phonons commonly participate and spectra thought to be due to four have been observed. The task is made somewhat easier because of the different temperature variation of some of the processes. In our discussion we have tacitly assumed that two phonons would be *created* and this certainly must be so at low temperatures. At higher temperatures, however, phonon concentrations will be present and one of the two phonons in a two-phonon process may be *emitted*, still conserving momentum. In this case we have

$$h\nu = \hbar\omega_1(q_1) - \hbar\omega_2(-q_1). \tag{103a}$$

In this case we must have $\omega_1 > \omega_2$. In general ω_1, will then correspond to an optical mode and ω_2 to an acoustic mode.

[87] *Proc. Roy. Soc.* A (1964) **281**, 274.

[88] See, for example, W. Cochran, *Adv. Phys.* (1960) **9**, 387; *The Dynamics of Atoms in Crystals* (Edward Arnold, 1973); A. A. Maradudin, E. W. Montroll and G. H. Weiss, *Solid State Physics* (Academic Press, 1963), suppl. 3; W. G. Spitzer, *Semiconductors and Semimetals* (Academic Press, 1967) **3**, 17; F. A. Johnson, *Proc. Roy. Soc.* A (1969) **310**, 79, 89, 101; *Optical Properties of Solids*, ed. E. D. Haidemenakis (Gordon and Breach, 1970), p. 331.

[89] *Infra-red Physics* (Clarendon Press, 1966).

[90] *Adv. Phys.* (1968) **18**, 545.

In order to obtain the temperature variation of the absorption we must take account of a very interesting effect which we shall consider in greater detail later in connection with laser action – namely stimulated emission, whose probability turns out to be the same as that for absorption. The phonon absorption process will be proportional to the number of phonons present which will be proportional to $F(\omega)$ where (see § 14.6)

$$F(\omega)=[\exp(\hbar\omega)/kT-1]^{-1}. \tag{104}$$

As we shall see, the probability of emission of a phonon is proportional to $1+F(\omega)$, the 1 arising from spontaneous emission. The total probability of absorption of a photon with emission of a single phonon will then be proportional to $\{1+F(\omega)\}-F(\omega)$ and so will be independent of temperature. The probability of emission of two phonons and absorption of a photon will be proportional to

$$\{1+F(\omega_1)\}\{1+F(\omega_2)\}-F(\omega_1)F(\omega_2)=1+F(\omega_1)+F(\omega_2)$$

while for emission of one phonon and absorption of one the probability is proportional to

$$\{1+F_1(\omega_1)\}F(\omega_2)-\{1+F(\omega_2)\}F(\omega_1)=F(\omega_2)-F(\omega_1).$$

The temperature variation for various combinations of three phonons is much more complex and has been given by F. A. Johnson (*loc. cit.*) who has achieved a good many of his identifications of phonon peaks by this means.

10.9 Photo-conductivity

The formation of pairs of free electrons and holes, through the absorption of radiation which causes band-to-band transitions, increases the number of charge carriers of both types. Let Δn be the number of extra electrons per unit volume and Δp the number of extra holes. If there is no differential trapping of electrons and holes we shall have $\Delta n=\Delta p$ (see § 7.3) and the conductivity σ will be given by

$$\sigma=e(n_0+\Delta n)\mu_e+e(p_0+\Delta p)\mu_h, \tag{105}$$

where n_0 and p_0 are the equilibrium carrier densities, giving a conductivity σ_0; we thus have

$$\frac{\Delta\sigma}{\sigma_0}=\frac{\sigma-\sigma_0}{\sigma_0}=\frac{\Delta n\mu_e+\Delta p\mu_h}{n_0\mu_e+p_0\mu_h}, \tag{106}$$

$$= \frac{(1+b)\Delta n}{n_0 b + p_0}, \tag{106a}$$

where $b = \mu_e/\mu_h$. We shall find that, for small intensities of illumination, $\Delta n \propto I$, the intensity of the incident radiation, so that $\Delta\sigma \propto I$.

Apart from exciton formation in indirect transitions, the long-wave limit of intrinsic photo-conductivity should coincide with the fundamental absorption edge. When we include excitons in the indirect transitions the long-wave limit of photo-conductivity should coincide at ordinary temperatures with the long-wave limit of exciton absorption since, because of their very small binding energy, the excitons are likely to become quickly dissociated and will then produce pairs of free electrons and holes. At very low temperatures, however, the photo-conductivity should start at a frequency slightly higher than that giving the long-wave limit of exciton absorption. In insulators this has been the main reason for supposing some of the absorption to be due to excitons.

Since, for most semiconductors, the long-wave limit of fundamental absorption lies below 1.5 eV, photo-conductivity occurs in the infra-red region of the spectrum. This has led to the use of semiconductors with small energy gaps such as the lead salts PbS, PbSe, PbTe and the inter-metallic compound InSb as infra-red detectors.[91] Photo-conductivity induced by band-to-band transitions is called intrinsic photo-conductivity to distinguish it from that due to impurities. Intrinsic photo-conductivity has been extensively studied[92] in a wide variety of substances. In many instances it has not been possible to work with pure single crystals and much of the available information has been derived from thin evaporated films. It is now well known that photo-conductive processes in such films are extremely complex. The intrinsic properties of the infra-red photo-conductors PbS, PbSe, PbTe have been discussed by R. A. Smith[93] in two review papers; those of Ge and Si have been described by E. Burstein, G. S. Picus and N. Sclar[94] and also by G. A. Morton, E. E. Hahn and M. L. Schultz.[95]

The creation of electron–hole pairs gives rise in certain conditions to a photo-voltage which appears across illuminated samples of semicon-

[91] R. A. Smith, F. E. Jones and R. P. Chasmar, *Detection and Measurement of Infra-red Radiation* (Clarendon Press, 2nd edition, 1968).
[92] See, for example, T. S. Moss, *Photo-conductivity in the Elements* (Butterworth, 1952); also *Atlantic City Photo-conductivity Conference* (John Wiley and Sons, and Chapman and Hall, 1956).
[93] *Advances in Physics* (*Phil. Mag.* Supplement) (1953) **2**, 321; *Physica* (1954) **20**, 910.
[94] *Proceedings of Atlantic City Photo-conductivity Conference* (John Wiley and Sons, and Chapman and Hall, 1956), p. 353. [95] *Ibid.* p. 556.

ducting material. This is due to inhomogeneities in the material or to surface effects, and we shall defer discussion of the various causes till later.

Ionization of impurities as discussed in § 10.7 also leads to photo-conductivity. For shallow donors and acceptors this photo-conductivity occurs in the far infra-red (~100 μm for group III and group V impurities in Ge) but at shorter wavelengths in the intermediate infra-red for deeper lying impurities such as Cu. Using intrinsic and impurity photo-conductivity a whole range of detectors have been developed covering the whole of the infra-red spectrum.[96]

We must now consider the processes of generation and recombination of the carriers added by illumination so as to determine the quantities Δn, Δp. Suppose we have a specimen with rectangular plane faces and of uniform thickness d, and that the radiation falls normally on one of the plane faces. We shall suppose that the radiation is monochromatic and that its frequency is sufficiently high to produce electron–hole pairs. Let I represent the intensity of the radiation, i.e. the incident power on unit area of surface. If I' is the intensity in the material then the power absorbed per unit volume per unit time is equal to $\alpha I'$, where α is the absorption coefficient. The number of electron–hole pairs produced per unit volume in unit time is then equal to $q\alpha I'/\boldsymbol{h}\nu$, where q is a constant known as the quantum efficiency; if each absorbed quantum produces *one* electron–hole pair, and the absorption due to causes other than band-to-band transitions is negligible, then $q=1$. The quantity I' depends on the thickness of the sample in a rather complicated way; if the sample is so thin that $\alpha d \ll 1$ we shall have interference between the radiation reflected from the front and back surfaces, provided these are plane and parallel, and the amount of radiation absorbed in the sample will vary rapidly with wavelength. Such interference effects are treated in most text-books on optics.[97] If the sample is several wavelengths thick or the surfaces are not very flat, the interference effects will not be marked and the *intensities* of the radiation reflected from the various surfaces may be added.

If RI is the intensity of the reflected radiation, and TI the intensity of the transmitted radiation, it may readily be shown that

$$R = R_s\left\{1+\frac{(1-R_s)^2\,e^{-2\alpha d}}{1-R_s^2\,e^{-2\alpha d}}\right\} \tag{107}$$

[96] See, for example, *Optical Properties of Solids*, ed. S. Nudelman and S. S. Mitra (Plenum Press, 1969); T. S. Moss, G. J. Burrell and B. Ellis, *Semiconductor Opto-electronics* (Butterworths, 1973).

[97] See, for example, R. A. Smith, F. E. Jones and R. P. Chasmar, *loc. cit.*

and
$$T=\frac{(1-R_s)^2 e^{-\alpha d}}{1-R_s^2 e^{-2\alpha d}}, \tag{107a}$$

where R_s is the reflection coefficient at the surface of an infinitely thick slab. The total amount of radiation absorbed by the sample in unit time is equal to $I(1-T-R)$, and if $\alpha d \ll 1$ the average rate of absorption per unit volume is equal to $I(1-T-R)/d$. When equations (107) and (107a) are valid, we have $(1-T-R)\to\alpha d$ when $\alpha d \ll 1$ and the radiation is uniformly absorbed throughout the sample at a rate αI per unit volume. The rate of creation of electron–hole pairs $\mathcal{R}$ is then given by the equation

$$\mathcal{R}=q\alpha I/h\nu \tag{108}$$

and the quantity I' referred to above is, in this case, equal to I. If, on the other hand, $\alpha d \gg 1$ we have, just inside the illuminated surface, $I' = I(1-R_s)$, there being no appreciable radiation reflected from the back surface; in this case we have

$$I'=I(1-R_s)\,e^{-\alpha x} \tag{109}$$

and the rate of pair creation $\mathcal{R}$ is given by the equation

$$\mathcal{R}=q\alpha I(1-R_s)\,e^{-\alpha x}/h\nu. \tag{110}$$

Under the conditions for which equations (107) and (107a) are valid, the general expression for $\mathcal{R}$ is

$$\mathcal{R}=\frac{q\alpha I(1-R_s)}{h\nu(1-R_s^2 e^{-2\alpha d})}\{e^{-\alpha x}+R_s \exp\left[-\alpha(2d-x)\right]\}. \tag{110a}$$

10.9.1 Uniform absorption rate

We shall first of all consider a thin sample for which $\alpha d \ll 1$ so that excess carriers are generated uniformly throughout the sample, and we shall calculate the excess carrier density and determine the magnitude of the photo-current when an electric field $\mathscr{E}$ is applied along the length of the sample. If we may neglect surface recombination the excess carrier density will be uniform, and the equation satisfied by the excess hole concentration Δp in n-type material will be (see § 7.4)

$$\frac{d\Delta p}{dt}+\frac{\Delta p}{\tau_p}=\mathcal{R}, \tag{111}$$

where, in this case, $\mathcal{R}$ is constant and given by equation (108). In the steady condition we have an excess carrier density Δp_0, given by the equation

$$\Delta p_0 = \mathcal{R}\tau_p = q\alpha I\tau_p/\boldsymbol{h}\nu. \tag{112}$$

Let us estimate the value of Δp_0 for a typical example; suppose that I is equal to 1 mW cm^{-2}; for $\lambda = 1\ \mu$m, $\boldsymbol{h}\nu \simeq 2\times 10^{-19}$ J, so that the light flux corresponds to 5×10^{15} quanta crossing one square centimetre per second; if we take $q = 1$, $\alpha = 100$ cm^{-1} and $\tau_p = 200\ \mu$s we have $\Delta p_0 = 10^{14}$ cm^{-3}. For an n-type semiconductor with $n_0 = 10^{14}$ cm^{-3} and $p_0 \ll n_0$, we have $\Delta\sigma/\sigma = (1+b)/b$ so that if $b = 2$ the conductivity will have been increased by a factor 5/2 by this light flux and, if the applied voltage does not change with the current, the photo-current will be 3/2 times the dark current. When τ_p is fairly large, it will be seen that large changes in conductivity may be brought about by quite moderate values of the light flux; we shall, however, usually be concerned with relatively small changes in conductivity so that $\Delta\sigma \ll \sigma$ and consequently $\Delta p_0 \ll n_0$ or p_0.

We shall first of all assume that the number of traps is so small that $\Delta n = \Delta p$ (see § 7.3); the change in conductivity $\Delta\sigma_0$ is then given by the equation

$$\begin{aligned}\Delta\sigma_0 &= \boldsymbol{e}\Delta p_0(1+b)\mu_h\\ &= \boldsymbol{e}\mathcal{R}\tau_p(1+b)\mu_h,\end{aligned} \tag{113}$$

and the photo-current i_p is given by

$$\begin{aligned}i_p &= \boldsymbol{e}w\,d\mathcal{R}\tau_p(1+b)\mu_h\mathcal{E}\\ &= \boldsymbol{e}qw\,d\alpha I\tau_p(1+b)\mu_h\mathcal{E}/\boldsymbol{h}\nu,\end{aligned} \tag{114}$$

where w is the width of the sample.

If $\Delta p \ll p$, so that τ_p is constant, the photo-current is proportional to I, while for large values of Δp the variation of i_p with I depends on the recombination law. If we have, for example, a recombination rate of the form Anp, as for radiative recombination, so that

$$\Delta p/\tau_p = A(np - n_0p_0),$$

we have Δp_0 given by the equation

$$\Delta p_0^2 + (n_0 + p_0)\Delta p_0 = \mathcal{R}/A, \tag{115}$$

n_0, p_0, being the equilibrium carrier concentrations; for values of I so large that $\Delta p \gg n_0$ or p_0 we have $\Delta p_0 \propto \mathcal{R}^{\frac{1}{2}}$ so that $i_p \propto I^{\frac{1}{2}}$. For recombination through traps we may write (see § 9.4)

$$\tau_p = \tau_0(1+a\Delta p)/(1+c\Delta p), \tag{116}$$

so that

$$(1+c\Delta p_0)\Delta p_0 = \mathscr{R}\tau_0(1+a\Delta p_0) \tag{117}$$

and we see that for both large and small values of Δp we have Δp_0, and hence i_p, proportional to I, but with a different constant of proportionality in each case.

Let us now suppose that at $t=0$ we have a steady condition and that the illumination is then suddenly switched off; for values of Δp sufficiently small to make τ_p constant we have at a subsequent time t

$$\Delta p = \Delta p_o \exp(-t/\tau_p). \tag{118}$$

Also we have

$$i_p = i_{p0} \exp(-t/\tau_p), \tag{119}$$

so that the photo-conductivity decays exponentially with time constant τ_p.

If $\Delta p = 0$ at $t=0$ and the illumination is switched on at $t=0$, held constant from $t=0$ to $t=t_0$ and switched off at $t=t_0$, we have

$$\begin{aligned} i_p &= i_{p0}[1-\exp(-t/\tau_p)] && (0 \leqslant t \leqslant t_0), \\ &= i_{p0}[1-\exp(-t_0/\tau_p)]\exp[-(t-t_0)/\tau_p] && (t \geqslant t_0). \end{aligned} \tag{120}$$

The variation of i_p with t is shown in Fig. 10.19; from such curves a direct measure of τ_p may be obtained. As we shall see, however, the

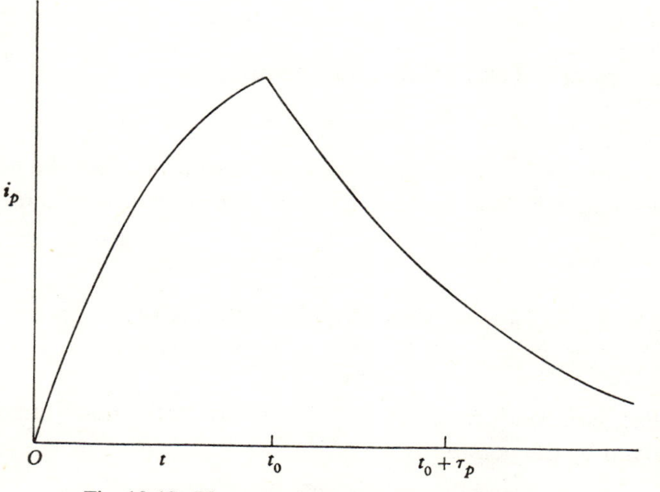

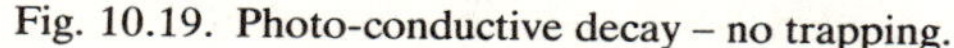

Fig. 10.19. Photo-conductive decay – no trapping.

effect of trapping may modify the form of the decay curve and lead to erroneous values of τ_p unless it is eliminated or taken into account.

10.9.2 Effect of trapping

If we have only recombination traps present, the analysis will be unchanged, except that if we have a high density of traps $\Delta n \neq \Delta p$, and the transient decay will not correspond to an exponential with time constant τ_p (see § 9.4). There is a good deal of evidence, however, that there exists in many materials another form of trap in which, when an electron has been captured, the trap is unlikely to capture a hole and so cause recombination. Such traps are sometimes called 'safe' traps to distinguish them from recombination centres.

We shall now have to introduce a number of time constants into the analysis. To be specific we shall consider holes in *n*-type material. Let τ_p represent the true minority carrier lifetime and τ_1 the average lifetime of a hole before it is caught in a 'safe' trap. Let τ_2 represent the mean time a hole spends in a 'safe' trap before being re-excited to the valence band. The equation satisfied by Δp is now

$$\frac{d\Delta p}{dt} = \mathscr{R} - \frac{\Delta p}{\tau_p} - \frac{\Delta p}{\tau_1} + \frac{\Delta N_s}{\tau_2}, \tag{121}$$

where ΔN_s is the excess hole concentration in the 'safe' traps; also we have

$$\frac{d\Delta N_s}{dt} = \frac{\Delta p}{\tau_1} - \frac{\Delta N_s}{\tau_2} \tag{122}$$

so that, in equilibrium, $$\Delta N_{s_0} = (\tau_2/\tau_1)\Delta p_0 \tag{123}$$

and $$\Delta p_0 = \mathscr{R}\tau_p. \tag{124}$$

The presence of the 'safe' traps does not therefore affect the number of *minority* carriers. We now have, from the condition of space-charge neutrality

$$\Delta n = \Delta p + \Delta N_s, \tag{125}$$

so that in equilibrium we have, from equation (123)

$$\Delta n_0 = \Delta p_0[1 + (\tau_2/\tau_1)]. \tag{126}$$

The change in conductivity σ_0 and the steady-state photo-current i_{p_0} are then given by the equations

$$\Delta\sigma_0 = e\mathscr{R}\tau_p[1 + b + b\tau_2/\tau_1]\mu_h, \tag{127}$$

$$i_{po} = ew\, d\mathcal{R}\tau_p[1+b+b\tau_2/\tau_1]\mu_h\mathcal{E}. \tag{128}$$

It will be seen that if $\tau_2 \gg \tau_1$ the traps will cause a very considerable increase in the photo-current and that, when this condition holds, an 'excess' hole spends considerably longer in a 'safe' trap than in the valence band; by comparing equation (128) with equation (114) an equivalent lifetime τ_p' may be defined by means of the equation

$$\tau_p' = \tau_p[1+b+b\tau_2/\tau_1]/(1+b). \tag{129}$$

If $\tau_2 \gg \tau_1$ then $\tau_p' \gg \tau_p$; the equivalent lifetime τ_p' does not, however, describe the transient behaviour.

The complete solution of equations (121) and (122) may be obtained, but is somewhat complex; we may, however, readily obtain an approximate solution if $\tau_2 \gg \tau_1$ and $\tau_2 \gg \tau_p$. If the illumination is suddenly cut off after the steady state has been established, the minority carrier concentration will drop rapidly with time constant τ_p, since initially we have $\Delta p/\tau_1 = \Delta N_s/\tau_2$; after this initial drop, we have approximately

$$\frac{d\Delta N_s}{dt} = -\frac{\Delta N_s}{\tau_2}, \tag{130}$$

so that ΔN_s decreases with time constant τ_2. The number of excess carriers (electrons) is now nearly equal to ΔN_s so we have $\Delta n \simeq \Delta N_s$ and

$$i_{po} = ew\, d\mathcal{R}\tau_p(b\tau_2/\tau_1)\exp(-t/\tau_2). \tag{131}$$

The trapped holes are slowly released, and while they are trapped they cause an excess of *majority* carriers equal to the number of trapped *minority* carriers.[98] The form of variation of photo-current after the illumination has been cut off is shown in Fig. 10.20; the initial fractional drop in the photo-current is equal to $(1+b)/[1+b+b(\tau_2/\tau_1)]$ and, being quite small if $\tau_2 \gg \tau_1$, may not be observed. Care must therefore be taken when photo-conductive decay is used to measure τ_p, that τ_2 is not measured instead.

The distribution of trapping centres in semiconductors is frequently much more complex than that we have discussed; there may, for example, be several kinds of traps present, and each of them may not even correspond to a single energy level, but have a distribution of levels. The simple situation we have discussed, nevertheless, brings out some important points in connection with lifetime measurements and has applications in one or two important cases. The 'safe' traps may also be

[98] See, for example, R. A. Smith, *Adv. Phys.* (1953) **2**, 321.

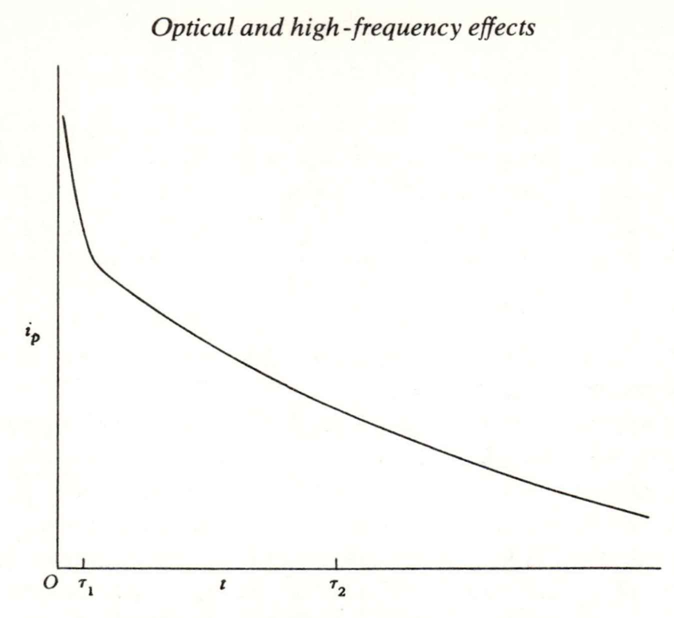

Fig. 10.20. Photo-conductive decay – effect of trapping.

located at the surface of the semiconductor; in this case a special treatment is required, as a space-charge is set up and plays an important part in the trapping process. The general problem of trapping and its effect on photo-conductivity has been discussed by A. Rose[99] and also by C. Herring.[100]

10.9.3 Effect of surface recombination

We have so far neglected surface recombination and assumed that the value of Δp was constant throughout the sample; this will not be so near the surface, unless the surface recombination velocity s is equal to zero. We have already discussed this problem in § 9.9, and found a solution corresponding to uniform absorption of radiation throughout the sample. If we neglect the effect of trapping, the only change introduced by the surface recombination is to replace τ_p in equation (112) by τ where

$$\frac{1}{\tau}=\frac{1}{\tau_p}+\frac{2s}{d} \tag{132}$$

[99] *Progress in Semiconductors* (Heywood, 1957) **2**, 109; *Proceedings of Atlantic City Photo-conductivity Conference* (John Wiley and Sons, and Chapman and Hall, 1956), p. 3.
[100] *Ibid.* p. 81.

provided that the thickness d is much less than the diffusion length L_p (see § 9.9). The ratio of the photo-conductive current to that when $s=0$ is equal to $1/(1+2s\tau_p/d)$, and if $s\tau_p \ll d$, the surface recombination has little effect on the magnitude of the photo-current; if, on the other hand, $s\tau_p \gg d$, the magnitude of the photo-current is independent of τ_p and depends only on surface conditions. It is, moreover, much less than for $s=0$ in the ratio $d/2s\tau_p$. When d is not small compared with L_p we have, using equation (73) of § 9.9

$$i_p = eqw\, d\alpha I\tau_p(1+b)\mu_h\mathscr{E}[1-f(s)], \tag{133}$$

where
$$f(s) = \frac{2s\tau_p(L/d)\sinh(d/2L_p)}{L_p\sinh(d/2L_p)+s\tau_p\cosh(d/2L_p)}.$$

If $d \gg L_p$ we have

$$f(s) \to \frac{2s\tau_p}{L_p+s\tau_p}\left(\frac{L_p}{d}\right), \tag{134}$$

so that $f(s) \to 2L_p/d$ for large values of s, and we see that the effect of surface recombination is small, as we should expect.

10.9.4 Non-uniform absorption rate

We must now consider the condition in which the thickness d is not small compared with the value of α^{-1}, so that the incident radiation is not absorbed at a uniform rate throughout the sample; the absorption rate, and hence the rate of creation of electron–hole pairs, decreases with the distance from the illuminated surface. The analysis for the general case is quite complicated, mainly due to the effect of radiation reflected from the non-illuminated surface and we shall therefore assume for simplicity that $\alpha d \gg 1$. In this case the intensity of I' of radiation inside the sample is given by equation (109). The equation satisfied by the excess carrier density Δp in an n-type sample, neglecting trapping effects, is now

$$D\frac{d^2\Delta p}{dx^2} - \frac{\Delta p}{\tau_p} = -\mathscr{R}\,e^{-\alpha x}, \tag{135}$$

where $\mathscr{R} = q\alpha I(1-R_s)/h\nu$. We may note that the diffusion constant D and diffusion length L will be the ambipolar diffusion constant and diffusion length (see § 7.7) unless $n \gg p$ when they will be equal to D_h and L_p, respectively.

Let us first consider a thick sample for which we also have $d \gg L$. The appropriate solution of equation (135) is

$$\Delta p = A\, e^{-x/L} - \tau_p \mathcal{R}\, e^{-\alpha x}/(L^2\alpha^2 - 1), \tag{136}$$

where A is a constant to be determined by the boundary condition at $x = 0$,

$$D\frac{\partial \Delta p}{\partial x} = s\,\Delta p. \tag{137}$$

This condition gives

$$DA/L - \alpha L^2 \mathcal{R}/(\alpha^2 L^2 - 1) = -sA + s\tau_p \mathcal{R}/(\alpha^2 L^2 - 1), \tag{138}$$

and we have therefore

$$\Delta p = \left(\frac{\mathcal{R}\tau_p}{\alpha^2 L^2 - 1}\right)\left(\frac{\alpha L^2 + s\tau_p}{L + s\tau_p}\, e^{-x/L} - e^{-\alpha x}\right). \tag{139}$$

The photo-current i_p is now obtained by integrating with respect to x; we obtain in this way

$$i_p = e\mathcal{E}(1+b)\mu_h w \int_0^\infty \Delta p\, dx \tag{140}$$

$$= \frac{eqwLI\tau_p\mu_h(1+b)(1-R_s)\mathcal{E}}{h\nu(L + s\tau_p)}\left(1 + \frac{s\tau_p}{L}\,\frac{1}{1+\alpha L}\right). \tag{141}$$

We note that, if $L\alpha \gg 1$, the excess holes penetrate to a much greater depth than the radiation and even if $\alpha \to \infty$ the penetration depth is of the order of L; for the latter condition we have $i_p = i_{p\infty}$, where

$$i_{p\infty} = \frac{eqwLI\tau_p\mu_h(1+b)(1-R_s)\mathcal{E}}{h\nu(L + s\tau_p)}. \tag{142}$$

We may therefore express i_p in terms of $i_{p\infty}$ by means of the equation

$$(i_p/i_{p\infty}) - 1 = \frac{s\tau_p}{L}\,\frac{1}{(1+\alpha L)}. \tag{143}$$

If $s\tau_p \gg L$, a large increase in i_p will take place as α decreases sufficiently to make $\alpha \simeq L^{-1}$; the reason for this is that when $s\tau_p \gg L$ the carriers generated by the radiation will recombine mainly at the surface if they are formed at a distance less than L from the surface, while if they are formed mainly at a distance greater than L, bulk recombination will predominate, with correspondingly longer lifetime. When α becomes small enough to enable an appreciable amount of radiation to pass through the sample, i_p begins to fall again and, when $\alpha d \ll 1$, i_p is proportional to α. The general case is somewhat complicated, but appropriate solutions may be obtained from equation (135) with suitable boundary conditions applied at $x = 0$ and $x = d$; analytical expressions have been given for Δp by O. Garreta and J. Grosvalet[101] and by H. B. de Vore[102] and a series of curves for numerical computation has been given by W. Gärtner.[103] The general form of the variation of the photo-current with wavelength is shown in Fig. 10.21 and compared

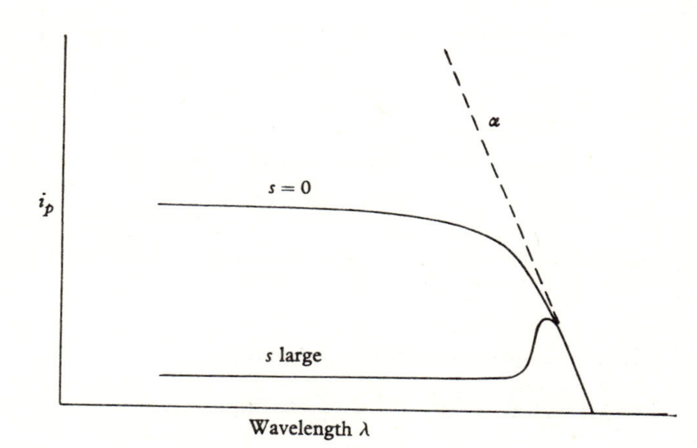

Fig. 10.21. Variation of photo-current with wavelength for large and small values of the surface recombination velocity s.

with the absorption coefficient α; it will be seen that, when s is small, the photo-current rises to a certain value and remains constant as α increases. When s is large, however, the photo-current passes through a maximum and reaches a steady value less than the corresponding value when $s = 0$; a curve obtained for InSb by D. W. Goodwin[104] and illustrating this effect is shown in Fig. 10.22.

When α is very large so that $\alpha L \gg 1$, but d is not much larger than L, we may obtain a solution by omitting the term on the right-hand side of equation (135) and replacing it by a boundary condition. Since all the holes are generated near the surface we may say that we have a hole current density at $x = 0$ equal to $-se\Delta p_0 + eIq(1 - R_s)/h\nu$, and at $x = d$

[101] *Progress in Semiconductors* (Heywood, 1956) **1**, 165.
[102] *Phys. Rev.* (1956) **102**, 86. [103] *Phys. Rev.* (1957) **105**, 823.
[104] *Report of Meeting on Semiconductors, April 1956* (London Physical Society), p. 137.

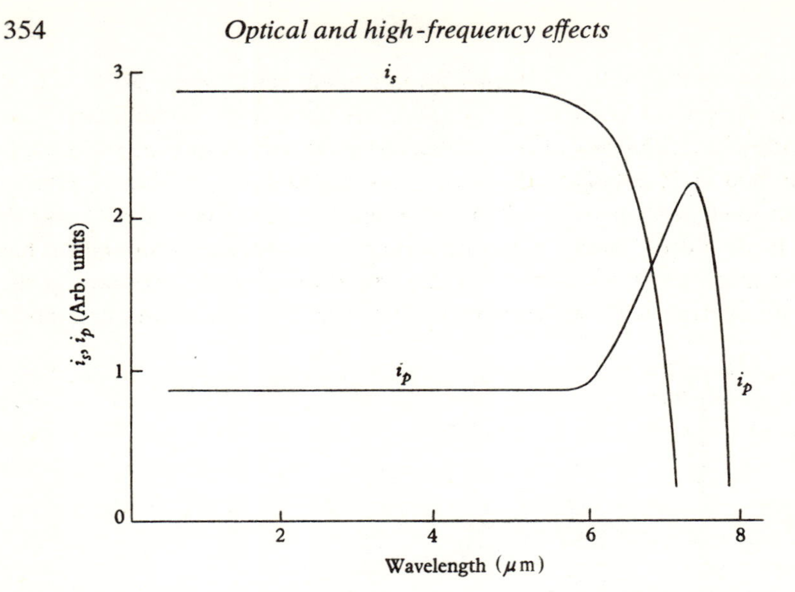

Fig. 10.22. Variation of photo-conductive and photo-magnetic currents with wavelength for InSb. (After D. W. Goodwin, *loc. cit.*)

equal to $se\Delta p_d$. The appropriate solution is now

$$\Delta p = A \cosh\{(d-x)/L\} + B \sinh\{(d-x)/L\}, \tag{144}$$

where A and B are determined by the two boundary conditions. On solving for A and B we obtain

$$\Delta p = \frac{Iq(1-R_s)\tau_p}{h\nu}\left[\frac{s\tau_p \sinh\{(d-x)/L\} + L \cosh\{(d-x)/L\}}{(L^2+s^2\tau_p^2)\sinh(d/L) + 2Ls\tau_p \cosh(d/L)}\right], \tag{145}$$

and the photo-current i_p is given by the equation

$$i_p = i_{p\infty}\frac{[L \sinh(d/L) + s\tau_p\{\cosh(d/L) - 1\}](L+s\tau_p)}{(L^2+s\tau_p^2)\sinh(d/L) + 2Ls\tau_p \cosh(d/L)}, \tag{146}$$

where $i_{p\infty}$ is given by equation (142); when $d \ll L$ (but $\alpha d \gg 1$) we have

$$i_p = \frac{eqw\, dI\mathscr{E}\tau_p\mu_h(1+b)(1-R_s)}{h\nu[2s\tau_p + d(1+s^2\tau_p^2/L^2)]}. \tag{147}$$

If $s\tau_p \ll L$ (but not $\ll d$) we have the same value for i_p as for uniform absorption, when the same amount of radiation is absorbed, i.e. equation (147) differs only from that for uniform absorption in that $I\alpha d$ is replaced by $I(1-R_s)$; if $s\tau_p$ is not small compared with L the photo-current is further reduced. In this case $s\tau_p \gg d$ and a very large reduction, as compared with the condition when $s = 0$, will be experienced. These formulae may be used to deduce values of L and s under appropriate conditions.

10.9.5 Two-photon photo-conductivity

So far, we have assumed that the free carriers produced as a result of absorption of radiation have in each case been due to single-photon absorption. However, we have seen in § 10.5.5, that simultaneous absorption of two photons of frequencies ν_1 and ν_2 will give rise to free carriers, and hence to intrinsic photo-conductivity provided $\boldsymbol{h}(\nu_1 + \nu_2) > \Delta E$. Similarly, for a filled donor impurity level at depth ϵ_d below the conduction band, impurity photo-conductivity would result if $\boldsymbol{h}(\nu_1 + \nu_2) > \epsilon_d$. A similar situation will also hold for empty acceptor levels which can receive an electron from the valence band. Both types of photo-conductivity have been observed, the former in a number of semiconductors.[105] The photo-conductivity of GaAs has been studied with high intensity laser radiation by S. Jayaraman and C. H. Lee[106] who considered, in particular, the effect of a strong impurity level.

The photo-conductive response due to two-photon absorption has been frequently used to obtain the two-photon absorption coefficients using a relationship similar to that given by equation (113) of § 10.9.1. This method has to be used with great care as has been shown by A. F. Gibson *et al.*[107] who have made very careful studies of the two-photon photo-conductivity as well as the two-photon absorption in InSb.

10.9.6 Intra-band photo-conductivity

An interesting and unusual type of photo-conductivity has been observed by A. F. Gibson and P. N. D. Maggs[108] in *p*-type Ge. This is due to the excitation of holes from the light-hole to the heavy-hole band. As we have seen in § 10.2 this causes absorption in the 10 μm region of the spectrum. That a photo-conductive effect should be produced is not at first sight obvious since no additional current carriers are created. The effect is, however, due to the different mobilities of the light and heavy holes. Gibson and Maggs (*loc. cit.*) have given a theoretical discussion of

[105] See, for example, C. C. Lee and H. Y. Fan, *Phys. Rev.* (1974) B**9**, 3502.
[106] *Appl. Phys. Lett.* (1972) **20**, 392.
[107] *Proc. Phys. Soc.* C (1976) **9**, 3259.
[108] *J. Phys.* D, *Appl. Phys.* (1974) **7**, 292.

the effect and compared the theory with their experimental results, finding very good agreement. The recovery time of the effect is very short and it is suggested that it might form the basis of a very fast infra-red detector.

10.10 The transverse photo-voltage

Because of the non-uniform distribution of excess minority carriers when the absorption coefficient α is not small compared with $1/d$ we shall have a diffusion current. When, as is usual, the diffusion constants for electrons and holes are different, an electric field must exist normal to the illuminated surface, if the total current normal to this surface is zero; the current density J_x is given by the equation

$$J_x = e(nb+p)\mu_h\mathscr{E}_x + e(b-1)D_h\frac{\partial \Delta p}{\partial x}. \tag{148}$$

Putting $J_x = 0$ and using the Einstein relationship we have, if $\Delta p \ll n_0$,

$$\mathscr{E}_x = -\left(\frac{kT}{e}\right)\frac{b-1}{nb+p}\frac{\partial \Delta p}{\partial x}. \tag{149}$$

The potential difference V_t between the illuminated and dark faces is therefore given by

$$V_t = \left(\frac{kT}{e}\right)\frac{(b-1)(\Delta p_0 - \Delta p_d)}{n_0 b + p_0}. \tag{150}$$

For the condition $\alpha d \gg 1$ and $d \gg L$ we have from equation (145)

$$V_t = \left(\frac{kT}{e}\right)\frac{qI\alpha L\tau_p(1-R_s)(b-1)}{(n_0 b + p_0)(L+s\tau_p)(1+\alpha L)} \tag{151}$$

and, if the voltage for very large values of α is $V_{t\infty}$, we have

$$V_t = V_{t\infty}\left(\frac{\alpha L}{1+\alpha L}\right). \tag{152}$$

The voltage V_t is known as the transverse photo-voltage, and the phenomenon is sometimes known as the Dember effect.

10.11 The photo-magnetic effect

When a slab of semiconducting material is illuminated on one side with radiation of wavelength somewhat shorter than that corresponding to the fundamental absorption edge, electron–hole pairs are formed near the surface, owing to the high value of the absorption coefficient ($\alpha \sim 10^4\ \mathrm{cm}^{-1}$). These charge carriers now diffuse into the sample and, if a magnetic field is applied parallel to the illuminated surface, the charge carriers will tend to drift to opposite sides of the sample, creating a photo-voltage, which, for not too large fields, is proportional to the magnetic field. This effect is known as the photo-magnetic effect and was first observed in Cu_2O by I. K. Kikoin and M. M. Noskov[109] and discussed by J. Frenkel;[110] it has been studied more recently[111] in high-purity single crystals of Ge and other materials.

When the magnetic field is applied parallel to the illuminated surface (z-direction) a current will be set up in a direction at right angles (y-direction). The hole and electron currents J_{hy}, J_{ey}, are given by (see § 5.2.2)

$$\left.\begin{aligned} J_{hy} &= \mu_h(ep\mathscr{E}_y + BJ_{hx}), \\ J_{ey} &= \mu_e(en\mathscr{E}_y - BJ_{ex}), \end{aligned}\right\} \qquad (153)$$

provided $\mu_h B \ll 1$, where B is the magnetic induction, and J_{hx} and J_{ex} are the hole and electron currents normal to the illuminated surface. As we have seen in § 10.10, $J_{ex} = -J_{hx}$, and we have

$$J_{hx} = ep\mu_h\mathscr{E}_x - eD_h\frac{\partial \Delta p}{\partial x}; \qquad (154)$$

on inserting the value of $\mathscr{E}_x$ given by equation (149) we obtain for the hole current

$$J_{hx} = -eD\frac{\partial \Delta p}{\partial x}, \qquad (155)$$

where D is the ambipolar diffusion coefficient.

In some of the early papers on the photo-magnetic effect the electric field in the y-direction, giving the open-circuit photo-magnetic voltage, was obtained by putting $J_{hy} + J_{ey} = 0$. It was, however, pointed out by W. van Roosbroeck[112] that this is incorrect as it leads to a value of $\mathscr{E}_y$ which depends on x. This cannot be so since, if B is constant, curl $\boldsymbol{\mathscr{E}} = 0$. Since we assume that $\partial\mathscr{E}_x/\partial y = 0$ we must have also $\partial\mathscr{E}_y/\partial x = 0$. To obtain $\mathscr{E}_y$

[109] *Phys. Z. Sowjet.* (1934) **5**, 586. [110] *Phys. Z. Sowjet.* (1935) **8**, 185.

[111] See, for example, T. S. Moss, L. Pincherle and Alice M. Woodward, *Proc. Phys. Soc.* B (1953) **66**, 743; also P. Aigrain and H. Bulliard, *C.R. Acad. Sci., Paris* (1953) **236**, 595.

[112] *Bull. Amer. Phys. Soc.* (1955) **30**, 10. See also L. Pincherle, *Proceedings of Atlantic City Photo-conductivity Conference* (John Wiley and Sons, and Chapman and Hall, 1956), p. 307.

we should use the condition that the total current in the y-direction is zero; this condition may be expressed in the form

$$\int_0^d (J_{ey} + J_{hy})\, dx = 0. \tag{156}$$

If $\Delta p \ll n_0$ or p_0 this becomes

$$\begin{aligned}\mathscr{E}_y &= -\frac{B(b+1)}{ed(n_0 b + p_0)} \int_0^d J_{hx}\, dx \\ &= \frac{BD(b+1)}{d(n_0 b + p_0)} \int_0^d \frac{\partial \Delta p}{\partial x}\, dx \\ &= -\frac{BD(b+1)}{d(n_0 b + p_0)} [\Delta p_0 - \Delta p_d].\end{aligned} \tag{157}$$

It is more usual to measure the short-circuit current due to the photo-magnetic effect; this current i_s is obtained by putting $\mathscr{E}_y = 0$ in equations (153), and integrating with respect to x. Thus we have

$$\begin{aligned}i_s &= w\mu_h(1+b)B \int_0^d J_{hx}\, dx \\ &= Dwe\mu_h(1+b)B[\Delta p_0 - \Delta p_d].\end{aligned} \tag{158}$$

For the conditions $\alpha d \gg 1$, $d \gg L$ we may take $\Delta p_d = 0$; also, from equation (139) we have

$$\Delta p_0 = \frac{\mathscr{R} L \tau_p}{(\alpha L + 1)(L + s\tau_p)} \tag{159}$$

$$= \frac{qI(1-R_s)\tau_p}{h\nu(L + s\tau_p)} \left(\frac{\alpha L}{1 + \alpha L}\right) \tag{160}$$

so that

$$i_s = \frac{we\mu_h(1+b)BqI(1-R_s)L^2}{h\nu(L + s\tau_p)} \left(\frac{\alpha L}{1 + \alpha L}\right). \tag{161}$$

If $i_{s\infty}$ is the value of i_s when $\alpha L \gg 1$, we have

$$\frac{i_s}{i_{s\infty}} = \frac{\alpha L}{1 + \alpha L}. \tag{162}$$

From equations (141) and (161) we may obtain an expression for the ratio of the photo-magnetic current per unit magnetic field to the photo-conductive current per unit electric field; we have

$$\frac{i_s/B}{i_p/\mathscr{E}} = \frac{\alpha L^2}{(1+\alpha L+s\tau_p/L)\tau_p} \tag{163}$$

and

$$\frac{i_{s\infty}/B}{i_{p\infty}/\mathscr{E}} = \frac{L}{\tau_p} = \frac{D^{\frac{1}{2}}}{\tau_p^{\frac{1}{2}}}. \tag{164}$$

Thus, if the diffusion constant D is known from mobility measurements, equation (164) enables τ_p to be determined from a measurement of the ratio $i_{s\infty}/i_{p\infty}$; this method is commonly used to enable τ_p to be determined for semiconductors for which it is too small for direct methods to be used. We note that i_s becomes small when αL is small, so that we have a small photo-magnetic effect unless the photons are mostly absorbed within a diffusion length of the illuminated surface. The photo-magnetic current does not show a maximum near an absorption edge as does the photo-conductive current, when s is large; it decreases steadily as α decreases and begins to fall off more rapidly than the photo-conductive current. The spectral variation of photo-magnetic and photo-conductive currents i_s, i_p obtained by D. W. Goodwin (see p. 354) for a specimen of InSb with a high value of surface-recombination velocity s is shown in Fig. 10.22; the maximum in the photo-conductive current is well illustrated in this example.

If $\alpha L \gg 1$ but d is not large compared with L we have, using equation (145),

$$i_s = i_{s\infty}\frac{[s\tau_p \sinh(d/L)+L\{\cosh(d/L)-1\}](L+s\tau_p)}{(L^2+s^2\tau_p^2)\sinh(d/L)+2Ls\tau_p\cosh(d/L)}. \tag{165}$$

In this case we have

$$\frac{i_s/B}{i_p/\mathscr{E}} = \frac{L}{\tau_p}\frac{s\tau_p \sinh(d/L)+L\{\cosh(d/L)-1\}}{L\sinh(d/L)+s\tau_p\{\cosh(d/L)-1\}} \tag{166}$$

and when $d \ll L$ we have the interesting result that

$$\frac{i_s/B}{i_p/\mathscr{E}} = s. \tag{167}$$

For small values of s the photo-magnetic effect is small if $d \ll L$; this gives a good method for checking that s is small for a thin strip, provided we are sure that $d \gg \alpha^{-1}$ and also that $d \ll L$.

A number of authors have given theoretical treatments of the photo-magnetic effect under various conditions. T. S. Moss, L. Pincherle and

Alice M. Woodward[113] have discussed the effect of strong illumination so that $\Delta p \simeq \Delta n \gg n_0$ or p_0; L. Pincherle[114] has discussed the effect of a strong magnetic field such that $B\mu_h \gg 1$; a comprehensive mathematical treatment together with a critical survey of the various approximations which have been made has been given by W. van Roosbroeck.[115] A numerical analysis leading to a series of curves which may be used to determine i_p and i_s under a wide variety of conditions and so enable L and τ_p to be determined from the variation of i_p and i_s with α and s, has been given by W. Gärtner.[116] The subject has been reviewed by O. Garreta and J. Grosvalet,[117] who give formulae for i_s and i_p for the condition in which αd is not very large, and also discuss the experimental verification of the formulae under various conditions. A good example of the use of the photo-magnetic effect to derive a large amount of data about the properties of semiconductors is given by D. W. Goodwin[118] in his study of InSb.

10.12 Photon drag

There is yet another photo-voltaic effect produced by a quite different mechanism from those already described. This is the so-called photon-drag effect. If a beam of radiation of wavelength about 10 μm (usually 10.6 μm from a CO_2 laser) is passed along a rod of Ge it is found that a photo-voltage appears along the rod. This is basically due to the transfer of momentum from the beam of photons to the electrons and holes in the rod. If the power in the beam is W per unit area there are $W/h\nu$ photons passing unit area per unit time. If a photon is absorbed it passes momentum $\hbar k$, where k is its wave-number, or $h\nu/c$. The momentum passed is therefore simply W/c per unit area. The absorption here is mainly due to free carriers, so that the momentum is passed to these. For simplicity let us consider what happens in a strongly n-type material, there being few holes present. The electrons then tend to be drawn down the beam and in a steady state this flow must be balanced by an electric field $\mathscr{E}$. If α is the absorption coefficient we must have $\alpha W/c = ne\mathscr{E}$ at each point of the rod. Because of absorption, W varies as $W_0 e^{-\alpha x}$ at distance x from one end and we have to perform an integra-

113 *Proc. Phys. Soc.* B (1953) **66**, 743.
114 *Proceedings of Atlantic City Photo-conductivity Conference* (John Wiley and Sons, and Chapman and Hall, 1956), p. 307.
115 *Phys. Rev.* (1956) **101**, 1713.
116 *Phys. Rev.* (1957) **105**, 823.
117 *Progress in Semiconductors* (Heywood, 1956) **1**, 166.
118 *Report of Meeting on Semiconductors, April 1956* (London Physical Society), p. 137.

tion over the length of the rod. If reflection losses are ignored then we have simply the open-circuit voltage $V = -W_0/cen$. In practice the situation is not so simple. This effect has been observed by A. F. Gibson, M. F. Kimmit and A. C. Walker[119] who have given more complex expressions for the voltage induced, taking account of the presence of both electrons and holes. It has also been observed by A. M. Danishevsky *et al.*[120] For n-type material the sign and magnitude of the photovoltage in Ge is more or less as expected but A. F. Gibson and A. C. Walker[121] have observed a reversal of sign as the temperature is changed for p-type Ge. It would be expected that in strongly p-type material the sign of the voltage would be opposite to that for n-type. Gibson and Walker have accounted for the sign reversal in terms of the degenerate band structure of Ge and scattering processes. The effect has also been observed in InAs by C. K. N. Patel.[122]

The photon-drag effect has been used by Gibson, Kimmit and Walker (*loc. cit.*) to provide an extremely fast infra-red detector. The carrier relaxation time would limit the speed to about 10^{-15} s but other effects reduce this somewhat.

10.13 Electroabsorption and electroreflectance

Measurements of the absorption and also of the reflectance of semiconductors at optical frequencies much higher than those corresponding to the fundamental absorption edge have been made and have shown up some higher absorption edges and discontinuities in reflectivity. Some of these have been interpreted in terms of band structure and indeed have helped in the clarification of calculated band structure schemes. The very high values of the absorption coefficient at such frequencies (usually in excess of 10^{-5} cm^{-1}) make measurements with unstrained single-crystal samples very difficult to obtain. The reflection coefficient naturally changes also at an absorption edge, but the change can be quite small and not easily observed. The subject has been reviewed by M. Cardona,[123] who has also discussed in some detail the results achieved. Optical properties of semiconductors in the visible and ultra-violet regions of the spectrum have also been reviewed by J. Tauc.[124]

[119] *Appl. Phys. Lett.* (1970) **17**, 75. [120] *Soviet Phys. JETP* (1970) **31**, 292.
[121] *J. Phys. C: Solid State Phys.* (1971) **4**, 2209.
[122] *Appl. Phys. Lett.* (1971) **18**, 25, 274.
[123] *Semiconductors and Semimetals* (Academic Press, 1967) **3**, 125.
[124] *Progress in Semiconductors* (Heywood, 1965) **9**, 87.

These difficulties have been overcome in a remarkable way by looking for the *change* in absorption and reflectivity when an electric field or pressure is applied. The *change* in reflectivity is easier to observe and this technique has been widely used to obtain a great deal of information about higher and lower bands in semiconductors. By using an alternating electric field or pressure and synchronous detection very small changes are readily observed.

The expectation of changes in reflectivity with electric field arose in the first instance from the prediction by W. Franz[125] and W. F. Keldysh[126] of absorption at photon energies just *below* the forbidden energy gap in the presence of an electric field. An electron excited to a state just below the gap in the forbidden energy band may yet pass into the conduction band by a tunnelling process (see § 14.4.3), the probability becoming much less as the energy deficit increases. The predicted absorption coefficient α_F for $h\nu < \Delta E$ is given approximately by

$$\alpha_F = \frac{Bhe\mathscr{E}}{\pi(512m_r)^{\frac{1}{2}}(\Delta E - h\nu)} \exp\left[-\frac{\pi(128m_r)^{\frac{1}{2}}(\Delta E - h\nu)^{\frac{3}{2}}}{3he\mathscr{E}}\right] \qquad (168)$$

where $\mathscr{E}$ is the electric field strength, m_r the reduced electron–hole mass and B the constant in equation (77) for the absorption without field in the case of a direct transition.

The approximations made in deriving equation (168) become invalid when $(\Delta E - h\nu)$ is less than about $(h^2e^2\mathscr{E}^2/64m_r)^{\frac{1}{3}}$. Clearly α_F does not become infinite as $h\nu \to \Delta E$. A considerable increase in α_F may however be expected as $h\nu \to \Delta E$.

The expressions for α_F obtained without these approximations and for values of $h\nu$ such that $h\nu > \Delta E$ are somewhat complex and involve Airy integrals.[127] When $h\nu > \Delta E$ the absorption coefficient α for a direct transition is given by

$$\alpha = \alpha_d[1 + F(\nu)] \qquad (169)$$

where α_d is the absorption coefficient when $\mathscr{E} = 0$ and $F(\nu)$ is an oscillatory function, the electric field sometimes adding to the absorption and sometimes subtracting from it. In this discussion we have neglected the effect of excitons which, as we have seen, modify the absorption near the fundamental edge.

[125] *Z. Naturforsch.* (1958) **13a**, 484. [126] *Soviet Phys. JETP* (1958) **7**, 788.
[127] See, for example, J. Callaway, *Phys. Rev.* (1963) **130**, 549; K. Tharmalingam, *ibid.*, 2204.

The general form of the absorption is shown in Fig. 10.23. It will be seen that there is a sharp rise in the *change* in absorption coefficient α to the low-frequency side of the forbidden energy gap and that for higher

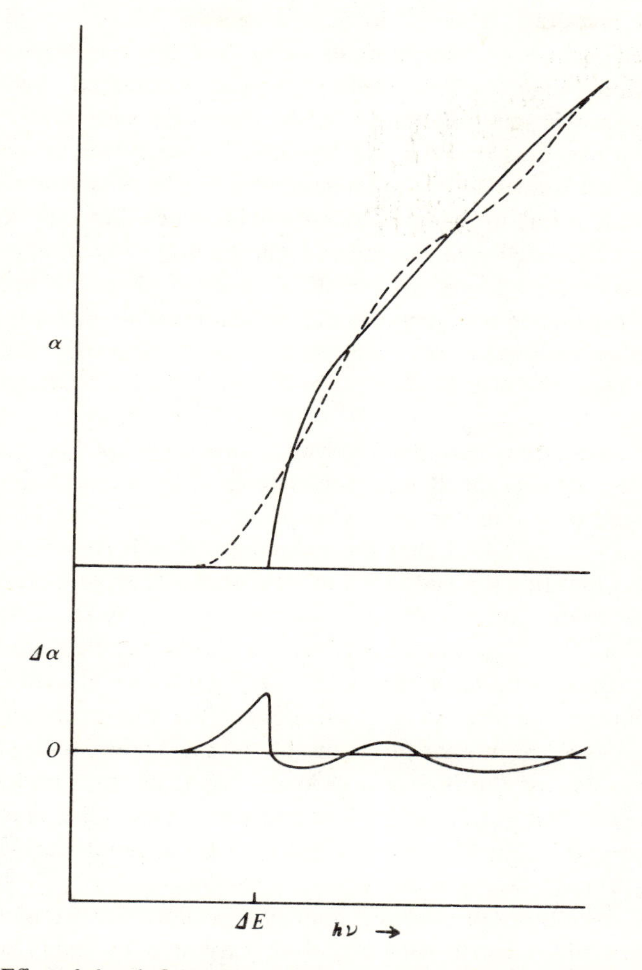

Fig. 10.23. Effect of electric field on absorption near fundamental absorption edge due to direct transition. The broken curve shows absorption coefficient with field on.

frequencies the change in α is oscillatory. This is typical of the changes observed but the precise form of $\Delta\alpha$ varies a great deal.

For an indirect transition these expressions must be modified to take account of the phonons emitted or absorbed. This situation has been treated by C. M. Penchina[128] and shows an oscillatory structure similar to that obtained for direct transitions.

The effect of steady electric fields on absorption has been observed in a number of semiconductors, for example by L. V. Keldysh, V. S. Vavilov and K. I. Bricin[129] in Si and by T. S. Moss[130] in GaAs. The observed changes in absorption, particularly to the long-wave side of the absorption edge, agree well with the calculated values. This phenomenon is generally known as the Franz–Keldysh effect.

T. S. Moss, G. J. Burrell and B. Ellis[131] have, however, pointed out that caution is needed in the interpretation of 'exponential tails' in the absorption as due to 'internal' electric fields, since they may also be due to other causes. Electroabsorption techniques have been applied by D. F. Blossey and P. Handler[132] to the detailed study of the fundamental absorption edge in a variety of semiconductors involving both direct and indirect transitions. In particular they have studied by this means the fine details of the modification of the absorption edge by the presence of excitons.

Even more important than the change in absorption due to an electric field is the consequent change in reflectivity. If the absorption coefficient is changed (and therefore the absorption index k) it will be seen from equation (11) of § 10.1 that the reflectivity R will also be changed. It turns out that this is a good deal more complex than would arise simply from the Franz–Keldysh effect as has been shown by D. E. Aspnes and N. Boltka.[133] A large change in reflectivity is also predicted to take place at any critical point for which $\Delta_k(E_1-E_2)=0$ where E_1 and E_2 refer to two different bands. As a consequence, the interpretation of electroreflectance spectra has proved to be the most powerful tool available for the study of the higher and lower bands of semiconductors. We should note that $\nabla_k(E_1-E_2)=0$ not only at points in the band structure for which both $\Delta E_1(\mathbf{k})=0$ and $\Delta E_2(\mathbf{k})=0$ but at points for which $E_1(\mathbf{k})$ and $E_2(\mathbf{k})$ have the *same* slope.

The observation of periodic changes of reflectivity on application of an alternating electric field has been exploited by many workers to obtain a great deal of experimental information on critical points in the band structure of a wide variety of semiconductors, particularly for

[128] *Phys. Rev.* A (1965) **138**, 924.
[129] *Proc. Vth Int. Conf. on Phys. of Semiconductors* (Czechoslovak Acad. of Sci., 1961), p. 824. [130] *J. Appl. Phys.* (1961) **32**, 2136.
[131] *Semiconductor Opto-electronics* (Butterworths, 1973), § 3.7.2.
[132] *Semiconductors and Semimetals* (Academic Press, 1972) **9**, 257.
[133] *Semiconductors and Semimetals* (Academic Press, 1972) **9**, 457.

energies in the range 1–6 eV. The electroreflectance spectrum of Ge obtained by M. Cardona, K. L. Shaklee and F. H. Pollak[134] is shown in Fig. 10.24. This is a fairly simple spectrum; usually such spectra are much more complex, showing lots of oscillations. The feature marked *A* is thought to correspond to a transition similarly marked in Fig. 13.1, which shows the band structure of Ge. Others marked *B*, *C*, *D* probably correspond to zone-edge transitions, but there is some doubt about their identification.

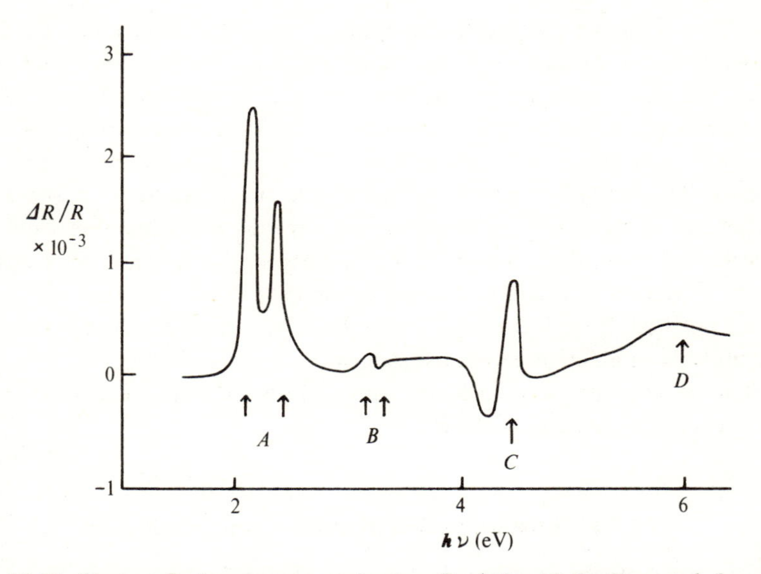

Fig. 10.24. Electroreflectance spectrum of *n*-type Ge. (After M. Cardona *et al.*, *loc. cit.*)

The experimental methods used for obtaining such spectra and their correlation with band structure have been discussed in detail by M. Cardona,[135] who with his collaborators has carried out a great deal of work on electroreflectance spectroscopy.

The different kinds of critical points, maxima, minima, saddle points, equal-slope points, etc., each produce a characteristic electroreflectance spectrum. These and also further developments in the techniques of measurement and correlation have been discussed in reviews by B. O.

[134] *Phys. Rev.* (1967) **154**, 696.
[135] *Semiconductors and Semimetals* (Academic Press, 1967) **5**, 125; *Solid State Phys.* (Academic Press, 1969) suppl. 11.

Seraphin,[136] who with his collaborators has also made extensive observations on and interpretation of these spectra. The various forms which the oscillatory spectra can take have also been listed and discussed by Y. Hamakawa, P. Handler and F. A. Germano.[137] Some more recent developments in technique and interpretation as well as further applications have been reviewed by Y. Hamakawa and T. Nishimo.[138]

10.13.1 Piezoreflectance

The fine details of piezoreflectance spectra are somewhat different from those obtained by means of electrically modulated reflectance, but apart from the fact that an alternating pressure is applied instead of an alternating electric field the techniques and interpretation are very similar. They have also been discussed in most of the references given above. The added pressure brings about small changes in the band structure. In particular, it modulates the energy difference between two bands and may also slightly shift the position in **k**-space of a critical point. These in turn cause changes in the reflectivity. The effect is again most marked at critical points in the band structure and these show up strongly as in electroreflectance spectra.

The basic theory and also experimental details have been given in an extensive review by I. Balslev.[139]

10.14 Emission of radiation from semiconductors

There is a general relationship between the radiation from a body in a small spectral interval and the absorption of the body in the same interval, known as Kirchhoff's law. If the body absorbs strongly it will emit strongly – if it is transparent it will not emit.[140] Strictly this applies only in thermal equilibrium and we shall usually be concerned with situations in which *excess* carriers have been injected either electrically or optically so that we are far from an equilibrium situation. However,

[136] *Optical Properties of Solids*, ed. E. D. Haidemenakis (Gordon and Breach, 1970), p. 213; *Semiconductors and Semimetals* (Academic Press, 1972) **9**, 1; *Optical Properties of Solids*, ed. F. Abeles (North Holland, 1972), p. 163.

[137] *Phys. Rev.* (1968) **167**, 709.

[138] *Optical Properties of Solids – New Developments*, ed. B. O. Seraphin (North-Holland, 1976), p. 255.

[139] *Semiconductors and Semimetals* (Academic Press, 1972) **9**, 403.

[140] See, for example, R. A. Smith, F. E. Jones and R. P. Chasmar, *Detection and Measurement of Infra-red Radiation* (Clarendon Press, 1968).

we should expect that a near-intrinsic semiconductor would radiate little at wavelengths just longer than that corresponding to the fundamental absorption edge. Indeed most of the recombination radiation is confined to a region of wavelengths just shorter, as we have seen in § 9.2.

10.14.1 Recombination radiation

The recombination radiation due to free electrons and holes is confined to a spectral region whose frequency ν is bounded at its lower limit by the condition $h\nu > \Delta E$, falling off rapidly at a rate which increases as the temperature is lowered (see § 9.2). The observed spectral form of the radiation is complicated by the presence of exciton recombination as has been shown by J. R. Haynes, M. Lax and W. F. Flood,[141] who have made a detailed study of the radiation spectrum of Si. Indeed it turns out that at a temperature of 83 °K 5/6 of the recombination radiation in pure Si is due to recombination through excitons (see § 9.5). These measurements were made with *excess* carriers injected into the sample either optically or electrically via a *p–n* junction. At lower temperatures the radiation is mainly due to exciton recombination. Even at 20 °K it was possible to measure the breadth of the main peak in the radiated spectrum and this corresponds well to the spread of energies one should expect from a Boltzmann distribution of translational velocities of the excitons in thermal equilibrium. It is not surprising that at such low temperatures the exciton binding energy in Si (0.0075 eV) would ensure that in thermal equilibrium a large proportion of the free holes and electrons would be bound to form excitons. Moreover this equilibrium between free electrons, holes and excitons should be rapidly established.

The situation is further complicated by the fact that the transition involved is an indirect one and four possible phonons T0, L0, TA, LA could be involved giving eight emission lines in all even for single phonon processes, four for emission and four for absorption. At low temperature the phonon absorption lines would be absent (see § 10.8). At low temperatures four lines are indeed seen but one, thought to be due to emission of a T0 phonon, predominates and so makes the analysis of line shape possible. The line shape will be seen from Fig. 10.25(*a*).

[141] *Proc. IVth Int. Conf. on Physics of Semiconductors: J. Phys. Chem. Solids* (1959) **8**, 392; *Proc. Vth Int. Conf. on Physics of Semiconductors* (Czechoslovak Acad. of Sci., 1961), p. 423.

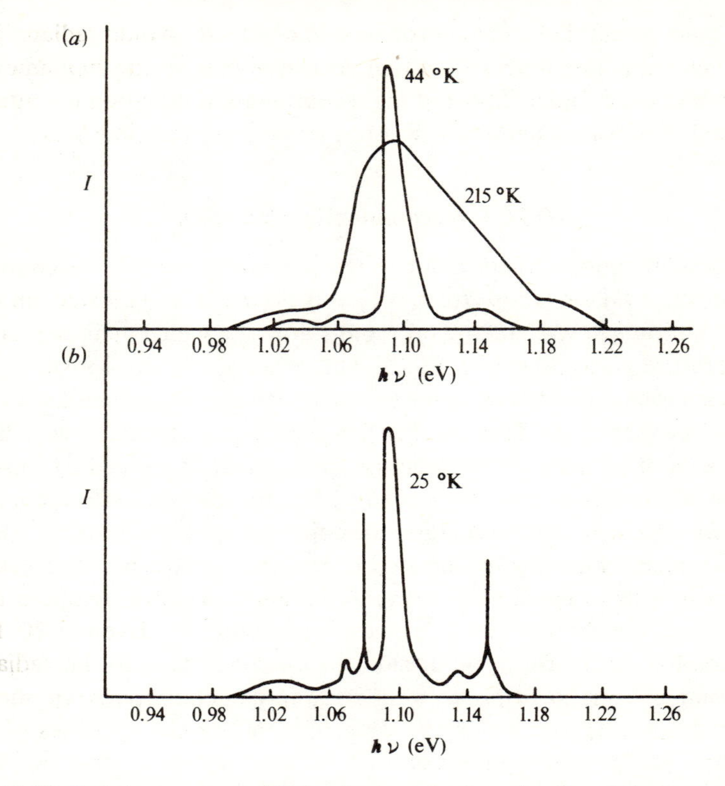

Fig. 10.25. Emission spectrum of Si showing intensity I as a function of photon energy $h\nu$. (*a*) Pure Si (after J. R. Haynes *et al., loc. cit.*); (*b*) Si with 10^{16} cm^{-3} As atoms (after J. R. Haynes, *loc. cit.*).

10.14.2 Emission from bound excitons

The experiments described in § 10.14.1 refer to pure Si. When J. R. Haynes[142] made measurements with Si doped with As atoms (8×10^{16} cm^{-3}) he found in addition to the former radiation a number of *very sharp* lines two of which were particularly strong (see Fig. 10.25(*b*)). One of these, near the former strong line, is attributed to emission from the decay of an exciton bound to an As donor with emission also of a T0 phonon. The other being at a frequency higher by that of the T0 phonon is due to emission *without* a phonon. Thus the change in crystal momentum in the transition is taken up by the crystal as a whole through the donor impurity. Since the excitons are *bound* before decay

[142] *Phys. Rev. Lett.* (1960) **4**, 361.

they have no translational motion and this accounts for the very sharp lines. The binding energy of the exciton to the As donor was found to be 0.0065 eV, about one tenth of the ionization energy of the donor.

Compound semiconductors present a much greater variety of possibilities than the element semiconductors and a very rich variety of emission spectra have been observed due to excitons bound to both donors and acceptors. D. G. Thomas and J. J. Hopfield,[143] for example, have found complex line spectra associated with excitons bound to donors in CdS and CdSe. Some of these lines have been interpreted by D. C. Reynolds[144] as due to the donors having been left in a variety of excited states after exciton decay.

The group III–V compounds, especially GaAs and GaP and their alloys, because of their technological importance as materials for photodiodes, have also been extensively studied. Doped with various impurities they give also a rich variety of narrow-line emission spectra arising from the decay of excitons bound to various imperfections.

A number of processes other than the binding of excitons to single donors or acceptors take place. Some of these we shall discuss in greater detail later (see §§ 14.1, 14.2).

Indeed the reciprocals of all the absorption processes discussed in the previous sections lead to emission. For example, electrons in un-ionized donors may fall to vacant levels in the valence band and so recombine with a hole, similarly electrons in the conduction band may recombine with a hole at an un-ionized acceptor. All the processes discussed so far are inefficient so far as light emission is concerned, most of the recombination taking place through non-radiative transitions.

A number of much more efficient processes have, however, been discovered. If we have both un-ionized donors and acceptors present recombination can take place through the exchange of an electron between them. Apart from the interaction energy, the energy involved in the transition will be $\Delta E - \epsilon_d - \epsilon_a$ where ϵ_d and ϵ_a are respectively the ionization energies of the donor and acceptor. However if they are separated by a distance R a Coulomb energy $-e^2/4\pi\epsilon R$ will have been created. The energy radiated $h\nu$ will then be given by

$$h\nu = \Delta E - \epsilon_d - \epsilon_a + e^2/4\pi\epsilon R. \qquad (170)$$

Since the electrons and holes are *bound*, the emission line for each value of R will be sharp but there will be a whole series of lines corresponding

[143] *Phys. Rev.* (1962) **128**, 2135.
[144] *Electronic Structure in Solids* ed. E. D. Haidemenakis (Plenum Press, 1969), p. 110.

to the various groups of sites at distances $R_1, R_2, \ldots$ in the crystal. Such series of sharp intense lines have been observed by D. G. Thomas, M. Gershenzon and F. A. Trumbore[145] for Si and Te pairs in GaP and also in a number of other semiconductors. P. J. Dean, C. H. Henry and C. J. Frosch[146] have observed intense emission spectra from GaP thought to be due to In–O and Cd–O pairs. These tend to congregate as nearest neighbour pairs so that most of the radiation is compressed into a strong narrow line.

Another type of centre which provides intense radiation including the well-known red light from GaAs photo-diodes is the isoelectronic centre, i.e. when one of the atoms is replaced by another of the same valency. Such centres have been studied by a large number of workers, including A. M. White *et al.*[147] who compared the GaAs spectra with those obtained with InP.

The theory of the emission from various types of centre has been treated in some detail by H. B. Bebb and E. W. Williams.[148] The large amount of experimental data now available on emission from GaAs has been reviewed and compared with experiments by E. W. Williams and H. B. Bebb.[149] J. C. Phillips has discussed the theory particularly as concerns GaAs and GaP.[150] An extensive review of narrow-line emission spectroscopy of semiconductors has also been given by P. J. Dean.[151]

10.14.3 Emission from free carriers

The 'black body' type of emission from a semiconductor will normally follow that of an ideal black body modulated by variations in the absorption and hence of the emissivity as indicated in § 10.14.1. A considerable part of this emission will come from the free carriers and also from the lattice vibrations. The latter we shall discuss in the next section. The free carrier emission may, however, be increased by injection of excess carriers and this has been observed by E. A. Ulmer and D. R. Frankl.[152] They were able, for example, to see the increased emission in the same part of the spectrum showing extra absorption due to excitation of holes from one band to another in the valence band (see § 10.2).

145 *Phys. Rev.* A (1964) **113**, 269. 146 *Phys. Rev.* A (1968) **168**, 812.
147 *J. Phys.* C (1972) **5**, 1727.
148 *Semiconductors and Semimetals* (Academic Press, 1972) **8**, 181.
149 *Ibid.* **8**, 321. 150 *Phys. Rev.* B (1970) **1**, 1545.
151 *Prog. Sol. State Chem.* (Pergamon Press, 1973) **8**, 1.
152 *Proc. IXth Int. Conf. on Physics of Semiconductors* (Nauka, 1968), p. 170.

10.14.4 Emission from lattice vibrations

The peak of the 'temperature' radiation from a body at room temperature occurs around 10 μm so that the lattice vibrations which contribute most of the absorption in this region of the spectrum must be mainly responsible. As we have seen (§ 10.8) these provide a very peaky absorption spectrum and so we should expect the emission to vary in a similar manner. This is found to be so, and this emission has been used by D. L. Stierwald and R. F. Potter[153] to study the lattice vibration spectra. Such studies are in several respects complementary to the absorption studies discussed in § 10.8 and give further information on the principal phonon frequencies. In particular the *reststrahl* bands stand out very clearly and have been located for a number of group III–V semiconductors. The form of the emission spectrum is quite complex due to the effect both of high absorption and high reflectivity caused by the interaction of the L0 and T0 phonons with the radiation (see §§ 8.3, 14.3) but may be analysed to give frequencies of both of these at $\mathbf{k} = 0$. The values found in this way agree very well with those found by a combination of absorption and reflection (see § 10.8).

[153] *Semiconductors and Semimetals* (Academic Press, 1967) **3**, 71.

11

Band structure and the effective-mass approximation

11.1 The effective-mass concept

In Chapter 2 we have briefly discussed the form of the energy as a function of the wave-vector **k** and also the form of the electron wave-functions corresponding to states having a particular value of the energy E. From the resemblance of the form of the energy function near extrema of the band structure to that for a free electron we concluded that free electrons and holes in a semiconductor could reasonably be regarded as particles having an effective mass, which in the simplest case would be scalar, but in general would be tensorial. These masses would only be independent of **k** near extrema and under certain conditions. The use of the effective-mass concept was strengthened by the equations of motion which again resemble those of classical particles. In fact the effective-mass hypothesis is a much more sophisticated concept, as we shall see. Already this has been indicated by the fact that we have applied *quantum theory* to these 'particles' to derive the binding energy of shallow impurities and excitons and to calculate their scattering probability. In retrospect this seems a highly illogical process, but as we shall see it can in fact be justified.

Before proceeding to this justification we must consider briefly how the energy function $E(\mathbf{k})$ is obtained from a knowledge of the crystal structure of the semiconductor. A great deal has been written on this subject and we shall later give some of the key references. A number of methods have been developed and each has its use under particular conditions, there being no completely general method which can be applied. We may say at once that although the general form of the band structure for a particular material is not too hard to obtain, it is a very difficult problem to calculate the complete structure of the conduction and valence bands, consisting in general of a number of overlapping bands. Nevertheless with passing years more and more success has been achieved. This is not only due to the development of better methods of calculation but also to the increasing power of electronic digital computers.

Here we shall only give a brief review of the principles involved. A much fuller discussion has been given by the author in *Wave Mechanics of Crystalline Solids* (*W.M.C.S.*), Chapters 6 and 11, and in a number of books referred to in the Preface. Particular methods are also described in great detail in a number of review articles referred to later.

11.2 Bloch and Wannier functions

The wave-function for an electron in a state in a perfect crystal corresponding to an energy $E(\mathbf{k})$ can be expressed in the form

$$\psi(\mathbf{r}) = U_k(\mathbf{r}) \exp [i\mathbf{k} \cdot \mathbf{r}] \tag{1}$$

where the function U_k has the periodicity of the crystal lattice.

The functions $\psi(\mathbf{r})$ are known as Bloch functions and are basic to any calculation of band structure. Their calculation has been discussed by many authors, a comprehensive treatment being given by J. M. Ziman.[1] Although such wave-functions describe the undisturbed motion of electrons (and they can be adapted for holes), *combination* of them is required to deal with localized disturbances due to imperfections, fields of force etc. For band-structure calculations the plain Bloch functions do not lead to rapidly converging approximations and, as we shall see, various methods have been devised to obtain better approximations.

When we are dealing with a sharply localized field of force, such as that due to an impurity, many Bloch functions would be needed, and another set of wave-functions derived from them is more appropriate. A function is defined for each unit cell in the crystal, that for the jth cell being, for a particular energy band, of the form

$$a_j(\mathbf{r}) = N^{-\frac{1}{2}} \Sigma U_k(\mathbf{r} - \mathbf{R}_j) \exp [i\mathbf{k}(\mathbf{r} - \mathbf{R}_j)] \tag{2}$$

where N is the number of unit cells in the crystal and $\mathbf{R}_j$ the position vector of the jth cell. It will be seen that $a_j(\mathbf{r})$ behaves in the jth cell in the same way as $a_l(\mathbf{r})$ behaves in the lth cell, and so may be written as $a(\mathbf{r} - \mathbf{R}_j)$. These wave-functions were introduced into the theory by G. Wannier,[2] and are generally known as Wannier functions.

[1] *Solid State Physics* (Academic Press, 1971) **26**, 1.

[2] *Phys. Rev.* (1937) **52**, 191.

11.3 Methods of band-structure calculation

A few exact solutions for motion in a one-dimensional periodic potential are available.[3] Moreover perturbation methods can be used to obtain solutions in a *small* three-dimensional periodic potential.[4] While these are useful in showing the general character of the energy bands and, in particular, the presence of a forbidden energy gap, they cannot be used for calculations of the band structure of real crystals. A tight-binding approximation[5] which, at the other extreme, can only be used when electrons are so tightly bound to the individual atoms of a crystal that there is very little overlap of the atomic wave-functions, is again useful only in certain very special circumstances for calculations on semiconductors.

11.3.1 The cellular method

In this method the crystal is divided up into unit cells centred on each atom of the crystal and an approximate wave-function calculated for the cell using the atomic potential. The main difficulty is in fitting boundary conditions at the surface of the cells, but various approximate methods have been devised. If $u(\mathbf{r}')$ is the wave-function where $\mathbf{r}'$ is the vector distance from the centre of the cell, the wave-function in the jth cell is written as

$$\psi(\mathbf{r}) = u(\mathbf{r} - \mathbf{R}_j) \exp[i\mathbf{k} \cdot \mathbf{r}] \tag{3}$$

which has the required Bloch form. This method, originally proposed by E. Wigner and F. Seitz[6] for monatomic crystals has been greatly extended by J. C. Slater[7] and his collaborators. It has also been applied to diatomic crystals such as PbTe with some success by Dorothy G. Bell *et al.*,[8] but has largely been superseded.

11.3.2 Linear combination of atomic orbitals (LCAO)

The LCAO method is an extension of the tight-binding approximation, to take into account the effect of a number of atomic wave-functions instead of just using one as in equation (3). The corresponding wave-function has the form

$$\psi(\mathbf{r}) = \sum_s \alpha_s u_s(\mathbf{r} - \mathbf{R}_j) \exp[i\mathbf{k} \cdot \mathbf{R}_j] \tag{4}$$

[3] *W.M.C.S.*, § 2.9.
[4] *W.M.C.S.*, §§ 6.1, 6.2.
[5] *W.M.C.S.*, § 6.3.
[6] *Phys. Rev.* (1933) **43**, 804.
[7] See, for example, *Handb. Phys.* (1956) **19**, 1.
[8] *Proc. Roy. Soc.* A (1953) **217**, 71.

the constants α_s being obtained by a variational method which minimizes the energy which is found from the latent roots of a so-called 'secular' determinant.

11.3.3 Plane-wave methods

The Bloch functions having the periodicity of the lattice may be expanded as a Fourier series[9] using the reciprocal lattice vectors $\mathbf{b}_n$. The wave-function of equation (1) may then be written in the form

$$\psi(\mathbf{r}) = \sum_n A_n \exp i(\mathbf{k} + 2\pi\mathbf{b}_n) \cdot \mathbf{r} \tag{5}$$

which may be interpreted as a sum of plane waves. The complete solution would require an infinite number of such plane waves, but an approximate solution could be obtained with a finite number s. The trouble with the method lies in the large number s required to give convergence where the wave-function is varying rapidly, i.e. near the atomic cores.

Two ways have been proposed for overcoming this difficulty and provide the two methods now in most general use for band-structure calculations. The first, the augmented plane-wave method (APW), uses the cellular method *inside* a sphere of radius R just smaller than the unit cell and plane waves in the region outside, fitting the wave-functions across the boundary. Since the rapidly varying wave-function near the atomic core is taken care of by the cellular method, only a small number of plane waves is required to obtain convergence.[10] The method was proposed by J. C. Slater[11] and has been expanded and developed by him and his collaborators. It has been applied now to a great many band-structure calculations with considerable success, but mainly to metals. A recent review has been given by J. O. Dimmock[12] and a book on the subject has been published by T. L. Loucks.[13]

The second method, the orthogonalized plane-wave method (OPW) was originally proposed by C. Herring[14] but has been developed by a considerable number of other workers. The wave-functions in the form of plane waves used for the valence electrons will not in general be

[9] *W.M.C.S.*, § 4.8. [10] *W.M.C.S.*, § 6.9.

[11] *Phys. Rev.* (1953) **92**, 603; *Quantum Theory of Molecules and Solids* (McGraw-Hill, 1965), Vol. 2.

[12] *Solid State Physics* (Academic Press, 1971) **26**, 104.

[13] *The Augmented Plane Wave Method* (Benjamin, 1967).

[14] *Phys. Rev.* (1940) **57**, 1169.

orthogonal to the more highly localized wave-functions of the electrons of the atomic core. The method then consists of adding to the plane waves linear combinations of atomic wave-functions to *make* them orthogonal.[15] In this way rapid convergence is again obtained with a small number of plane waves. The method has been successfully used to calculate the band structure of a number of semiconductors. The development and application of the method has been discussed by T. O. Woodruffe.[16] A more recent review has been given by T. C. Collins, R. N. Euwema and D. J. Stukel.[17]

11.3.4 The pseudopotential

A method of band-structure calculation originally introduced by J. C. Phillips and L. Kleinman[18] and further developed by M. H. Cohen and V. Heine[19] is really a consequence of the OPW method. By using the orthogonalized plane waves a modification[20] can be made of the crystal potential which effectively removes the rapidly varying part near the atomic cores. Its Fourier expansion then requires only a few values of the reciprocal lattice vector $\mathbf{b}_n$ and the Fourier coefficients may readily be calculated. These may be chosen to fit a few values of the energy known from experiments. The method is most frequently used to determine the full band structure when a few values of the energy are known. It is found that with only three parameters the band structure of monatomic semiconductors like Si and Ge may be found, agreeing fairly well with a full OPW calculation. More importantly (since there are far more diatomic semiconductors) use of six parameters reproduces well OPW calculations for materials like GaAs and so may be expected to give quite good results for other compound semiconductors without the labour of a full OPW calculation.[21]

An extensive recent review of the development of the method has been given by V. Heine[22] and its application to the fitting of experimental data to obtain the band structure of a number of semiconductors has been discussed by M. L. Cohen and V. Heine.[23]

[15] *W.M.C.S.*, § 6.10.
[16] *Solid State Physics* (Academic Press, 1957) **4**, 367.
[17] *Optical Properties of Solids*, ed. E. D. Haidemenakis (Gordon and Breach, 1970), p. 81.
[18] *Phys. Rev.* (1959) **116**, 287.
[19] *Phys. Rev.* (1961) **122**, 182.
[20] *W.M.C.S.*, § 6.11.
[21] See, for example, J. C. Phillips, *Solid State Physics* (Academic Press, 1966) **18**, 55; *Bonds and Bands* (Academic Press, 1973).
[22] *Solid State Physics* (Academic Press, 1970) **24**, 1.
[23] *Solid State Physics* (Academic Press, 1970) **24**, 38.

11.3.5 The $\mathbf{k}\cdot\mathbf{p}$ method

The only other method used to make an estimate of band structure which we shall mention is the so-called $\mathbf{k}\cdot\mathbf{p}$ method. In this, the differential equation for the quantities $U_k(\mathbf{r})$ is used, rather than that for the full Bloch wave-functions in equation (1). Treating the magnitude of the wave-vector $\mathbf{k}$ as small the equation is solved by perturbation methods to give approximate solutions valid near $\mathbf{k}=0$. It may also be adapted to give solutions valid near any critical point $\mathbf{k}=\mathbf{k}_m$. The energy is expressed in the form[24]

$$E_n(\mathbf{k})=E_{n0}+\frac{\hbar^2k^2}{2m}+\frac{\hbar^2}{m^2}\sum_{m\neq n}\frac{(\mathbf{k}\cdot\mathbf{p}_{mn})(\mathbf{k}\cdot\mathbf{p}_{nm})}{E_{n0}-E_{m0}} \tag{6}$$

where E_{n0} and E_{m0} are band-edge energies and $\mathbf{p}_{nm}$ is a momentum matrix element calculated with $k=0$ and so obtainable by the cellular method. A large contribution from the terms in the sum is obtained only when $E_{m0}\simeq E_{n0}$. The method is most useful when we have a small gap between two bands, when only one term of the sum need be used. In this case we may write by appropriate choice of axes

$$E(\mathbf{k})=\frac{\hbar^2}{2}\sum_{r=1}^{3}\frac{k_r^2}{m_r} \tag{7}$$

where for the top band

$$\frac{1}{m_r}=\frac{1}{m}+\frac{2|\mathbf{p}_{12}^r|^2}{m^2(E_1-E_2)} \tag{8}$$

and for the lower band

$$\frac{1}{m_r}=\frac{1}{m}-\frac{2|\mathbf{p}_{12}^r|^2}{m^2(E_1-E_2)}. \tag{8a}$$

If $|\mathbf{p}_{12}^r|$ is the same for all values of r it will be seen that the top band has a minimum at $\mathbf{k}=0$ and the bottom band a maximum provided $2|\mathbf{p}_{12}|^2>m(E_1-E_2)$.

The method has been used by E. O. Kane[25] to discuss the band structure of Si and Ge near $k=0$ and also that of numerous III–V compounds. It has been treated in some detail by T. P. McLean.[26]

[24] *W.M.C.S.*, § 6.13.
[25] *J. Phys. Chem. Solids* (1956) **1**, 245; *ibid.* (1959) **8**, 38; *Semiconductors and Semimetals* (Academic Press, 1966) **1**, 75.
[26] *Rendiconti della Scuola Internazionale di Fisica Enrico Fermi*, Corso XXII (Academic Press, 1963) p. 483.

The method also provides an interpolation scheme in the same manner as the pseudopotential method and has been used extensively as such by M. Cardona and F. H. Pollak.[27]

When the energy gap ΔE is small, an expression for E of the form (7) is valid only for very small values of k. Kane (*loc. cit.*) has used the $\mathbf{k}\cdot\mathbf{p}$ method to obtain an expression for the deviation from 'parabolicity' in the form

$$E=-\tfrac{1}{2}\Delta E+\frac{\hbar^2k^2}{2m}\pm\frac{1}{2}\left[\Delta E^2+\frac{4\hbar^2k^2|\mathbf{p}_{12}|^2}{m^2}\right]^{\frac{1}{2}}. \tag{9}$$

This reduces to equations (8) and (8*a*) for small values of k. Equation (9) may also be written in terms of the effective mass m_e for the conduction band

$$E=-\tfrac{1}{2}\Delta E+\frac{\hbar^2k^2}{2m}+\tfrac{1}{2}\Delta E\left[1+\frac{2\hbar^2k^2}{\Delta E}\left(\frac{1}{m_e}-\frac{1}{m}\right)\right]^{\frac{1}{2}}. \tag{9a}$$

If $m_e \ll m$ this may be simplified to

$$E=-\tfrac{1}{2}\Delta E+\tfrac{1}{2}\Delta E\left(1+\frac{2\hbar^2k^2}{m_e\,\Delta E}\right)^{\frac{1}{2}}. \tag{9b}$$

Equation (9*b*) gives an indication of the deviation of the conduction band from 'parabolicity'.

11.4 Band-structure calculations

As indicated above, a large number of calculations on a great variety of semiconductors has now been carried out using the various methods. Extensive reviews of this work are available, each with many references. Of the more recent reviews those by J. Callaway,[28] M. L. Cohen and V. Heine,[29] and F. Herman[30] cover the semiconductor field fairly thoroughly. It is difficult to say at present which method has proved most effective. The APW method has been used mainly for metals and the OPW method seems to be favoured for semiconductors particularly; as pointed out by F. Herman (*loc. cit.*) the pseudopotential method, being a derivative, is very suitable for fitting in the complete band structure when a 'first principles' calculation has given the energy for a small number of critical points in the Brillouin zone.

[27] *Phys. Rev.* (1966) **142**, 530.
[28] *Solid State Physics* (Academic Press, 1958) **7**, 100.
[29] *Solid State Physics* (Academic Press, 1970) **24**, 38.
[30] *Electronic Structure in Solids*, ed. E. D. Haidemenakis (Plenum Press, 1969) p. 41.

It has been shown by L. Pincherle[31] that although numerical values of various energies in the band structure depend on the atomic core potentials, the general *form* of the bands may be deduced from consideration of the symmetry properties of the material in question by a judicious use of group theory. He has also made a critical comparison of the various methods and has given a very clear and readable account of their special fields of application.[32] The relative merits of the various methods together with their relationship with bond theory has been discussed by J. C. Phillips.[33]

11.5 The effective-mass approximation

Having determined the band structure of a semiconductor near an extremum, usually the bottom of the conduction band and top of the valence band, we may obtain the elements of the effective-mass tensor. In this section we shall first of all assume these to be constant indicating that the energy function $E(\mathbf{k})$ is a quadratic function of the components of $\mathbf{k}$. We shall now indicate how it is possible to describe quantum-mechanically the motion of particles having such an effective-mass tensor and under what conditions this is justified. This seems to have first been discussed by G. Wannier[34] using the functions bearing his name which we have already introduced (§ 11.2). We write the Bloch wave-functions as before in the form

$$b_n(\mathbf{r}) = U_{nk}(\mathbf{r}) \exp [i\mathbf{k} \cdot \mathbf{r}], \tag{10}$$

the index n referring to the nth energy band. The corresponding Wannier functions $a_n(\mathbf{r})$ are then defined as in § 11.2. It may then be shown that the wave-function arising from an additional potential $V(\mathbf{r})$ in the crystal may be expressed quite generally in the form[35]

$$\psi(\mathbf{r}) = \sum_{nj} A_n F_n(\mathbf{R}_j) a_n(\mathbf{r} - \mathbf{R}_j) \tag{11}$$

where the quantities $F_n(\mathbf{R}_j)$ are derived from a series of algebraic difference equations involving the matrix elements of the potential with the Wannier functions. In the situation in which we may restrict the summation over n to a single band (generally $\hbar^2 k^2/m \ll \Delta E$) we may

[31] *Rendiconti della Scuola Internazionale di Fisica Enrico Fermi*, Corso XXII (Academic Press, 1963) p. 1.
[32] L. Pincherle, *Electronic Energy Bands in Solids* (Macdonald, 1971).
[33] *Bonds and Bands in Semiconductors* (Academic Press, 1973).
[34] *Phys. Rev.* (1937) **52**, 191. [35] *W.M.C.S.*, § 11.2.3.

under certain conditions (which turn out to be essentially that $V(r)$ does not change appreciably over a lattice distance, i.e. is 'slowly variable') replace these *difference* equations by a *differential* equation for a function $F(\mathbf{r})$. This differential equation has a surprisingly simple form. If $E(\mathbf{k})$ is the energy function (which we assume known) we have

$$E(-i\nabla)F(\mathbf{r})+V(\mathbf{r})F(\mathbf{r})=EF(\mathbf{r}) \tag{12}$$

where E is the energy of a stationary state wave-function given by

$$\psi=AU_0(\mathbf{r})F(\mathbf{r}) \tag{13}$$

where A is a normalizing factor and $U_0(\mathbf{r})$ is the periodic part of the Bloch function for $k=0$. We thus see that $F(\mathbf{r})$ acts as a slowly varying modulating function, the variation within each cell being given by $U_0(\mathbf{r})$.

If we have a scalar effective mass m_e, so that $E(\mathbf{k})=\hbar^2k^2/2m_e$, $F(\mathbf{r})$ satisfies the equation

$$\nabla^2F(\mathbf{r})+\frac{2m_e}{\hbar^2}[E-V(\mathbf{r})]=0 \tag{14}$$

which is just the Schrödinger equation for a particle of mass m_e.

More generally if m_{rs} is the effective mass tensor, $F(\mathbf{r})$ satisfies the equation

$$\sum_{rs}\frac{1}{m_{rs}}\frac{\partial^2F}{\partial x_r\,\partial x_s}+\frac{2}{\hbar^2}[E-V(\mathbf{r})]F=0. \tag{15}$$

This is not quite a Schrödinger equation but is what we should expect for a classical particle whose Hamiltonian is

$$\frac{1}{2}\sum_{rs}\frac{1}{m_{rs}}p_rp_s.$$

Equation (14) gives the basis of our previous calculation of the energy of a hydrogenic impurity centre and (15) its modification for ellipsoidal constant energy surfaces. It is also the basis of the use of a quantum-mechanical calculation of the scattering of electrons and holes in semiconductors.

A modification is required when we have multiple minima. If the minimum is at $\mathbf{k}=\mathbf{k}_0$ the appropriate wave-function is given by

$$\psi=AU_{k_0}\exp[i\mathbf{k}_0\cdot\mathbf{r}]F(\mathbf{r}). \tag{16}$$

The wave-function for M multiple minima is then expressed in the form

$$\psi=\sum_{r=1}^{M}A_rF_r(\mathbf{r})U_{k_{0r}}\exp[i\mathbf{k}_{0r}\cdot\mathbf{r}], \tag{17}$$

the constants A_r being determined by a variational procedure. Such wave-functions have been used by W. Kohn[36] who has discussed in some detail the use of the effective-mass approximation in the calculation of the energy levels of shallow donors, although a different set of base functions is used. For degenerate bands, as in the valence band of Si and Ge, some modification is required. This has also been discussed by W. Kohn (*loc. cit.*) and by K. Mendelson and D. R. Schultz.[37]

A somewhat different approach to the effective-mass approximation has been made by T. P. McLean,[38] who has adapted the $\mathbf{k} \cdot \mathbf{p}$ approximation to study this problem. The final results are very similar but the method enables some rather more complex situations to be treated.

11.5.1 Application to impurity levels and excitons

The application of the effective-mass approximation and its extension to deal with deeper impurities for which the 'slowly varying' potential is no longer applicable have been discussed extensively by A. M. Stoneham.[39] A. M. Stoneham and A. H. Harker[40] have made calculations for a variety of centres including those involving more than a single atom. Such calculations involve application of electronic digital computers and although not so complex as band-structure calculations are nevertheless quite complicated.

Stoneham (*loc. cit.*) has made an extensive review of the experimental data on energy levels of impurities in a variety of semiconductors and has compared these with values obtained by calculation. The subject has also been treated in some detail by A. G. Milnes.[41] A very detailed comparison of experiment and theory, in the particular case of donors in GaAs, has been made by H. R. Fetterman, *et al.*[42] The most recent calculations show extremely good agreement between theory and experiment, except for some of the deeper imperfections in Si and Ge for which there are still some discrepancies which are not fully accounted for. As we have seen, the agreement between theory and experiment is very good for shallow donors in Si and Ge, except for the ground states for which the range of forces due to the 'central core' is too short

[36] *Solid State Physics* (Academic Press, 1957) **5**, 257.
[37] *Phys. Stat. Solidi* (1969) **31**, 59.
[38] *Rendiconti della Scuola Internazionale de Fisica Enrico Fermi,* Corso XXII (Academic Press, 1963) p. 479. [39] *Theory of Defects in Solids* (Clarendon Press, 1975).
[40] *J. Phys. C, Solid State Phys.* (1975) **8**, 1102, 1109.
[41] *Deep Impurities in Semiconductors* (Wiley, 1973).
[42] *Phys. Rev. Lett.* (1971) **26**, 975.

for the simple effective-mass approximation to be valid. Calculations by M. Jaros[43] have however brought theory and experiment very close by taking into account spatial variations of effective mass and dielectric constant due to the 'central core'.

A modification of the effective-mass approximation required to deal with a two-particle system has been used to calculate the binding energy of excitons. We have already discussed this in § 10.6.

11.6 Magnetic quantization

An important application of the effective-mass approximation is to give a quantum-mechanical treatment of the motion of electrons (and holes) in a magnetic field. A magnetic field may be regarded as a slowly varying perturbation to the crystalline forces and so should provide a favourable situation for application of the effective-mass approximation. A difficulty arises, however, in that one cannot represent a magnetic field by a scalar potential – we require a vector potential. However, it is well known how the Schrödinger wave equation has to be modified to take account of a magnetic field, and it is not unreasonable to suppose that the effective-mass wave equation should be modified in the same way.

We have already seen that a condition for magnetic quantization is that the field be high enough to make $\omega_c\tau$ somewhat greater than 1, where ω_c is, as usual, the cyclotron frequency and τ the scattering mean free time. This can also be expressed as the condition that $B\mu$ should be somewhat greater than 1, where B is the magnetic induction and μ the mobility.

The wave equation for a free electron in a magnetic field is simply obtained by replacing the momentum operator $\mathbf{p} = -i\hbar\nabla$ by $\mathbf{p} - e\mathbf{A}$, where A is the vector potential for the magnetic field so that $\mathbf{B} = \nabla \times \mathbf{A}$. For an electron having a scalar effective mass m_e the equation for the modulating function $F(\mathbf{r})$ would then be expected to be

$$\frac{1}{2m_e}(i\hbar\nabla + e\mathbf{A})^2 F(\mathbf{r}) = EF(\mathbf{r}). \tag{18}$$

This equivalence which we have written down intuitively can be formally established, as has been shown by W. Kohn.[44] The more general equation replacing equation (12) which we should expect to be

$$E(-i\nabla - e\mathbf{A}/\hbar)F(\mathbf{r}) = EF(\mathbf{r}) \tag{19}$$

[43] *J. Phys.* C (1971) **4**, 1162. [44] *Phys. Rev.* (1959) **115**, 1460.

is more difficult to justify, as has been shown by Kohn (*loc. cit.*). For ellipsoidal constant energy surfaces we should have

$$\sum_{r=1}^{3} \frac{1}{2m_r}\left(i\hbar\frac{\partial}{\partial x_r}+e\mathbf{A}_r\right)^2 F(\mathbf{r})=EF(\mathbf{r}). \tag{20}$$

If the magnetic field is along the z-axis the vector potential $\mathbf{A}$ may be taken as $(0, Bx, 0)$ and equation (18) reduces to

$$\frac{\partial^2 F}{\partial x^2}+\frac{\partial^2 F}{\partial z^2}+\left(\frac{\partial}{\partial y}-\frac{ieBx}{\hbar}\right)^2 F+\frac{2m_e EF}{\hbar^2}=0, \tag{21}$$

an equation solved by L. Landau[45] in discussing the quantum theory of free electrons in a magnetic field. We shall discuss the solution of this equation and its application to semiconductors in § 12.5.

[45] *Z. Phys.* (1930) **64**, 629.

12

Effect of high electric and magnetic fields on transport and optical properties

12.1 Modification of distribution function

When we discussed the transport properties of semiconductors in Chapters 5 and 6 we tacitly assumed that the distribution function giving the electron (or hole) population of the various allowed energy states was not very different from the equilibrium distribution function $f_0(\mathbf{k})$, whether given by the 'classical' form for the non-degenerate condition or by the Fermi–Dirac function. That the distribution should be slightly changed was seen to be necessary to provide for electric and thermal currents, but the change amounted to the superposition of small drift velocities on the otherwise largely unmodified function. (This corresponds to replacing the distribution function $f(\mathbf{k})$ in the small terms of the Boltzmann equation by the distribution function $f_0(\mathbf{k})$ (see *W.M.C.S.* § 10.5).) This is rather like the effect of a gentle wind on the molecular distribution of velocities in the air. A small magnetic field introduces some 'eddies' but again does not greatly upset the equilibrium distribution. In the present section we shall consider the effect of both electric and magnetic fields that are so high that this condition no longer holds.

Firstly we shall consider an electric field without the presence of a magnetic field. As the field increases we shall see that the mean energy of the electrons is increased so that they are no longer in thermal equilibrium with the crystal lattice. At first this departure may be represented by an 'effective temperature' T_e of the electrons, generally higher than the lattice temperature T but with the same *form* of distribution, together with a drift term. As the field is increased further the distribution function begins to depart from the equilibrium form, considerably more fast electrons being produced. When these have enough energy to generate electron–hole pairs by impact ionization a great increase of excess carriers rapidly builds up and a breakdown or avalanche occurs. As we shall see, electrons in the conduction band may also be excited from the lowest minimum to higher minima with quite drastic results.

The modification of the distribution function by a high magnetic field is in some ways more dramatic even when we have no electric field. The phenomenon of magnetic quantization sets in strongly when $B\mu \gg 1$, as we have indicated in § 10.4.2 and, as we shall see, this drastically changes the *equilibrium* distribution function $f_0(\mathbf{k})$ since it greatly modifies the allowed energy levels. We should note the difference between this and the situation created by a high electric field. For the latter a *steady* state may be possible (but not in the avalanche condition) but it is far from an equilibrium condition when the field is large. For a large magnetic field, however, in the absence of an electric field, an equilibrium situation can be established since no energy is being fed into the system. As we shall see, the new distribution function $f_0(\mathbf{k})$ modifies the transport properties when even small electric fields and thermal gradients are applied, and modifies drastically the optical properties. We must now consider how these changes in the distribution function $f(\mathbf{k})$ take place.

12.2 Energy exchange between electrons and lattice

We have assumed when considering scattering of electrons by impurities and also by the crystal lattice that such scattering is nearly elastic and that only small amounts of energy are exchanged in scattering processes. For lattice scattering, we saw that an electron either gained an amount of energy $\hbar\omega_s$ or lost an amount $\hbar\omega_s$, where $\omega_s/2\pi$ is the frequency of the lattice wave causing the scattering. If the probability of scattering with loss were the same as that with gain there would be no net exchange of energy with the lattice. That this is not so is indicated by the fact that the ohmic heating of a crystal is due to electrons under the influence of an electric field giving more energy to the lattice in scattering processes than they receive from it. In a steady state the energy gained by the electrons (or holes) from the electric field just balances that lost to the lattice.

If the mean energy of the electrons is appreciably increased by an electric field this will affect the mobility, and the current density will no longer be proportional to the field. The mobility then must be regarded as a function of the applied field. If the current density J is not proportional to the electric field $\mathscr{E}$ we have a deviation from Ohm's law. Such deviations are common in semiconductors, as we have seen, but are generally due to surface effects or to inhomogeneities such as junctions. The effect which we now consider is a bulk effect and may take place in a perfectly homogeneous semiconductor.

It was pointed out by W. Shockley[1] that departures from Ohm's law of this kind would be much more readily produced in semiconductors than in metals. In a semiconductor a change in mean energy of the order of $\boldsymbol{k}T$ ($\frac{1}{40}$ eV at room temperature) would represent a large change in the mean energy $\frac{3}{2}\boldsymbol{k}T$, while in a metal it would hardly be noticed, the mean energy being several electron volts.

We must now consider in more detail the energy gain due to scattering and to the electric field. For the moment we shall deal only with lattice scattering, since impurity scattering is highly elastic and is important in this connection only at very low temperatures. The energy gain δE due to scattering, with emission or absorption of a phonon, is given approximately by equation (18) of § 8.4. For our purpose, however, this equation is not sufficiently accurate and we must obtain expressions valid to the second power in (u/v), where u is the velocity of sound in the crystal and v the velocity of the electron before scattering. Let ϵ_p be the energy of the phonon emitted or absorbed. Then the angle through which the electron is scattered is the same only to the first power in (u/v) for emission and absorption. Let θ' be the angle of scattering when a phonon is absorbed and θ the angle when a phonon is emitted. Then from equations (9) and (17) of § 8.4 we may readily show, neglecting higher powers of (u/v) than the second, that

$$\epsilon_p = 2m_e v^2(u/v)\sin\tfrac{1}{2}\theta'[1+(u/v)\sin\tfrac{1}{2}\theta'] \tag{1}$$

and

$$\epsilon_p = 2m_e v^2(u/v)\sin\tfrac{1}{2}\theta[1-(u/v)\sin\tfrac{1}{2}\theta]. \tag{2}$$

Equations (1) and (2) represent a better approximation to δE $(=\epsilon_p)$ than equation (18) of § 8.4.

In order to estimate the average gain in scattering we must know the relative probabilities of scattering with emission and with absorption. It is a well-known[2] result of quantum theory that these are in the ratio

[1] *Bell Syst. Tech. J.* (1951) **30**, 990.

[2] This follows in exactly the same way as the well-known result for photons derived in § 14.6.

$(N_p+1)/N_p$, where

$$N_p = \frac{1}{\exp(\epsilon_p/\boldsymbol{k}T)-1}. \tag{3}$$

Normalizing the probabilities p_e, p_a for emission and absorption so that their sum is unity, we have if $\epsilon_p \ll \boldsymbol{k}T$

$$p_e = \tfrac{1}{2}(1+\epsilon_p/2\boldsymbol{k}T), \tag{4}$$

$$p_a = \tfrac{1}{2}(1-\epsilon_p/2\boldsymbol{k}T). \tag{5}$$

The contribution to the loss in energy due to emission of a phonon with scattering of an electron through an angle θ into a solid angle $d\omega$ is

$$\frac{\epsilon_p}{2}(1+\epsilon_p/2\boldsymbol{k}T)\frac{d\omega}{4\pi}. \tag{6}$$

To obtain a similar expression for scattering through an angle θ with absorption, we must first make a correction for the change of angle, i.e. we must replace ϵ_p by $\epsilon_p \sin(\frac{1}{2}\theta)/\sin(\frac{1}{2}\theta')$, and $d\omega$ by $\sin\theta\, d\theta\, d\omega/\sin\theta'\, d\theta'$. The mean gain in energy $\overline{\delta E}(\theta)$ is then given by

$$\overline{\delta E}(\theta)\, d\omega = \frac{d\omega}{8\pi}\left(-\epsilon_p(1+\epsilon_p/2\boldsymbol{k}T)+\epsilon_p(1-\epsilon_p/2\boldsymbol{k}T)\frac{\sin\frac{1}{2}\theta \sin\theta\, d\theta}{\sin\frac{1}{2}\theta' \sin\theta'\, d\theta'}\right). \tag{7}$$

From equations (1), (2) it will be readily seen that we have approximately

$$\sin\tfrac{1}{2}\theta' = \sin\tfrac{1}{2}\theta[1-2(u/v)\sin\tfrac{1}{2}\theta] \tag{8}$$

so that

$$\sin\tfrac{1}{2}\theta' \sin\theta'\, d\theta' = \sin\tfrac{1}{2}\theta \sin\theta\, d\theta[1-8(u/v)\sin\tfrac{1}{2}\theta]. \tag{9}$$

Inserting these expressions in equation (7) and also substituting for ϵ_p in terms of θ we have, neglecting powers of (u/v) higher than the second,

$$\overline{\delta E}(\theta)\, d\omega = 8m_e u^2\left(1-\frac{m_e v^2}{4\boldsymbol{k}T}\right)\sin^2\tfrac{1}{2}\theta(d\omega/4\pi). \tag{10}$$

Averaging over all angles we have then for the average energy gain $\overline{\delta E}$ on scattering

$$\overline{\delta E} = 4m_e u^2\left(1-\frac{E}{2\boldsymbol{k}T}\right), \tag{11}$$

where $E=\frac{1}{2}m_e v^2$ is the kinetic energy of the electron before scattering. We note that if $E > 2\boldsymbol{k}T$ the electron will, on the average, lose energy.

Let us now assume that the electrons have a Maxwellian distribution of velocities corresponding to a temperature T_e so that the number $N(v)\,dv$ with velocities between v and $(v+dv)$ is given by

$$N(v)\,dv = A \exp(-m_e v^2/2\mathbf{k}T_e)v^2\,dv, \tag{12}$$

the constant A being a normalizing factor. When $T_e = T$ this corresponds to the distribution in thermal equilibrium when there is no degeneracy. Equation (11) could now be used to average the energy gain δE over all values of the velocity v. The quantity of greater interest, however, is the rate of gain of energy by the electrons as a result of collisions. For an electron of velocity v the mean collision rate is $\tau^{-1} = v/l$, where l is the constant mean free path due to lattice scattering. The mean rate at which an electron gains energy is $v\,\overline{\delta E}/l$ and the average rate for all the electrons $(dE/dt)_s$ is given by the equation

$$\left(\frac{dE}{dt}\right)_s = 4m_e u^2 l^{-1}\int_0^\infty v^3\left(1-\frac{m_e v^2}{4\mathbf{k}T}\right)\exp(-m_e v^2/2\mathbf{k}T_e)\,dv$$
$$\div \int_0^\infty v^2 \exp(-m_e v^2/2\mathbf{k}T_e)\,dv = 8u^2(2\pi m_e \mathbf{k}T_e)^{\frac{1}{2}}(T-T_e)/\pi l T. \tag{13}$$

It can be seen that in thermal equilibrium, when $T = T_e$, $(dE/dt)_s$ is equal to zero, i.e. there is on the average no gain of energy by the electrons due to collisions.

When an electric field $\mathscr{E}$ is applied, however, the electrons gain energy at a rate $(dE/dt)_f$ given by

$$\left(\frac{dE}{dt}\right)_f = \text{force} \times \text{average velocity}.$$

If we define the mobility μ for field strength $\mathscr{E}$ by means of the relationship

$$\text{average velocity} = \mu\mathscr{E},$$

we have

$$\left(\frac{dE}{dt}\right)_f = e\mu\mathscr{E}^2. \tag{14}$$

For a steady state we must have

$$\left(\frac{dE}{dt}\right)_f + \left(\frac{dE}{dt}\right)_s = 0, \tag{15}$$

and by means of this equation we may obtain an expression for the effective temperature T_e as a function of the electric field $\mathscr{E}$.

In order to express μ as a function of the new average velocity or of T_e we note that the relaxation time is still given by $\tau = l/v$. When we average over the velocity distribution as in § 5.3 we obtain instead of equation (98)

$$\mu = \frac{4el}{3(2\pi m_e kT_e)^{\frac{1}{2}}} = \mu_0\left(\frac{T}{T_e}\right)^{\frac{1}{2}}, \tag{16}$$

where μ_0 is the small-field mobility. The mobility μ has therefore the form

$$\mu = B/TT_e^{\frac{1}{2}}, \tag{17}$$

where B is the same constant as in equation (99) of § 5.3.2. Inserting the values of $(dE/dt)_s$ and $(dE/dt)_f$ in equation (15), we obtain a quadratic equation for the effective temperature T_e of the electrons of the form

$$(T_e/T)^2 - (T_e/T) = (3\pi/32)(\mu_0\mathscr{E}/u)^2, \tag{18}$$

an equation obtained by W. Shockley (*loc. cit.*).

When the drift velocity $\mu_0\mathscr{E}$ is much less than the velocity of sound u we see that T_e is nearly equal to T, the temperature corresponding to the lattice vibrations. For electrons in Ge with $\mu_0 = 0.39\ \mathrm{m^2\,V^{-1}\,s^{-1}}$ for $T = 300\ °\mathrm{K}$ and $u = 5.4 \times 10^3\ \mathrm{m\,s^{-1}}$ we have $\mathscr{E}\mu_0 \simeq u$ when $\mathscr{E} = 10^4\ \mathrm{V\,m^{-1}}$. Deviations from a linear relationship between $\mathscr{E}$ and $(\mu\mathscr{E})$ might therefore be expected to occur at reasonably small values of the field. Equation (18) may be solved to give

$$T_e/T = \tfrac{1}{2}[1 + \{1 + (3\pi/8)(\mu_0\mathscr{E}/u)^2\}^{\frac{1}{2}}], \tag{19}$$

and if we have $\mathscr{E} \ll u/\mu_0$ we may write equation (19) in the approximate form

$$T_e = T[1 + (3\pi/32)(\mu_0\mathscr{E}/u)^2]. \tag{20}$$

For this condition $T_e > T$, and $(T_e - T)$ is proportional to $\mathscr{E}^2$. Thus we see that the effective temperature of the electrons is raised slightly above the equilibrium value, but that this is a second-order effect, so that the usual assumption that $T_e = T$ is justified. Inserting the value of T_e given by equation (16) in equation (20) we have to the same approximation

$$\mu = \mu_0[1 - \alpha\mathscr{E}^2], \tag{21}$$

where $\alpha = 3\pi\mu_0^2/64u^2$. Thus we see that $(\mu - \mu_0)$ is also proportional to $\mathscr{E}^2$ and that, provided $\mathscr{E} \ll \mu_0/u$, it is a good approximation to assume that μ is independent of $\mathscr{E}$ and equal to μ_0.

The quadratic departure of μ from constancy as $\mathscr{E}$ is increased from very small values has been verified experimentally for n-type Ge by J. B. Arthur, A. F. Gibson and J. W. Granville,[3] although the value found for α is smaller than that predicted by equation (21). The above analysis is based on a model which assumes a scalar effective mass m_e, but it is thought that the use of a tensor effective mass, as is required for Ge, would not greatly affect the mean rate of energy exchange between the electrons and the lattice at ordinary temperatures.[4] We must therefore assume that there is some other scattering mechanism present which speeds up the energy exchange. This is unlikely to be impurity scattering since this is highly elastic. Its effect is mainly to change the variation of τ with v, which alters the expression for $(dE/dt)_s$. Both ionized impurity scattering and a mixture of this and lattice scattering have been considered by J. B. Gunn.[5] For ionized impurity scattering alone the value of α in equation (21) is negative, so that the mobility at first increases as $\mathscr{E}$ is increased, but from the experimental conditions it is clear that the small values of α observed cannot arise from a fortuitous balance between lattice and impurity scattering. The discrepancy seems likely to arise from the neglect of scattering by the optical and higher acoustic modes as shown by a much more elaborate treatment of the scattering processes. These have been examined in great detail by Esther M. Conwell[6] who, in a supplementary volume of the series *Solid State Physics* devoted entirely to the subject, has extended the treatment of all the scattering processes which we have discussed in Chapter 8 to higher electric fields. She has also discussed in detail their application to Ge and Si as well as to other semiconductors. As we shall see later, at very high fields the effect of higher valleys is important, but not for determining the quantity α which gives the variation at moderate electric fields. The subject has also been treated from a slightly different point of view by E. G. S. Paige,[7] a whole volume of *Progress in Semiconductors* being devoted to it.

12.3 Hot electrons

As $\mathscr{E}$ is further increased we should expect to have significant departures of μ from constancy. When $\mathscr{E} \gg \mu_0/u$, we may express equation

[3] *J. Electron.* (1956) **2**, 145.
[4] See, for example, M. Shibuya, *Phys. Rev.* (1955) **99**, 1189.
[5] *Progress in Semiconductors* (Heywood, 1957) **2**, 213.
[6] *Solid State Physics* (Academic Press, 1967) Suppl. 9.
[7] *Progress in Semiconductors* (Heywood, 1964) **8**.

(19) in the approximate form

$$T_e/T = (3\pi/32)^{\frac{1}{2}}\mu_0\mathscr{E}/u \tag{22}$$

and inserting this value in equation (16) we have

$$\mu = (32/3\pi)^{\frac{1}{4}}(\mu_0 u)^{\frac{1}{2}}\mathscr{E}^{-\frac{1}{2}}. \tag{23}$$

The mean drift velocity $v_d = \mu\mathscr{E}$ is given by

$$v_d = (32/3\pi)^{\frac{1}{4}}(\mu_0 u\mathscr{E})^{\frac{1}{2}} \tag{24}$$

and increases as $\mathscr{E}^{\frac{1}{2}}$ rather than as $\mathscr{E}$, when Ohm's law is obeyed.

Equation (24) will not hold for indefinitely large values of $\mathscr{E}$, since other effects set in which produce a marked modification at high values of $\mathscr{E}$. So far, we have not considered scattering by the optical modes of the lattice vibrations. These require an energy $\boldsymbol{h}\nu_0$ to excite them, where ν_0 is the frequency corresponding to $K = 0$ for the optical mode, so that when the value of v_d is such that $\frac{1}{2}m_e v_d^2$ is comparable with $\boldsymbol{h}\nu_0$, they will begin to be excited. In this case a large fraction of the kinetic energy of an electron may be emitted, and the process is highly inelastic. If we assume that the optical modes are not excited thermally to an extent that the probability of absorption is comparable with that of emission, we see that as the energy of an electron approaches $\boldsymbol{h}\nu_0$ its velocity is likely to be reduced to a small value by exciting an optical phonon. If τ_0 is the relaxation time for this process we have, neglecting absorption of optical phonons,

$$\left(\frac{dE}{dt}\right)_s = -\boldsymbol{h}\nu_0/\tau_0. \tag{25}$$

We have also

$$\left(\frac{dE}{dt}\right)_f = \boldsymbol{e}^2\tau_0\mathscr{E}^2/m_e \tag{26}$$

and inserting these values in equation (15) gives

$$\tau_0 = (\boldsymbol{h}\nu_0 m_e)^{\frac{1}{2}}/\boldsymbol{e}\mathscr{E}. \tag{27}$$

From this we may calculate the saturation drift velocity v_s which is given by

$$v_s = \boldsymbol{e}\tau_0\mathscr{E}/m_e = (\boldsymbol{h}\nu_0/m_e)^{\frac{1}{2}}. \tag{28}$$

Thus the current density through the sample will saturate at a value J_s given by

$$J_s = n\boldsymbol{e}v_s = n\boldsymbol{e}(\boldsymbol{h}\nu_0/m_e)^{\frac{1}{2}}. \tag{29}$$

This effect was predicted theoretically by W. Shockley (*loc. cit.*) and observed experimentally in Ge by E. J. Ryder[8] and later by J. B. Gunn[9] using a rather more elaborate experimental technique. Gunn found the current density constant for fields between 4.5 and 9.0×10^5 V m^{-1}, corresponding to a value of v_s of 5.2×10^4 m s^{-1}. The corresponding values obtained by Ryder and by Arthur, Gibson and Granville (*loc. cit.*) were respectively 6.2×10^4 m s^{-1} and 6.5×10^4 m s^{-1}. The value of v_s given by equation (28) is 1.7×10^5 m s^{-1} if we take $m_e = \boldsymbol{m}/5$ and $\boldsymbol{h}\nu_0 = kT_0$, where $T_0 = 550$ °K. Thus the agreement between experiment and the simple theory of optical mode scattering is somewhat better than for the low-field values, but the theoretical value of v_s is too high by a factor of about two. This is perhaps not surprising when we consider the simplifying assumptions we have made.

The form of the curve showing the variation of drift velocity v_d with electric field $\mathscr{E}$ obtained by Gunn is shown in Fig. 12.1. It will be seen

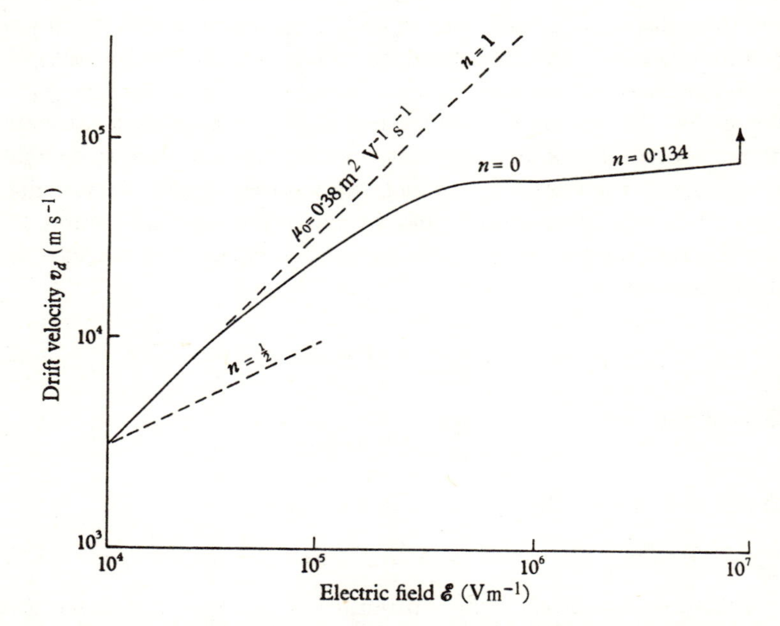

Fig. 12.1. Variation of drift velocity v_d with electric field $\mathscr{E}$. The number n represents a variation of the form $v_d \propto \mathscr{E}^n$. (After J. B. Gunn, *loc. cit.*)

that for values of $\mathscr{E}$ greater than 9×10^5 V m^{-1} the drift velocity begins to increase again. This means that some electrons reach velocities

[8] *Phys. Rev.* (1953) **90**, 766.

[9] *J. Electron.* (1956) **2**, 87.

greater than that required to excite the optical phonons and is what we should expect, since, although the probability of being scattered with emission of an optical phonon is high as soon as $\frac{1}{2}m_e v^2 > h\nu_0$, it is not exactly equal to unity, and there is a small probability that some electrons will reach much higher energies in high fields. There is indeed good experimental evidence that this is so, since light is observed to be emitted from certain types of junctions in semiconductors[10] at which there exist strong electric fields of the order of 10^7 V m^{-1}; the light is supposed to arise from these fast electrons recombining with holes. Evidence for the emission of fast electrons from such junctions in Si has also been obtained.[11]

Using a numerical method (Monte Carlo method) for solving the Boltzmann equation (see § 12.4) W. Fawcett and E. G. S. Paige[12] have made a more elaborate calculation of the variation of mobility with electric field and find a variation very similar to that shown in Fig. 12.1 at room temperature. As we shall see later, the presence of higher valleys in the ⟨100⟩ directions modifies the form of the curve for lower temperatures.

When the mean electron velocity greatly exceeds the thermal velocity corresponding to the lattice temperature T the electrons have come to be known as 'hot' electrons. In this case the distribution of velocities is far from Maxwellian as has been shown by W. Fawcett and E. G. S. Paige (*loc. cit.*). When the distribution is nearly Maxwellian but with raised effective temperature T_e the electrons are known as 'warm'.

From the form of the curve in Fig. 12.1 it will be seen that a sudden and very large increase in v_d appears to take place at a field of the order of 8×10^6 V m^{-1}. What is observed is a sudden increase in current. This marks the onset of a new phenomenon, and is due to a sudden increase in the number of carriers which hitherto we have assumed to be constant. This increase is due to ionization of the atoms of the crystal and provides more evidence for the existence of some electrons with energies considerably in excess of $h\nu_0$. Only a few electrons are needed as this is an avalanche process, each electron producing a number of secondaries which in turn may be accelerated to produce more secondaries. The minimum energy which an electron must possess to produce a hole–electron pair is somewhat in excess of ΔE, the forbidden energy gap; this is because momentum as well as energy must be conserved.

[10] R. Newman, *Phys. Rev.* (1955) **100**, 700; A. G. Chynoweth and K. G. McKay, *ibid.* (1956) **102**, 369.
[11] J. A. Burton, *Phys. Rev.* (1957) **108**, 342.
[12] *J. Phys.* C: *Solid State Phys.* (1971) **4**, 1801.

The minimum value if the electron and hole masses are equal is $\frac{3}{2}\Delta E$, which for Ge corresponds to an energy of the order of 1.05 eV and is greatly in excess of $h\nu_0$.

An excellent review of the electron lattice energy interchange has been given by E. G. S. Paige[13] with particular reference to Ge and Si but does not include the later work referred to above.

This sudden increase in the number of hot electrons takes place when the optical mode scattering is insufficient to transfer the energy gained by the electrons from the field to the lattice and the breakdown situation sets in. This process has been treated theoretically by H. Fröhlich and B. V. Paranjape.[14] It has also been discussed in some detail by J. Yamashita.[15] The work on hot electrons has also been reviewed by H. Pötzl.[16] To describe the modified distribution function they introduced a 'displaced' Maxwellian function of the form

$$f(\mathbf{k}) = A \exp\left[-\hbar^2|\mathbf{k}-\mathbf{q}|^2/2m_e kT\right] \tag{30}$$

where $\mathbf{q}$ is a vector parallel to the electric field and of magnitude $m_e\mu\mathscr{E}/\hbar$ so that the drift velocity is just $\hbar\mathbf{q}/m_e$. They showed, treating optical-mode scattering as the main transfer mechanism, that at a certain field, which they were able to calculate, breakdown occurs. For GaAs for example this is about 3.5×10^5 V m^{-1}, rather lower than for Ge because of the smaller value of m_e. For this material, however, other effects come into operation before breakdown occurs.

It is interesting to note that to the first order in $\mathscr{E}$ the distribution function given in (30) is just equal to the modified function we have used in § 7.3 equation (23).

12.4 Transferred electron effects

When the effective temperature of the electrons in the lowest minimum of the conduction band rises sufficiently, some of them will be transferred to a higher minimum. Assuming that we can define an effective temperature T_e this will begin to take place rapidly as kT_e approaches $\Delta E_1 - \Delta E$, where ΔE_1 is the value of the next smallest energy gap at a minimum. For semiconductors for which $\Delta E_1 - \Delta E$ is small such a transfer will take place at relatively low electric fields, generally before an avalanche condition sets in. For example in GaAs the next minimum

[13] *Progress in Semiconductors* (Heywood, 1964) **8**.
[14] *Proc. Phys. Soc.* B (1956) **69**, 21.
[15] *Progress in Semiconductors* (Heywood, 1960) **4**, 63.
[16] *Electronic Materials*, ed. N. B. Hannay and U. Colombo (Plenum Press, 1973), p. 89.

is only 0.35 eV above the central minimum at $\mathbf{k}=0$. This effect, known as electron transfer, as expected, occurs very strongly in this semiconductor and has a marked effect in the variation of mobility with electric field. In GaAs, being partially ionic, polar optical-mode scattering is dominant at high temperatures and in particular at the effective temperatures created by a strong electric field. These, as we shall see, can reach several thousand degrees K for fields of the order of 10^6 V m^{-1} and 10^4 °K for 10^7 V m^{-1}. It is clear then that for such fields kT_e is comparable with or greater than $\Delta E_1 - \Delta E$ and a lot of electrons will be excited up to the higher minima. Were it not for a curious feature this would not drastically affect the voltage–current charactistic. This feature is that the effective mass is much greater and the mobility considerably less at these higher minima. Consequently when a lot of electrons are transferred we should expect the current to drop rapidly as the field is increased. Expressing the current density in the form $J = nev_d$ so that v_d is a kind of average drift velocity we obtain a variation of v_d against the electric field $\mathscr{E}$ like that shown in Fig. 12.2. It will be seen that v_d (and hence J) passes through a sharp maximum, falls rapidly and then increases quite slowly. The falling part of the curve corresponds to a negative resistance. Such a negative resistance characteristic had been long sought to provide a means of counteracting the positive resistance of a circuit and so providing a solid-state oscillator.

If n_1, μ_1 are the electron concentration and mobility for the central minimum and n_2, μ_2 the same for the higher minima, the current density J will be given by

$$J = e(n_1\mu_1 + n_2\mu_2)\mathscr{E} = nev_d \tag{31}$$

where $n = n_1 + n_2$, so that

$$v_d/\mathscr{E} = (n_1\mu_1 + n_2\mu_2)/(n_1 + n_2). \tag{32}$$

When n_2 is small $v_d = n\mathscr{E}\mu_1$ and when n_1 is small $v_d = n\mathscr{E}\mu_2$. If μ_1 is considerably greater than μ_2, it requires a rather sudden decrease in n_1 with increase of $\mathscr{E}$ to produce the maximum shown in Fig. 12.2, otherwise the *slope* of the $v_d/\mathscr{E}$ curve just decreases. So the turning point may be expected to occur not far below the critical field for breakdown where the effective temperature in the central 'valley' rises rapidly. In GaAs this is just over 3×10^7 V m^{-1} while, as we have seen, the breakdown field would be at 3.5×10^7 V m^{-1}. The value of T_e for the electrons in the central minimum is also shown in Fig. 12.2 and confirms this.

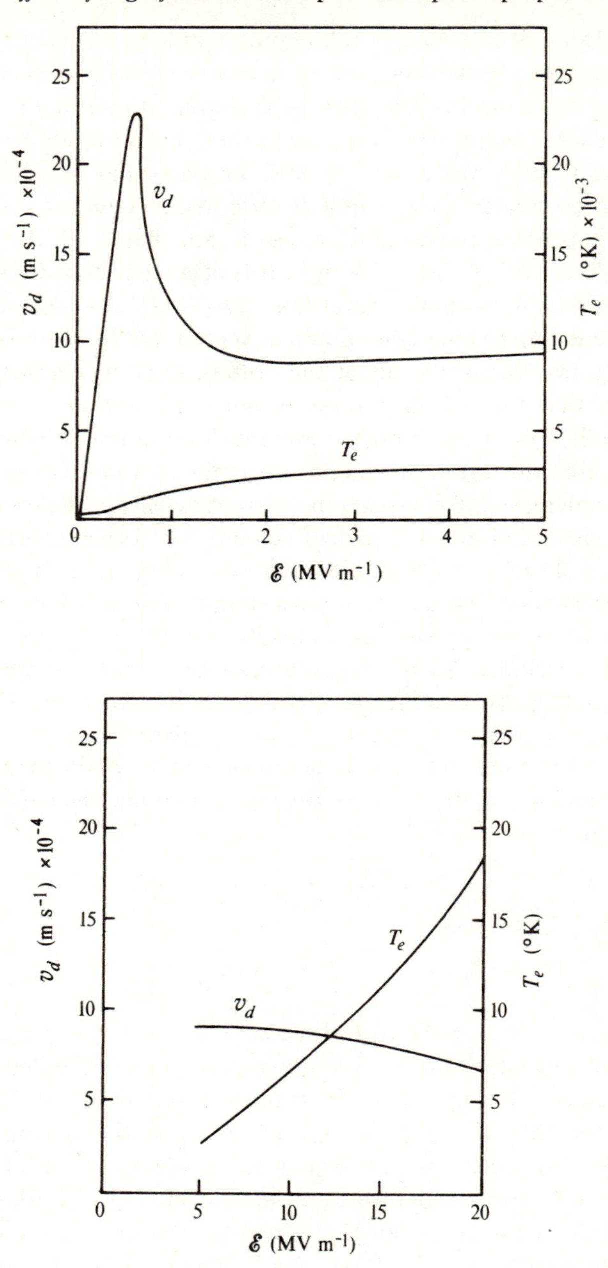

Fig. 12.2. Drift velocity v_d and effective temperature T_e in the lowest conduction band minimum of GaAs as a function of electric field $\mathscr{E}$. (After P. N. Butcher, *Rept. Prog. Phys.* (1967) **30**, 97.)

It turns out that the steady state in which the negative resistance holds over all the space between two electrodes is unstable as was first pointed out by B. K. Ridley[17] although the assumptions on which his analysis was based turned out to be not quite correct. This instability was also studied by J. B. Gunn[18] who found that when an electric field is applied to a short length of GaAs high enough to produce the negative slope, oscillations are maintained, the frequency of which is proportional to the length of the sample. This effect, now known as the Gunn effect, provides the basis of microwave oscillators and has thus stimulated a great deal of research on the phenomenon of electron transfer.

It is clear that regions of some kind are driven across the space between electrodes at more or less constant velocity and arrive periodically at the anode. These are high-field regions where most of the electrons have been driven into the low-mobility minima. They start at the cathode and drift across with velocities of the order of the peak value of v_d which for GaAs is about 2×10^5 m s^{-1}. When they reach the anode they are reformed at the cathode and the process is repeated. With a drift velocity of 2×10^5 m s^{-1} a length of 20 μm is required for an oscillator of frequency 10 GHz, or a wavelength of 3 cm.

The main reason for the instabilities described by B. K. Ridley (*loc. cit.*) and J. B. Gunn (*loc. cit.*) is not difficult to find. Since the effective mobility is negative beyond the maximum, the differential resistivity ρ_d is also negative. The quantity τ_0 in § 7.3 equation (28), equal to $\epsilon\rho_d$, will now be negative so that any space-charge will *grow* exponentially rather than decay. This means that an electron accumulation at the cathode will grow and an electron depletion at the anode will grow. The details of this growth, which has been discussed by many authors, are very complex. A review of the situation has been given by C. Hilsum.[19]

If we have a region in which nearly all the electrons are in the low-mobility minima, the field being $\mathscr{E}_2$, next to a region with most of the electrons in the centre minimum, with field $\mathscr{E}_1$, in order to conserve current we must have $\mathscr{E}_1\mu_1=\mathscr{E}_2\mu_2$ so that $\mathscr{E}_2/\mathscr{E}_1=\mu_1/\mu_2$. For GaAs $\mu_1\simeq8000$ cm^2 V^{-1} s^{-1} and $\mu_2\simeq150$ cm^2 V^{-1} s^{-1}, so that $\mathscr{E}_2\simeq60\,\mathscr{E}_1$. Such an extreme situation does not in fact occur but it is clear that high-field regions will be readily created even when n_2 is somewhat less than n_1. We may also note that because of the higher effective mass, 0.35 $\boldsymbol{m}$ in the higher minima as against 0.067 $\boldsymbol{m}$ in the central minimum,

[17] *Solid State Commun.* (1963) **1**, 88. [18] *I.B.M. J. Res. Devel.* (1964) **8**, 141.
[19] *Electronic Structures in Solids*, ed. E. D. Haidemenakis (Plenum Press, 1969), p. 360.

together with the fact that there are three ⟨100⟩ minima, there are about thirty times more states per unit energy in the higher levels.

Domains will not form under all conditions. For example it is clearly necessary that the transit time should exceed the space-charge relaxation time $\mathscr{E}/ne\overline{\mu_d}$ where $\overline{\mu_d}$ is the negative differential mobility ($\sim 3000\ \mathrm{cm^2\ V^{-1}\ s^{-1}}$). For GaAs the condition is that the product nl should have a value greater than about $2\times 10^{11}\ \mathrm{cm^{-2}}$. For $l = 20\ \mu\mathrm{m}$ this means that $n > 10^{14}\ \mathrm{cm^{-3}}$ which is readily achieved.

The calculation of the effective temperature is now much more complex than for a single minimum. One method which has been used is to assume an effective temperature T_s for each minimum and to use a 'displaced' Maxwellian distribution with displacement vector $\mathbf{q}_s$. The drift velocity for the sth minimum is then given by $\hbar\mathbf{q}_s/m_s$ where m_s is the effective mass for the minimum. By applying certain conservation conditions, the effective temperature (assumed to be the same in all the higher minima) is calculated. This method seems first to have been used by C. Hilsum[20] who made the simplifying assumption that the mobility is constant in the central and higher valleys, and showed that a characteristic of the general shape of that shown in Fig. 12.2 could be obtained. Clearly this is much too simple a model since we should expect the mobilities to be field-dependent at fields high enough to excite an appreciable fraction of electrons into the higher minima.

A much more elaborate calculation was made by P. N. Butcher and W. Fawcett,[21] also using displaced Maxwellian distributions but taking various scattering mechanisms into account. The curves shown in Fig. 12.2 are based on this calculation for GaAs. It will be seen that for the central valley T_e reaches the very high value of $1.8\times 10^4\ ^\circ\mathrm{K}$ at a field of $2\times 10^7\ \mathrm{V\ m^{-1}}$. They also made calculations for InP and CdTe and found $v_d/\mathscr{E}$ characteristics of similar shape.

Although the use of displaced Maxwellian distributions give the general features of the voltage–current characteristic it is clearly desirable to calculate the form of the distribution *ab initio* from Boltzmann's equation. This is a formidable task but can be done using modern computational methods (Monte Carlo) as has been shown by W. Fawcett, A. D. Boardman and S. Swain[22] who have applied the technique to GaAs. They have shown that at the higher fields the displaced Maxwellian distributions are not very good and that there are more electrons at higher energies than they indicate. The technique has also been

[20] *Proc. Inst. Rad. Engrs N.Y.* (1962) **50**, 185.
[21] *Proc. Phys. Soc.* (1965) **86**, 1205.
[22] *J. Phys. Chem. Solids* (1970) **31**, 1963.

applied to Ge, to determine the effect of the higher minima, by E. G. S. Paige.[23] The form of the $J/\mathscr{E}$ characteristic is similar to Fig. 12.1, i.e. with no maximum followed by a negative slope for $T = 300\ °K$ but a negative slope develops for lower temperatures. The effect is, however, not nearly so marked as for GaAs, mainly due to the much stronger polar scattering in the latter, and the larger ratio of the mobilities.

A different approach has been adopted by Esther M. Conwell and M. O. Vassell[24] based on the assumption of nearly elastic collisions but also solving the Boltzmann equation. Surprisingly the results of the calculation do not differ a great deal from those obtained by the Monte Carlo method.

The method is also discussed in the review by Esther M. Conwell already referred to (p. 390). Reviews of the theory of the drift velocity–field characteristic have also been given by P. N. Butcher.[25]

A great deal of device development has taken place as a result of the discovery of the effects of electron transfer and especially of the Gunn effect. These have been recently reviewed by C. Hilsum[26] who has also considered devices based on avalanche breakdown. The Gunn effect and its applications to the development of microwave oscillators and other electronic devices has been described in detail in a book by G. S. Hobson.[27]

12.4.1 The acoustoelectric effect

An extreme case of interaction between electrons and the lattice takes place when the lattice vibrational energy is excited by means of externally applied ultra-sonic waves. This takes place in piezoelectric materials where pressure changes can induce quite strong electric fields which interact with electrons and holes. It was observed by H. D. Nine[28] that the ultra-sonic attenuation in CdS crystals varied with conductivity changes induced by illumination with light. Such changes were also observed by A. R. Hutson, J. H. McFee and D. L. White[29] when an external electric field was applied. It was found that when the field was high enough to cause the drift velocity of electrons to exceed the wave velocity of the ultra-sound, energy was *given* to the waves and instead of attenuation, amplification was obtained. A theory of these interactions

[23] *I.B.M. J. Res. Dev.* (1969) **13**, 562.
[24] *Phys. Rev.* (1968) **166**, 797.
[25] *Rept. Prog. Phys.* (1967) **30**, 97; *Electronic Structures in Solids*, ed. E. D. Haidemenakis (Plenum Press, 1969), p. 366.
[26] *Proc. XIIIth Int. Conf. on Phys. of Semiconductors* (Tipographia Marves, 1976), p. 74.
[27] *The Gunn Effect* (Clarendon Press, 1974).
[28] *Phys. Rev. Lett.* (1960) **4**, 359.
[29] *Phys. Rev. Lett.* (1961) **7**, 237.

was developed by A. R. Hutson and D. L. White[30] to account for this effect, which became known as the acoustoelectric effect. This theory deals only with the 'small-signal' condition. For applied electric fields beyond the value which produces amplification the current tends to saturate as in Fig. 12.1.

Clearly when amplification of the ultra-sonic waves is possible, large amounts of power may be built up. When the power exceeds a certain value instabilities occur and the electron energy distribution deviates markedly from its equilibrium form with the formation of domains rather similar to those which we have discussed in connection with the Gunn effect (see § 12.4) which results in a strong interaction of the electric field in microwaves with the electrons and holes.

A good deal of device development has arisen from the acoustoelectric effect. Since the instabilities propagate with fairly constant speed, near that of ultra-sound in the crystal, the effect has been used to produce delay lines and similar devices. A review of this work and also of the small-signal and large-signal theories has been given by A. Many and I. Balberg.[31]

12.5 Quantum theory of motion of electrons in a high magnetic field

The effective-mass approximation as discussed in § 11.6 is very appropriate for description of the motion of electrons (and holes) in a strong external magnetic field.

When we have a scalar effective mass m_e we may write the equation for the function $F(\mathbf{r})$ defined in § 11.5 in the form (see § 11.6)

$$\frac{1}{2m_e}(i\hbar\nabla + e\mathbf{A})^2 F(\mathbf{r}) = EF(\mathbf{r}). \tag{33}$$

For a magnetic field of induction B along the z-axis the vector potential $\mathbf{A}$ may be taken as $(0, Bx, 0)$ since the vector $\mathbf{B}$ is given by $\nabla \times \mathbf{A}$.

Solutions of equation (33) for this vector potential were first given by L. Landau[32] for free electrons and may at once be adapted for electrons in semiconductors. The function $F(\mathbf{r})$ may be expressed as

$$F(\mathbf{r}) = \phi(x') \exp[ik_y y + ik_z z] \tag{34}$$

with a change of variable $x' = x - \hbar k_y / eB$.

[30] *J. Appl. Phys.* (1962) **33**, 40, 2547.
[31] *Electronic Structure in Solids*, ed. E. D. Haidemenakis (Plenum Press, 1969), p. 385.
[32] *Z. Phys.* (1950) **64**, 629.

$\phi(x')$ satisfies the well-known harmonic oscillator equation

$$\frac{d^2\phi}{dx'^2}+\frac{2m_e}{\hbar^2}[E'-\tfrac{1}{2}m_e\omega_c x'^2]\phi=0 \tag{35}$$

where ω_c is the cyclotron frequency eB/m_e and

$$E'=E-\hbar^2k_z^2/2m_e. \tag{36}$$

The solution of equation (36) is only appropriate if E' has the values $\hbar\omega_c(n+\frac{1}{2})$ where n is an integer. The energy E is therefore given by

$$E=\hbar^2k_z^2/2m_e+(n+\tfrac{1}{2})\hbar\omega_c. \tag{37}$$

The discrete levels given by different values of n are known as Landau levels. This looks very different from the allowed energy in the absence of a magnetic field

$$E=\hbar^2(k_x^2+k_y^2+k_z^2)/2m_e. \tag{38}$$

The form of the energy is illustrated in Fig. 12.3.

The energy levels given by (38) would, of course, be quantized into a set of closely spaced levels by boundary conditions for a finite crystal but

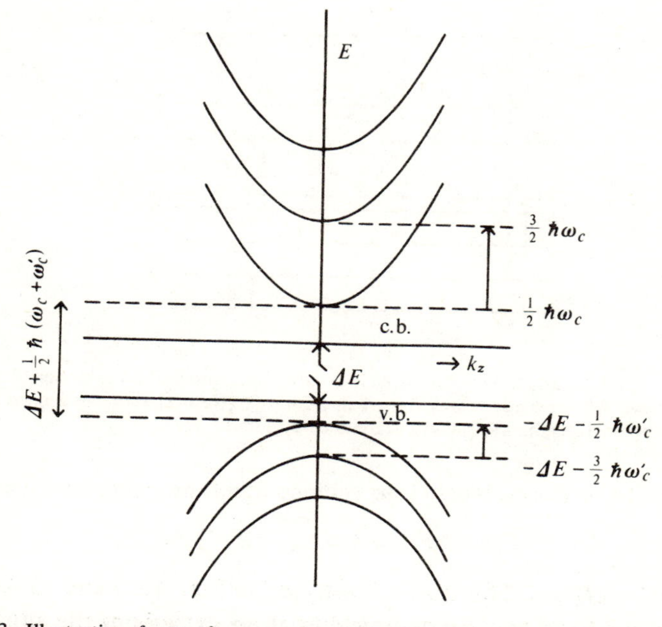

Fig. 12.3. Illustrating form of energy in conduction and valence bands in a strong magnetic field.

apart from counting the levels we have neglected this separation and can still continue to do so. We note that the kinetic energy of the electrons is now represented only by their motion parallel to the magnetic field and this is unaffected. The x, y motion consists of quantized orbits which are circles, as we have seen, when we have a scalar effective mass. The energy therefore consists of a series of discrete values above each of which there is the kinetic energy of motion parallel to the magnetic field. As we shall see, the one-dimensional character of this distribution has a very marked effect on the density of states.

For the moment, however, let us ignore the kinetic energy and consider the nature of the discrete states. They form a 'ladder' of equally spaced levels, the spacing between them being $\hbar\omega_c$. The bottom of the conduction band is no longer an allowed state, the lowest energy in the band being $\frac{1}{2}\hbar\omega_c$ above the bottom of the band. This is illustrated in Fig. 12.4.

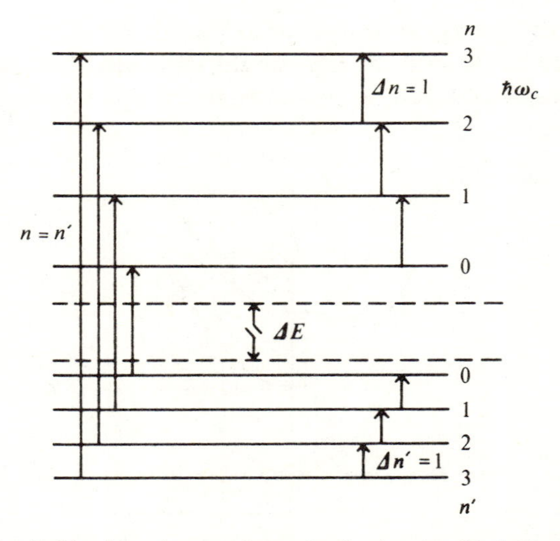

Fig. 12.4. Landau 'ladders' for electrons in conduction band and holes in non-degenerate valence band. Allowed transitions $\Delta n = 1$ and $\Delta n' = 1$ give cyclotron resonance for electrons and holes. Inter-band transitions are allowed for $n = n'$.

For holes in a non-degenerate valence band the energy is given by

$$E = -\Delta E - \hbar^2 k_z^2/m_h - (n' + \tfrac{1}{2})\hbar\omega'_c \qquad (39)$$

where $\omega'_c = eB/m_h$. This is also illustrated in Figs. 12.3 and 12.4.

When we have two bands touching at an extremum the situation is much more complex. This situation has been treated by J. M.

Luttinger.[33] When we may use the heavy-hole and light-hole approximation we might expect to get two 'ladders', the spacing being given by $e\hbar B/m_{h1}$ and $e\hbar B/m_{h2}$. The situation is more complex when we take electron spin into account, as we shall see, and in fact we get four 'ladders', for which the numbering of the levels is not so simple as when there is no degeneracy at $\mathbf{k}=0$.

12.5.1 Density of states in high magnetic fields

When we consider the density of states for the one-dimensional distribution of values of k_z we find an important difference from the three-dimensional distribution. Writing E_k for $\hbar k_z^2/2m_e$ we find that $N(E_k)$ is proportional[34] to $E_k^{-\frac{1}{2}}$, and not to $E^{\frac{1}{2}}$ as for a three-dimensional distribution. This means that at $k_z=0$ we have an infinitely high density of states. We have seen in § 10.4.2 that this only holds strictly if we may neglect collisions over a period $2\pi/\omega_c$, the condition being $B\mu \gg 1$. This condition still holds, for the above analysis has been made on the assumption that the electron or hole is free to move in the crystal.

When we take the kinetic energy in the x- and y-directions into account we see that as $B\mu$ becomes much greater than 1 the levels corresponding to k_x and k_y bunch into the discrete levels which are therefore highly degenerate. This is illustrated in Fig. 12.5. We may

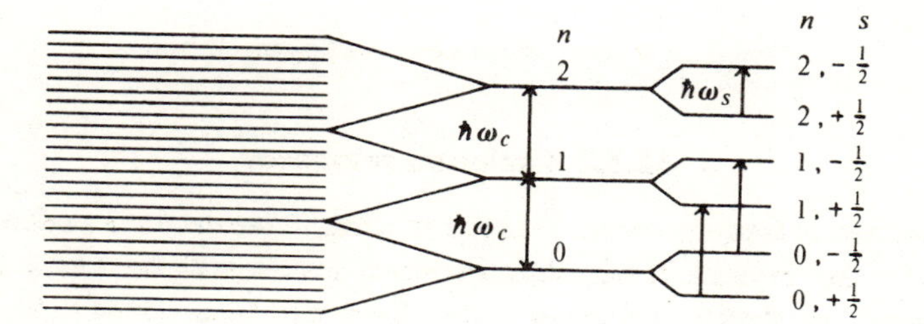

Fig. 12.5. Illustrating grouping of levels into a Landau 'ladder' and further splitting by spin.

evaluate this degeneracy and count up the levels in each Landau level and check that they add up to the same value as before. When the magnetic field is small and there are many Landau levels in a small interval δE we can check that we again get the same number as without a magnetic field.[35]

[33] *Phys. Rev.* (1956) **102**, 1030. [34] *W.M.C.S.*, p. 29. [35] *W.M.C.S.*, § 12.3.1.

To get an expression for the density of states in terms of the energy E we must sum over the Landau levels and we obtain for unit volume

$$N(E)=\frac{eB(2m_e)^{\frac{1}{2}}}{4\pi^2\hbar^2}\sum_n[E-(n+\tfrac{1}{2})\hbar\omega_c]^{-\frac{1}{2}}. \tag{40}$$

This is shown in Fig. 12.6. It may readily be shown that when the spacing between the Landau levels is small and we replace this sum by an integral we simply obtain the value for $N(E)$ given by equation (12a) of § 4.2.

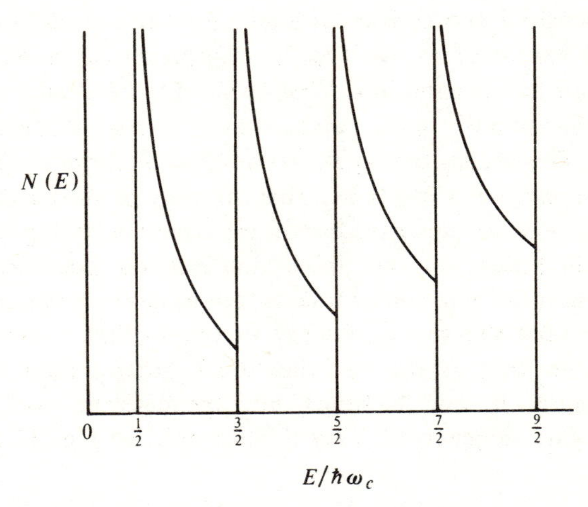

Fig. 12.6. Density of states in a high magnetic field.

12.5.2 Cyclotron resonance

If radiation of frequency equal to the spacing between the Landau levels falls on the crystal, transitions between levels will be excited. The harmonic oscillator selection rule $\Delta n=\pm 1$ (see *W.M.C.S.* § 1.7) indicates that *only* transitions between neighbouring levels should take place. (Because of coupling much weaker transitions with $\Delta n=2$ are possible and have been seen.) The transitions with $\Delta n=\pm 1$ take place at frequency $\omega_c/2\pi$ and are simply cyclotron resonance. Quantum theory therefore gives the same result (apart from the weak $\Delta n=2$ transitions) as the semi-classical theory given in § 10.4.1.

When we have ellipsoidal constant-energy surfaces with effective masses m_1, m_2, m_3, equation (35) may be generalized to give

$$\sum_{r=1}^{3}\frac{1}{2m_r}\left(i\hbar\frac{\partial}{\partial x_r}+e\mathbf{A}_r\right)^2F(\mathbf{r})=EF(\mathbf{r}) \tag{41}$$

where A_r are the components of the vector potential along the three principal axes. If B is along one of the axes, say axis 3, this equation, by means of a simple transformation, may be solved as before to obtain[36]

$$E = \hbar^2 k_z^2/2m_3 + \hbar eB(n+\tfrac{1}{2})/(m_1 m_2)^{\frac{1}{2}}. \tag{42}$$

The cyclotron resonance angular frequency ω_3 is therefore equal to $eB/(m_1 m_2)^{\frac{1}{2}}$. Similarly when the field is along the axes 1 and 2 the angular frequencies ω_1 and ω_2 are respectively equal to $eB/(m_2 m_3)^{\frac{1}{2}}$ and $eB/(m_1 m_3)^{\frac{1}{2}}$. By a further transformation of co-ordinates it can readily be shown that when the magnetic field is in the direction (l, m, n) the cyclotron frequency ω_c is given by [36]

$$\omega_c^2 = l^2\omega_1^2 + m^2\omega_2^2 + n^2\omega_3^2. \tag{43}$$

This is the same result as given in equation (61) of § 10.4.1 but now obtained quantum-mechanically.

If the energy is not given exactly by equation (38) the spacing of the Landau levels will not be equal and there will be a broadening of the cyclotron resonance. Collisions also cause broadening. As we shall see later the change in spacing between the levels of the Landau 'ladder' may be shown more readily by optical absorption techniques involving transistions between a hole 'ladder' and an electron 'ladder'. The above equations for electrons apply equally well to holes but do not take account of the warping of the valence band due to degeneracy at $\mathbf{k}=0$ (see § 11.6).

The interaction between this warping and the electron spin produces a great deal of fine structure in the spectrum for the valence band. This has been treated theoretically by J. M. Luttinger and W. Kohn.[37] It has been studied experimentally by R. C. Fletcher, W. A. Yager and F. R. Merritt[38] and in rather more detail by J. J. Stickler, H. J. Zeiger and G. S. Heller[39] who have resolved most of the fine structure for Ge by using a wavelength of 2.0 mm at 1.2 °K. The fine structure of the hole spectrum for Te has also been resolved by K. J. Button *et al.* using far infra-red radiation from lasers (see § 13.4).

A full discussion of cyclotron resonance including the factors determining line shape and interaction with other effects such as electron spin has been given by B. Lax and J. G. Mavroides[40] and by K. J. Button.[41]

[36] *W.M.C.S.*, § 11.4. [37] *Phys. Rev.* (1955) **97**, 869; *ibid.* (1956) **102**, 1030.
[38] *Phys. Rev.* (1955) **100**, 747. [39] *Phys. Rev.* (1962) **127**, 1077.
[40] *Rendiconti della Scuola Internazionale di Fisica Enrico Fermi*, Corso XXII (Academic Press, 1963), p. 240; *Solid State Physics* (Academic Press, 1960) **11**, 261.
[41] *Optical Properties of Solids*, ed. E. D. Haidemenakis (Gordon and Breach, 1970), p. 253.

12.6 Splitting of levels due to electron spin

In addition to the condensation of the energy levels into a Landau 'ladder' a strong magnetic field has another effect. So far, we have largely ignored the splitting of the energy levels in a magnetic field due to electron spin. A free electron has energy $\pm\beta B$ in a magnetic field of induction B where β is the Bohr magneton $e\hbar/2\boldsymbol{m}$. For electrons in atoms and also for electrons and holes in solids this has to be modified. The energy is generally written $\pm\frac{1}{2}g\beta B$ where g is a number known as the Landau splitting factor and has the value 2 for a free electron. Because of the strong spin–orbit coupling in semiconductors the value of g may be very different from 2 and may even be negative.

When we have a scalar effective mass m_e the value of the g factor as given by B. Lax, L. M. Roth and S. Zwerdling[42] is

$$g = 2\left[1 + \left(\frac{m_e - \boldsymbol{m}}{m_e}\right)\left(\frac{\Delta E_s}{3\Delta E + 2\Delta E_s}\right)\right] \tag{44}$$

where, as usual, ΔE_s is the spin–orbit splitting. For InSb, g has the value -40. For semiconductors such as Ge and Si the expressions for g are much more complex and have been discussed by L. M. Roth and B. Lax.[43]

To the energy given by equation (37) we must add the spin term so that

$$E = \hbar^2 k_z^2/2m_e + (n + \tfrac{1}{2})\hbar eB/m_e \pm \tfrac{1}{2}g\beta B \tag{45}$$

so that each Landau level is doubled and we have two 'ladders'. This is shown in Fig. 12.5. A transition between two states with the same value of n can take place if the spin is reversed, leading to a frequency of absorption $\omega_s/2\pi$ where $\hbar\omega_s = g\beta B$. This is the spin resonance frequency (see § 10.7.1).

12.7 Effect of magnetic field on inter-band transitions

We must now consider the effect of magnetic quantization on inter-band transitions. We shall first of all deal with the simplest possible case, that of a direct transition for two non-degenerate 'spherical' bands with maximum and minimum at $\mathbf{k} = 0$. The Landau level structure is shown in Fig. 12.4 for which we have assumed that the holes have effective mass comparable with that of the electrons. For transitions between such

[42] *Proc. IVth Int. Conf. on Phys. of Semiconductors, J. Phys. Chem. Sol.* (1958) **8**, 311.
[43] *Phys. Rev. Lett.* (1959) **3**, 217.

bands it is found that the selection rule $\Delta n = 0$ operates so that the lowest energy for which a transition can take place is no longer given by $h\nu_0 = \Delta E$ but by

$$h\nu_0 = \Delta E + \tfrac{1}{2}\hbar eB\left(\frac{1}{m_e} + \frac{1}{m_h}\right) \tag{46}$$

corresponding to $n = 0$. The value of the low-frequency limit of absorption therefore increases as B is increased. This shows up as a shift of the fundamental absorption edge to shorter wavelengths as the field is increased. The effect is sometimes known as the 'magnetic band shift'. (See § 10.4.3.) For higher values of n we shall have, on writing m_r for the reduced mass of the electron and hole,

$$h\nu_n = \Delta E + (n + \tfrac{1}{2})\hbar eB/m_r. \tag{47}$$

Because of the spread of kinetic energy we should not expect to have sharp absorption lines. We should expect $\Delta k_z = 0$, corresponding to $\Delta \mathbf{k} = 0$ for direct transitions without magnetic field, but this would still give continuous absorption corresponding to different initial values of k_z. However, because of the very large value of the density of states at $k_z = 0$ we should have a sharp maximum in absorption at $k_z = 0$ especially at low temperatures. Instead therefore of a line absorption spectrum one with a series of maxima is found as shown in Fig. 12.7. When the

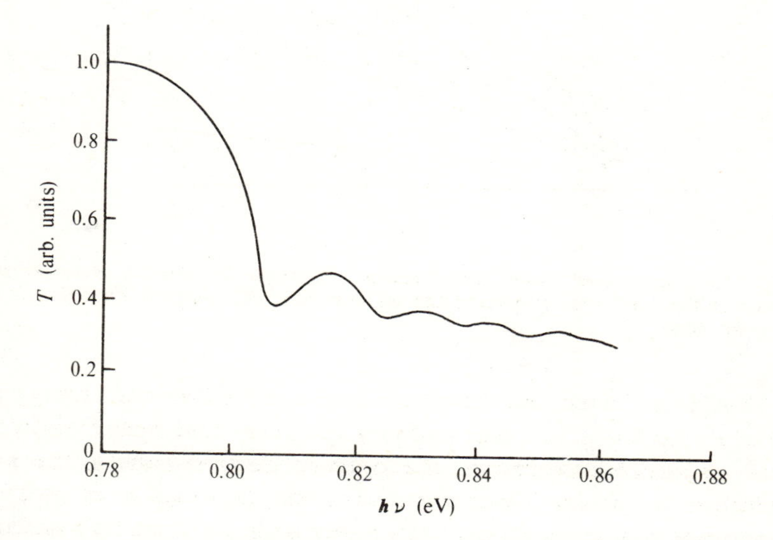

Fig. 12.7. Transmission T as a function of quantum energy $h\nu$ for $B = 3.6$ T. (After S. Zwerdling and B. Lax, *loc. cit.*)

positions of these maxima are plotted as a function of B we should expect to obtain a series of straight lines converging on $h\nu = \Delta E$ and having slopes $(n+\frac{1}{2})\hbar eB/m_r$. This is exactly what is found, although for higher values of n and B they are generally not quite straight, but this gives additional information about the non-parabolicity of the bands. The measurements of S. Zwerdling and B. Lax[44] on the direct transition at $\mathbf{k}=0$ for Ge are shown in Fig. 12.8.

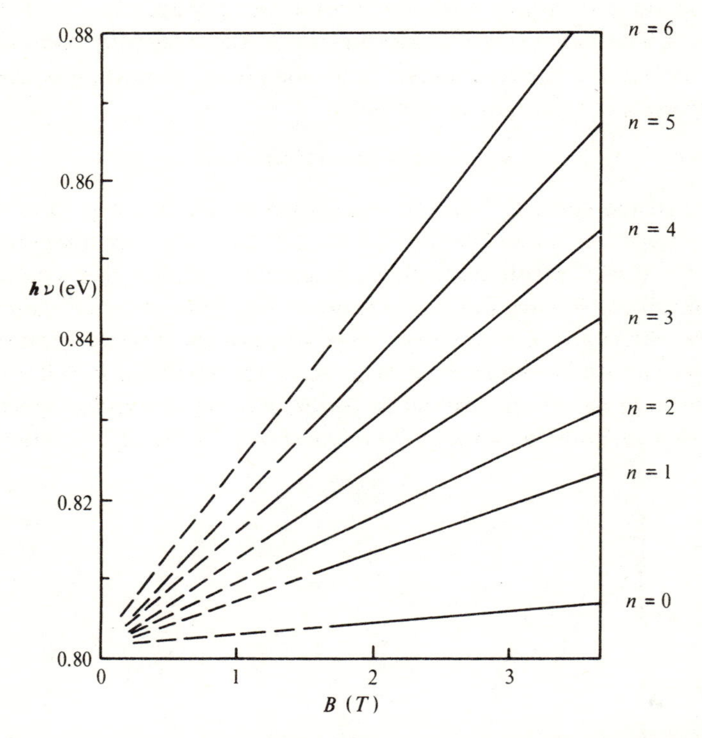

Fig. 12.8. Variation of transmission maxima with magnetic induction B for different values of the index n for combining Landau levels ($\Delta n = 0$). (After S. Zwerdling and B. Lax, *loc. cit.*)

This gives a direct and accurate measure of the forbidden energy gap (0.80 eV) at $\mathbf{k}=0$, and also, knowing the heavy hole mass from cyclotron resonance experiments, the effective electron mass at the $\mathbf{k}=0$ minimum is $0.036\boldsymbol{m}$. This cannot, of course, be found from cyclotron resonance measurements since this minimum is not normally populated.

[44] *Phys. Rev.* (1957) **106**, 51.

Similar measurements have now been carried out for a great number of semiconductors. The absorption maxima (or transmission minima) seem to show up more readily than the cyclotron resonance and the method has been used to obtain effective masses and accurate values of ΔE for a variety of materials with 'direct' gaps, such as InSb or GaAs, even at room temperature. For semiconductors with very narrow gaps, such as InSb and alloys like $Pb_{1-x}Sn_xTe$, a complication arises due to the interaction of the valence and conduction bands. This has been discussed by C. Pidgeon[45] who has used the method even for semimetals like HgTe.

When higher resolution is used to study the spectrum corresponding to direct transition in Ge at $\mathbf{k}=0$ fine structure is found. The valence band in Ge is degenerate at $\mathbf{k}=0$ and this causes interaction between the spin terms in the energy and the 'Landau' terms. This situation has been discussed in § 12.5.2 in connection with cyclotron resonance and shows up too in the optical spectra.

When we have the non-degenerate situation the spin is unchanged in the transition. Now, however, we have four separate Landau 'ladders' for the valence band and two for the conduction band.

The system of levels has been discussed in detail by L. M. Roth, B. Lax and S. Zwerdling[46] who show that while the conduction band levels are identified by the Landau quantum number n and the spin numbers $s=\pm\frac{1}{2}$, the valence band requires n and a magnetic quantum number j which takes the values $\frac{3}{2}, \frac{1}{2}, -\frac{1}{2}, -\frac{3}{2}$. The 'ladders' with $j=\pm\frac{1}{2}$ correspond to the light-hole band and can be identified as 'spin-up' and 'spin-down' states as for the conduction band. Those with $j=\pm\frac{3}{2}$ have their spin states mixed with the 'orbital' angular momentum and cannot be regarded as 'spin-up' or 'spin-down', but there is a probability of spin being either 'up' or 'down'. This as we shall see later has some important consequences for other semiconductors which show the same behaviour (see § 14.7).

Measurements of the absorption spectrum of Ge by Roth, Lax and Zwerdling (*loc. cit.*) were compared with their theoretical predictions which indicated that the $s=-\frac{1}{2}$ 'ladder' links with the $j=-\frac{1}{2}$ and $j=\frac{3}{2}$ 'ladders', selection rule for n being $\Delta n=0$, while the $s=\frac{1}{2}$ 'ladder' links with the $j=\frac{1}{2}$ and $-\frac{3}{2}$ 'ladders' with the selection rule $\Delta n=-2$. By this means they were able to account for all the lines in the spectrum. The 'ladder' scheme and allowed transitions are shown in Fig. 12.9. It will be

[45] *Electronic Structure in Solids*, ed. E. D. Haidemenakis (Plenum Press, 1969), p. 47.
[46] *Phys. Rev.* (1959) **114**, 90.

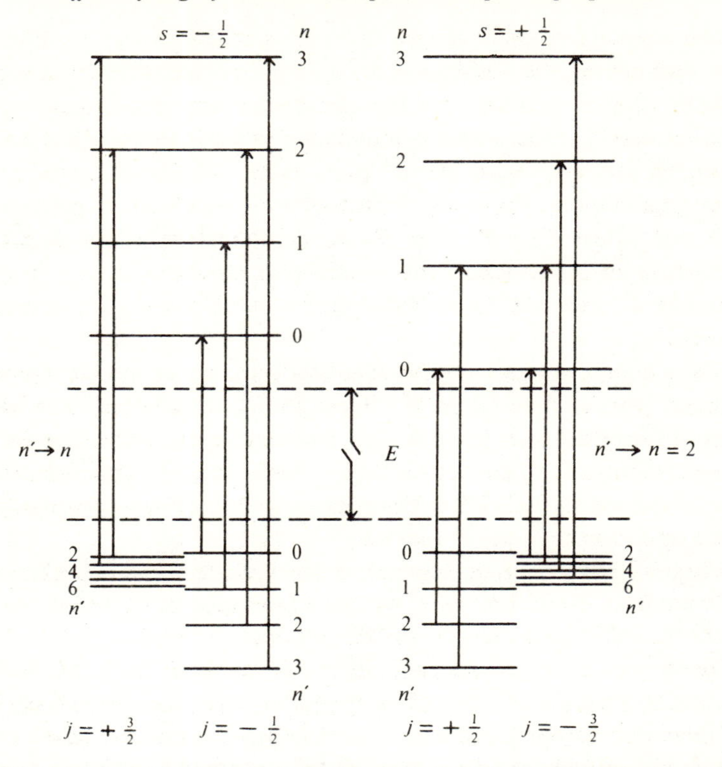

Fig. 12.9. Landau 'ladders' for conduction band and valence band degenerate at $\mathbf{k}=0$, showing spin splitting and allowed transitions. (After L. M. Roth *et al.*, *loc. cit.*)

noticed that the $n=0$ and $n=1$ states are missing from the $j=\pm\frac{3}{2}$ 'ladders'.

For indirect transitions such as the one across the lowest forbidden energy gap in Si and Ge a somewhat different form of spectrum is predicted because of the extra freedom of choice of phonon energy required for the conservation of crystal momentum. Instead of maxima we now have a series of steps in the absorption as shown by L. M. Roth, B. Lax and S. Zwerdling (*loc. cit.*) and found experimentally by S. Zwerdling, B. Lax, L. M. Roth and K. J. Button.[47] The positions of these steps may also be analysed to give information about the forbidden energy gap and band structure near the extrema.

Extensive reviews of these and other magnetic effects have been given by B. Lax (see p. 405) and also by J. G. Mavroides.[48]

[47] *Phys. Rev.* (1959) **114**, 80.
[48] *Optical Properties of Solids*, ed. F. Abeles (North Holland, 1972), p. 351.

12.7.1 Zeeman effect in impurity absorption

The effect of a strong magnetic field on the absorption spectrum of impurities in semiconductors is very similar to that in atomic spectroscopy except that, while in the latter the Zeeman splitting of the energy levels is small compared with the energies themselves, now it can be comparable and no longer can be regarded as a small perturbation. The splitting shows up the multiplicities of the levels and their magnetic characteristics and provides a great deal of further information on the electronic structure of impurity centres. There is a very rich variety in the spectra of such centres and this has also been discussed in some detail by B. Lax in his review of magneto-optical effects (see p. 405).

Using Fourier-transform spectroscopy in the far infra-red R. Kaplan[49] has been able to study both absorption and photo-conductivity due to 'hydrogenic' impurity levels in GaAs and InSb and to measure directly the change in impurity energy levels in a strong magnetic field, confirming the theoretical treatment of such levels (see § 11.5.1). The subject has also been reviewed by H. Hasegawa.[50]

12.8 Transport phenomena in high magnetic fields

We have already discussed some aspects of electrical transport phenomena in a strong magnetic field in § 5.3.5 and have found that the Hall constant R for electrons is given simply by $R = -1/ne$ rather than by $-r/ne$, where r is the constant defined by equation (111) of § 5.3.3. Various other predictions were made on the basis of the analysis, for example that there would be no longitudinal magneto-resistance for a semiconductor such as Si, with field along one of the $\langle 100 \rangle$ directions. This would also be true for a semiconductor with spherical constant-energy surfaces. Moreover, for such a semiconductor it was shown that, if degenerate, there would be no transverse magneto-resistance, but that there would be in the non-degenerate condition. Expressions were found for the limit ρ_∞ to which the resistivity would be expected to tend for very high fields. For example for a non-degenerate n-type semiconductor with spherical constant-energy surfaces it was shown that $\rho_\infty/\rho_0 = (32/9)\pi$ for pure lattice scattering which is really quite a small effect; but when impurity scattering predominates $\rho_\infty/\rho_0 = (32/3)\pi$.

[49] *Optical Properties of Solids*, ed. E. D. Haidemenakis (Gordon and Breach, 1970), p. 301.

[50] *Physics of Solids in High Magnetic Fields*, ed. E. D. Haidemenakis (Plenum Press, 1969).

More generally $\rho_\infty/\rho_0 = \langle 1/\tau\rangle\langle\tau\rangle$. For mixed conduction the ratio predicted for ρ_∞/ρ_0 is quite complex, depending on the ratio n/p and on the mobility ratio b (see § 5.3.5 equation (168)). The magneto-resistance disappears, i.e. $\rho_\infty = \rho_0$ when $n = p$ as for an intrinsic semiconductor.

Of all these predictions only that for the high-field Hall constant turns out to be confirmed by experiment when the magnetic field is so high that $\omega_c\tau \gg 1$, and even here small deviations are found. These however, are of no great importance in determining the carrier concentration, and high field measurements of R are used to give n or p (or indeed $n-p$ since $R \to 1/\boldsymbol{e}(p-n)$ for mixed conduction). That most of these predictions are not borne out by experiment is hardly surprising since in their derivation magnetic quantization and its drastic effect on the distribution function f_0 was not taken into account as it must be if $\omega_c\tau \gg 1$.

As we have seen, the proper starting point of a calculation to find the effect of a small electric field in the presence of a large magnetic field would be to use for f_0 the distribution function f_{0B} obtained by using the expression (37) for the energy. This would give

$$f_{0B}^{-1} = \exp\{[E' + (n+\tfrac{1}{2})\hbar\omega_c - E_F]/\boldsymbol{k}T\} + 1 \tag{48}$$

where $E' = \hbar^2 k_z^2/2m_e$ for a scalar effective mass m_e. To obtain expressions for the electric or heat currents we should have to sum over the index n as well as average over k_z. Moreover, the Boltzmann equation on which the averaging process was based in § 5.3.5 is not strictly valid in these conditions and the transport equations are much more complex.

There is, however, one situation in which the sum over n may be avoided, known as the extreme quantum condition. If $\boldsymbol{k}T \ll \hbar\omega_c$ virtually all the electrons in the conduction band will be in the lowest Landau level and we need only consider $n = 0$. We then have in the non-degenerate condition

$$f_{0B} = A \exp(-E'/\boldsymbol{k}T) \tag{49}$$

and the number of states per unit interval of E' is $CE'^{-\frac{1}{2}}$, where A and C are constants that may be readily evaluated (*W.M.C.S.* § 12.3.1). For conduction parallel to the magnetic field we may carry through an analysis similar to that given in § 5.3.1. We easily find that the current density J is given by the same expression as before but with the average $\langle\tau\rangle$ replaced by

$$\bar{\tau} = \int_0^\infty \tau(E')E'^{\frac{1}{2}} \exp(-E'/\boldsymbol{k}T)\, dE' \bigg/ \int_0^\infty E'^{\frac{1}{2}} \exp(-E'/\boldsymbol{k}T)\, dE'. \tag{50}$$

We see therefore that in this situation the weighting function $E^{\frac{3}{2}} \exp(-E/kT)$ is replaced by $E'^{\frac{1}{2}} \exp(-E'/kT)$.

Even if we know $\tau(E)$ in the absence of a magnetic field we yet do not know $\tau(E')$ without a recalculation of τ. This is generally very difficult. If however τ is a function of the velocity v it is perhaps not unreasonable to assume that if $\tau(E)=AE^{-s}$ then $\tau(E')=A'E'^{-s}$. With this assumption we readily find that

$$\frac{\rho_\infty}{\rho_0}=\frac{\bar{\tau}}{\langle\tau\rangle}=\frac{3-2s}{3}. \tag{51}$$

For pure lattice scattering, with $s=\frac{1}{2}$, we have $\rho_\infty/\rho_0=\frac{2}{3}$; while for ionized impurity scattering, with $s=-\frac{3}{2}$, ρ_∞/ρ_0 becomes negative but now we cannot satisfy the condition $\omega_c\tau \gg$ for small values of E'. All that we can reasonably conclude from this simple analysis is that the amount of impurity present will strongly affect the high-field longitudinal magneto-resistance. The main conclusion, however, is that unless $\tau(E')$ is independent of energy we do have a longitudinal magneto-resistance effect and that the prediction of the former analysis is incorrect. We may note that for pure lattice scattering the resistivity at high fields is *decreased*.

We have, so far, not considered conductivity transverse to the magnetic field. In the extreme quantum condition this would be expected to be quite small, the paths of electrons and holes being tightly bound into circles in the high field. Indeed this leads us to a further difficulty. In the effective-mass approximation we required that the 'external' field should vary little over a lattice spacing and indeed a steady high magnetic field satisfies this condition. It is also, however, a condition that the 'modulation function' $F(\mathbf{r})$ should not vary much over a lattice spacing. Clearly this will not be so if an electron is confined to move in a circle whose radius is comparable with a lattice spacing. There is therefore also an *upper* limit to the magnetic field for which the effective-mass approximation is valid.

From equation (62) of § 10.4.2 we see that when $n=1$ the radius of the circle in which an electron moves is independent of its effective mass and is equal to $(\hbar/eB)^{\frac{1}{2}}$. For $B=10$ T this is about 10^{-6} cm and so is still quite a bit larger than the lattice spacing, so this is not likely to be too serious a restriction. This calculation based on 'classical' quantum theory may be made more precisely in terms of the wave-functions which are solutions of equation (33) (*W.M.C.S.* p. 449) and the same result obtained. For a material like InSb with its small electron effective mass and consequently high mobility, reaching values of the order of $100\ \mathrm{m^2\,V^{-1}\,s^{-1}}$, even with $B=1$ T we have $B\mu=\omega_c\tau=100$ so that we are in the region of magnetic quantization and also in the condition for which the effective-mass approximation is good. The extreme quantum condition when all the electrons are in the $n=0$ Landau level may be expressed as $\boldsymbol{k}T \ll \hbar\omega_c$. For InSb with $B=1$ T and $m_e \simeq 0.01\boldsymbol{m}$, $\hbar\omega_c$ is about 10^{-2} eV. This means that if $T<12$ °K a condition of extreme quantization may be established with $B=1$ T and at 1.2 °K with $B=0.1$ T. InSb would therefore be an excellent material for measurements in the extreme quantum condition. These are, however, complicated by the phenomenon of carrier freezeout (see § 12.9).

As we have already indicated, a proper treatment of transport, both electrical and thermal, in a very strong magnetic field is very complex, mainly because we have to abandon the simple Boltzmann equation. A quantum-mechanical treatment based on the use of the density matrix formalism has been given by P. N. Argyres[51] and also by E. N. Adams and T. D. Halstein.[52] They find that the expression we have given for the high-field Hall constant is correct apart from small oscillatory terms which are particularly marked as the degenerate condition is approached, i.e. they are of greater significance for highly-doped semiconductors. This is similar to the de Haas–van Alphen effect in metals (*W.M.C.S.* § 12.6). A somewhat different treatment has been given by R. Kubo, S. J. Miyake and N. Hashitsume[53] who have also reviewed the previous work on transport in high magnetic fields. A still more elaborate treatment has been given by J. Hajdu[54] who has also considered the limitations and range of application of the previous theories. Various formulae are given for the transport coefficients valid under a variety of different conditions.

It is hardly surprising that there are few measurements of thermal transport coefficients with high magnetic fields since they are difficult to make and in any case are not very important at low temperatures where the quantum conditions can be most readily established. The available data have been compared with theory by S. M. Puri and T. H. Geballe.[55] More measurements have been made of electrical conductivity in extreme quantum conditions but have not always been unambiguous as we have already pointed out.

[51] *Phys. Rev.* (1958) **109**, 1115. [52] *J. Phys. Chem. Solids* (1959) **10**, 254.
[53] *Solid State Phys.* (Academic Press, 1965) **17**, 269.
[54] *Electronic Structure in Solids*, ed. E. D. Haidemenakis (Plenum Press, 1969), p. 305.
[55] *Semiconductors and Semimetals* (Academic Press, 1966) **1**, 203.

For example L. J. Neuringer[56] has reported measurements on the magneto-resistance of InSb with carrier concentrations in the range 5×10^{13} cm^{-3}–10^{16} cm^{-3} at liquid helium temperatures where the extreme quantum condition is obtained. The resistivity rises by nearly 10^6 as the field is increased from 0.1 T to 20 T. This effect is, however, largely due to carrier freezeout (see § 12.9). A saturation of both transverse and longitudinal magneto-resistance of PbTe has been reported by G. Suryan and S. Nagabhushana[57] using pulsed fields up to 20 T. Negative magneto-resistance has been observed in GaAs at high fields by S. Askenaye, J. P. Ulmet and J. L. Leotin.[58] This could not be due to lattice scattering in the conditions of their experiments where impurity scattering prevails. Negative magneto-resistance at high fields has also been reported in Se and fairly heavily doped Ge by Y. V. Shmartsev *et al.*[59] A curious behaviour of compensated GaAs at liquid He temperatures was observed by B. M. Vul, N. V. Kotelnikova, E. I. Zavoritskaya and I. D. Voronova[60] in which a sudden increase in conductivity takes place rather like that due to a Mott transition (see § 13.14) causing a sudden decrease in electron mobility. Localization of electrons in potential wells arising from fluctuations in impurity concentration also is thought to cause a decrease in conductivity in fields of the order of 5–15 T.

Clearly the tight confinement of the electrons by the magnetic field tends to make any form of localization more likely. There is, however, yet a great deal to be clarified before these phenomena are fully understood. Some of the experimental work is reviewed and compared with theory by A. C. Beer[61] in a volume of *Solid State Physics* devoted entirely to transport in magnetic fields. The subject has also been discussed in detail by L. M. Roth and P. N. Argyres.[62]

12.9 Carrier freezeout

In a number of semiconductors the binding energy of electrons in donors is so small that even at liquid helium temperatures the donors remain largely ionized. This also applies to holes and acceptors but

[56] *Proc. IXth Int. Conf. on Phys. of Semiconductors* (Nauka, Moscow, 1968), p. 715.
[57] *Proc. XIIIth Int. Conf. on Phys. of Semiconductors* (Tipographia Marves, 1976), p. 1121. [58] *Sol. State Commun.* (1969) **7**, 979.
[59] *Proc. XIIth Int. Conf. on Phys. of Semiconductors* (Pol. Sci. Publ., 1972), p. 410.
[60] *Proc. XIIIth Int. Conf. on Phys. of Semiconductors* (Tipographia Marves, 1976), p. 1188. [61] *Solid State Physics* (Academic Press, 1963) Suppl. 4.
[62] *Semiconductors and Semimetals* (Academic Press, 1966) **1**, 159.

because of the generally larger effective mass of the hole this situation tends to be less common than for electrons. However, the lead chalcogenides all exhibit this phenomenon for both donors and acceptors. For InSb shallow donors (not accurately identified) remain ionized at very low temperatures. It is found, however, that on application of a high magnetic field while the material is kept at a temperature of a few degrees K the carriers can be frozen into the shallow impurity centres. The behaviour of InSb is particularly striking and has been described by R. J. Sladek[63] and by E. H. Putley.[64] Two separate mechanisms cause the freezeout. The impurity level, if consisting of a donor having a single nearly-free electron, will have its energy split by the magnetic field, one of the split levels moving up in energy and going into the conduction band while the other is lowered. The gap between this lower level and the bottom of the conduction band increases as the magnetic field is increased and may become sufficient to bind the electron at low temperature. The amount of the splitting is readily calculable from the splitting of the ground state of the hydrogen atom with appropriate change of effective mass.

The second effect is due to the contraction of the donor electron wave-function. Again this is known for the hydrogen atom and may be estimated with appropriate scaling factor. If the impurity concentration is high enough for the wave-functions of neighbouring impurity atoms to overlap, the binding energy may be decreased to vanishing point. With the contraction of the wave-function on application of a strong magnetic field the overlap may be considerably decreased with consequent increase in the binding energy.

The theory of these effects has been given by Y. Yafet, R. W. Keyes and E. N. Adams[65] and gives a good description of what takes place in the case of simple 'hydrogenic' impurities.

This phenomenon should not be confused with magneto-resistance. It results from a change in the *number* of carriers and not in their transport properties, although the two effects are inter-related. Indeed a failure to appreciate the effects of carrier freezeout has been one of the main causes of confusion in the interpretation of magneto-resistance measurements in high magnetic fields at low temperature. The subject has been extensively reviewed by E. H. Putley,[66] who has used the

[63] *J. Phys. Chem. Solids* (1958) **5**, 157.
[64] *Proc. Phys. Soc.* (1960) **76**, 802.
[65] *J. Phys. Chem. Solids* (1956) **1**, 137.
[66] *Semiconductors and Semimetals* (Academic Press, 1966) **1**, 289.

photo-conductivity produced by freeing of the magnetically bound electrons in InSb as the basis of a sensitive detector of far infra-red radiation.

13

Semiconducting materials

13.1 Materials used in solid-state devices and in basic research

In this chapter we shall consider briefly some of the basic properties of a few semiconductors which, because of their scientific or technological importance, have been intensively studied. Of the element semiconductors only Ge, Si and Te fall into this class. Se hardly does for, although it is technologically important, it is its amorphous form, which acts more like an insulator, that is mostly used in the photocopying industry. Grey tin has also been studied but because of the great difficulty of providing pure samples has not received nearly so much attention as the others. Si and Ge must be regarded as the basic semiconductors. Of the two, Ge has been more used for fundamental physical studies. It is easier to process and purify because of its lower melting point (937 °C) as compared with Si (1420 °C). Ge was used for most of the first solid state electronic devices such as transistors but has now been replaced almost entirely by Si which is much more abundant. The complex integrated circuits of modern electronics are almost all fabricated from Si. Te is interesting in itself, being non-isotropic as a conductor in single crystal form, but has not, so far, had much use in solid-state devices.

Of binary compound semiconductors two groups have come into prominence. Of the II–VI group CdS has long been studied because of its use in light meters while PbS, PbSe and PbTe have been studied in some detail because of their use in infra-red detectors. Of the III–V compounds InSb, GaAs and GaP have been most intensively studied although all nine compounds involving Al, In, Ga, combined with Sb, As, P have received some attention. Alloys of these as well as the pure compounds have also been used. We shall mainly be concerned with these materials but shall mention one or two others. More complex semiconductors such as ternary compounds are beginning to receive some attention but have not yet found widespread technological use. However, their variety is so great that it will be surprising if some compound does not emerge to meet a technological need. Already such compounds are being used on a small scale in non-linear optics and we shall briefly mention some of those whose properties are better known.

13.2 Preparation and measurement of semiconductor materials

The two main methods of preparation of crystals of semiconductor materials are growth from the melt and epitaxial growth. Two methods for the former are used: the so-called Stockbarger method in which the molten material is lowered very slowly through a small temperature difference which causes it to solidify, and the Czochralski method in which, using a seed crystal to start with, the growing crystal is pulled slowly vertically from the melt. A variation of this is the horizontal-zone method in which the 'pulling' is done by moving a 'boat' containing the seed and melt through a temperature gradient.

Epitaxial growth is achieved by making the desired crystal grow on another by deposition, either from a gas or liquid phase. Controlled impurity inclusion can more readily be achieved by this means but large crystals are not so easily obtained. It is, however, finding increasing use in modern electronic device development.

So much has been written on the preparation and purification of semiconducting materials and the methods differ so much for each material that it would take far too much space to give an adequate account of these. Fortunately there are several excellent books on the subject dealing with most of the materials of technological interest.[1] A great variety of methods for determining the basic properties of semiconducting materials is also used. Some of these we have indicated in the text while discussing these properties. They fall into three main groups: (1) theoretical calculations, (2) electrical and magnetic measurements, and (3) optical and magneto-optical measurements.

[1] See, for example, P. F. Kane and G. B. Larrabee, *Characterization of Semiconducting Materials* (McGraw-Hill,,1975); *Electronic Materials*, ed. N. B. Hannay and U. Colombo (Plenum Press, 1973); M. S. Brooks and J. K. Kennedy, *Ultra-purification of Semiconductors* (Macmillan, 1962); R. L. Parker, 'Crystal growth mechanisms' in *Solid State Physics* (Academic Press, 1970) **25**, 152; *The Art and Science of Growing Crystals*, ed. J. J. Gilman (Wiley, 1963); *Crystal Growth* (series), ed. W. Bardsley, D. J. T. Heurle and J. B. Mullin (North Holland, 1973).

Again there are several books[2] dealing extensively and in detail with the procedures used in each case. Because of the different precautions that have to be taken we cannot adequately discuss these in the space available. The methods used for determining the various parameters are in some cases discussed and are generally indicated in the compilations of material properties to which we shall later make reference.

Broadly speaking, forbidden energy gaps are determined by a combination of theory and measurement, the most accurate values being now obtained by measurement of optical absorption and emission both with and without a magnetic field. Effective masses are determined by cyclotron resonance, Faraday rotation or magneto-optical absorption. Mobilities are determined by Hall constant and conductivity measurements combined with theory of electron and hole scattering, when direct drift methods cannot be used. Impurity levels are obtained both by Hall-constant measurements and by optical absorption and emission, the latter two being generally now favoured. Minority-carrier lifetimes are found by observing the decay of injected carriers from either photo-conductivity or related experiments although several ingenious indirect ways are also used. We have selected a few semiconductors of different types for more detailed description so as to bring out the rich variety of properties which these materials possess.

Various compilations of the basic properties of semiconducting materials have been made in which a large number of references have been given. Fairly standard properties, such as thermal conductivity, lattice constant or density, are unlikely to be changed by further work but some of the more difficult quantities to obtain, such as depth of impurity levels or effective masses, may be changed by more up-to-date measurements and for these the latest available data should be consulted. Most of the values quoted are taken from the following: *Materials used in Semiconductor Devices*, ed. C. A. Hogarth (Wiley–Interscience, 1965); *Selected Constants relative to Semiconductors* (Pergamon Press, 1961); *Handbook of Electronic Materials* (IFI–Plenum, 1971–72) Vols. 2–7; *Electronic Materials*, ed. N. B. Hannay (Plenum Press, 1973); *Semiconductors and Semimetals*, ed. R. K. Willardson and A. C. Beer (Academic Press, 1966–68), Vols. 1–4.

A great deal of information on a wide variety of semiconducting materials may be obtained from the 'materials' sections in the series of

[2] See for example, P. F. Kane and G. B. Larrabee, *loc. cit.*; W. R. Runyan, *Semiconductor Measurements and Instrumentation* (McGraw-Hill, 1975).

biennial conferences on the physics of semiconductors to which we have frequently referred.

13.3 Silicon and germanium

We treat the two materials Si and Ge together because their properties are very similar but also show some interesting differences. Both are 'indirect' semiconductors, both have the top of their valence band at $\boldsymbol{k} = 0$. Si has the bottom of its conduction band in the ⟨100⟩ directions at an internal point of the Brillouin zone ($0.8\ k_{max.}$) and so has six minima, while Ge has the bottom of its conduction band at the points in the ⟨111⟩ directions at the edge of the zone and so has four minima. Near the bottom of the conduction band both have their constant-energy surfaces in the form of ellipsoids of revolution (see § 2.3). The valence band is degenerate at $\boldsymbol{k} = 0$ and has a spin–orbit split-off band below it (see § 2.3). The form of the bands for Ge and Si are shown schematically in Figs. 13.1, 13.2 together with some values of the separations between

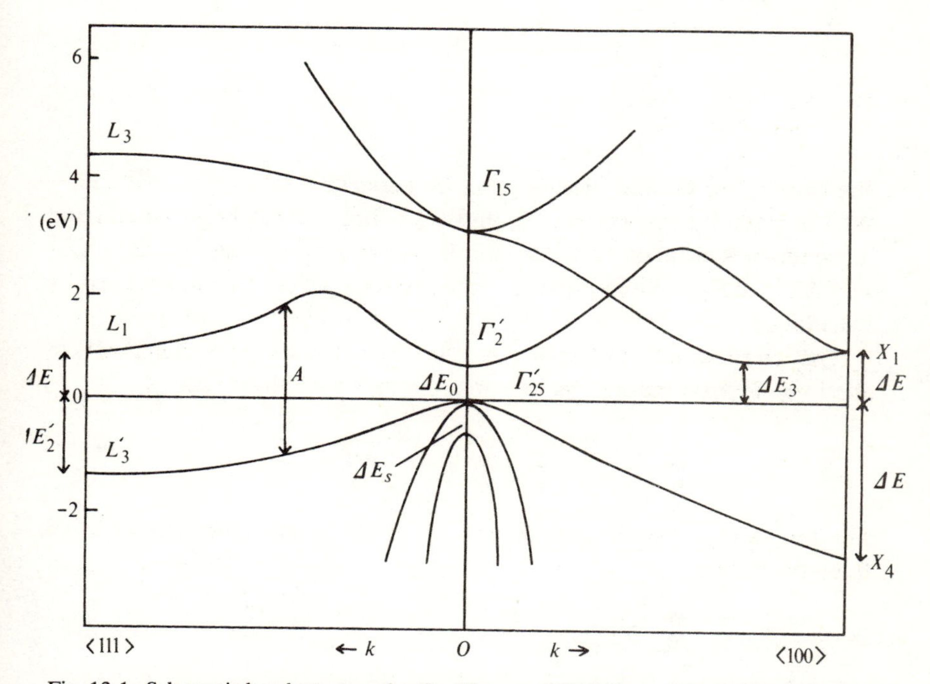

Fig. 13.1. Schematic band structure for Ge. The gaps (300 °K) are $\Delta E = 0.665$ eV, $\Delta E_0 = 0.805$ eV, $\Delta E_s = 0.28$ eV, $\Delta E_1 = 1.2$ eV, $\Delta E_1' = 3.1$ eV, $\Delta E_2' = 1.45$ eV, $\Delta E_3 = 0.845$ eV. A indicates a transition for which $\Delta_k(E_e - E_v) = 0$ (cf. Fig. 10.24).

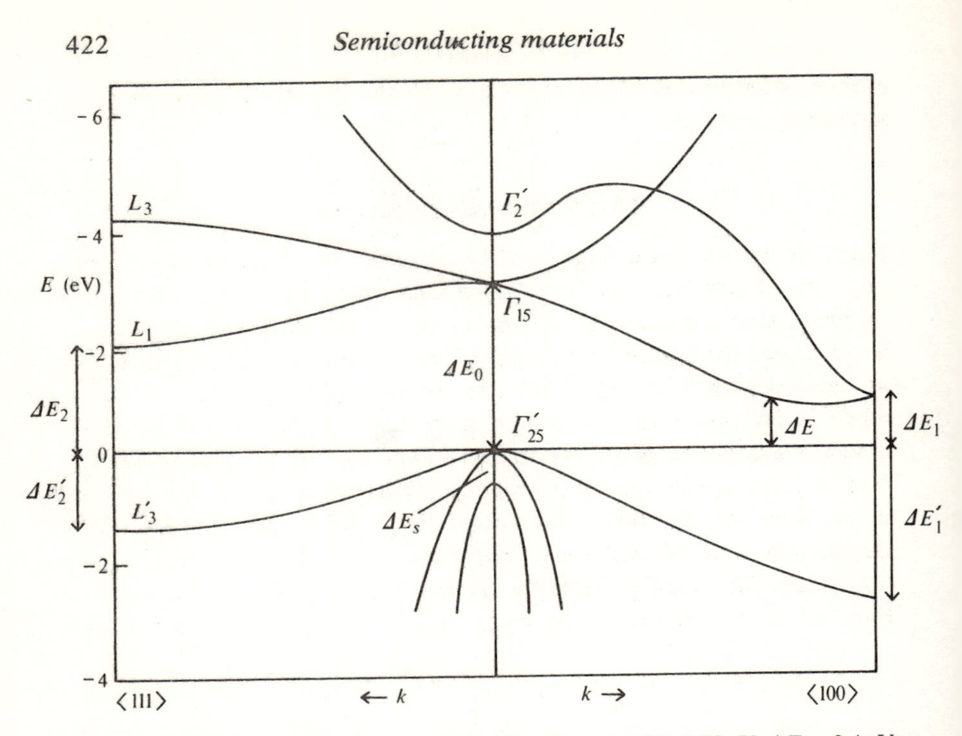

Fig. 13.2. Schematic band structure for Si. The gaps are $\Delta E = 1.12$ eV, $\Delta E_0 = 3.4$ eV, $\Delta E_s = 0.035$ eV, $\Delta E_1 = 1.2$ eV, $\Delta E'_1 = 3.1$ eV, $\Delta E_2 = 1.9$ eV, $\Delta E'_2 = 2.2$ eV.

the bands. For Ge the quoted gaps are experimental except that $\Delta E_1 + \Delta E'_1$ is given by experiment, E_1 and E'_1 being separately obtained from a pseudo-potential calculation.[3] Similarly for Si the values of $\Delta E_2 + \Delta E'_2$ and $\Delta E_1 + \Delta E'_1$ are experimental, the separate values being obtained by calculation.

The longitudinal and transverse effective masses are given in terms of the free electron mass $\boldsymbol{m}$ as follows, together with their ratio K:

	m_l/m	m_t/m	K
Ge	1.64	0.082	2.0
Si	0.98	0.19	5.1

For the valence band the constants A, B, C in equation (25) of § 2.3 have the values:

	A	B	C
Ge	13.1	8.3	12.5
Si	4.0	1.1	4.1

[3] See, for example, D. Brust, J. C. Phillips and F. Bassani, *Phys. Rev. Lett.* (1962) **9**, 94; D. Brust, M. L. Cohen and J. C. Phillips, *ibid.* (1962) **9**, 389 for much more detailed band structure.

In terms of the approximation that we have heavy and light holes of masses m_{h1}, m_{h2} and a 'split-off' hole of mass m_{h3} we have (see § 2.3):

	m_{h1}/m	m_{h2}/m	m_{h3}/m
Ge	0.28	0.044	0.077
Si	0.49	0.16	0.245

The variation of the forbidden energy gap ΔE with temperature is shown for Ge in Fig. 13.3 and for Si in Fig. 13.4. These were obtained by optical absorption measurements by G. G. Macfarlane *et al.*[4] It will be seen that for a large range of temperature ΔE varies linearly with T but tends to a constant value as $T \to 0$.

The mobilities of electrons and holes in very pure Ge and Si are due to lattice scattering and inter-valley scattering and are given below.

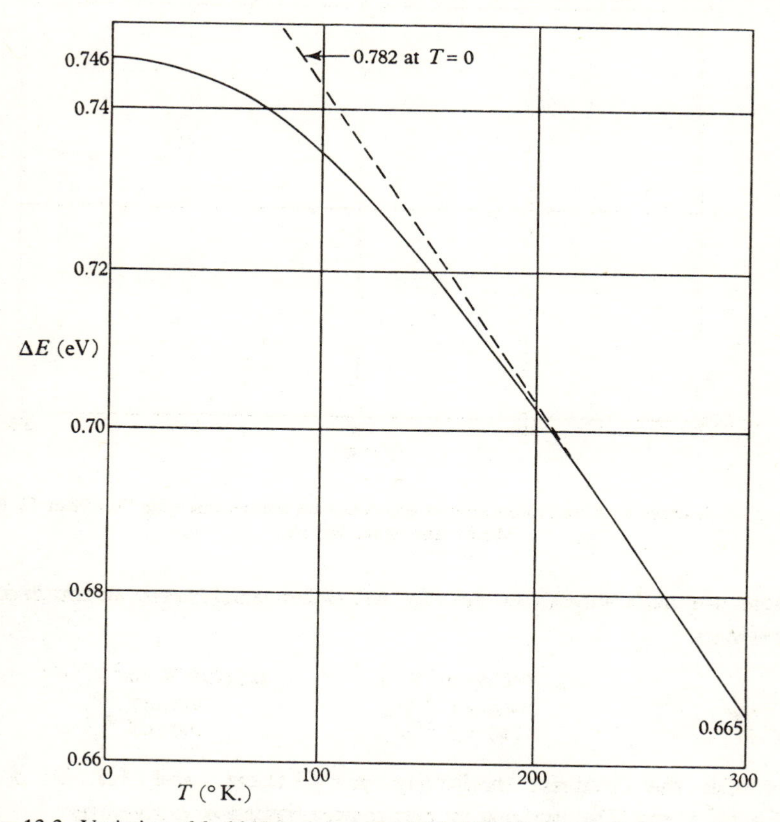

Fig. 13.3. Variation of forbidden energy gap ΔE with temperature for Ge. (After G. G. Macfarlane *et al., loc. cit.*)

[4] *Phys. Rev.* (1957) **108**, 1377; *ibid.* (1958) **111**, 1245.

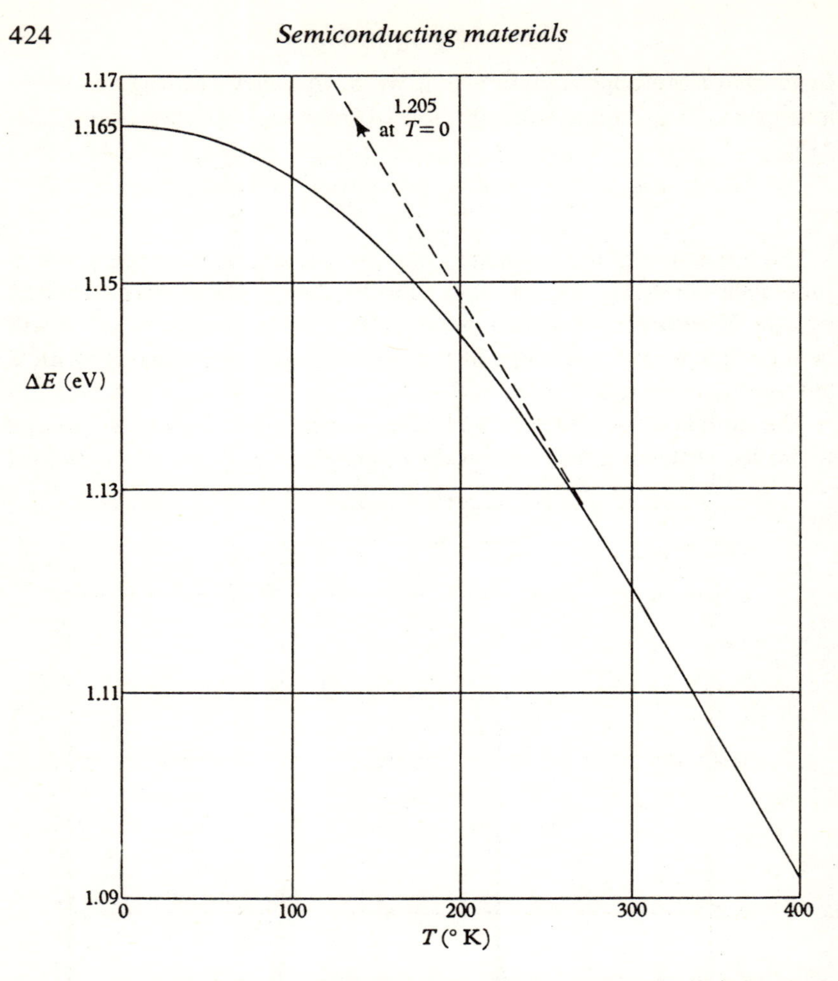

Fig. 13.4. Variation of forbidden energy gap ΔE with temperature for Si. (After G. G. Macfarlane *et al.*, *loc. cit.*)

These are drift mobilities but do not differ much from conductivity mobilities:

	μ_{de} (290 °K) cm^2 V^{-1}s^{-1}	μ_{dh} (290 °K) cm^2 V^{-1} s^{-1}
Ge	3900 ($\propto T^{-1.6}$)	1900 ($\propto T^{-1.3}$)
Si	1350 ($\propto T^{-2.5}$)	480 ($\propto T^{-2.7}$)

For Ge the intrinsic resistivity $\rho_i = 47\ \Omega$ cm, and for Si $\rho_i = 2.5 \times 10^5\ \Omega$ cm. The intrinsic carrier concentration n_i is given by:

Ge	$n_i = 2.25 \times 10^{13}$ cm^{-3}	($T = 290$ °K)
Si	$n_i = 7 \times 10^9$ cm^{-3}	($T = 290$ °K)

Intrinsic Ge at room temperature is commonplace but intrinsic Si will be very difficult to achieve. The variation of n_i with T is given in § 4.2.

Some of the more common impurity levels in Si and Ge are given below. The values marked with a + denote the amount *above* the top of the valence band in eV and those marked with a − the amount *below* the conduction band in eV:

Group III acceptors	B	Al	Ga	In	Tl
Ge	+0.0104	+0.0102	+0.0108	+0.0112	+0.0100
Si	+0.044	+0.069	+0.073	+0.155	+0.26
Group V donors	**P**	**As**	**Sb**		
Ge	−0.0120	−0.0127	−0.0096		
Si	−0.044	−0.049	−0.039		

	Li	Cu	Ag	Au	Zn	Cd
Ge	−0.0093	+0.045	+0.13	+0.05	+0.03	+0.05
		+0.32	−0.39	+0.15	+0.09	+0.16
		−0.26	−0.029	−0.20		
				−0.04		
Si	−0.033			+0.39	+0.092	
				−0.30	+0.30	

Ge alloys in all proportions with Si although it is difficult to prepare large single crystals if the Si:Ge ratio is between 15% and 95%. The properties of Ge/Si alloys have been reviewed by F. Herman, M. Glicksman and R. A. Parmenter.[5] The electron mobility of pure material falls from that of pure Ge to a minimum of about 500 $cm^2 V^{-1} s^{-1}$ at 15% Si and rises again to that of pure Si.

Molten Si and Ge behave like metals, the resistivity rising with increasing temperature. On melting the resistivity of Ge drops by a factor of about 20 to about 6.5×10^{-5} Ω cm. Its properties have been studied by C. A. Dominicali.[6]

13.4 Other element semiconductors

Besides Ge and Si, other element semiconductors which have been studied are Se, Te and grey tin (α-Sn). B shows some semiconducting properties but has not been studied in detail. Allotropic forms of other elements, such as S, P and I, may be semiconductors but have only been studied in this respect in the form of evaporated layers.[7]

[5] *Progress in Semiconductors* (Heywood, 1957) **2**, 1.
[6] *J. Appl. Phys.* (1957) **28**, 749.
[7] T. S. Moss, *Photoconductivity of the Elements* (Butterworths, 1952).

The properties of Se have been described by A. Jenkins.[8] The fact that zone refining methods appear to be ineffective with this material has made its purification very difficult. Te on the other hand can be zone refined and has been studied with controlled amounts of impurity, particularly with a view to clarifying various mechanisms of carrier recombination.[9] Its band structure, and in particular its anisotropy, are of considerable theoretical interest.

Cyclotron resonance has now been applied to this semiconductor after many unsuccessful attempts, success being mainly due to the higher frequencies now available. It has been observed both for electrons in the conduction band (thermally excited) and for holes in the valence band, in spite of the fact that Te is normally *p*-type and no doping has yet produced *n*-type. This has enabled the electron mass m_e for the conduction band to be determined, its value being $0.135\boldsymbol{m}$. The cyclotron resonance for holes shows a complex structure, there being several peaks in the spectrum. This is due to the interaction of the two branches of the band, degenerate at the maximum, with the electron spin. The details of this spectrum have been studied by K. J. Button *et al.*[10] using far-infra-red radiation from HCN and D_2O lasers at wavelengths of 0.337 mm and 0.1186 mm. The band structure of PbTe has been discussed by K. J. Button[11] in the light of these measurements and found to be in reasonably good agreement with band-structure calculations.

Grey tin (α-Sn) has been studied for some time (see T. S. Moss, *loc. cit.*) and, mainly from measurements of photo-conductivity and Hall effect, was thought to be a semiconductor with a forbidden energy gap of about 0.08 eV.[12] As a result of more recent magneto-optical experiments grey tin is now thought to be a semimetal with zero forbidden energy gap. This band structure was suggested by S. Groves and W. Paul[13] and supported by experiments. More recent experiments using magneto-optical techniques have confirmed it.[14] The schematic form of the band structure is shown in Fig. 13.5. The lower branch of the Γ_8^+ band, degenerate at $\boldsymbol{k} = 0$, corresponds to the valence band, and the

[8] *Materials used in Semiconductor Devices*, ed. C. A. Hogarth (Wiley Interscience, 1965), p. 49.

[9] J. S. Blakemore, *Semiconductor Statistics* (Pergamon Press, 1967).

[10] *Phys. Rev. Lett.* (1969) **23**, 14.

[11] *Optical Properties of Solids*, ed. E. D. Haidemenakis (Gordon and Breach, 1970), p. 253.

[12] G. Busch and R. Kern, *Solid State Physics* (Academic Press, 1960) **11**, 1.

[13] *Proc. VIIth Int. Conf. on Phys. of Semiconductors* (Dunod, 1964), p. 41.

[14] C. R. Pidgeon, in *Electron Structure in Solids*, ed. E. D. Haidemenakis (Plenum Press, 1969), p. 47.

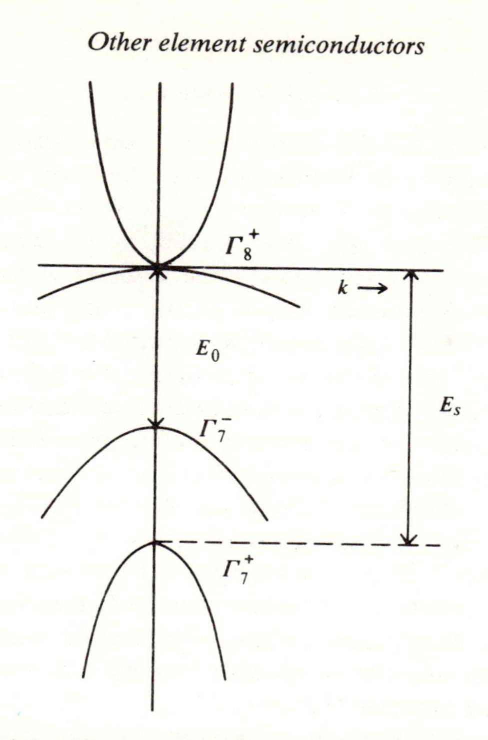

Fig. 13.5. Schematic band structure for α-Sn showing zero forbidden energy gap typical of a semimetal. ΔE_s is the spin–orbit splitting.

upper branch to the conduction band. The valence band is full but there is zero energy gap between it and the conduction band. Grey tin is therefore a semimetal (see § 13.8) and not a semiconductor. The value given by Pidgeon (*loc. cit.*) for the gap ΔE_0 between the next higher band at $\boldsymbol{k} = 0$ is 0.413 eV and this is much too large to correspond to the previously determined gap of 0.08 eV. It is not at present clear why the careful Hall-constant and other measurements reported by Busch and Kern (*loc. cit.*) should show an activation energy of 0.08 eV. This is certainly small but well within the accuracy of their experimental determination. It is possible that the methods of preparation led to the inclusion of impurities from which electrons could be excited across the zero-energy gap and this could lead to an apparent activation energy. Another possibility is that the crystals were strained. It has been shown by A. W. Ewald,[15] who has also reviewed the work on α-Sn, that strain can open up a gap between the valence and conduction band. However, the strain required to produce a gap of 0.08 eV would be rather large.

[15] *Helv. Phys. Act.* (1968) **41**, 795.

13.5 III–V compounds

The properties of the III–V compound semiconductors were first studied intensively by H. Welker and his colleagues. Their properties have been reviewed by H. Welker and H. Weiss,[16] and also by O. Madelung.[17] They have also been treated by C. Hilsum and A. C. Rose-Innes,[18] and in great detail in the volumes of the series *Semiconductors and Semimetals* (see p. 420). InSb, and especially the problems involved in crystal growth and purification, have been treated by K. F. Hulme[19] in some detail. In particular, the difficulty of avoiding the growth of mirror-like facets when pulling from the melt is discussed. These facets lead to uneven distribution of dopants. InSb has also been the main vehicle for study of growth striations, another manifestation of uneven impurity distributions. These have been studied by A. F. Witt and H. C. Gatos.[20] Naturally, having two elements, the III–V semiconductors are more difficult to prepare in very pure form or with small controlled amounts of impurity than the element semiconductors. Fortunately there do not appear to be any great problems regarding the stoichiometric ratio of the two elements, the compounds forming with very nearly equal numbers of each constituent atom.

There are nine III–V binary compounds resulting from the combination of In, Ga, Al, with Sb, As, P. InSb has the smallest forbidden energy gap of all the III–V semiconductors: 0.165 eV at 300 °K. The variation with temperature is very similar to that for Si and Ge, being linear over a large range of temperatures but tending to a constant value (0.23 eV) near 0 °K. For InSb the optical transition corresponding to the gap is 'direct'. The very steep rise in absorption is shown in Fig. 10.7.

Of the nine III–V binary compound semiconductors five have 'direct' gaps and four 'indirect'. The values of the minimum gap ΔE (shown as D or I) are given in Table 13.1. We note the high mobility of electrons in InSb. At low temperatures this can exceed 10^6 cm^2 V^{-1} s^{-1}.

InSb approaches very closely to an ideal semiconductor having a very small scalar effective mass for electrons in the conduction band with no other band minima close. The band structure of InSb is shown schematically in Fig. 13.6. The notation for the various maxima and minima is now conventional and as for Si and Ge is derived from group theory.

[16] *Solid State Physics* (Academic Press, 1956) **3**, 1.
[17] *Physics of III–V Compounds* (Wiley, 1964).
[18] *Semiconducting III–V Compounds* (Pergamon Press, 1961).
[19] *Materials used in Semiconductor Devices*, ed. C. A. Hogarth (Wiley Interscience, 1965), p. 115.
[20] *J. Electrochem. Soc.* (1966) **113**, 808.

TABLE 13.1 *Properties of III–V semiconductors* (300 °K)

	ΔE (eV)	m_e/m	m_{h1}/m	μ_e ($cm^2\,V^{-1}\,s^{-1}$)	μ_h ($cm^2\,V^{-1}\,s^{-1}$)
InSb	0.18D	0.014	0.4	78 000	750
InAs	0.35D	0.022	0.4	28 000	450
GaSb	0.72D	0.044	0.23	44 000	700
InP	1.33D	0.078	—	5 200	150
GaAs	1.42D	0.065	0.5	8 900	400
AlSb	1.62I	—	—	200	—
AlAs	2.13I	—	—	—	—
GaP	2.26I	—	—	200	100
AlP	2.43I	—	—	—	—

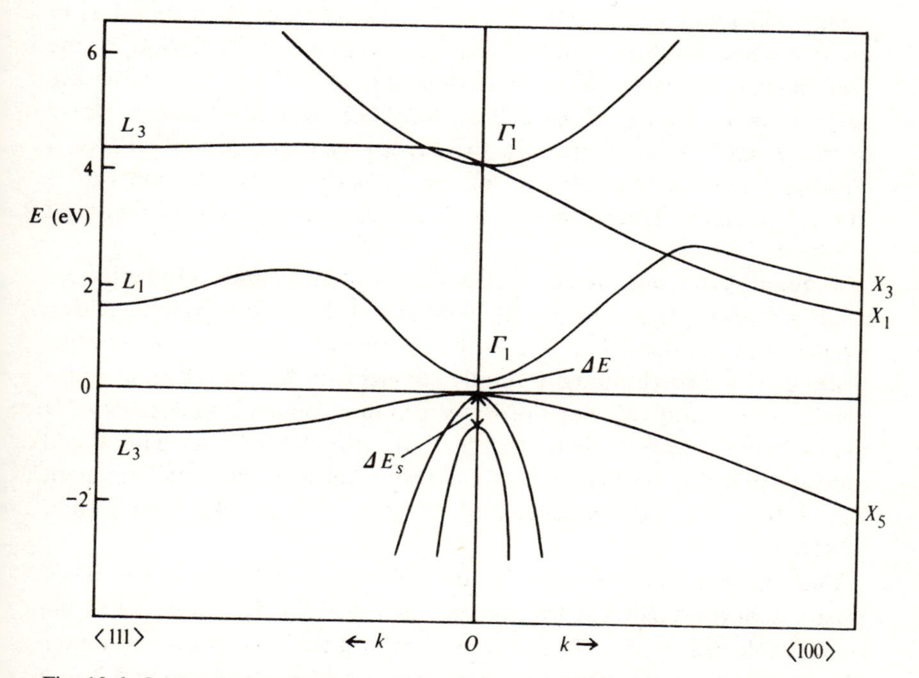

Fig. 13.6. Schematic band structure for InSb. The gap $\Delta E = 0.18$ eV (at 300 °K), $\Delta E_s = 0.9$ eV.

This schematic is not strictly correct. When a further refinement is made in the calculation of the form of the bands for InSb near $\boldsymbol{k} = 0$ a splitting is found which makes the heavy-hole band rise slightly, as k increases from zero, before falling. The maximum rise is however, only of the order of 1 meV as shown by C. Pidgeon.[21] Also, although 'spherical',

[21] *Electronic Structure in Solids*, ed. E. D. Haidemenakis (Plenum Press, 1969), p. 47.

there are deviations from the strictly quadratic behaviour of $E(\boldsymbol{k})$ because of the very small electron effective mass.

The band structure of GaAs is very similar but there is a subsidiary minimum in the conduction band only 0.35 eV above the minimum at $\boldsymbol{k}=0$. The values given for the principal energy gaps are experimental, the others being derived by calculation.[22]

Values of some of the other gaps for the remaining III–V semiconductors and also variation with temperature are given in some detail in *Handbook of Electronic Materials* Vol. 7 (see p. 419).

These materials all alloy with each other to produce ternary III–V semiconductors such as GaP_xAs_{1-x}. The variation of the properties of these semiconductors with x has been tabulated in detail in the *Handbook of Electronic Materials* referred to above. The band structure and general properties of these alloys have been discussed in some detail by D. Long.[23] It is interesting to see, for example, how GaP_xAs_{1-x} changes from a semiconductor having an 'indirect' gap to one with a 'direct' gap as x changes from 0 to 1, the cross-over point being about $x=0.44$.

A quantity of some interest in III–V binary compounds is the effective ionic charge e^*. This measures to what extent the binding is ionic and to what extent covalent. If $e^*=\boldsymbol{e}$ each ion has just one electronic charge, while if $e^*=0$ the binding is wholly covalent, as for the element semiconductors Si and Ge. For the III–V group of binary compounds $e^*/\boldsymbol{e}$ varies from 0.33 for GaSb to 0.68 for InP with 0.5 for GaAs. This would indicate that the binding in GaAs is about half ionic and half covalent (see § 3.2). (For determination of e^* see § 8.3, p. 242; also § 14.3, p. 458.)

One other quantity which is quite anomalous for some of the III–V binary semiconductors is the g-factor (see §12.6). For InSb g has the remarkable value of -40 (it varies somewhat with impurity content). This means that spin splitting in a magnetic field is very large and is one of the factors which make InSb a suitable material for use in magnetically tunable lasers (see §12.6). For InAs $g=-12$ and for GaSb $g=-6$. For the other compounds g is usually positive and somewhat less than the free electron value 2.

The energies of impurity levels in these materials and indeed their precise nature are not nearly so well known as for Si and Ge. This is

[22] See, for example, M. L. Cohen and T. K. Bergstresser, *Phys. Rev.* (1966) **141**, 789, for much more detailed band structures.

[23] *Semiconductors and Semimetals* (Academic Press, 1966) **1**, 143.

hardly surprising because of the much greater variety of possibilities which can occur; e.g. we may have two kinds of lattice vacancies and two kinds of interstitial atom together with various combinations of these. There are also different ways of substitution.

The group II elements Zn and Cd appear to substitute for the group III atoms in all these compounds. In InSb they give rise to shallow acceptors with ionization energy about 0.0075 eV. S, Se and Te from group VI substitute for the group V atoms and give donors. Because of the small effective mass in InSb, according to the simple 'hydrogen' model, they would have ionization energy of only 0.0007 eV and very large Bohr radii. Except in very small concentrations they would overlap and their energies would merge with the conduction band. Un-ionized impurities of this kind have only been observed in strong magnetic fields which lower their binding energy (see §12.9). Cu, Ag and Au give double acceptors in InSb while substitution by other group III or group V atoms seems to lead to neutral impurities. In GaAs they give acceptors 0.034 eV above the valence band while Ge gives donors 0.006 eV below the conduction band. Cu and Ag act as acceptors but tend to form complexes with other imperfections.

A good deal of information is available on deep-lying impurities in GaP through the study of sharp-line spectra associated with excitons trapped at such impurities[24] (see §§10.14, 14.1). For example, *pairs* of impurities – one a donor and one an acceptor, such as Te (D) and Si (A) – form such a double impurity. O and N form deep-lying impurities in both GaAs and GaP, generally in combination with other impurities.

13.6 II–VI compounds

The II–VI compounds which have been studied in greatest detail are the lead chalcogenides PbS, PbSe and PbTe. They are so similar that we shall consider them together. Their properties have been treated in some detail by E. H. Putley[25] and by W. W. Scanlon.[26]

A good deal of confusion has arisen over their band structure, since the first calculations showed PbTe to be an 'indirect' semiconductor.[27] Later calculations by J. B. Conklin, L. E. Johnson and G. W. Pratt[28] and

[24] See, for example, M. Gershenzon, *Semiconductors and Semimetals* (Academic Press, 1966) **2**, 289.
[25] *Materials used in Semiconductor Devices* (Wiley Interscience, 1965), p. 71.
[26] *Solid State Physics* (Academic Press, 1959) **9**, 83.
[27] Dorothy G. Bell *et al.*, *Proc. Roy. Soc.* A (1953) **217**, 71.
[28] *Phys. Rev.* (1965) **137**A, 1282.

by P. J. Lin and L. Kleinman,[29] taking account of relativistic effects on the wave-functions arising from the heavy atoms, showed that for all three compounds both the lowest minima in the conduction band and the highest minima in the valence band are at the zone edge in the ⟨111⟩ directions. There are therefore four maxima and four minima and these are all 'direct' semiconductors, but *not* with the minimum gap at $\boldsymbol{k} = 0$. The constant-energy surfaces near the bottom of the conduction band and also near the top of the valence band are ellipsoids of revolution. They are therefore different from Si and Ge and also from the III–V semiconductors and form a very interesting and unique group of materials. They all have the sodium chloride crystal structure.

A schematic energy band diagram for PbTe is shown in Fig. 13.7. A much more detailed band structure is given by Lin and Kleinman (*loc. cit.*) including the rather complex group theoretical notation for the maxima and minima. The values for the forbidden energy gap ΔE are

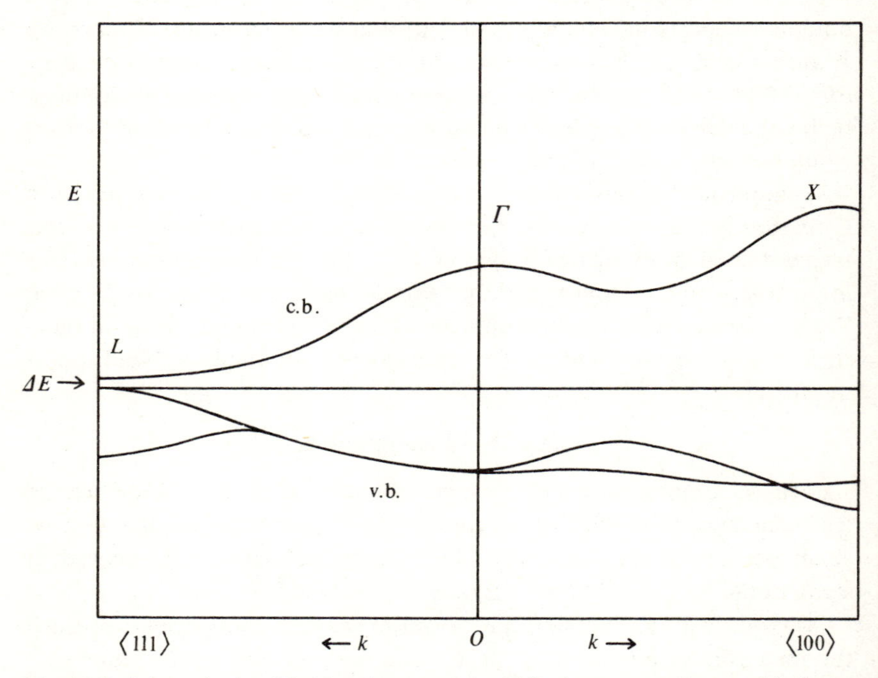

Fig. 13.7. Schematic simplified band structure for PbTe (not accurately to scale). Structures for PbS and PbSe are similar.

[29] *Phys. Rev.* (1966) **142**, 478.

given in Table 13.2 for different temperatures. It should be noted that the values of ΔE for these materials *decrease* as the temperature is lowered, in contrast to the behaviour of Si and Ge and the III–V compounds, and that PbSe is out of order, having a smaller value of ΔE

TABLE 13.2 *Forbidden energy gaps* ΔE *of lead chalcogenides*

	T (°K)	ΔE (eV)
PbS	290	0.41
	77	0.31
	4	0.29
PbSe	290	0.27
	77	0.17
	4	0.15
PbTe	290	0.32
	77	0.22
	4	0.19

than PbTe. They are also sensitive to strain in the crystal as has been shown by D. L. Mitchell, E. D. Palik and J. N. Zemel[30] who have also obtained by magneto-optical absorption studies the effective masses. These differ slightly from values obtained by others. T. S. Moss, G. J. Burrell and B. Ellis[31] have analysed the different measurements and give the values quoted in Table 13.3 for a temperature of 77 °K. The

TABLE 13.3 *Effective masses for PbS, PbSe and PbTe*

		m_T	m_L	K
PbS	c.b.	0.10	0.11	1.1
	v.b. (holes)	0.10	0.11	1.1
PbSe	c.b.	0.05	0.08	1.6
	v.b. (holes)	0.05	0.08	1.6
PbTe	c.b.	0.025	0.22	8.7
	v.b. (holes)	0.025	0.22	8.7

only material to have a large value of K is PbTe, the departure from 'spherical' form for PbS being quite small. The above values would appear to indicate that the band parameters for the conduction and valence bands are the same. At 77 °K this appears to be very nearly so but differences do appear at lower temperatures. For example, while K for the conduction band in PbTe is 10 at 4 °K, it is 14 for the valence band.

[30] *Proc. VIth Int. Conf. on Phys. of Semiconductors* (Dunod, 1964), p. 325.
[31] *Semiconductor Opto-electronics* (Butterworths, 1973).

The anomalous order of the forbidden energy gap of PbSe has been discussed by a number of authors but has finally been explained by R. L. Bernick and L. Kleinman[32] who have shown that the symmetry properties of the conduction band of PbTe is different from those of PbS and PbSe. With this adjustment the calculated effective masses are also in better agreement with experiment than those derived in previous calculations.

Mitchell, Palik and Zemel (*loc. cit.*) also obtained values for the g-factor as follows: PbS 10.0 (c.b.), 8.5 (v.b.); PbSe 22 (c.b.), 22 (v.b.).

Data on impurity levels in these semiconductors are scanty, mainly on account of the difficulty of obtaining them with strictly stoichiometric composition. Excess Pb atoms act as very shallow donors whereas excess S, Se or Te act as shallow acceptors in PbS, PbSe or PbTe respectively. These problems have been treated in some detail by J. Bloem and F. A. Kröger,[33] who have shown that for PbS at its maximum melting point there are 6×10^{18} cm^{-3} Pb atoms in excess. Bi acts as a donor and may be used partially to compensate p-type material while Ag and Cu act as acceptors when they substitute for Pb. On the other hand they may act as donors when in an interstitial position.

The equilibrium between vacancies (V_c^+) and added Bi has also been studied by Bloem and Kröger (*loc. cit.*) who have shown that under certain circumstances the number of vacancies is just equal to the number of added Bi atoms a situation known as valence compensation.

All three compounds have high refractive indices. In the near infrared they are PbS 4.1, PbSe 4.85 and PbTe 5.65. They increase rather slowly as the frequency is lowered but at extremely low frequencies a great increase has been reported especially for PbTe, there is however some doubt about the precise values. W. Cochran[34] gives a value 17.6 for PbTe. A recent review by R. Dalven[35] has discussed in detail the properties of this very interesting group of semiconductors.

The group of semiconductors CdS, CdSe and CdTe has also been studied, though less fully than the lead chalcogenides. The values of ΔE are respectively 2.4 eV, 1.75 eV and 1.51 eV. CdS has been used as a typical polar semiconductor for investigations of the properties of vacancies, both anion and cation, and of their relationship with the number of free carriers of each type. The physical chemistry of these materials and also of the lead compounds has been studied in detail by

[32] *Solid State Commun.* (1970) **8**, 569. [33] *Z. Phys. Chem.* (1956) **7**, 1.
[34] *Phys. Lett.* (1964) **13**, 193.
[35] *Solid State Physics* (Academic Press, 1973) **28**, 179.

F. A. Kröger, H. J. Vink and J. van den Boomgard.[36] CdS has also been widely used for the investigation of exciton spectra (see § 10.6). As a result very accurate values for the energies of some of the impurity levels are known but the defects have not always been unambiguously identified.

CdS is also the material for which the electro-acoustic effect has mainly been studied (see § 12.4.1). The use of the material for electro-acoustic delay lines in addition to its widespread use in light meters has stimulated its examination as a semiconductor.

A number of other semiconducting compounds involving S, Se and Te have been examined, but in much less detail. In particular, Bi_2Te_3 has received a good deal of attention as a thermo-electric material. Its properties have been described by H. J. Goldsmid.[37] The properties of the compounds CdSb and ZnSb have been reviewed by G. R. Blackwell.[38]

A review of the properties and methods of preparation of a number of polar semiconductors, but of the lead chalcogenides in particular, has been given by W. W. Scanlon[39] who has also discussed the formation of vacancies and deviations from stoichiometry. A more recent review of the II–VI compound semiconductors has been given in a collection of papers edited by D. G. Thomas.[40]

13.7 Ternary and quaternary compounds

All the compounds discussed so far have been binary compounds, except for mixed crystals which are really alloys. The possibility of having ternary compound semiconductors, or even quaternary compound semiconductors, has been appreciated for some time, but very little progress has been made in this direction, mainly owing to the difficulties of making and purifying the compounds. In view of the difficulties of obtaining a stoichiometric balance when two elements are involved, it is not surprising that this should be even more difficult when three are present. Such compound semiconductors have been discussed by H. Hahn *et al.*[41] and by C. H. L. Goodman and R. W. Douglas,[42] who have described the principles on which the making of more complex

[36] *Z. Phys. Chem.* (1954) **203**, 1.
[37] *Materials used in Semiconductor Devices*, ed. C. A. Hogarth (Wiley Interscience, 1965), p. 165. [38] *Ibid.*, p. 199.
[39] *Solid State Physics* (Academic Press, 1959) **9**, 83.
[40] *II–VI Semiconducting Compounds*, ed. D. G. Thomas (Benjamin, 1967).
[41] *Z. Anorg. Allg. Chem.* (1953) **271**, 153. [42] *Physica* (1954) **20**, 1107.

compound semiconductors are based; these depend on balanced valencies and the formation of closed shells of eight electrons, as for the binary compounds. Several compounds of this type were made and shown to be semiconductors, but very little detailed information on their properties is available as yet; examples are $CuFeS_2$, $CuInSe_2$, $AgInSe_2$. The first of these belongs to a class of materials known as chalcopyrites which have long been known to show rectification at a metal point contact; it is also stated by Goodman and Douglas to be transparent in the infra-red at wavelengths greater than 2 μm and so presumably has a forbidden energy gap of about 0.5 eV; it has a crystal structure very similar to the diamond structure. In this structure the Cu atom is monovalent and the Fe atom trivalent, the four electrons thus provided, together with the twelve from S_2, making two closed groups of eight. Quaternary compounds of the form Cu_2FeSnS_4, known as stannites, should also show semiconducting properties, and on this analogy it is suggested by Goodman and Douglas that compounds like $Cu_2CdSnTe_4$ should be semiconductors; new compounds made and studied by Goodman and Douglas include $CuInTe_2$, $CuInSe_2$ and $CuAlTe_2$. These show marked rectification characteristics, have optical energy gaps of about 0.9 eV and electron mobilities of the order of $300\ cm^2\ V^{-1}\ s^{-1}$. Attempts to pull single crystals of these materials were not successful.

Clearly, when we include ternary and quaternary compounds the number of possible semiconducting compounds becomes very great. Some suggestions for extending the principle of balanced valency discussed above have been put forward by E. Mooser and W. B. Pearson.[43]

After the discovery of their non-linear optical properties a large number of ternary compounds have been investigated, their choice being based on the principle of balanced valency. Of these, the chalcopyrite compounds have received most attention. These take two main forms which may be expressed as 245_2 and 136_2, the full numbers representing the group in the periodic table and the subscripts the number of atoms in the compound. The former may be regarded as an extension of the III–V compounds and the latter as an extension of the II–VI compounds. Examples of the 245_2 compounds are as follows:

$CdGeAs_2$	$\Delta E = 0.55$ eV
$CdSnP_2$	$\Delta E = 1.15$ eV
$ZnGeP_2$	$\Delta E = 2.2$ eV

[43] *Report of Meeting on Semiconductors*, April 1956 (London Physical Society), p. 65.

Examples of the 136_2 compounds are

$AgInSe_2$	$\Delta E = 1.2$ eV
$AgInS_2$	$\Delta E = 2.0$ eV
$AgGaS_2$	$\Delta E = 2.7$ eV
$CuAlS_2$	$\Delta E = 3.5$ eV

An extensive literature has rapidly grown up on these materials. The properties of some of them have been reviewed by C. Hilsum.[44] The use of these materials in opto–electronics has been discussed by R. C. Smith[45] who has listed values of ΔE for a number of both types of chalcopyrite and whether the gap is 'direct' or 'indirect'. He has also discussed their optical properties and, in particular, use of their non-linear characteristics for optical frequency mixing.

One of the interesting features of these materials is that many of them have 'direct' gaps and so should be useful for luminescence applications and laser action (see § 14.6). Indeed a number of them have shown such action.

The problems of crystal growth which proved too difficult to overcome when these materials were first studied have now to some extent been overcome, and large single crystals of fairly good optical quality have been produced in a number of laboratories. The techniques now used have been described by R. S. Feigelson.[46] In particular, he has discussed the growth of crystals of $CdGeAs_2$, $AgGaSe_2$ and $AgGaS_2$ for optical applications. The growth of high-quality crystals of $AgGaS_2$ has also been described by M. Matthes *et al.*[47]

In a book devoted entirely to the subject of ternary chalcopyrite semiconductors J. L. Shay and J. H. Wernick[48] have dealt with the known properties of a large number of compounds. Much less is known of the electrical properties of these materials than of their optical properties but a good deal of information is rapidly becoming available. Many of the 245_2 compounds have room-temperature electron or hole mobilities of the order of $100\ cm^2\ V^{-1}\ s^{-1}$ or less but a few such as $CdGeAs_2$ have considerably higher mobilities of the order of $1500\ cm^2\ V^{-1}\ s^{-1}$. The 136_2 compounds appear to have rather lower mobilities of the order of $1–20\ cm^2\ V^{-1}\ s^{-1}$, apart from $AgInSe_2$ which

[44] *Electronic Materials*, ed. N. B. Hannay and U. Colombo (Plenum Press, 1973), p. 69.
[45] *J. de Phys.* (1975) **36**, C3–89.
[46] *J. de Phys.* (1975) **36**, C3–57.
[47] *J. de Phys.* (1975) **36**, C3–105.
[48] *Ternary Chalcopyrite Semiconductors – Growth Electronic Properties and Applications* (Pergamon Press, 1975). See also p. 227.

has $\mu_e \simeq 750\ \mathrm{cm^2\ V^{-1}\ s^{-1}}$. The electrical properties as derived from Hall-effect, conductivity and other measurements are described in considerable detail by Shay and Wernick (*loc. cit.*). A whole volume (**36**, C3) of the *Journal de Physique* (1975) is devoted to a series of papers on this class of semiconductor. (See also p. 227.)

The band structure of these compounds shows interesting differences from those of the corresponding III–V and II–VI semiconductors, due mainly to the fact that the tetrahedral bonding is slightly distorted and also that the unit cell is twice as large as for the binary compounds. The use of pressure to augment the deviation further from tetrahedral symmetry has helped to clarify the rather complex structure which results. Electro-reflectance spectroscopy has also been effectively used to identify a number of critical points in the band structure. These techniques have been discussed by R. A. Bendorius *et al.*[49] and also by A. Shileika.[50] Band-structure calculations have been reported by a number of authors including Y. I. Polygalov, A. S. Poplavnoi and A. M. Ratner[51] and also L. Paseman *et al.*[52] While these are not fully in line with the experimental findings they have helped greatly in filling in the structure between the critical points so that we now have a very good picture of the band-structure of a number of these compounds.

13.8 Very-narrow-gap semiconductors

For many years a search has been made for semiconductors with a forbidden energy gap of the order of 0.1 eV or lower. Until 1959 the only material which showed such a low gap was α-Sn and it now turns out that this may not be an 'intrinsic' gap (see § 13.4). The search was largely prompted by the need for infra-red detectors whose response extended to wavelengths larger than 10 μm. The only simple compound semiconductor with a gap approaching 0.1 eV is InSb whose gap at room temperature is 0.18 eV. There are, however, semimetals with zero or very small negative energy gaps which alloy with semiconductors with forbidden energy gaps somewhat larger than 0.1 eV. It occurred to W. D. Lawson and some of his colleagues that by alloying a little of a semimetal with a semiconductor the forbidden energy gap of the latter might be lowered. This turned out to be so and was demonstrated[53] by

[49] *Proc. XIth Conf. on Phys. of Semiconductors* (Polish Sci. Publ., 1972), p. 838..
[50] *Surf. Sci.* (1973) **37**, 730.
[51] *J. de Phys. Supplt* (1975) C3–129. [52] *Phys. Stat. Solidi* (1976) **77**, 527.
[53] W. D. Lawson *et al.*, *J. Phys. Chem. Solids* (1959) **9**, 325.

alloying HgTe, a semimetal, with CdTe, a semiconductor with $\Delta E \approx$ 1.5 eV. Both the device development and study of this very interesting alloy has been held up by the difficulty of growing high quality single crystals of uniform composition. Crystals of better quality than formerly available have been prepared by B. E. Bartlett, J. Deans and P. C. Ellen,[54] and have been used not only for the study of the material, but as the basis for infra-red detectors[55] for medical and other applications involving the viewing of bodies at or near that temperature for which the maximum radiation is at a wavelength of about 10 μm. Although there is a zero energy gap in HgTe at $\boldsymbol{k} = 0$, the valence band rises slightly for larger values of $\boldsymbol{k}$ and there is an overlap or negative energy gap. As CdTe is added, the conduction band rises, but it is not until about 10% CdTe has been added that it clears the valence band and there is a true forbidden energy gap.

According to J. L. Schmit and E. L. Stelzer[56] the forbidden energy gap varies nearly linearly, from −0.1 eV to 1.5 eV for $Hg_xCd_{1-x}Te$ as x varies from 0 to 1.

Various measurements on effective masses have been made at different concentrations of Hg. These have been reviewed by T. S. Moss, G. J. Burrell and B. Ellis.[57]

Another method for obtaining a semiconductor with very small and controllable forbidden energy gap was suggested by J. O. Dimmock, I. Melngailis and A. J. Strauss[58] as a result of studies of the band structure of the PbS group of semiconductors and of SnTe (see § 13.6). They noticed that the bands for PbTe and SnTe were very similar but the properties of the lowest conduction and highest valence band were *inverted*. This is illustrated in Fig. 13.8, where we show a simplified version of the bands using the usual group theory notation. It will be seen that whereas the L_6^- band in PbTe is the conduction band it is the valence band in SnTe. Similarly in PbTe the L_6^+ band is the valence band and in SnTe the conduction band. It is found that the two compounds alloy and we should therefore expect the valence and conduction bands to cross over at some composition, each becoming inverted and then separating. At a critical composition, which turns out to be about 62% SnTe at room temperature but 34% at 4.2 °K, the forbidden energy gap is zero. This is illustrated in Fig. 13.9.

[54] *J. Matls. Sci.* (1969) **4**, 266. [55] B. E. Bartlett *et al., Infrared Phys.* (1969) **9**, 35.
[56] *J. Appl. Phys.* (1969) **40**, 4865.
[57] *Semiconductor Opto-Electronics* (Butterworths, 1973), p. 398.
[58] *Phys. Rev. Lett.* (1966) **16**, 1193.

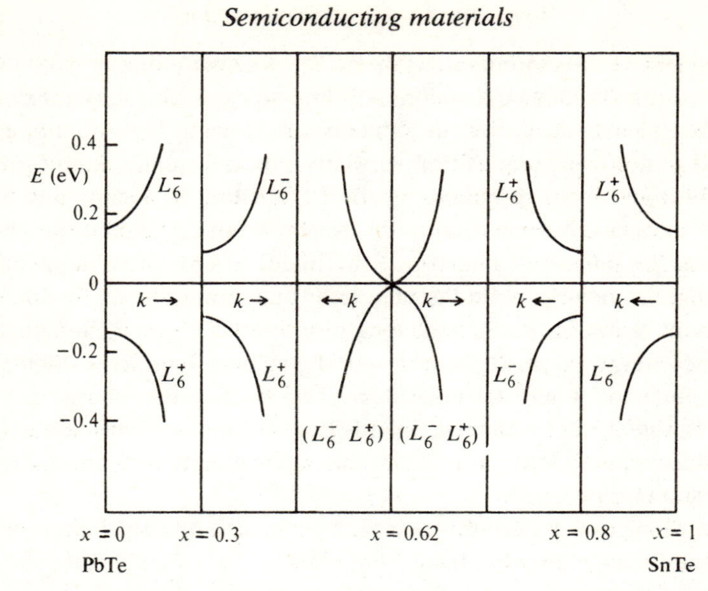

Fig. 13.8. Illustrating changeover from L_6^- band to L_6^+ as x goes from 0 to 1 in $Pb_{1-x}Sn_xTe$. (After J. O. Dimmock *et al.*, *loc. cit.*)

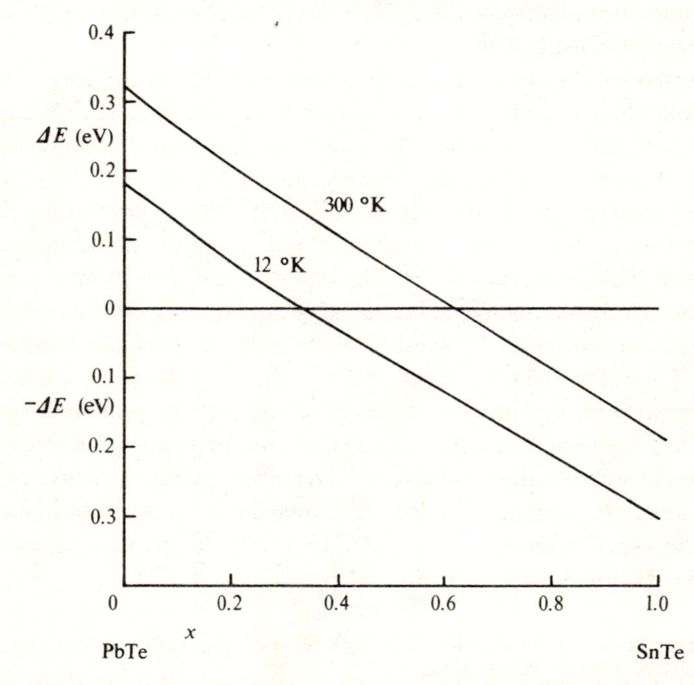

Fig. 13.9. Variation of forbidden energy gap ΔE with composition for $Pb_{1-x}Sn_xTe$. (After J. O. Dimmock *et al.*, *loc. cit.*)

As we have seen, for pure PbTe the forbidden energy gap decreases as the temperature is lowered. For SnTe, as for most other semiconductors, the gap increases as the temperature is lowered. This is consistent with the variation indicated in Fig. 13.9.

Good single crystals of the material can be grown and by careful control of the Te content both n-type and p-type material can be prepared. Both electron and hole mobilities are high as has been shown by L. Melngailis and T. C. Harman.[59] For material having 17% SnTe at 77 °K, μ_n and μ_h are both in excess of 30 000 cm^2 V^{-1} s^{-1}. Rather small effective masses for such material have been found by B. Ellis and T. S. Moss[60] using cyclotron resonance measurements, both holes and electrons having m_{eT} and m_{hT} of the order of $0.014\boldsymbol{m}$ with $K = 7$. These may also be derived theoretically from a study of the band structure close to the L point. Such a study has been made by J. O. Dimmock using the $\boldsymbol{k} \cdot \boldsymbol{p}$ method (see § 11.3.5).[61] The use in infra-red detector has been described by T. S. Moss, G. J. Burrell and B. Ellis.[62]

Because of the small effective electron mass in these materials, although the conduction band may still be 'spherical' it will deviate considerably from the simple 'parabolic' form for the energy as a function of $\boldsymbol{k}$. The effect of such deviations on transport properties has been discussed in some detail by W. Zawadski.[63]

A general review of the properties of these interesting new materials has been given by various authors in *Physics of Semimetals and Narrow-Gap Semiconductors.*[64]

13.9 Oxide semiconductors

A number of metallic oxides behave as semiconductors with rather large forbidden energy gaps. Of these Cu_2O and ZnO have been most studied, apart from the transition metal oxides such as NiO which pose some fascinating theoretical problems and consequently have received more attention. These generally have incomplete d-shells and so one should expect them to show metallic behaviour. Some oxides, notably ReO_3, do just this, showing a steady increase of resistivity with

[59] *Appl. Phys. Lett.* (1968) **13**, 180. [60] *Phys. Stat. Solidi* (1970) **41**, 531.
[61] Physics of Semimetals and Narrow-Gap Semiconductors, ed. D. L. Carter and R. T. Bates (Pergamon Press, 1971), p. 319.
[62] *Semiconductor Opto-Electronics* (Butterworths, 1973), p. 390.
[63] *Adv. Phys.* (1974) **23**, 435.
[64] *Physics of Semimetals and Narrow-Gap Semiconductors*, ed. D. L. Carter and R. T. Bates (Pergamon Press, 1971), Chapters 6–8.

temperature and having a value only about ten times greater than the noble metals. Pure NiO on the other hand behaves like an insulator with $\Delta E = 3.7$ eV. Much more striking, however, is the behaviour of oxides like VO_2 and V_2O_3 which both show metallic conduction at high temperatures, but the conductivity drops suddenly by more than 10^6 at a critical temperature. For the lower temperatures they then behave as semiconductors (see Fig. 13.10).

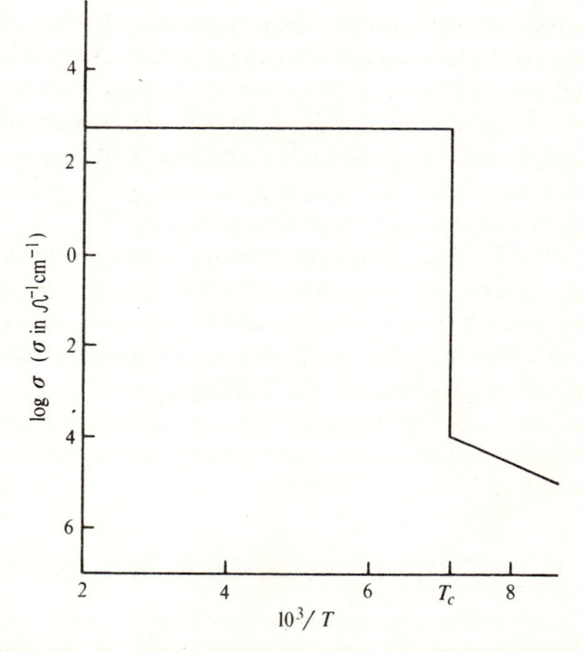

Fig. 13.10. Variation of conductivity σ with temperature T (°K) of a transition-metal oxide having a metal–semiconductor transition. The values shown are approximately those for V_2O_3.

This very odd behaviour is apparently in complete disagreement with the theory based on Bloch wave-functions in a crystal. One can understand why a material which has filled bands should behave as a metal, because of overlap of these bands, but it is not easy to see how a material with partially filled bands, as the *d*-bands for these materials must be, should behave as a semiconductor or even as an excellent insulator like MnO.

Several attempts have been made to resolve this paradox. N. F. Mott[65] has suggested that one should use localized wave-functions rather than Bloch type when their overlap from neighbouring atoms is small. The usual treatment deals inadequately with correlation effects between electrons. If one were to neglect these in solid hydrogen one would get a metallic phase, and indeed such a phase has been predicted at high pressures when atoms come closer together. At normal pressures the correlation effects cause the electrons and atoms to be paired to form H_2 molecules so that solid H_2 is a molecular crystal. Mott suggested that similar correlation effects should cause the electrons in the d-bands of oxides like V_2O_3 to be localized, an activation energy being required to release them for conduction. At a critical temperature, however, this would break down and a metallic state would set in suddenly. Such a transition has come to be known as a Mott transition. The evidence that such transitions do in fact take place in the transition metal oxides is now thought to be questionable, other possibilities having come to light to explain the sudden transition.

It has been found that in both V_2O_3 and VO_2 a change in crystal structure takes place, and the resultant change in band structure can account for the change from metal to semiconductor. Whether this change in crystal structure is driven by a Mott transition is another and still open question. The effect of the change in crystal structure in both cases is to produce a unit cell of twice the size and hence a Brillouin zone of half the size. Also the d-band is split, as has been shown by D. Adler and H. Brooks.[66]

This is illustrated in Fig. 13.11. The split d-band corresponds to the low-temperature situation, the transition being what is known as a first-order phase change and takes place suddenly over a very small temperature range. This would indicate that at temperatures above the transition temperature the d-band is half filled, the Fermi level being near the middle of the band and so the material would be expected to show metallic conduction. At temperatures below the transition temperature a finite forbidden energy gap would appear and the lower band would be completely filled so that the material would behave like a semiconductor. Moreover, the transition temperature would be expected to vary with pressure and this variation as well as the general

[65] *Can. J. Phys.* (1956) **34**, 1356; *Phil. Mag.* (1961) **6**, 287.
[66] *Phys. Rev.* (1967) **155**, 826.

behaviour of the conductivity and magnetic properties of VO_2 and V_2O_3 have been studied by J. Feinlieb and W. Paul.[67]

The schematic band structure shown in Fig. 13.11 is very much simplified. Each section of the split *d*-band is double and higher and

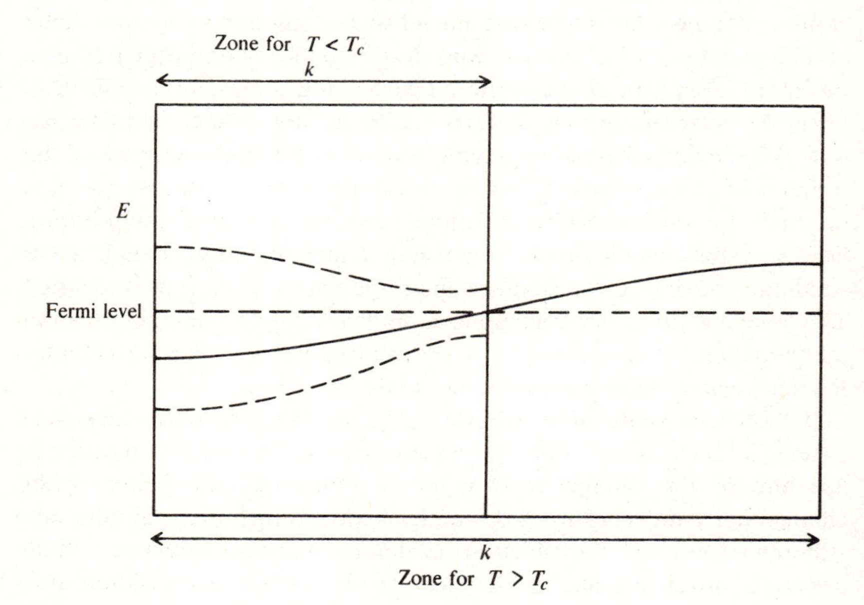

Fig. 13.11. Showing development of a forbidden energy gap through reduction of Brillouin zone by change in crystal structure at $T = T_c$.

lower bands are not too far off, but not near enough to lead to the observed value of activation energy. The experimental data have been carefully compared with the theory of Adler and Brooks (*loc. cit.*) by D. Adler *et al.*[68] and found to agree very well with the predictions based on the effect of the crystallographic change. That the transition is due to magnetic ordering is shown to be less likely. These problems, including that of NiO, have been discussed at some length by J. Feinlieb.[69]

This theory does not explain the semiconducting properties of NiO. Such an explanation almost certainly requires a band-structure calculation involving more than the *d*-band, taking into account covalent bonding between the *d*-electrons and those of the O^{2-} ions. Attempts to

[67] *Phys. Rev.* (1967) **155**, 841. [68] *Phys. Rev.* (1967) **155**, 851.
[69] *Electronic Structure in Solids*, ed. E. D. Haidemenakis (Plenum Press, 1969), p. 231.

do this have been made by J. Feinlieb and D. Adler.[70] The conductivity and magnetic properties of the metal oxides have been reviewed by D. Adler.[71] Reviews have also been given by S. Doniach[72] and by I. G. Austin and R. Gamble.[73]

The compounds $SrTiO_3$ and $BaTiO_3$, while normally insulators, become semiconducting with appropriate doping procedures. Their band structure is in some ways similar to that of the transition metal oxides and has been compared with these by H. P. R. Frederikse.[74]

13.10 Refractory semiconductors

While most of the semiconductors we have discussed operate satisfactorily in electronic devices at room temperature, those with values of the forbidden energy gap ΔE of the order of 1 eV or less lose the properties given to them by insertion of controlled amounts of impurity as the temperature is increased. There are a number of applications in which it is essential that the material should still retain its desired properties at much higher temperatures and this implies that it should have a considerably higher forbidden energy gap, and also high melting point. A search has therefore been made for semiconductors having not only these properties but also having the capability of being made controllably *n*-type or *p*-type so as to form *p–n* junctions etc. Such materials are known as refractory semiconductors.

Most work on materials of this kind has been carried out on SiC which at one time was thought might provide an alternative to Si for high-temperature operation. Also, this material showed strong light emission and so had possibilities for use in opto-electronic applications. The results of a great deal of work in this respect are disappointing. Some of the difficulty arises because SiC occurs in a number of allotropic forms. Impurity concentration also appears to be very difficult to control. The value of ΔE is about 2.8 eV as determined by W. J. Choyke and L. Patrick[75] from a study of the fundamental absorption edge.

Attention has more recently turned to the III–V compounds with higher forbidden energy gaps such as AlP, AlAs and InP (see §13.5).

[70] *Phys. Rev. Lett.* (1968) **21**, 1010.
[71] *Solid State Physics* (Academic Press, 1968) **21**, 1.
[72] *Advan. Phys.* (1969) **18**, 819.
[73] *Conduction in Low-Mobility Materials*, ed. N. Klein, D. S. Tannhäuser and M. Pollak (Taylor and Francis, 1971), p. 17.
[74] *Electronic Structure in Solids*, ed. E. D. Haidemenakis (Plenum Press, 1969), p. 259.
[75] *Phys. Rev.* (1957) **105**, 1721.

Some compounds of N with group V elements provide materials of this kind including AlN, GaN and InN. Some compounds of B are also refractory semiconductors such as BN, BP and BAs. A lot of work has still to be done before the properties of these compounds are as well understood as those of the III–V compounds. Also although a start has been made on controlled 'doping' a great deal remains to be done before their properties can be controlled as for the III–V compounds with smaller values of ΔE.

Refractory semiconductors have been treated in some detail by Y. V. Shmartsev, Y. A. Valov and A. S. Borshchevskii.[76]

13.11 Superconducting semiconductors

A number of semiconductors have been reported as showing superconductivity at temperatures below 1 °K. There has in some instances, however, been some doubt cast on the interpretation of these observations. For example all three lead chalcogenides PbS, PbSe and PbTe have at various times been reported as being superconducting when highly non-stoichiometric and *n*-type, thus having a high concentration of electrons. This apparent superconductivity is now thought to be due to lead filaments in the material which become superconducting.

Superconductivity has been fairly well established for GeTe by R. A. Hein *et al.*[77] and also in SnTe. Both of these also show deviations from stoichiometry leading to carrier concentrations of the order of 10^{21} cm^{-3}.

The best-studied superconducting semiconductor is however the oxide $SrTiO_3$, which in the condition of oxygen defect shows superconductivity[78] at carrier concentrations in the range 5×10^{18} cm^{-3}–10^{21} cm^{-3}. A number of other mixed oxides have been investigated, such as WO_3, but these require addition of a few per cent of alkali metals in order to become superconducting. These and other materials have been discussed by H. P. R. Frederikse.[79]

A number of semiconductors become superconducting under high pressure. They then behave as metals rather than as semiconductors (see H. P. R. Frederikse *loc. cit.*). Examples are InSb, InTe and the element Te.

[76] *Refractory Semiconductor Materials* (Consultants Bureau, N.Y. 1966).
[77] *Phys. Rev. Lett.* (1964) **12**, 320.
[78] J. F. Schooley *et al.*, *Phys. Rev. Lett.* (1965) **14**, 305.
[79] *Electronic Structure in Solids*, ed. E. D. Haidemenakis (Plenum Press, 1969), p. 270.

The theory of superconductivity as applied to semiconductors has been given by V. L. Gurevich, A. I. Larkin and Y. A. Firsov,[80] and by M. L. Cohen.[81] The former deals primarily with polar semiconductors while the latter is more general and is in good agreement with the observations, particularly in giving the variation of transition temperature with carrier concentration.

13.12 Magnetic semiconductors

A number of materials of interest mainly for their magnetic properties are semiconductors. Most of these are oxides or chalcogenides and we have already discussed some of them in §13.9. NiO, for example, is antiferromagnetic as is CoO and FeO. They are low-mobility semiconductors (see §13.14). Various garnets and spinels which are ferrimagnetic have semiconducting properties with low values of mobility but not so low as for the antiferromagnetic oxides – of the order of $10^{-2}\ cm^2\ V^{-1}\ s^{-1}$. $FeCr_2S_4$ and Cr_2S_3 are semiconductors, again with fairly low mobility. More striking are a few ferromagnetic materials which are semiconductors, of which those most studied are EuO, EuS, EuSe and $CdCr_2Se_4$.

All these materials have either transition metal atoms with $3d$-electrons or rare-earth metals with $4f$-electrons. They have been treated in some detail by C. Haas.[82] The subject has also been reviewed by I. G. Austin and D. Elwell[83] who discuss the difficulties encountered in preparing and purifying high-quality single crystals of these materials. Minute quantities of ferromagnetic impurities such as Fe in some of the materials can have a marked effect and mask their intrinsic properties.

13.13 Organic semiconductors

A number of organic substances show semiconductivity. There has been considerable interest in these. Typical of the organic semiconductors is anthracene, which has received most detailed study. Others which have received some attention are the pthalocyanines. Work on these materials has been impeded by the difficulty of preparing pure compounds to anything like the standards available for the inorganic semiconductors. The compounds tend to be mixed with small quantities of related

[80] *Soviet Phys. Solid State* (1962) **4**, 131.
[81] *Phys. Rev.* A (1964) **134**, 442.
[82] *Electronic Materials*, ed. N. B. Hannay and U. Colombo (Plenum Press, 1973), p. 169.
[83] *Contemp. Phys.* (1970) **11**, 455.

compounds which may, however, have rather different properties. Another difficulty has been the preparation of good quality crystals of appreciable size, since these materials are normally soft and very fragile. Without great precautions crystals tend to be severely strained.

In spite of this, a good deal of information on the semiconducting properties of some of these materials has been obtained, mainly through the study of the variation of their conductivity with temperature and by measurements of their optical absorption, the latter not requiring large single crystals.

The properties of organic semiconductors have been discussed in some detail by H. Inokuchi and H. Akanto[84] and also by F. Gutman and L. E. Lyons.[85] Their behaviour under pressure has been described by H. G. Drickamer.[86]

13.14 Low-mobility semiconductors

We have already discussed briefly a number of instances in which the mobility of electrons and holes in semiconductors is low, or indeed very low: for example in the impurity conduction found at very low temperatures, in some of the oxides discussed in § 13.9 and in most of the organic semiconductors. In the oxides mobilities as low as 10^{-6}–10^{-8} cm^2 V^{-1} s^{-1} have been found but these are exceptional and confer insulating rather than semiconducting properties. Mobilities of the order of 1–10^{-2} cm^2 V^{-1} s^{-1} are more common. In amorphous materials mobilities are normally not greater than this, but we shall postpone discussion of liquid and amorphous semiconductors to Chapter 15.

In considering the low-mobility crystalline semiconductors it is important to realise that the theory based on Bloch waves in a crystal is not really applicable and one must use the ideas associated with hopping as briefly outlined in § 5.4 and which we shall discuss further in connection with amorphous semiconductors (see §§ 15.1, 15.2).

It is not difficult to see why the energy band theory, which has been so successful in elucidating the properties of high-mobility semiconductors, fails when the mobility is less than about 1 cm^2 V^{-1} s^{-1}. If we use the relationship $\mu = e\tau/m_e$, which does not depend on crystal structure or band theory, we find that if we take $m_e = \boldsymbol{m}$ and $\mu = 1$ cm^2 V^{-1} s^{-1} then $\tau = 1.6 \times 10^{-16}$ s.

[84] *Solid State Physics* (Academic Press, 1961) **12**, 93.
[85] *Organic Semiconductors* (Wiley, 1967).
[86] *Solid State Physics* (Academic Press, 1965) **17**, 1.

Since there will be an uncertainty in the electron's state in a time of order τ, we can express the uncertainty δE in energy through the Uncertainty Principle as $\delta E\tau \sim \hbar$. Inserting the value of τ we find that $\delta E \simeq 1$ eV. An uncertainty of energy of this amount makes band structure meaningless. Moreover the concept of mean free path also becomes meaningless (see § 15.1).

Low-mobility semiconductors and the problems associated with the determination of their properties have been discussed by A. F. Ioffe and A. R. Regel.[87] The limitations of band theory, especially as applied to narrow bands with high effective mass, have been discussed by H. Fröhlich and G. L. Sewell.[88] Transport properties of low-mobility semiconductors, including the hopping process, have been discussed by M. J. Klinger.[89] For a further discussion of the hopping process and references see § 5.4 and also our discussion of amorphous semiconductors (§§ 15.1, 15.2).

13.15 Other semiconductors

We have discussed only a few of the great variety of materials which are now known to behave as semiconductors. Many, but not all, of these are rather complex compounds, usually oxides or chalcogenides. Binary compounds have been fairly fully explored since they are limited in number. Yet only a few have been studied in anything like the detail of those we have discussed. As we have seen (§ 13.7) a start has been made with some ternary compounds and here there is a still greater variety of compounds to choose from. When one gets to quaternary and higher compounds the number available is simply enormous. A few have been studied but not in depth and it will indeed be surprising if there should not emerge one or two new materials with rather special properties. It must, however, be realised that the more elements in the compound the harder it will be to process to the degree of perfection required in semiconductor technology.

A number of lists of compounds together with such information as is available, usually only on the forbidden energy gap, have been prepared.[90]

A rather curious group of semiconducting materials have their minority-carrier lifetime τ_n (or τ_p) less than the dielectric relaxation

[87] *J. Phys. Chem. Solids* (1959) **8**, 6; *Prog. in Semiconductors* (Heywood, 1960) **4**, 237.
[88] *Proc. Phys. Soc.* (1959) **74**, 643.
[89] *Proc. Vth Int. Conf. on Phys. of Semiconductors* (Inst. of Phys. Lond., 1962), p. 205.
[90] See references in § 13.1; also B. Pamplin, *New Scientist* (1974) **64**, 739.

time τ_0 (see § 7.3). Their properties were discussed by W. Van Roosbroeck and H. C. Casey[91] and they have come to be known as relaxation semiconductors. As we have seen in § 7.3, τ_0 is usually of the order of 10^{-12} s and $\tau_n \gg \tau_0$ leading to the condition of space-charge neutrality. When this is no longer so, space-charge can build up and majority carrier depletion can come about.

Very little experimental work on these materials has been carried out since the original paper which suggested that highly compensated GaAs might behave in this way. The subject has recently been reopened by C. Popescu and H. K. Henisch,[92] who have considered the effects resulting from the injection of minority carriers into these materials. They have also investigated the effect of trapping centres at which these minority carriers may be held and their effect on recombination rates.

[91] *Proc. Xth Int. Conf. on Phys. of Semiconductors* (USAEC Publications, 1970), p. 832.
[92] *Phys. Rev.* (1975) B **11**, 1563; *J. Phys. Chem. Solids* (1975) **37**, 47; *Phys. Rev.* (1976) B **14**, 517.

14

Some special topics

14.1 Excitonic molecules

We have discussed in §§ 10.6 and 10.14 the attachment of single excitons to various impurity centres to form exciton complexes. The striking characteristic of these complexes is the radiative decay of the exciton when bound, to give strong emission in the form of narrow lines. Although it had been predicted in 1958 by M. A. Lampert[1] that a pair of excitons could combine to form an excitonic molecule, rather like H_2, it was not until 1966 that emission from such a molecule was observed by J. R. Haynes[2] in the spectrum of optically excited Si. Emission from Ge due to excitonic molecules was later found by C. Benoit à la Guillaume, F. Salvan and M. Voos,[3] and also in a variety of II–VI semiconductors such as CdS and CdSe as well as materials like CuBr. This work and subsequent theoretical investigations has been reviewed by E. Hanamura.[4]

Haynes estimated that the binding energy between two excitons to form a molecule would be about 2 meV, deducing this by comparison with the H_2 molecule. To a first approximation it might appear that the molecular binding energy W_M is related to the single-exciton binding energy W_{ex} by the simple expression $W_M/W_{ex} = W_{H_2}/W_H = 0.33$. However, because of the similarity of the effective masses of the holes

[1] *Phys. Rev. Lett.* (1958) **1**, 450.

[2] *Phys. Rev. Lett.* (1966) **17**, 860. There is now some doubt as to whether the radiation observed by Haynes does in fact come from excitonic molecules. It has been suggested that it is more likely to be due to condensed excitons (see § 14.2). The same may be true of the radiation from Ge. The evidence for radiation from excitonic molecules in CuBr and such materials is, however, much stronger (private communication, T. I. Galkina).

[3] *Proc. Xth Int. Conf. on Phys. of Semiconductors* (U.S. Atomic Energy Commission, 1970), p. 516.

[4] *Optical Properties of Solids – New Developments*, ed. B. O. Seraphin (North Holland, 1976), p. 81.

and electrons, the problem of calculating the binding energy is not exactly the same as for a H_2 molecule. O. Akimoto and E. Hanamura[5] showed, however, that such a molecule would be stable for all values of the hole/electron mass ratio. For a many-valley semiconductor such as Ge or Si the calculation of the binding energy is quite complicated. This has been carried out by O. Akimoto[6] and others and is discussed by E. Hanamura (*loc. cit.*). The binding energy of the molecule appears to be proportional to the exciton binding energy, but the constant of proportionality varies vetween 0.01 and 0.3 depending on the ratio m_e/m_h, being largest for small values of the ratio (as would be expected from the analogy with the hydrogen molecule) and also largest when the anisotropy is greatest. It appears therefore that the bi-exciton is energetically metastable but whether there are other groups of excitons which are still more stable is another question to which we shall return later. It appears that in materials such as CuCl and CdS the conditions are more favourable than in Si and Ge for the formation of single molecules rather than groups, and that is probably why the excitonic molecule has been observed more readily in these materials.

It has been proposed by J. S. Wang and C. Kittel[7] that because of the multi-valley structure of Si and Ge molecules Ex_8 and Ex_{12} should exist and be more stable than Ex_2. There seems, however, to be no good experimental evidence at present for the existence of these molecules. On the other hand, as we shall see in § 14.2, much larger groups of excitons can gather together to form droplets of macroscopic size.

There is evidence that the bi-exciton can also be bound to a neutral impurity such as N in GaP. For example J. L. Merz, R. A. Faulkner and P. J. Dean[8] have observed sharp lines in the photo-luminescence of GaP which they interpret as due to the radiative decay of bi-excitons bound to N impurities.

14.2 Condensation of excitons into electron–hole drops

Excitons being zero-spin particles (the electron and hole having their spins anti-parallel as in the lowest state of the hydrogen atom), one might expect them to condense to form a dense formation like a liquid (Bose condensation) at low temperature. This would be analogous to liquid H_2. There is some evidence (N. Nakata *et al.*[9]) that such conden-

[5] *J. Phys. Soc. Japan* (1972) **33**, 1537; *Solid State Commun.* (1962) **10**, 253; *ibid.* (1962) **11**, xiii.
[6] *J. Phys. Soc. Japan* (1973) **35**, 973.
[7] *Phys. Lett.* (1972) **42**A, 189.
[8] *Phys. Rev.* (1969) **188**, 1228.
[9] *J. Phys. Soc. Japan* (1975) **38**, 903.

sations may occur in highly polar materials like CuCl and CuBr. For materials such as Si and Ge, however, it appears that a lower-energy state is obtained when the individual excitons break up and form a dense electron–hole plasma. This would be analogous to liquid metallic hydrogen if it existed. There is in fact a critical temperature T_c below which a condensation of molecules would change over to a 'metallic' complex, this being a kind of Mott transition as discussed in § 13.9.

So far, such a transition has not been observed, most work on exciton condensations having been carried out at the temperature of liquid helium, estimated to be well below T_c. That excitons should condense to form a 'metallic' phase was suggested by L. V. Keldish[10] in 1968. The phenomenon has since been observed and studied by many experimenters and has indeed provided one of the most vigorous fields of research in solid-state physics in recent years.

In calculating the binding energy per exciton, the 'metallic' state is assumed to be in equilibrium with a 'gas' of free excitons. Various attempts have been made to obtain this binding energy as a function of the mean separation r of the particles in the plasma, for example by W. F. Brinkman *et al.*[11] and also by M. Combescot and P. Nozières.[12] The main result of these calculations is to show that for Ge the binding energy per exciton has a well-defined maximum at a value of r equal to r_m, the value being 1.7 meV relative to the free-exciton binding energy of 3.6 meV (see § 10.6, p. 333). It is interesting to note that if the calculation is carried out for a simple semiconductor with equivalent isotropic electron and hole masses only a very shallow minimum is obtained. The value of r_m corresponds to an exciton density n_0 of $1.8 \times 10^{17}\ \text{cm}^{-3}$. A number of other calculations have been made and these have been reviewed by M. Voos and C. Benoit à la Guillaume.[13]

In most of the experimental work on the 'metallic' phase the exciton density required has been produced by illumination with laser radiation but in some instances electron injection from an n^+–p junction has been used. The first evidence for the formation of electron–hole droplets came from the observation of additional broadened lines to the long-wave side of the free-exciton emission lines in Ge and Si, by Y. E. Pokrovsky and K. I. Svistunova,[14] respectively. The evidence for a

[10] *Proc. XIth Int. Conf. on Phys. of Semiconductors* (Nauka, 1968), p. 1303.
[11] *Phys. Rev. Lett.* (1972) **28**, 1961; *Phys. Rev.* (1973) B**7**, 1508; *ibid.* (1973) B**8**, 1570.
[12] *J. Phys. C: Solid State Phys.* (1972) **5**, 2369.
[13] *Optical Properties of Solids – New Developments*, ed. B. O. Seraphin (North Holland, 1976), p. 143.
[14] *JETP Lett.* (1969) **9**, 261; *Soviet Phys. Semicond.* (1970) **4**, 409.

'metallic' phase comes mainly from the line shape, which corresponds to a value of n_0 of the order of 2×10^{17} cm^{-3}.

Evidence for the existence of electron–hole droplets and an estimate of their size has been obtained in a very simple way. If the droplets are created in the *n*-type region of a *p–n* junction, when one of them reaches the junction it will break up in the strong electric field there. A large current pulse should then be produced which should show up in the current produced by the junction photo-voltage. This kind of experiment was first carried out successfully by V. M. Asnin, A. A. Rogachev and N. I. Sablina[15] but has been repeated under a variety of conditions by a number of others (see M. Voos and C. Benoit à la Guillaume, *loc. cit.*). These experiments indicate droplet radii in Ge of about 1–10 μm. The effect shows up at a temperature of about 2 °K but not at 15 °K, indicating that at the higher temperature the droplets have evaporated to form a free-exciton gas. This is also confirmed by the spectroscopic measurements.

A beautiful experiment has been carried out by Y. E. Pokrovsky and K. I. Svistunova,[16] by V. S. Bagaev *et al.*,[17] and by a number of others who have used Rayleigh scattering of 3.39 μm radiation from a He–Ne laser to show up the presence of exciton–hole droplets and to measure their radii. Values found are again in the region 2–8 μm. V. S. Bagaev *et al.* (*loc. cit.*) found that the drop radius *increases* with temperature, varying from about 4 μm to 10 μm as T is increased from 2.6 °K to 3.2 °K.

By studying far infra-red emission and absorption by droplets V. S. Vavilov, V. A. Zayats and V. N. Murzin,[18] by relating the maximum absorption to emission at $h\nu = 8.7$ meV, have estimated that $n_0 = 2\times10^{17}$ cm^{-3}, in good agreement with other measurements.

While most of the experiments have shown droplets of size 1–10 μm a few very large drops have been reported, for example by B. J. Feldman.[19] There appears to be some doubt, however, whether these are nearly spherical collections of many much smaller droplets. If an inhomogeneous stress pattern is applied to Ge a localized region can be produced which favours condensation relative to the rest of the crystal. By this means C. D. Jeffries, J. P. Wolfe and R. S. Markiewicz[20] have produced drops of radius 0.3 mm with $n_0 = 0.7\times10^{17}$ cm^{-3}.

[15] *JETP Lett.* (1970) **11**, 99. [16] *JETP Lett.* (1971) **13**, 212.
[17] *Soviet Phys. – Solid State* (1974) **15**, 2179.
[18] *Proc. Xth Int. Conf. on Phys. of Semiconductors* (USAEC Publ., 1970), p. 509.
[19] *Phys. Rev. Lett.* (1974) **33**, 359.
[20] *Proc. XIIIth Int. Conf. on Phys. of Semiconductors* (Tipographia Marves, 1976), p. 879.

The application of stress tends to lower the binding energy, as shown both by the calculations discussed above and by experiments. V. S. Bagaev, T. I. Galkina and O. V. Gogolin[21] found a reduction in the binding energy as shown by the shift of the peak in the emission spectrum due to exciton condensation. They also found that the emission line splits in a high magnetic field of the order of 10 T due to the spin splitting of the electron and hole energies.

Various values for the lifetime of droplets have been given. On removal of the excitation they decay in times of the order of 40 μs. With the larger droplets, having lower exciton density, Jeffries Wolfe and Markiewicz (*loc. cit.*) have found decay times of the order of 200 μs. Small droplets tend to diffuse but there is a very large variation in the values reported for the diffusion length. Under certain conditions, however, they appear to be able to spread to distances of the order of 1 mm. A. S. Alekseev, V. S. Bagaev and T. I. Galkina[22] have shown that diffusion is aided by a stress field.

While most of the experimental work has been carried out on Ge, some has been done with Ge–Si alloys. Both theoretically and experimentally the binding energy is less for these. The 'indirect' gap and anisotropy of the electron effective mass appear to make Ge particularly suitable for droplet formation. 'Direct' gap semiconductors seem to be unfavourable, but R. F. Lekeny and J. Shah[23] have reported the observation of electron–hole droplets in CdS and CdSe. The binding energy in CdS is stated to have the surprisingly high value of 13 meV while for CdSe it is less than 2 meV.

14.3 Polarons and polaritons

Polarons and polaritons, although similarly named, are quite different entities. Polarons are free electrons or holes moving in a material polarized by their presence and carrying this polarization with them as they move. Polaritons are excitations which come about through the interaction of light or infra-red radiation with other excitations such as phonons or excitons. We shall briefly consider polarons first of all.

In a polar medium a free electron will repel the negative ions and attract the positive ions, thus polarizing the material round about it. For a polar crystal the range of this polarization caused by the Coulomb field

[21] *Proc. Xth Int. Conf. on Phys. of Semiconductors* (USAEC Publ., 1970), p. 500.
[22] *Soviet Phys. – JETP* (1972) **36**, 536.
[23] *Proc. XIIIth Int. Conf. on Phys. of Semiconductors* (Tipographia Marves, 1976), p. 394.

of the electron can be quite large. As the electron moves it will carry this polarization with it and this will affect its effective mass. The theory of this effect was given by H. Fröhlich, H. Pelzer and S. Zienau[24] as long ago as 1950 but has been updated and reviewed by H. Fröhlich.[25]

When the interaction parameter α which we have defined in equation (28) of § 8.6 is small an expression for the modified effective mass m_e^* is given by the expression

$$m_e^* = m_e(1+\tfrac{1}{6}\alpha). \tag{1}$$

For the III–V compound semiconductors α is quite small so that the effective mass is not much affected. For PbS $\alpha = 0.28$ and again the effect is fairly small, although in this case the theory is not strictly valid. For strongly polar crystals α is larger still and a more sophisticated theory is necessary. The polaron is a manifestation of the interaction of an electron with the optical phonons and has been discussed from this point of view by P. G. Harper, J. W. Hodby and R. Stradling.[26] A whole volume of papers on the subject has been published dealing not only with alkali halides, for which the concept of the polaron was initially developed, but also with polar semiconductors.[27]

We now turn to a brief discussion of polaritons. As a first example we consider the interaction of photons with the T0 mode of the lattice vibrations. As we have seen, the energy $\hbar\omega$ of the T0 mode near $\boldsymbol{k} = 0$ does not vary much with k and we represent ω in Fig. 14.1 as a horizontal line. For the photon, on the other hand, ω is equal to $\epsilon_0^{\frac{1}{2}}kc/\epsilon^{\frac{1}{2}}$ (we use here $\boldsymbol{k}$ as the wave-vector both for photon and phonon). The phonon and photon dispersion curves intersect when $k = \epsilon^{\frac{1}{2}}\omega_{\mathrm{T0}}/c\epsilon_0^{\frac{1}{2}}$. As is general in quantum theory, such curves do not cross but behave as shown by the broken curves in Fig. 14.1. For small and large values of k the photon and phonon behave as separate entities, but near $k = \epsilon^{\frac{1}{2}}\omega_{\mathrm{T0}}/c\epsilon_0^{\frac{1}{2}}$ they are strongly coupled and form what is known as a polariton.

In fact Fig. 14.1 is an over-simplification of the situation for polar compounds since we have two distinct optical phonons at $\boldsymbol{k} = 0$, the T0 and L0 phonons. The situation which generally exists in crystals with some polar binding is shown in Fig. 14.2 where, according to the

[24] *Phil. Mag.* (1950) **41**, 221.

[25] *Polarons and Excitons*, ed. C. G. Kuper and G. D. Whitfield (Oliver and Boyd, 1962), p. 1.

[26] *Rept. Prog. Phys.* (1973) **36**, 1.

[27] *Polarons in Ionic Crystals and Polar Semiconductors*, ed. J. T. Devreese (North Holland, 1972).

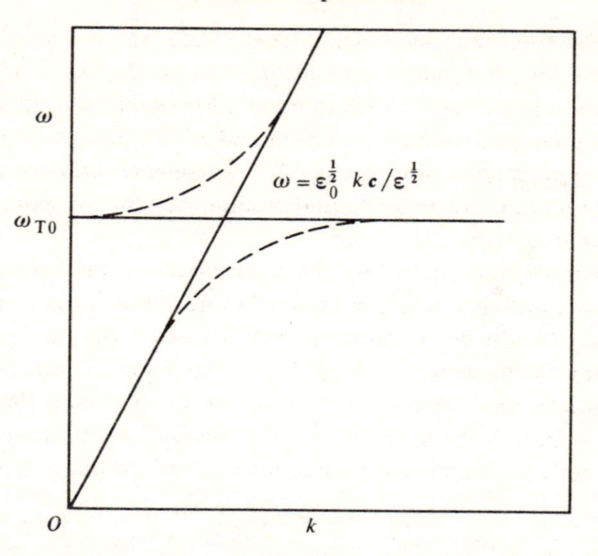

Fig. 14.1. Illustrating interaction of a photon with a T0 phonon to form a polariton. The dashed curves show the polariton dispersion.

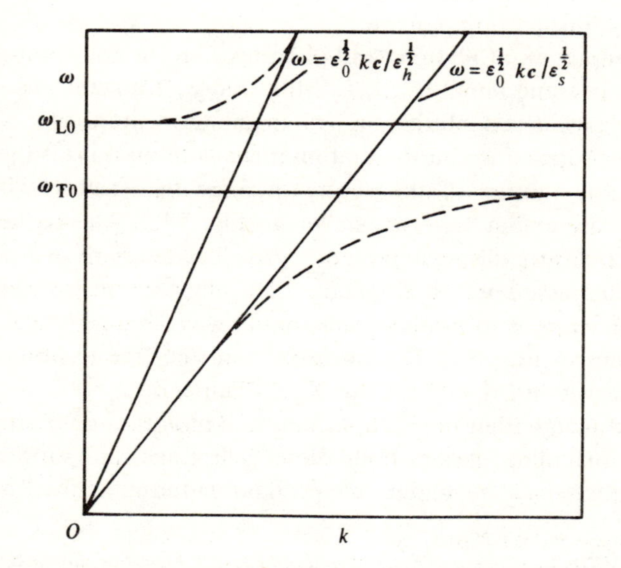

Fig. 14.2. Interaction of a photon with T0 and L0 phonons in a polar crystal. The dashed curves show the polariton dispersion.

Lyddane–Sachs–Teller relationship (see § 8.3), $\omega_{LO}^2 = \epsilon_s \omega_{TO}^2/\epsilon_h$, ϵ_s and ϵ_h being the low- and high-frequency permittivities. For small values of k we have a photon and an L0 phonon, with which it hardly interacts. For large values of k we have a photon and a T0 phonon and here again the interaction is small; but when $\epsilon_0^{\frac{1}{2}}kc/\epsilon^{\frac{1}{2}}$ is near to ω_{TO} or ω_{LO} we get strong interaction and two separate frequencies of the polariton for a given value of k.

The polariton thus shows up the disturbance in the refractive index between ω_{TO} and ω_{LO}, leading to the strong reflection and absorption associated with the *reststrahl* frequency to which we have referred in § 8.3. Using the polariton concept D. L. Mills and E. Burstein[28] have given a simple and elegant derivation of the Lyddane–Sachs–Teller relationship and its connection with the ionicity of the material. They have also derived the polariton dispersion relationship, in the form

$$\omega^2 = \frac{1}{2}\left(\frac{c^2k^2\epsilon_0}{\epsilon_h} + \omega_{TO}^2 + \omega_p^2\right) \pm \frac{1}{2}\left[\left(\frac{c^2k^2\epsilon_0}{\epsilon_h} - \omega_{TO} - \omega_p\right)^2 + \frac{4c^2k^2\epsilon_0\omega_p^2}{\epsilon_h}\right]^{\frac{1}{2}} \tag{1}$$

with
$$\omega_{LO}^2 = \omega_{TO}^2 + \epsilon_0\omega_p^2/\epsilon_h = (\epsilon_s/\epsilon_h)\omega_{TO}^2. \tag{2}$$

ω_p is the ion plasma frequency $e^*(N/M_r\epsilon_0)^{\frac{1}{2}}$ (cf. equation (45) of § 10.3), where e^* is the effective ionic charge, M_r the reduced mass and N the number of ions per unit volume.

The interaction of photons and phonons to form polaritons has been discussed in some detail by D. L. Mills and E. Burstein (*loc. cit.*) who have also considered interaction with magnetic excitations.

Another form of excitation that interacts with photons is the exciton. Through the motion of the centre of mass the exciton will have a parabolic dispersion curve as shown in Fig. 14.3. The dashed curves show the polariton dispersion relationship. The decay of such polaritons has been discussed by J. J. Hopfield.[29] The effect of the lower part of the dispersion curve is to produce emission at wavelengths longer than the low-frequency cut off of the emission from the free exciton. This has been observed in CdS at 12 °K by W. C. Tait *et al.*[30]

A great many other interactions between photons and excitations are possible, including plasma oscillations (plasmons) and surface waves. Many of these have been identified by light scattering of the Raman type

[28] *Rept. Prog. Phys.* (1974) **37**, 817.
[29] *Proc. VIIIth Int. Conf. on Phys. of Semiconductors, J. Phys. Soc. Japan*, Suppl. (1966) **21**, 77.
[30] *II–VI Semiconducting Compounds*, ed. D. G. Thomas (Benjamin, 1967), p. 370.

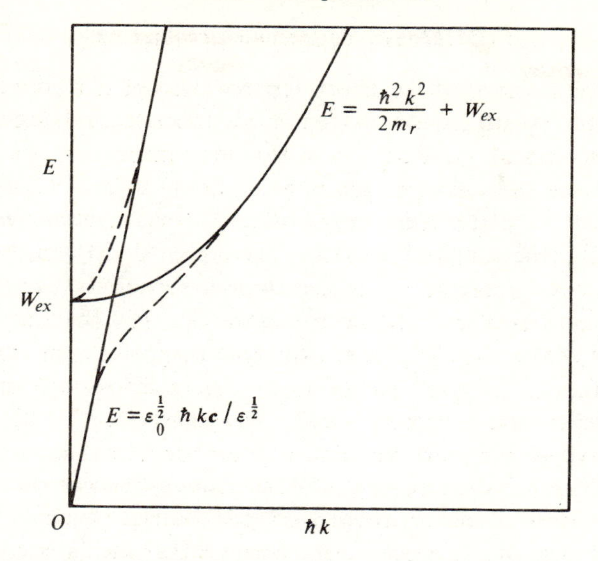

Fig. 14.3. Interaction of a photon with a free exciton to form a polariton. The dashed curves show the polariton dispersion.

in which the energy of the polariton is either taken from or added to the energy of the scattered light quantum[31] (see §14.7). The subject has been treated extensively in a series of papers published in book form.[32]

There are other interactions of this type. One is similar to a polariton but involves radiation, usually in the microwave region of the spectrum, free electrons or holes and optical phonons. This is the interaction between cyclotron resonance and optical phonons. When the cyclotron frequency approaches the frequency of a long-wave optical phonon there is a strong interaction between the two. This shows up as a splitting of the cyclotron absorption line and also a sudden broadening of the line. A simple theoretical treatment and review of the experimental work on this interesting phenomenon has been given by R. F. Wallis.[33]

Interaction between phonons and plasmons has also been studied by means of Raman scattering. This interesting interaction has been described by A. L. McWhorter[34] and by P. M. Platzman and P. A. Wolff.[35]

[31] See, for example, C. H. Henry and J. J. Hopfield, *Phys. Rev. Lett.* (1965) **15**, 964.
[32] *Polaritons*, ed. E. Burstein and F. de Martini (Pergamon Press, 1974).
[33] *Optical Properties of Solids*, ed. E. D. Haidemenakis (Gordon and Breach, 1970).
[34] *Electronic Structure in Solids*, ed. E. D. Haidemenakis (Plenum Press, 1969), p. 350.
[35] *Solid State Physics* (Academic Press, 1973) Suppl. 13.

14.4 Heavily-doped semiconductors

For many years most of the effort in preparation of semiconductors for experimental investigations and for use in electronic and optical devices, apart from crystal growing, consisted in elimination of undesired impurities and the uniform insertion of those desired. Generally the quantities of the latter were very small and special techniques, such as first 'doping' the material and then using this 'doped' material as the impurity to be inserted, were developed. Generally the donor and acceptor concentrations used were a good deal less than the quantities N_c and N_v defined in § 4.2 but still large compared with the intrinsic concentration n_i at room temperature. The main use of much higher concentrations was to form n^+- and p^+-junctions (see § 7.10); transport in the heavily-doped sections led to no new problems so was not much studied. With the development of Esaki diodes (see below) and semiconductor lasers, however, heavily-doped material became a sensitive and active part of a device and in recent years much greater attention has been paid to the transport and optical properties of heavily-doped materials, i.e. materials for which $p \geqslant N_v$ or $n \geqslant N_c$. Indeed, several books have been devoted to these materials alone (see below).

For such heavily-doped materials we can no longer use the simple exponential form given by § 4.3, equations (29), (30), for the concentrations n and p and have to use expressions derived in a similar way but based on the Fermi–Dirac form of the distribution function. How this is done is indicated in § 4.2, equation (19). We shall also have to modify the expressions for the average values of such quantities as $\langle \tau \rangle$ and this leads to much more complex expressions for the transport coefficients, particularly those for the thermo-magnetic effects. We shall not write down expressions for these transport coefficients. The full analysis leading to their derivation has been given in a number of review articles and books, including those by V. L. Bonch-Bruevich[36] and by V. I. Fistul.[37] We shall simply consider one or two *new* phenomena which have come to light as a result of investigations on the properties of heavily-doped semiconductors.

14.4.1 Screening by free carriers

One immediate effect of the increased electron and hole concentrations is to modify the energy levels due to donor and acceptor impurity

[36] *The Electronic Theory of Heavily Doped Semiconductors* (Elsevier, 1966).
[37] *Semiconductors and Semimetals* (Academic Press, 1966) **1**, 101; *Heavily Doped Semiconductors* (Plenum Press, 1969).

centres. This applies particularly to the 'hydrogenic' type of impurities which we have discussed in § 3.4.1. For impurities whose energy levels lie further from the conduction or valence bands the forces acting on the 'extra' electron or hole are of shorter range and are less affected by the free electrons due to other impurities. For the 'hydrogenic' type the Coulomb field is screened by the other free electrons and instead of moving in a potential $-e^2/4\pi\epsilon r$ the potential energy $V(r)$ is given by

$$V(r) = -\frac{e^2}{4\pi\epsilon r}\exp\left(-\frac{r}{r_0}\right) \tag{3}$$

where r_0 is the so-called Debye radius given by

$$r_0^2 = \epsilon kT/ne^2. \tag{4}$$

This we have already obtained in connection with ionized impurity scattering (see § 8.9). In such a potential there may be a small number of bound states or none, and if r_0 is small enough the impurity will no longer be able to bind an electron. The electrons will therefore all be free and will form an impurity band. The conditions for this have been discussed by L. Pincherle[38] and also by V. L. Bonch-Bruevich (*loc cit.*). Experimental investigations including the use of high carrier concentrations injected into lightly-doped semiconductors have been carried out by V. B. Glasko and A. G. Mironov[39] and also by V. L. Bonch-Bruevich.[40] These give verification of the theory of screening but also indicate that there are further problems when the doping is very heavy. The general result is, however, that for impurities such as group V donors in Ge, which have ionization energies of the order of 0.01 eV when n is less than $10^{16}\,\text{cm}^{-3}$, this energy steadily decreases as n increases, becoming zero as n approaches N_c (i.e. about $10^{19}\,\text{cm}^{-3}$).

The properties of Ge when heavily doped have been described in some detail by J. I. Pankove.[41]

14.4.2 The Burstein–Moss effect

Another interesting physical effect arising from heavy doping is the change in the position of the fundamental absorption edge as the electron or hole concentration becomes large. The absorption edge for

[38] *Proc. Phys. Soc.* A (1951) **64**, 663.
[39] *Soviet Phys. Solid State* (1962) **4**, 241.
[40] *Soviet Phys. Solid State* (1961) **3**, 558.
[41] *Progress in Semiconductors* (Heywood, 1965) **9**, 47.

n-type InSb was observed by E. Burstein[42] and by T. S. Moss[43] to be particularly sensitive to doping and moved to shorter wavelengths as the doping concentration was increased. The explanation of this phenomenon was given independently by Burstein and by Moss and the effect is generally known as the Burstein–Moss effect.

Because of the very light electron mass in InSb its conduction band states fill up rapidly as n is increased, the Fermi level moving above the bottom of the conduction band. This is illustrated in Fig. 14.4. The

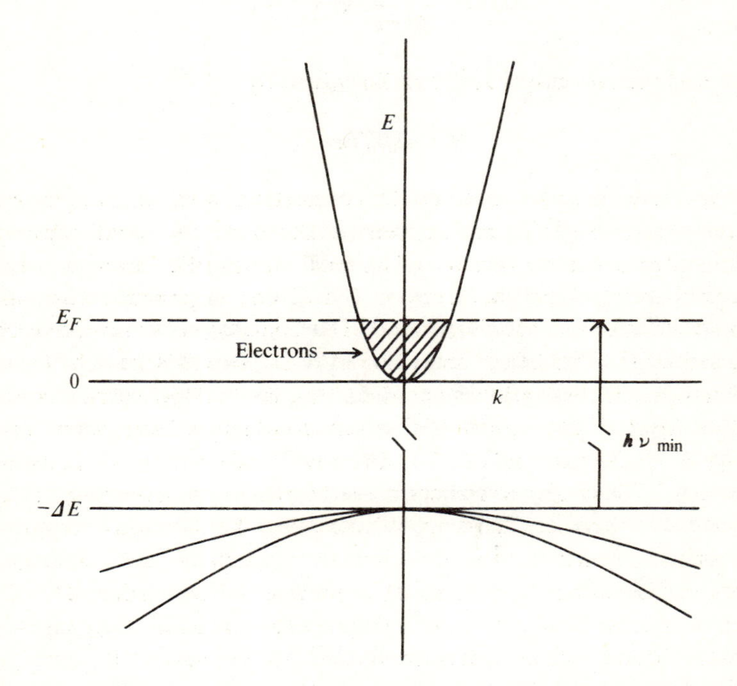

Fig. 14.4. Effect of heavy doping in InSb on low-frequency limit of fundamental absorption edge.

states below the Fermi level, being occupied, are not available for transitions from the top of the valence band so that the lowest energy transition is not to the foot of the conduction band, as for lightly-doped materials, but to the Fermi level. As this rises with increasing values of n so the absorption edge moves to shorter wavelengths. For *p*-type InSb, because of the heavier hole masses, the effect is much less marked.

[42] *Phys. Rev.* (1954) **93**, 632.

[43] *Proc. Phys. Soc.* B (1954) **67**, 775.

The effect has been observed in a number of semiconductors including n-type Ge. For this the detailed theory has been given by J. I. Pankove and P. Aigrain[44] and compared with experiment. For values of n in excess of 10^{19} cm^{-3} an increase in slope of the absorption edge was also found, thought to be due to the effect of scattering processes at these high carrier concentrations. These may remove the need for phonons to conserve energy in an indirect transition such as that in Ge.

14.4.3 Tunnelling in heavily-doped semiconductors

One of the most interesting new effects due to heavy doping is the marked change in the shape of the voltage–current characteristic of a p–n junction when both the n-type and p-type sections are heavily doped. This was first discovered by L. Esaki[45] and such junctions are generally known as Esaki diodes. The voltage–current characteristic of such a junction is shown for the 'forward' direction in Fig. 14.5.

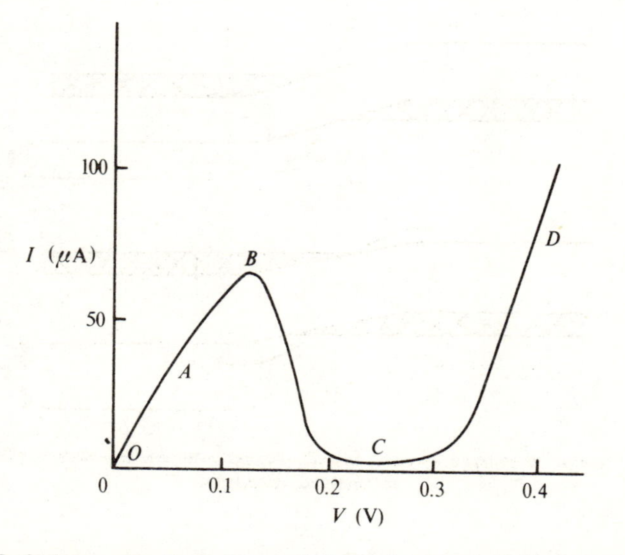

Fig. 14.5. Typical current–voltage characteristic (I/V) in 'forward' direction for heavily-doped Si p–n junction (Esaki diode). The regions marked *O, A, B, C, D* correspond to Fig. 14.6 *o, a, b, c, d* respectively.

The point O corresponding to zero bias will give the band condition as illustrated in Fig. 14.6(o). This should be compared with Fig. 7.8 for a

[44] *Phys. Rev.* (1962) **126**, 956.

[45] *Phys. Rev.* (1958) **109**, 603.

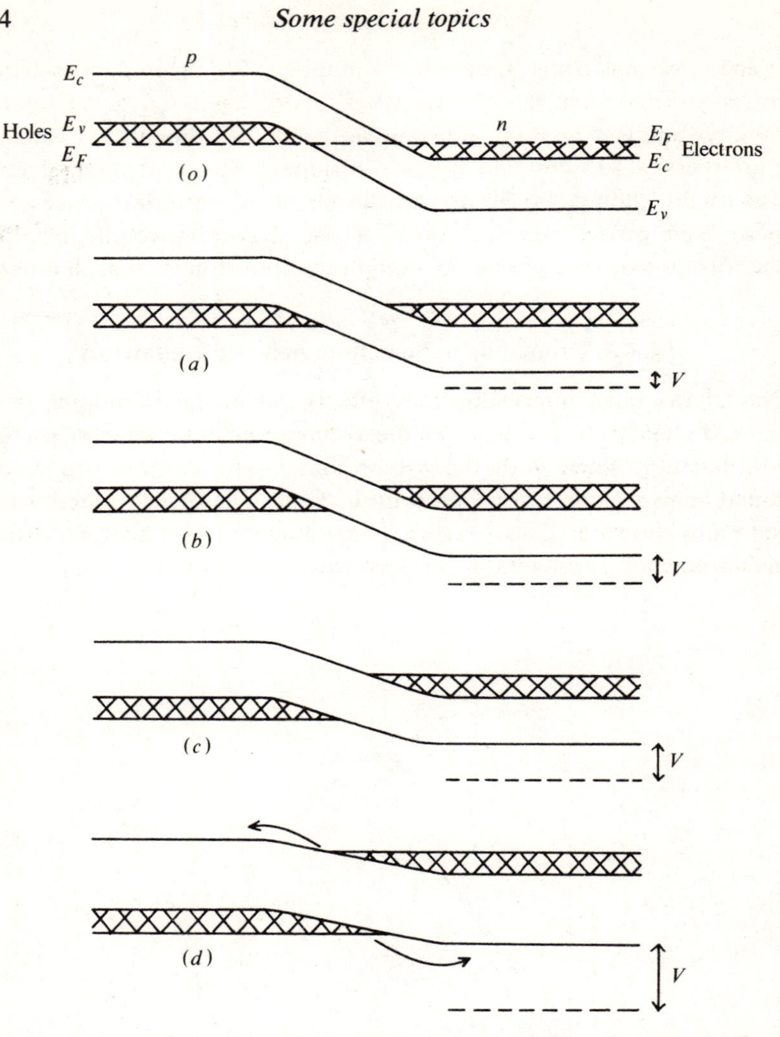

Fig. 14.6. Variation of band levels in heavily-doped p–n junction with applied voltage V in 'forward' direction.

lightly-doped junction. It will be seen that the energy difference between E_c or E_v in the n and p parts far from the junction is now *greater* than the forbidden energy gap ΔE because of the degeneracy of the n and p sections. This implies that excitation of carriers over the potential hump, when a voltage is applied, will be less than in a lightly-doped junction. Passage of carriers between the two sides does, however, take place by tunnelling through the junction region, provided that states of equal energy are *available*.[46] In the equilibrium condition most of the electron states in the *p*-type region opposite electron states in the *n*-type are occupied (empty of holes) and only a small amount of exchange will take place. As soon as a voltage is applied, however, some occupied electron states in the *n* side will be opposite empty electron states (filled hole states) in the *p* side. This is illustrated by Fig. 14.6(*a*) and leads to a steady increase in the forward current as at *A* in Fig. 14.5. This increase continues until the holes in the *p* section fully overlap with the electrons in the *n* section, as shown in Fig. 14.6(*b*). A further increase in *V* leads to a decrease in this overlap as shown in Fig. 14.6(*c*). The current *I* therefore reaches a maximum shown at *B* in Fig. 14.5, then falls rapidly. It is this negative slope in the *I*/*V* curve that gives the Esaki diode its practical importance. Because the electron and hole concentrations do not cut off sharply at the Fermi level there is some overlap of their occupied states causing some current at *C* which continues through a flat minimum until the voltage is sufficiently increased for the normal *p–n* junction mechanism of thermal excitation over the decreasing potential hump to set in as illustrated in Fig. 14.6(*d*). This causes the rapid increase in current as shown at *D* in Fig. 14.5.

Tunnel diodes have been made and studied with a variety of semiconductors besides Ge and Si, for example GaSb and GaAs. Their characteristics are very similar, apart from changes in the position of the maximum at *B* and minimum at *C* in Fig. 14.5.

A theory of the tunnelling process has been given by L. V. Keldish[47] and elaborated by T. Takeuti and H. Funada.[48] These theories make quantitative the simple description given above and also introduce some finer points which we have omitted in our description of the phenomenon.

Since the tunnelling process does not depend much on the temperature and since the Fermi level in a highly-degenerate situation is also insensitive to temperature variations, the voltage–current characteristic up to the minimum will be very much the same at low temperatures. Here, however, a new phenomenon is observed. Instead of a smooth characteristic as shown in Fig. 14.5, sudden increases in tunnelling current are observed at voltages which have been identified with various excitations, especially phonons, and this has led to a new subject, *tunnelling spectroscopy*, which we shall discuss in § 14.4.4. The

[46] See *W.M.C.S.* §§ 2.10, 6.5.

[47] *Zh. Exp. Teor. Fiz.* (1957) **33**, 994.

[48] *J. Phys. Soc. Japan* (1965) **20**, 1854.

subject of tunnelling in p–n junctions has been reviewed by L. Esaki[49] and a series of papers on various aspects of the subject published in a book *Tunneling Phenomena in Solids.*[50]

14.4.4 Electron tunnelling spectroscopy

When the characteristics of a Si tunnel diode were examined by L. Esaki and Y. Miyahara[51] and by N. Holonyak *et al.*,[52] four discontinuities were found in the I/V characteristic which corresponded closely in energy to the discontinuities in the absorption spectrum of Si caused by phonon assistance of the indirect transitions (see § 10.5.3). These were interpreted as phonon-assisted tunnelling. It is not surprising that such phonon assistance should aid the tunnelling process in an indirect-gap semiconductor, since crystal momentum must be exchanged in passing from the maximum of the valence band to the minimum of the conduction band.

From these observations has grown the subject of tunnelling spectroscopy. By observing the *slope* of the I/V characteristic directly, i.e. by measuring dI/dV as a function of V, these discontinuities show up much more clearly and this technique is now generally employed; even the second derivative d^2I/dV^2 is displayed directly to show up even more fine structure.

A typical I/V characteristic for a tunnel diode at liquid helium temperature is shown in Fig. 14.7. It should be compared with Fig. 14.5 which represents the behaviour at room temperature.

The theory of phonon-assisted tunnelling has been treated by L. V. Keldish,[53] by E. O. Kane[54] and by L. Kleinman.[55] The subject has also been reviewed by C. B. Duke[56] who deals with a rather wider range of solids, including metals.

Metal–semiconductor junctions and metal oxide–semiconductor junctions have also shown up discontinuities in the I/V characteristics and are widely used to study phonons and other excitations. These include Landau levels (see § 12.5.1) which show up when the junction is operated in a high magnetic field. This is to be expected because of the

[49] *Electronic Structure in Solids*, ed. E. D. Haidemenakis (Plenum Press, 1969), p. 1.
[50] *Tunneling Phenomena in Solids*, ed. E. Burstein and S. Lundquist (Plenum Press, 1969).
[51] *Solid State Electron.* (1960) **1**, 13.
[52] *Phys. Rev. Lett.* (1959) **3**, 167.
[53] *Soviet Phys. – JETP* (1958) **7**, 665.
[54] *J. Appl. Phys.* (1961) **32**, 83.
[55] *Phys. Rev.* (1965) **140**, A637; *Tunneling Phenomena in Solids*, ed. E. Burstein and S. Lundquist (Plenum Press, 1969), p. 181.
[56] *Solid State Physics* (Academic Press, 1969) Suppl. 10.

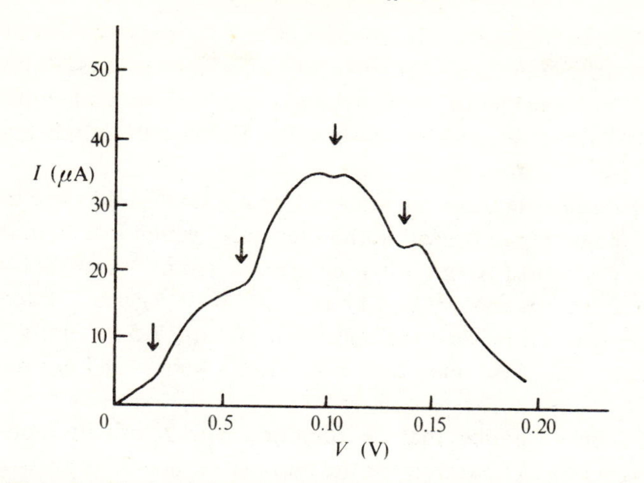

Fig. 14.7. Discontinuities in *I*/*V* characteristic of tunnel diode at liquid He temperature due to phonon-assisted tunnelling.

drastic change produced in the density of states. Plasma oscillations (see § 10.3) also show up as discontinuities in the *I*/*V* characteristic and even certain impurity levels cause sudden changes in the tunnelling current (L. Esaki, *loc. cit.*). The subject has recently been treated in detail by E. L. Wolf[57] who has given a review of the theory and experimental results on a wide range of materials.

14.5 High-pressure effects

A technique which has shed a great deal of new light on the band structure of solids is the application of high pressures. This is particularly so for semiconductors, where the electronic properties depend very markedly on the band structure and, in particular, on the value of the forbidden energy gap. That the latter should change significantly with application of pressure is to be expected since it depends sensitively on the atomic spacing which will change as pressure is applied. At first sight it might appear that the energy gaps would increase with pressure since a closer spacing of the crystal atoms usually means tighter binding. However, this is not found to be so and where we have several gaps due to multiple minima in the conduction band some increase and some decrease with pressure. For semiconductors with 'direct' gaps ΔE generally increases with pressure but there are some notable exceptions,

[57] *Solid State Physics* (Academic Press, 1975) **30**, 2.

namely Te and the lead chalcogenides. The different behavior of minima at $\boldsymbol{k} = (000)$, and in the $\langle 111 \rangle$ and $\langle 100 \rangle$ directions in crystals having the diamond and zincblende structures has enabled several doubts to be resolved in the relationship between the III–V compounds and Si and Ge.

Before discussing these we must indicate a small difference in units of pressure measurements used in the literature, which is now quite extensive. The units most frequently used are 10^3 kg cm^{-2} and kbar, and they are related approximately as 1 kbar $\simeq 1.02 \times 10^3$ kg cm^{-2}. Except when high accuracy of pressure measurement is required the units are practically the same. (We may also note that 1 kbar = 987 normal atmospheres.)

Quite marked effects may be obtained with relatively modest pressures, say up to 15 kbar, but for the more striking effects pressures up to about 500 kbar are required. While it is possible to get fairly evenly distributed hydrostatic pressure up to about 100 kbar, above this there is some uncertainty. The techniques for obtaining various pressures and their measurement have been described in detail in a number of books on the subject.[58]

Using pressures up to 12 kbar, which are quite readily available with fairly simple equipment, R. W. Keyes[59] has shown that the conductivity of p-type InSb ($N_a \simeq 10^{17}$ cm^{-3}) falls rapidly with pressure at 85 °C while at 0 °C it remains fairly constant. This shows that at 85 °C and zero pressure it is intrinsic, the change in conductivity being mainly due to the increase of ΔE with pressure, while at 0 °C it is extrinsic, the conductivity being due mainly to ionized acceptors. Studies of this kind have given much information on conduction processes in a variety of semiconductors. The variation of deformation potentials (see § 8.5) with pressure has led to the identification of the frequently complex relationships between them. The basic theory of such pressure effects and a review of experimental work, mainly on transport properties under pressure, has been given by R. W. Keyes.[60] Detailed studies of Ge and Si have been described in a review article by W. Paul and H. Brooks.[61]

Perhaps the most important quantities to observe under pressure are the forbidden energy gap ΔE and various other gaps in the band structure. These may be obtained by use of optical absorption, which

[58] See, for example, *Solids Under Pressure*, ed. W. Paul and D. M. Warschauer (McGraw-Hill, 1963).
[59] *Phys. Rev.* (1955) **99**, 490.
[60] *Solid State Physics* (Academic Press, 1960) **11**, 149.
[61] *Progress in Semiconductors* (Heywood, 1963) **7**, 135.

does not require the use of electrical contacts, the latter frequently leading to difficulties at high pressure. A large amount of work has been done on this subject, mainly in the higher pressure ranges, by W. Paul and colleagues and by H. G. Drickamer and colleagues. W. Paul and D. M. Warschauer[62] have reviewed this work in *Solids Under Pressure*. Work in the very high pressure range has been described by H. G. Drickamer.[63]

For Ge the value of the indirect gap at low pressures increases with pressure, the rate being 7.5×10^{-3} eV kbar^{-1}, while that for Si decreases at a rate of 2×10^{-3} eV kbar^{-1} as determined by T. E. Slykhouse and H. G. Drickamer.[64] However, at a pressure of about 50 kbar the rate of increase for Ge becomes zero and at higher pressures tends to the same value as for Si, namely -2×10^{-3} eV kbar^{-1}. This is illustrated in Fig. 14.8. Further experiments on III–V compounds have confirmed that this

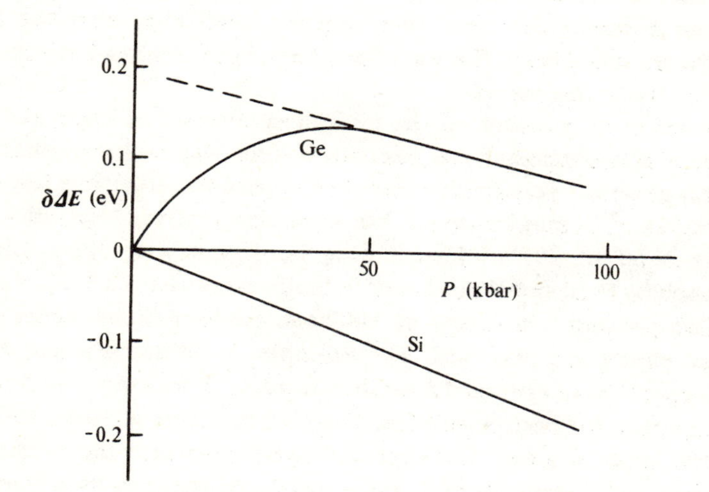

Fig. 14.8. Change $\delta\Delta E$ in forbidden energy gap ΔE as a function of pressure P for Ge and Si. (After T. E. Slykhouse and H. G. Drickamer, *loc. cit.*)

behaviour is due to the fact that the gap at (111) increases with pressure and that near (100) decreases. It also appears that the rate, with a few exceptions, is nearly the same for Si and Ge and for III–V compound semiconductors. This indicates that the (100) band edge (X-point) in Ge

[62] *Solids Under Pressure*, ed. W. Paul and D. M. Warschauer (McGraw-Hill, 1963), p. 179.
[63] *Solid State Physics* (Academic Press, 1965) **17**, 1.
[64] *J. Phys. Chem. Solids* (1958) **7**, 210.

falls below the (111) band edge (L-point) at a certain pressure and above this pressure Ge has a band structure similar to that of Si. The transition pressure is given as 35 kbar by Paul and Brooks (*loc. cit.*) and this agrees well with the measurements of Slykhouse and Drickamer (*loc. cit.*).

The direct gap for Ge is found to increase with pressure at a rate of about 15×10^{-3} eV kbar^{-1} and a similar rate is found for a number of direct-gap III–V semiconductors, such as InSb and GaAs, with rates of about 12×10^{-3} eV kbar^{-1}. An interesting changeover from direct to indirect gap was observed for GaSb by A. L. Edwards and H. G. Drickamer.[65] For low pressures the rate is about 12×10^{-3} eV kbar^{-1} but at a pressure of 18 kbar it changes to 7.3×10^{-3} eV kbar^{-1}, typical of a (111) minimum. This would indicate that the (000) minimum had risen above the (111) minimum and at higher pressures the gap would be 'indirect'. At a considerably higher pressure of about 50 kbar the rate of increase becomes zero and then negative, indicating that the (100) minima are now lowest. Similar effects have been observed in a number of other III–V compounds.

The effect of pressure on the ionization energies of impurities in a variety of semiconductors has been studied for 'shallow' impurities. The effect is generally very small, rates of change of the order of a few times 10^{-5} eV kbar^{-1} being observed. For some deeper-lying impurities rates of the order of 10^{-3} eV kbar^{-1} have been observed. The subject is discussed by Paul and Warschauer in their review (*loc. cit.*).

When pressures in excess of 100 kbar are used some rather more drastic effects are observed. For example, S. Minomura and H. G. Drickamer[66] have observed a sudden decrease in resistance by 5 orders of magnitude for both Si and Ge. The former occurs at about 190 kbar and the latter at about 120 kbar. At higher pressures their resistance increases with temperature as for a metal. At the transition temperature, for each, a change in crystal structure has been observed by J. C. Jamieson.[67] The structure changes from diamond to tetragonal, being similar to that of metallic tin. The change is therefore similar to the change of grey tin to white tin, the change in this case being from semimetal to metal (see §13.8).

Similar transitions have been observed for a number of III–V semiconductors, and also for some II–VI semiconductors such as ZnTe, by G. A. Samara and H. G. Drickamer.[68] For the III–V compounds the one

[65] *Phys. Rev.* (1961) **122**, 1149.
[66] *J. Phys. Chem. Solids* (1963) **23**, 451.
[67] *Science* (1963) **139**, 762.
[68] *J. Phys. Chem. Solids* (1963) **23**, 457.

having the highest transition pressure is GaAs which does not become metallic until the pressure reaches 250 kbar.

Various other properties of semiconductors change with pressure, including mobility. For example, M. I. Nathan, W. Paul and H. Brooks,[69] have studied the effect of pressure on inter-valley scattering (see § 8.7). They found that in Ge the mobility at room temperature drops by a factor of 5 between zero pressure and 30 kbar and were able to account very accurately for the form of the variation, which was not linear.

Variation with pressure of a number of other properties including refractive index have been discussed in the reviews to which we have already made reference by R. W. Keyes, by W. Paul and H. Brooks, by W. Paul and D. M. Warschauer and by H. G. Drickamer (see p. 469).

14.6 Laser action in semiconductors

In our treatment of recombination radiation in § 9.2 we considered only the spontaneous emission of radiation in a transition from a higher to a lower level, the rate of emission being assumed independent of the local intensity of radiation of the same frequency. It was, however, pointed out as long ago as 1917 by A. Einstein[70] that one must take account of another process, stimulated emission, in which the rate of emission is proportional to the radiation energy density per unit frequency interval. Unless the energy density is high this process may generally be neglected in the visible and near infra-red region of the spectrum; but it is of vital importance when circumstances are such that the radiation intensity can grow, leading to laser action, the resonant generation of radiation. Lasers are now well known as sources of intense, coherent and nearly monochromatic radiation over a wide range of spectral frequencies. Their basic physics and practical development have been described in a vast number of papers and in several books. We shall only be concerned with laser action in semiconductors and, as for electronic devices based on these, we must forgo description of laser development and concentrate on the basic physics of laser action in semiconductors. The physical principles of lasers have been treated in a number of books.[71]

We have already met the contribution of stimulated emission in the case of phonon emission, when we used the fact that phonon absorption

[69] *Phys. Rev.* (1961) **124**, 391. [70] *Phys. Z.* (1917) **18**, 121.
[71] See, for example, R. Loudon, *The Quantum Theory of Light* (Clarendon Press, 1973); M. Sargent, M. O. Scully and W. E. Lamb, *Laser Physics* (Addison-Wesley, 1974).

is proportional to N_p the phonon density, but phonon emission is proportional to $(1+N_p)$, the term 1 corresponding to spontaneous emission and N_p to stimulated emission (see §§ 8.7, 12.2). Einstein (*loc. cit.*) showed by means of a thermodynamic argument that the probability of stimulated emission is the same as the probability for absorption and introduced a coefficient B to describe it, and another coefficient A to describe the spontaneous emission. These are known as the Einstein coefficients, and their values may be readily derived by means of a simple calculation.

Let us first of all consider two well-defined levels whose energies are E_1 and E_2, with $E_2 > E_1$, connected by radiation of angular frequency ω such that $\hbar\omega = E_2 - E_1$. Let us suppose that there are N_1 systems occupying level 1 and N_2 occupying 2, so that $N_2/N_1 = \exp(-\hbar\omega/\boldsymbol{k}T)$. By definition of the Einstein coefficients we must have, if $W(\omega)$ is the energy density per unit angular frequency range,

$$dN_2/dt = -AN_2 - BN_2 W(\omega) + B'N_1 W(\omega). \tag{5}$$

In thermodynamic equilibrium $dN_2/dt = 0$ so that

$$W(\omega) = \frac{AN_2}{B'N_1 - AN_2} \tag{6}$$

$$= \frac{A}{B' \exp(\hbar\omega/\boldsymbol{k}T) - B}. \tag{6a}$$

In equilibrium this should reduce to the Planck's formula for the radiation intensity as given in § 9.2, but expressed in terms of ω, and this will only be so if $B = B'$, i.e. if the probability of spontaneous emission is the same as for absorption for each system. Moreover

$$A/BW(\omega) = \exp(\hbar\omega/\boldsymbol{k}T) - 1 = N_p^{-1} \tag{7}$$

where N_p is the photon occupation number. The same argument applies to phonons giving the result we have quoted above. The value of A may be obtained from Planck's formula but we shall not require it.

We have found these relationships for an equilibrium situation, but the *probabilities* may be applied to radiation not in thermal equilibrium. Also, these relationships may be deduced more rigorously from the quantum theory of radiation (see, for example, R. Loudon, *loc. cit.*). There are difficulties when $W(\omega)$ represents strictly monochromatic radiation for which the density per unit frequency interval is infinite.

These may, however, be overcome using the quantum theory of radiation.

Let us now consider the situation in which the ratio of N_2/N_1 is not that corresponding to thermodynamic equilibrium and $W(\omega)$ also does not correspond to thermal radiation. Neglecting for the moment the spontaneous emission which introduces a loss into the system, the rate of absorption of radiation will be proportional to $B(N_1-N_2)$. We may thus define an absorption coefficient $\bar{\alpha}$ given by

$$\bar{\alpha}=\alpha(N_1-N_2)/N \tag{8}$$

where $N=N_1+N_2$. This reduces to the normal absorption coefficient α when $N_2 \ll N_1$, a situation which usually holds for optical frequencies except under conditions of high external excitation. If, however, by some means N_2 can be made greater than N_1 then the absorption coefficient becomes negative and we have *gain* instead of loss. If this gain exceeds the loss due to spontaneous emission and other causes, the radiation intensity would increase as a beam progresses through such a medium and this constitutes laser action. The process of making $N_2>N_1$ by external means is termed 'pumping'.

A semiconductor into which a large number of electron–hole pairs have been injected through contacts, or by other means such as external radiation, provides a natural system of just this kind. The upper states correspond to electrons in the conduction band and the lower states to holes at the top of the valence band, and so are empty of electrons. It is therefore not surprising that laser action can take place under appropriate conditions. The system is, of course, much more complex than the simple two-level system we have discussed, but the basic principles are unchanged. On introducing the density of states N_e for electrons and N_h for holes one may express the generation rate G for electron levels with $E \simeq E_2$, and hole levels with $E \simeq E_1$, in the form (see § 9.2)

$$G=BN_e(E)N_h(E)P_e(E_1)P_h(E_2)W(\omega) \tag{9}$$

where $P_e(E)$ and $P_h(E)$ are the occupation probabilities defined in § 4.1. The recombination rate R is given by

$$R=BN_eN_hP_h(E_1)P_e(E_2)W(\omega)+AN_eN_hP_h(E_1)P_e(E_2) \tag{10}$$

the constant B in (9) and (10) being the same.

In a steady state these will be equal. On equating them we have

$$BW(\omega)[\exp(\hbar\omega/kT)-1]=A \tag{11}$$

which is just equation (7). Again, if we express the generation rate and recombination rate in terms of probabilities per electron–hole pair we find that the ratio of the probability for stimulated emission to that for spontaneous emission is given by equation (7).

We have not, so far, mentioned the very important property of the stimulated emission, namely that it has the same direction, polarization and frequency as the radiation creating it and is coherent with it, while the radiation arising from spontaneous emission has random direction and polarization and is incoherent.

When we have large departures from thermal equilibrium the *probabilities* still apply and the considerations leading to negative values of the absorption coefficient, i.e. to a positive *gain* coefficient, still hold. Let us consider for a moment a semiconductor in which a highly degenerate electron and hole concentration has been created, the conduction band being filled up to the quasi-Fermi level E_{Fe} and the valence band empty above a level $E_{Fh}=-\Delta E-E'_{Fh}$.

This situation is illustrated in Fig. 14.9. It will be seen that absorption will only take place for photon energies greater than $\Delta E+E_{Fe}+E'_{Fh}$. This is similar to the Burstein–Moss effect, but for the latter the term E_{Fh} does not apply since the effect refers to conditions in which either the conduction band is filled to level E_F or the valence band is empty above level E_F (see § 14.4.2). Emission may, however, take place for all frequencies in the band $\Delta E<h\nu<\Delta E+E_{Fe}+E'_{Fh}$. Since the lower level is empty this corresponds to complete population inversion and so should lead to a considerable gain (negative α). Under suitable circumstances quite large values of g $(=-\bar{\alpha})$ are found, up to 100 cm^{-1} or more, so that semiconductors are relatively easy materials in which to induce laser action. The spectrum of the emitted radiation would be rather broad if no further constraints were put on the system. As for all laser systems for which the gain is positive over a wide range of frequencies, a resonant optical cavity is used to increase the effective path length of the radiation in the material and to limit the frequency range over which laser action takes place. For semiconductor lasers this frequently consists simply of two parallel cleaved faces of a single crystal. Although laser action has been observed in uniform material strongly excited by means of an external laser, by far the most common

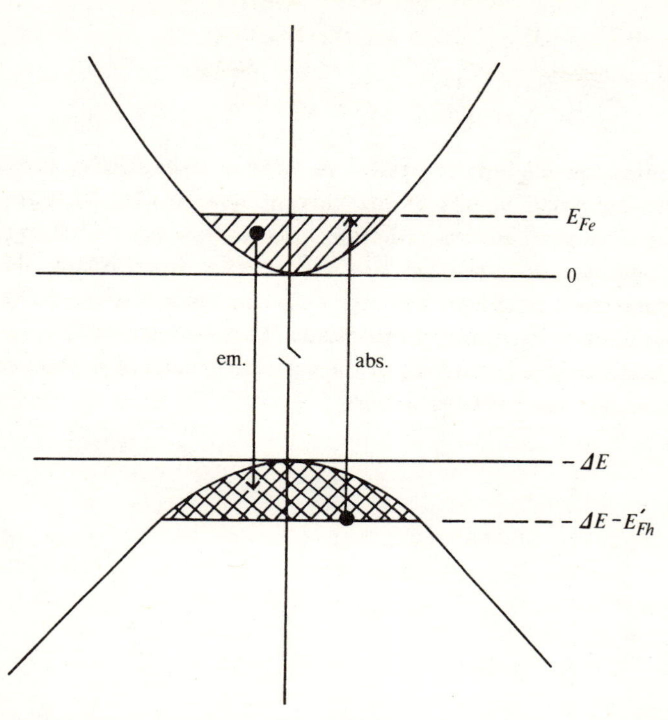

Fig. 14.9. Absorption and stimulated emission from high density of electrons and holes created by injection.

method of excitation is simply by injection from a p–n junction. The active region then consists of the narrow junction region where there are both holes and electrons, terminated by two parallel cleaved faces. In more recent developments heterojunctions have been used. These as we have already discussed (§ 7.14) have the advantage of low free-carrier absorption and can be arranged with material having a lower refractive index on either side of the active region so as to confine the radiation to the cavity.

Let us suppose that the cavity has length L and that the reflectivities of the two faces are R_1 and R_2. If a loss α per unit length results from absorption due to other causes and loss of radiation while the gain due to stimulated emission is g, the intensity I after one transit across the cavity of length L and back of radiation of intensity I_0 will be given by

$$I = I_0 R_1 R_2 \exp(2gL - 2\alpha L). \tag{12}$$

The condition for laser action will therefore be

$$g > \alpha - \frac{1}{2L} \ln (R_1 R_2). \tag{13}$$

When injection current is passed through a laser diode, emission of radiation for small values of the current will be due to spontaneous emission of recombination radiation. This is how light-emitting diodes (LED) operate. If conditions are suitable for laser action, when the diode current is increased beyond a certain value the intensity of the emitted radiation increases dramatically. This is illustrated in Fig. 14.10. At the same time a narrowing of the emitted spectrum is observed, also indicating the onset of laser action.

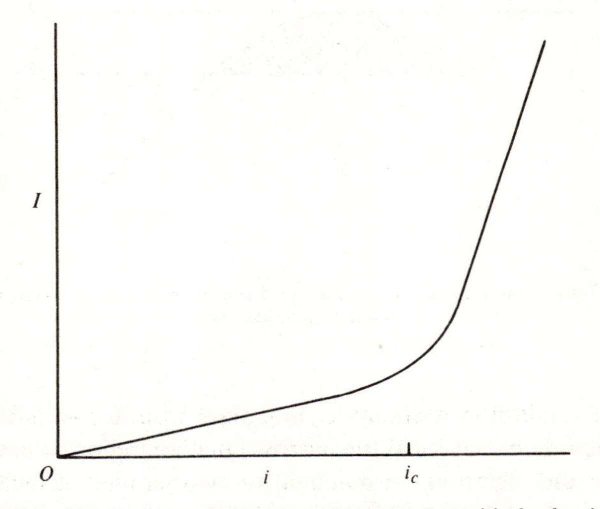

Fig. 14.10. Illustrating large increase in output intensity I at critical value i_c of injection current in a diode laser.

Since the radiative recombination is, as we have seen, a relatively inefficient process, only a small amount of the energy used to inject electrons reappears as radiation. The rest appears as heat and this limits the injection current unless pulsed operation is used. This was done in all the early work.

Although we have discussed laser action in terms of band-to-band transitions it is clear that it can also take place through impurity centres. Some of these which produce strong radiation through the recombination of trapped excitons we have discussed in § 10.14. In this case the conduction band acts rather like the broad band of states to which electrons are pumped in a gas laser before falling into the active upper state.

14.6.1 Laser materials

The earliest work on semiconductor lasers as reported by R. N. Hall *et al.*,[72] by M. I. Nathan *et al.*[73] and by T. M. Quist *et al.*,[74] used GaAs as the active material, mainly because it has a 'direct' gap which, as we have seen, leads to a higher recombination rate than when phonon-assisted transitions are involved, but also because the technique of making LEDs with this material had been developed; pulsed operation at 77 °K was used. The lead chalcogenides also have *direct* gaps at the zone edges in the ⟨111⟩ directions and were thought to be suitable for studies of laser action. The early work on these materials especially by C. E. Hurwitz, A. R. Calawa and R. H. Rediker,[75] used electron beam pumping, but diode operation was also found to be possible. Optical pumping using a ruby laser was also used by N. G. Basov *et al.*[76] to produce laser action in GaAs, CdTe, CdSe and CdS.

With more recent developments in technique, diode lasers may be made using a variety of materials including the lead chalcogenides and alloys of them, alloys such as $Pb_{1-x}Cd_xS$ and various alloys of III–V compounds. The narrow-gap materials $Pb_{1-x}Sn_xTe$ and $Pb_{1-x}Sn_xSe$ have also been used (see §13.8). By use of these materials, diode lasers have been fabricated providing spectral coverage from about 0.6 μm to 32 μm. These lasers are also tunable, each over a narrow spectral range, by varying the injection current. This, in turn, varies the temperature and hence the gap and lasing frequency. Lasers of this kind have been described in some detail by A. Mooradian[77] and by I. Melngailis and A. Mooradian,[78] together with their application to molecular spectroscopy.

GaAs lasers with $Ga_{1-x}Al_xAs$ heterojunctions can now be made to operate continuously at room temperature but the lead-salt lasers have to be cooled to 77 °K for such action. Lasers of this type are, however, usually operated under pulsed conditions. Various other methods of tuning have been used, including application of pressure and magnetic field, but use of current change is by far the simplest. Very narrow

[72] *Phys. Rev. Lett.* (1962) **9**, 366.
[73] *Appl. Phys. Lett.* (1962) **1**, 62.
[74] *Appl. Phys. Lett.* (1962) **1**, 91.
[75] *J. Quantum Electron.* (1965) **1**, 102.
[76] *Proc. VIIIth Int. Conf. on Phys. of Semiconductors* (Phys. Soc. Japan, 1966) p. 277.
[77] *Very High Resolution Spectroscopy*, ed. R. A. Smith (Academic Press, 1976), p. 75.
[78] *Laser Applications to Optics and Spectroscopy*, ed. S. F. Jacobs *et al.* (Addison-Wesley, 1975) Vol. 2, p. 1.

emission lines with width of the order of 50 kHz have been observed (A. Mooradian, *loc. cit.*) with continuous power outputs of a few mW.

Earlier reviews of semiconductor laser development have also been given by H. J. Queisser[79] and by M. H. Pilkuhn.[80]

14.7 Spin-flip scattering and lasers

Another new phenomenon leading to laser applications is a form of Raman scattering which has come to be known as 'spin-flip' since, in the process of scattering of light from a conduction band electron, the spin of the electron is reversed. If the electron is in a magnetic field the spin energy $g\beta B$ (see § 12.6) is either taken from or given to the energy of the scattered quantum, resulting in either a decrease or increase in frequency of an amount $g\beta B/\boldsymbol{h}$. The former is known in the terminology of Raman scattering as a Stokes shift and the latter as an anti-Stokes shift. Transitions of this kind turn out to be forbidden for electrons in free space but not for electrons in the conduction band of a semiconductor with a degenerate valence band.

Other forms of Raman scattering from the electrons in a semiconductor in a magnetic field arise from transitions between Landau levels in the conduction band. The Raman scattering process takes place through a two-stage transition in some ways similar to the indirect inter-band transitions, although in this case phonons are not involved. A transition is induced by the incident radiation from an initial state i to an intermediate state t and then from the state t to a final state f. The intermediate state is occupied for so short a time that energy need not be conserved, exactly as for the intermediate state in indirect inter-band transitions discussed in § 10.5.3. Energy will be conserved between the quanta of incident and scattered radiation and the initial and final states. The intermediate state may be above the initial and final states, below them or even between them. The three situations are illustrated in Fig. 14.11. The intermediate state must be a real one to which an *allowed* transition can take place. If $E_i < E_f$ the frequency of the scattered radiation ν_s will be less than that of the incident radiation ν_i since

$$\boldsymbol{h}\nu_s = \boldsymbol{h}\nu_i - (E_f - E_i) \qquad (14)$$

and we shall have a Stokes shift; if $E_i > E_f$ we shall have an anti-Stokes shift.

[79] *Electronic Materials*, ed. N. B. Hannay and U. Colombo (Plenum Press, 1973), p. 41.
[80] *Phys. Stat. Solidi* (1968) **25**, 9.

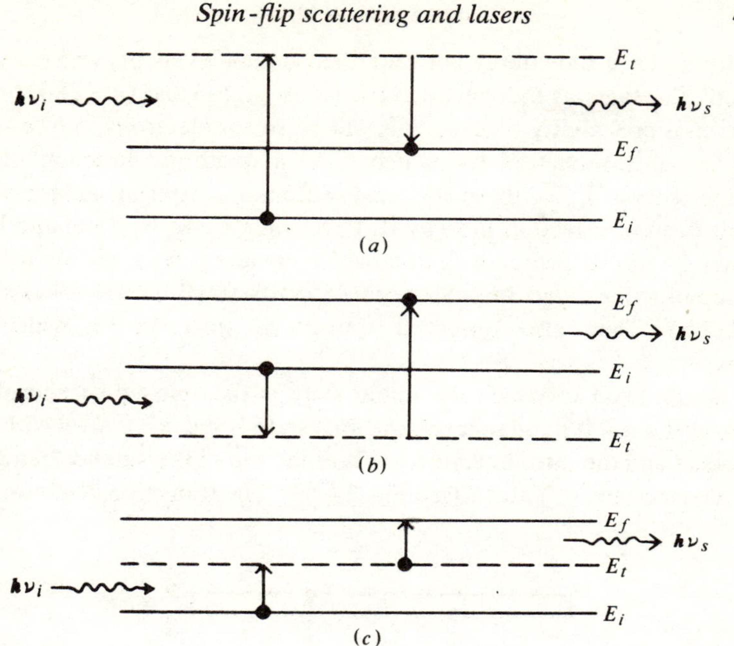

Fig. 14.11. Raman scattering as a two-stage process through an intermediate state. (*a*) $E_t > E_f > E_i$; (*b*) $E_t < E_i < E_f$; (*c*) $E_i < E_t < E_f$. ($h\nu_s = h\nu_i - E_f + E_i$.)

The situation in which the state i is the Landau level (n) and the state f the level $(n+2)$, while the intermediate state is the Landau level $(n+1)$, all in the conduction band, has been treated theoretically by P. A. Wolff.[81] This is illustrated in Fig. 14.12. It is found that only scattering with $\Delta n = 2$ is allowed. No change in spin is allowed in such transitions.

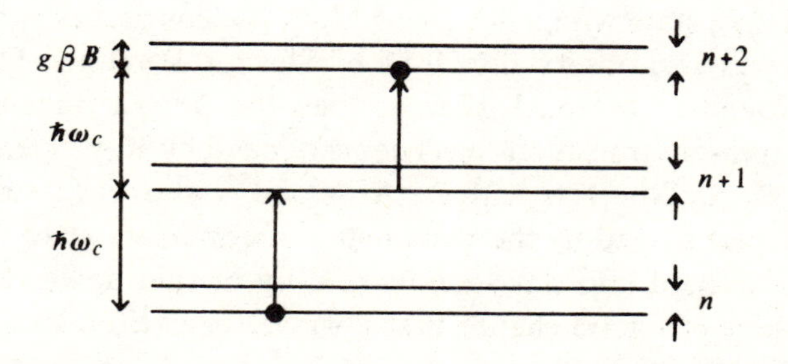

Fig. 14.12. Scattering with $\Delta n = 2$ transition through intermediate Landau level with no change of spin. ($h\nu_s = h\nu_i - 2\hbar\omega_c$.)

[81] *Phys. Rev. Lett.* (1966) **16**, 225.

Moreover, if the energy between the Landau levels (n) and ($n+1$) is exactly the same as that between the levels ($n+1$) and ($n+2$), then the transition probability is zero. This will be so for electrons in free space and in semiconductors for which E is a quadratic function of the components of $\boldsymbol{k}$; for these this kind of Raman scattering cannot occur. It has been observed in InSb by R. E. Slusher, C. K. N. Patel and P. R. Fleury,[82] and its presence is due to the deviation from parabolicity in the conduction band of InSb because of its small effective mass (see § 11.3.5). They also observed transitions with $\Delta n = 1$ which are unexpected.

The situation in which the initial state is the spin-up or spin-down state of the $n = 0$ Landau level, the final state is the $n = 0$ level with spin reversed and the intermediate state is at the top of the valence band, has been treated by Y. Yafet[83] (see Fig. 14.13). The transition is allowed for

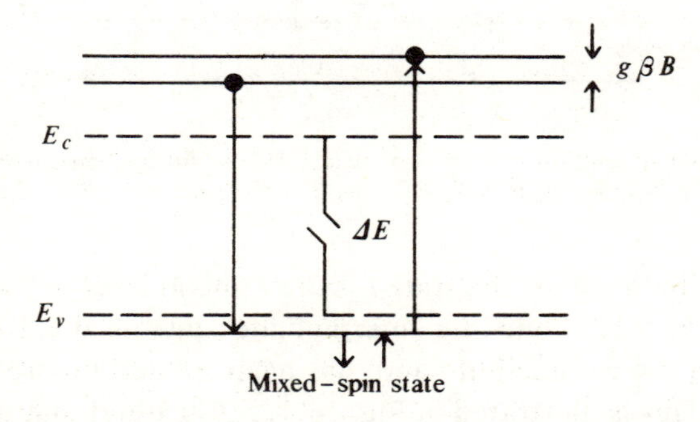

Fig. 14.13. Scattering with spin reversal through intermediate mixed-spin state in valence band. ($h\nu_s = h\nu_i - g\beta B$.)

the mixed-spin state with $n = 1$ (see § 12.7), such as occurs in InSb. This transition was also observed in InSb by Slusher, Patel and Fleury (*loc. cit.*) and found to be much sharper than the $\Delta n = 2$ transitions. The theory of spin-flip transitions has been extended by R. B. Dennis, C. R. Pidgeon, B. S. Wherrett and R. A. Wood[84] who have stressed the important part played by the mixed-spin intermediate state. When the electron is excited into this state its spin-up or spin-down character is lost and there is a good chance that it will be re-excited into either the

[82] *Phys. Rev. Lett.* (1967) **18**, 77.
[83] *Phys. Rev.* (1966) **152**, 858.
[84] *Proc. Roy. Soc.* (1972) A**331**, 203.

spin-up or spin-down $n = 0$ state in the conduction band. It is when it is re-excited with a *change* of spin that we get spin-flip scattering. InSb is rather unique and has proved particularly suitable for study of this kind of scattering which is so strong that the *scattered* radiation may be readily made to show laser action.

For such lasers, the input radiation of frequency ν_i may be considered as the pump and the scattered radiation and its augmentation by stimulated emission as the output. This may be readily separated from the input by filtering. InSb is again a particularly favourable material because of its very large g ($\sim$40) (see § 13.5) which means that the frequency shift will be much greater than for materials with normal g values. Laser action with the spin-flip scattered radiation was first observed by C. K. N. Patel and E. D. Shaw,[85] using transverse scattering, and soon after by R. L. Allwood *et al.*[86] and A. Mooradian, S. R. J. Brueck and F. A. Blum,[87] using forward scattering. The latter, although giving more formidable problems in separating the pump and laser radiation, produces a higher intensity of laser radiation.

Since the frequency shift is proportional to the magnetic induction B, the spin-flip laser provides a continuously tunable source of radiation and so has excited considerable interest.

CO_2 lasers with output in the region of 10.6 μm and CO lasers at 5.3 μm have mainly been used as pump sources. The latter has the advantage of producing a near-resonance between the initial and intermediate states which greatly enhances the Raman scattering.

Many materials other than InSb have been tried as a base for the spin-flip laser but none, apart possibly from $Hg_xCd_{1-x}Te$ (see § 13.8), have given nearly such intense spin-flip scattering.

Because of the interest in obtaining a continuously tunable narrow-line source in the infra-red a great deal of work has been stimulated by the discovery of spin-flip scattering, and a large literature has rapidly developed on the subject. Further theoretical developments have been reviewed by J. F. Scott.[88] The physical problems associated with the development of spin-flip lasers and their application to spectroscopy have been discussed by S. D. Smith[89] and by A. Mooradian.[90]

[85] *Phys. Rev. Lett.* (1970) **24**, 251.
[86] *J. Phys. C: Solid State Phys.* (1970) **3**, L186. [87] *Appl. Phys. Lett.* (1970) **17**, 481.
[88] *Laser Applications to Optics and Spectroscopy* (Addison-Wesley, 1975) Vol. 2, p. 123.
[89] *Very High Resolution Spectroscopy*, ed. R. A. Smith (Academic Press, 1976), p.13.
[90] *Ibid.*, p. 82.

14.8 Raman scattering from lattice vibrations

The scattering of light by the phonons in the lattice vibrations is exactly analogous to the usual Raman scattering from the normal vibrations of molecules. The techniques are a little more difficult but have been greatly aided by the availability of lasers with their highly monochromatic output and high intensity. These techniques have been described in some detail by A. Mooradian.[91] The basic theory of Raman scattering from the lattice vibrations has been given by R. Loudon[92] who has also reviewed the experimental results and information obtained by this means on the phonon frequencies. This information has proved to be a valuable addition to that obtained by absorption spectroscopy (see § 10.8) and by neutron scattering. In particular, because some of the selection rules which operate with this kind of scattering are different from those for absorption, identification of some of the phonons has been made more definite.

[91] *Science* (1970) **169**, 20.

[92] *Adv. Phys.* (1964) **13**, 423.

15

Amorphous semiconductors

15.1 New theoretical concepts

That the theoretical ideas on which we have based the treatment of the properties of crystalline semiconductors are inadequate to deal with amorphous semiconductors has long been appreciated. In a review article in 1960 A. F. Ioffe and A. R. Regel,[1] for example, drew particular attention to the fact that band structure theory based on Bloch waves is inapplicable when the mobility is less than $10\ \mathrm{cm^2\ V^{-1}\ s^{-1}}$ and not likely to be very good if it falls below $100\ \mathrm{cm^2\ V^{-1}\ s^{-1}}$ (see § 13.14). This it certainly does in any amorphous semiconductor examined so far. We have already drawn attention to the fact that a different kind of conductivity, which has been variously described as 'hopping' and 'diffusion', must be applicable even in crystalline semiconductors for which the mobility is low and we have described briefly some of the theoretical ideas used for this kind of conductivity. Some of these ideas can be taken over to describe in general terms transport properties of amorphous semiconductors, but here completely new problems arise.

Some of the objections raised by Ioffe and Regel against the theoretical ideas which have been so successful in providing a very detailed description of the physical processes that determine the properties of high-mobility semiconductors now seem hardly to be justified. Their main objection that they do not include *all* semiconducting materials seems to be a bit hard to maintain. From the work done in the past sixteen years, both experimental and theoretical, it is quite clear that crystalline semiconductors with mobilities in excess of say $100\ \mathrm{cm^2\ V^{-1}\ s^{-1}}$ are quite different materials from the low-mobility semiconductors and also from the *same* materials in amorphous form.

[1] *Progress in Semiconductors* (Heywood, 1960) **4**, 237.

For the latter the mobilities are *very much* lower than for the crystalline form. For example, amorphous Ge at room temperature has an electron mobility of about 0.15 $cm^2 V^{-1} s^{-1}$ as against 4000 $cm^2 V^{-1} s^{-1}$ for good crystalline Ge. As well as explaining the low mobilities found in glassy materials like As_2S_3 the new theoretical ideas must also explain why the mobility of a material like Ge falls by more than four orders of magnitude on becoming amorphous. They must also explain why some semiconductors (not Ge) retain some of their semiconducting properties on melting.[2]

That the theory of electron energy bands based on Bloch waves cannot explain these matters is hardly a justified criticism so long as its range of applicability is realized. In the past sixteen years it has continued to give an increasingly accurate account of the varied and beautiful phenomena associated with high-mobility semiconductors and to predict new phenomena which have subsequently been observed. It has also formed a sound basis for the design of electronic and opto-electronic devices. What more can one ask from a theory? Perhaps the fact that without a great deal of modification the basic theory can also deal with most of the properties of metals has led one to expect that it should be applicable to all materials. It may be that some such unified theory will emerge eventually but we do not have it now.

In the circumstances, theorists have tried to develop an alternative approach and to introduce new ideas which go some way towards explaining the basic properties of amorphous semiconductors, including liquids and glasses. A great deal of progress has been made but one has to admit that there are still many theoretical uncertainties and alternatives which have not yet been resolved. A great deal has been written on the subject and there are now about as many review articles and books on amorphous semiconductors as there are on the crystalline materials. For this reason we have felt it appropriate to deal with the subject only briefly. To do otherwise would require a considerable expansion of the present volume which is neither possible nor desirable. We have therefore concentrated on trying to bring out the essential differences between crystalline and non-crystalline semiconductors and stressing those properties which justify including the latter in the class of materials called semiconductors.

As we have seen in § 1.3, a decrease in conductivity with increase in temperature cannot always be taken as a clear indication that a material

[2] For a detailed discussion of liquid semiconductors see J. Enderby, in *Amorphous and Liquid Semiconductors*, ed. J. Tauc (Plenum Press, 1974), p. 361.

is *not* a semiconductor. Most of the amorphous semiconductors do, however, have negative temperature coefficients of resistance. More definitive, we saw, was the existence of a forbidden energy gap which generally shows up in optical absorption measurements. For amorphous materials the existence of a gap seems also fundamental but, as we shall see, we have to change our ideas about the kind of gap we have. In § 3.2 we considered ideas based on chemical binding as an alternative approach to the physical picture of a semiconductor and in § 3.3 we saw that these ideas could be used to develop a better definition of what we mean by a semiconductor.

While the concept of mean free path cannot be taken over for amorphous materials these ideas about binding can be, since they do not depend, as does the former concept, on long-range order. The bond picture is largely based on nearest-neighbour interactions – longer-range interactions acting as perturbations. The concept of a semiconductor as a material in which the electronic bonds are 'saturated', i.e. involve pairs of electrons, is still sound. For a semiconductor, an energy of not much more than 1 eV would be required to produce a positive hole, where an electron is deficient at an atomic site, and an excess electron when an extra electron is near an atomic site. The binding energy would show up as an increase in absorption in the infra-red. The 'fundamental' absorption is, as we should expect because of varying environment of a particular atom in amorphous material, not so sharp as for a single crystal.

15.2 Electronic states

We do not have to look far for the reason why the Bloch type of electronic states are strongly modified in an amorphous material. The scattering by the disordered arrangement of the atoms makes the continuous phase of such waves meaningless. Nevertheless it has been shown by theoretical work, mainly by P. W. Anderson,[3] N. F. Mott,[4] M. H. Cohen[5] and their colleagues, that so-called extended states can exist, i.e. states for which the wave-function extends through a region of macroscopic dimensions. In addition, bound states in which the wave-function is localized in a region whose dimensions are not greatly in excess of the inter-atomic spacing also exist. Moreover the two types for a given energy cannot co-exist. There is therefore a sharp division

[3] *Phys. Rev.* (1958), **109**, 1492. [4] *Advan. Phys.* (1967) **16**, 49.
[5] M. H. Cohen, H. Fritzsche and S. R. Ovshinsky, *Phys. Rev. Lett.* (1969) **22**, 1065.

between the two. It has been shown that the extended states can form an energy band rather similar to those for a crystal and that the localized states occur towards the edges of the bands, extending them beyond where they would cut off in crystalline material. This is illustrated in Fig. 15.1. The energies marked E_v and E_c correspond to the band edges of

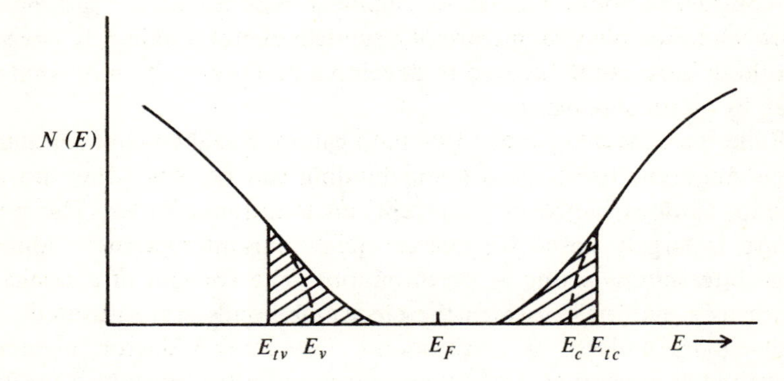

Fig. 15.1. Illustrating tails to bands occupied by localized states (shaded areas).

crystalline material. Those marked E_{tv} and E_{tc} correspond to the sharp division between localized and extended states near what would be the top of the valence and bottom of the conduction band respectively. The regions occupied by localized states are shown shaded.

It has been shown in the particular case of group IV semiconductors that, provided no broken bonds (generally known as 'dangling' bonds) exist, there will exist a forbidden energy gap.[6] For practical material, however, it seems that this ideal condition is unlikely to be attainable and that the tails of the bands may overlap so that there exist localized states throughout the space between E_c and E_v as shown in Fig. 15.2(*a*) (see also Fig. 15.3). The nature of these states, and in particular their relationship to dangling bonds, has been discussed by N. F. Mott, A. E. Davis and R. Street.[7] In this sense the energy gap is closed. However, conductivity in the localized states is forbidden and the mobility corresponding to them is zero. Thus instead of an energy gap we have a mobility gap ΔE_μ as shown in Fig. 15.2(*b*). It will be seen that ΔE_μ is a little bigger than ΔE, the forbidden energy gap for crystalline material.

[6] See, for example, D. Weaire, *J. Non-Cryst. Solids* (1971) **6**, 181; M. F. Thorpe and D. Weaire, *Phys. Rev.* (1971) B**4**, 3518; V. Heine, *J. Phys.* C: *Sol. State Phys.* (1971) **4**, L221.

[7] *Phil. Mag.* (1975) **32**, 961.

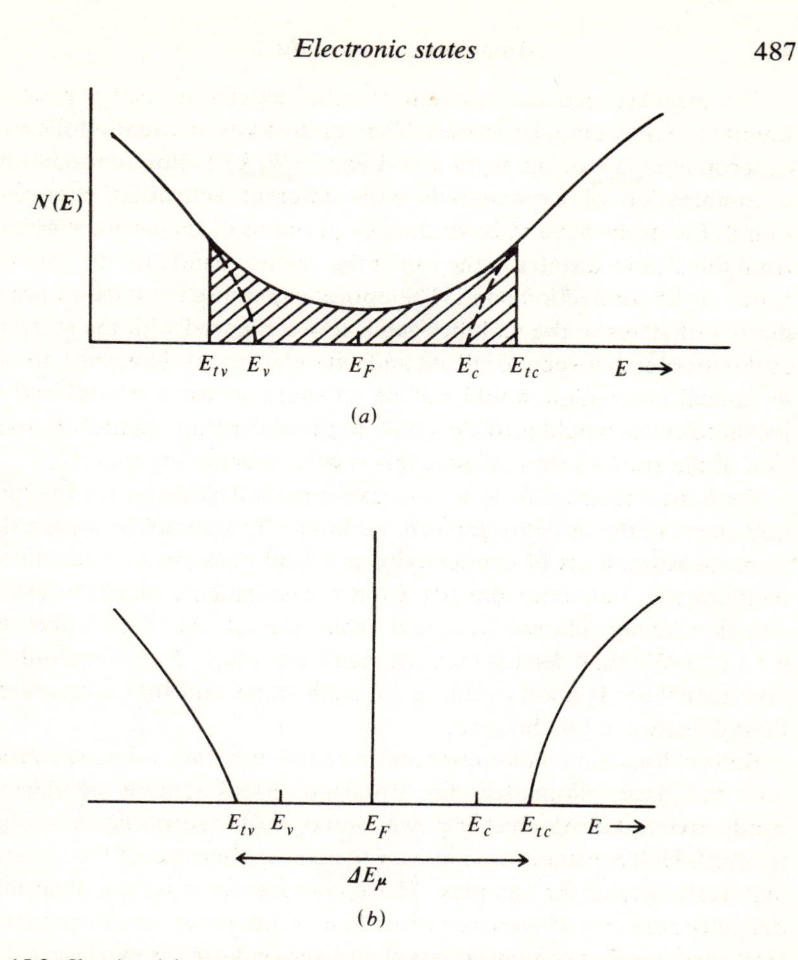

Fig. 15.2. Showing (*a*) overlap of tails occupied by localized states with no forbidden energy gap, (*b*) the mobility gap ΔE_μ.

Many theoretical papers have been written on these ideas. These have been reviewed and extended in a comprehensive treatment by N. F. Mott and E. A. Davis[8] who have also discussed the interpretation of the considerable amount of experimental data. The theoretical and experimental work has also been reviewed recently in series of papers by a number of authors[9] and in several books.[10]

[8] *Electronic Processes in Non-crystalline Materials* (Clarendon Press, 1971).

[9] *Electronic and Structural Properties of Amorphous Semiconductors*, ed. P. G. Le Comber and J. Mort (Academic Press, 1973); *Amorphous and Liquid Semiconductors*, ed. J. Tauc (Plenum Press, 1974); W. E. Spear, *Adv. Phys.* (1974) **23**, 523.

[10] A. I. Gabanov, *Quantum Theory of Amorphous Semiconductors* (Consultants Bureau, N.Y., 1965).

The mobility gap can account for the activation energy generally found for conduction processes. The conductivity σ usually follows an experimental law of the form $\sigma = A \exp(-W/kT)$ although sometimes a combination of exponentials with different activation energies is found. Electrons have to be excited by phonons of the lattice vibrations from the localized states at the top of the valence band into the extended states of the conduction band. The optical gap is also explained since the density of states in the mobility gap is low compared with the steep rise at the division between localized and extended states. However, the rise in optical absorption would not be so sharp as for a crystal and the localized states would provide a 'tail' to the absorption on the long-wave side of the rise. As we shall see, this is what is generally found.

We must enquire now as to how experimental evidence for the localized states in the mobility gap is to be found. This comes to some extent from measurements of conductivity and Hall constant as a function of temperature, but more directly from measurements of photo-electric emission under illumination and from use of the field effect (see § 7.11.1). We shall discuss measurements for a-Si in § 15.4 and only say now that there is good evidence for such states and that a measure of their density can be obtained.

Before discussing measurements on actual materials we must consider how the basic parameters are obtained. Measurement of electrical conductivity, thermo-electric power and optical absorption are straightforward. Hall constants are not easy to measure because of the generally high resistivity of the samples. There is moreover a serious theoretical difficulty here in that there is a good deal of theoretical evidence that the Hall constant R is somewhat less than $1/ne$ or $1/pe$ for extrinsic n-type and p-type material. One may obtain a Hall mobility μ_H from the product $R\sigma$ but its interpretation and its relationship to the drift mobility μ_d is not as clear as for high-mobility crystalline material. Fortunately the drift mobility may be obtained by means of a technique developed mainly by W. E. Spear.[11] A very short pulse of electrons is injected optically or by electron bombardment into one face of a thin specimen. The arrival of these at the opposite face is observed with a fast oscilloscope and, knowing the thickness of the specimen, from the time-delay μ_d is obtained.

The arrival pulse is usually followed by a 'tail' indicating that carriers get trapped on their way through the material.

[11] *Adv. Phys.* (1974) **23**, 523.

15.3 Lattice vibrations

Although the band structure for the lattice modes of vibration will be absent for the amorphous material there will be similarities with the crystal modes. For the long-wave acoustic modes irregularities in the atomic arrangement should average out and the ω/k relationship will depend on the elastic constants. Any difference will merely show up as a change in these and we may still speak of long-wave acoustic phonons. Again the long-wave optical phonons will depend very much on the local atomic arrangement. Indeed, in his original work on scattering from solids Raman considered only the modes of the local 'molecular' arrangement. For compounds whose unit is a molecular group such as As_2S_3 this is a very good approximation and, as would be expected, the Raman spectrum and lattice absorption due to the optical modes does not differ much between amorphous and crystalline forms. For materials like Si and Ge on the other hand the local tetrahedral group of atoms is much more strongly coupled to its neighbours. In this case the two-phonon spectra are similar but without the fine structure of the lattice vibration spectrum for crystalline material. One-phonon absorption can occur in amorphous material since the crystal momentum conservation law no longer holds.

15.4 Amorphous Si and Ge

Because of their open structure, the group IV semiconductors Si and Ge behave rather differently from the glassy amorphous semiconductors such as As_2S_3 on which a great deal of experimental work has been carried out. Moreover they are available in very pure form and their properties as crystalline materials are well understood. They have therefore also been extensively studied in amorphous form and provide an excellent opportunity to compare the properties of the two forms. We shall therefore consider them first of all before discussing the more complex compound amorphous semiconductors.

The amorphous form of Si and Ge consists of groups of four atoms tetrahedrally bonded. The main difference between the amorphous form and the crystalline form is that in the former the tetrahedra are randomly oriented with respect to each other. This common arrangement of nearest neighbours is very important in ensuring that their properties are not too dissimilar, as pointed out by Ioffe and Regel (see p. 483). On melting, this tetrahedral co-ordination is lost. A much more dense packing having eight nearest neighbours results, and Si and Ge become metallic as they do at very high pressures (see § 14.5).

There are two views of the random orientation. One considers all orientations to be possible and the other considers only two, that as in the diamond structure and that as in the wurtzite structure. These have been discussed by R. Grigorovici[12] who shows that, as well as the hexagonal rings of atoms characteristic of the crystalline structure, pentagonal rings occur in the amorphous structure, taking up all pairs of bonding electrons. There are no dangling bonds, unless they are broken thermally, optically or otherwise or are brought about by structure defects.

Amorphous films of the order of a few μm thick may be prepared by various methods including evaporation, sputtering, and by decomposition of gases by means of an electrical discharge. Until recently it was thought that one of the very large differences between amorphous and crystalline Si and Ge is that while the latter are, as we have seen, very sensitive to quite small amounts of group III or group V dopants, the former are very insensitive, requiring concentrations of the order of 10^{20} cm^{-3} to make an appreciable difference to the conductivity. This was thought to be due to the fact that an excess or defect electron could be taken up by the variation of bonding in the defect structure or in other bonds. The impurity states could also be taken up by unfilled localized states, the Fermi level for undoped material being located near the middle of the mobility gap (see Fig. 15.2). Recent work by W. E. Spear[13] has shown, however, that the existence of many states in the gap depends on the method of preparation, and that films can be prepared with fewer states in the gap than had previously been found. As a result he was able to influence the conductivity with concentrations of both n-type and p-type impurities very much smaller than previously used. He was moreover able to form p–n junctions and this is a great step forward in the technology of amorphous semiconductors, opening up the possibility of using them for large-area solar generators of electric power.

The films were deposited by means of a glow discharge in silane or germane and the impurities introduced by means of a few parts per million of phosgene or diborane. By this means the intrinsic conductivity of the order of $10^{-12}\ \Omega^{-1}$ cm^{-1} can be increased both with n-type and p-type material to about $10^{-2}\ \Omega^{-1}$ cm^{-1}. Room temperature mobilities

[12] *Amorphous and Liquid Semiconductors*, ed. J. Tauc (Plenum Press, 1974), p. 45.

[13] *Proc. XIIIth Int. Conf. on Phys. of Semiconductors* (Tipographia Marves, 1976), p. 515.

of about 0.1 $cm^2\,V^{-1}\,s^{-1}$ in a-Si are not much changed by doping but are decreased at lower temperatures.

Measurements by W. E. Spear and P. G. Le Comber,[14] using the field effect (see § 7.11.1), of the density of states in the mobility gap have shown that near the minimum it is about $10^{17}\,cm^{-3}\,eV^{-1}$ in contrast to a value of about $10^{19}\,cm^{-3}\,eV^{-1}$ generally supposed to exist for Si and Ge amorphous films. Details of the method of calculation of $N(E)$ from the field effect measurements are given by A. Madan, P. G. Le Comber and W. E. Spear.[15] Measurements on films specially prepared for doping are shown in Fig. 15.3. The density of states $N(E)$ shows an interesting structure and is clearly more complex than that, shown in Fig. 15.2, obtained by the superposition of two 'tails'.

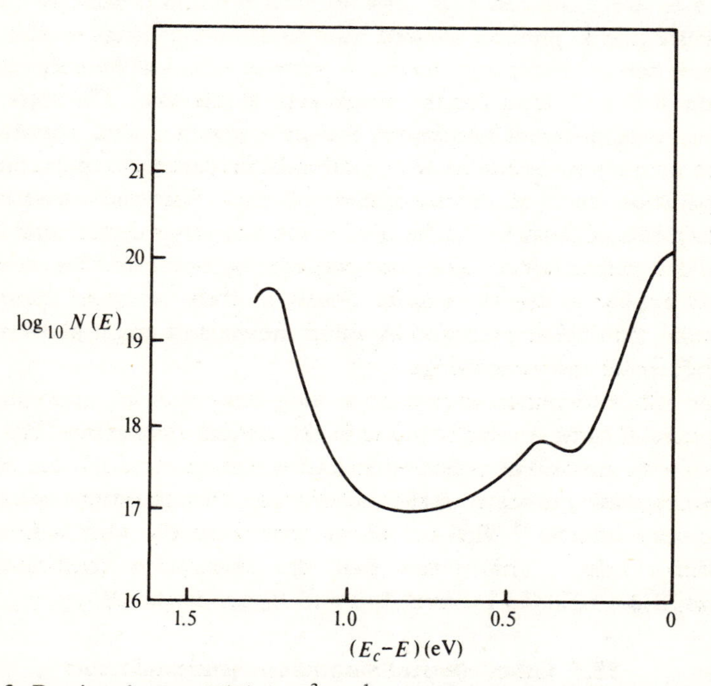

Fig. 15.3. Density of states $N(E)$ ($cm^{-3}\,eV^{-1}$) in mobility gap for a-Si. (After W. E. Spear, *loc. cit.*)

15.4.1 Optical properties of a-Si and a-Ge

The optical absorption of a-Si and a-Ge taken over a wide spectral range, say from 0.5 eV to 1.5 eV, as measured by T. M. Donovan *et al.*[16]

[14] *Phil. Mag.* (1976) **33**, 935.
[15] *J. Non-Cryst. Solids* (1976) **20**, 259.
[16] *Phys. Rev.* (1970) B**2**, 397.

does not differ very greatly from that of the crystalline materials but lacks the structure found in the latter. The rapid rise at the fundamental absorption edge starts at lower frequency, for example at 0.55 eV for Ge, the absorption coefficient rising rapidly to about 10^3 cm^{-1} at 0.6 eV and more slowly to about 10^5 cm^{-1} at 1.5 eV. The reflection spectrum also is similar but again without the peaks which have been associated with critical points in the band structure as in Fig. 10.24. This we should expect since the band structure no longer applies in the amorphous material. Measurements out to 9 eV by W. E. Spicer and T. M. Donovan[17] on a-Ge show that in the range 5–9 eV there is not much difference between a-Ge and crystalline Ge, the absorption of a-Ge being slightly higher. When the fundamental absorption edges of a-Ge and a-Si are examined with high spectral resolution none of the fine structure due to phonon assisted transitions and excitons is seen, but a gradual rise in absorption starting at various values of $h\nu$ generally less by about 0.1 eV than for the single-crystal material. The shift varies greatly with different specimens and this, together with variations in other properties, indicates that materials prepared by sputtering and evaporation are not representative of truly 'intrinsic' material but contain lots of flaws in the form of voids and other disturbances to an ideal disordered lattice. The films prepared by Spear and his colleagues would appear to approach more closely to truly 'intrinsic' amorphous material than those prepared by other means and might show a more characteristic absorption edge.

For all amorphous materials a long-wave tail of approximately exponential form is usually found in the optical absorption. This again varies with method of preparation and is almost certainly due to electrons originating in states in the mobility gap, though other explanations have been offered.[18] This tail obeys approximately what is known as Urbach's rule.[19] This states that the absorption coefficient $\alpha = A \exp[a(h\nu - E_0)/kT]$ where A, a and E_0 are constants.

15.5 Other element amorphous semiconductors

Experiments with a variety of other element semiconductors in amorphous form have been carried out including Se, Te and B. Of these only Te has been purified to anything like the state achieved for Si and Ge. Because it is highly anisotropic, having in crystal form a strongly marked

[17] *J. Non-Cryst. Solids* (1970) **2**, 66.

[18] J. Tauc, in *Amorphous and Liquid Semiconductors*, ed. J. Tauc (Plenum Press, 1974), p. 159.

[19] F. Urbach, *Phys. Rev.* (1953) **92**, 1324.

c-axis, it is difficult to form into amorphous layers. Even so, some detailed comparisons between the amorphous and crystal forms have been made (see J. Tauc, *loc. cit.*). Apart from the difficulty of purification of Se, it exists in two crystalline forms, monoclinic and triclinic, and these can be intermixed. Because of its widespread use in photo-copying machines this material has received a good deal of attention. Its optical absorption has been frequently measured but this seems to vary a good deal with method of preparation. It seems generally to lie roughly between the absorption curve for the two crystalline forms.[20] Alloys of Se and Te have also been studied.

Drift mobilities for Se have been obtained in a number of laboratories and seem to agree quite well. The room temperature values for holes and electrons as given by F. K. Dolezalek and W. E. Spear[21] are $0.14\ \text{cm}^2\ \text{V}^{-1}\ \text{s}^{-1}$ and $6\times10^{-3}\ \text{cm}^2\ \text{V}^{-1}\ \text{s}^{-1}$. They fall off at lower temperatures, decreasing exponentially and so indicating an activated process. The Hall mobility on the other hand, while having about the same value at room temperature, varies very much less with temperature, suggesting that the drift mobility is governed by trapping. This has been discussed by H. Fritzsche[22] who has also compared the experimental data obtained in a number of laboratories. The data for pure Te are scanty. Alloying Se with 1% Te drops the drift mobility of holes at room temperature by a factor of ten.

The optical properties of amorphous films are very similar to those of Si and Ge, the absorption edge being shifted to lower frequencies as compared with crystalline material. The situation is much more complex, however, because of the two crystalline forms of Se and the different behaviour of Te for radiation polarized parallel with and perpendicular to the c-axis. The subject has been discussed in some detail by J. Tauc.[23]

15.6 Other simple compound amorphous semiconductors

As well as for the group IV semiconductors, amorphous films of some of the III–V compounds including InSb, GaAs and GaSb have been prepared and studied.[24] As for Ge and Si, the conductivity falls much more

[20] A. E. Davis, in *Electronic and Structural Properties of Amorphous Semiconductors*, ed. P. G. Le Comber and J. Mort (Academic Press, 1973), p. 425.
[21] *J. Non-Cryst. Solids* (1970) **4**, 97.
[22] *Amorphous and Liquid Semiconductors*, ed. J. Tauc (Plenum Press, 1974), p. 211.
[23] *Ibid.*, p. 159.
[24] J. Stuke, *Proc. Int. Conf. on Low-Mobility Matls* (Eilat, 1971), p. 193.

rapidly with temperature than for crystalline material, the rate depending on the annealing temperature of the deposited films, in which form the material was studied. The difficulties in obtaining really intrinsic material seem to be even greater than for Ge and Si.

Very little seems to be known of the properties of II–IV compound semiconductors in amorphous form.

15.7 Glassy semiconductors

A great variety of glassy materials showing semiconducting properties have been examined (see general refs. in § 15.2). Of these the simplest are the chalcogenide glasses such as As_2S_3, As_2Se_3 and As_2Te_3. Because of their relatively simple structure a good deal of the fundamental work on semiconducting glasses has been done on these, and on As_2S_3 in particular. Very much more complex glasses, such as $Te_{81}Ge_{51}A_4$ where A is normally an element from group V of the periodic table have also received a good deal of attention of late because of their use in switches and memory devices (see § 15.7.2). The simpler glasses, however, give a better chance of unravelling the rather complex physical problems which these materials present and we shall restrict our discussion to them.

The picture of these materials which emerges from the considerations of § 15.2 is that they differ mainly from the group IV semiconductors in having very much fewer states in the mobility gap. Indeed, it may well be that the localized states are restricted to small 'tails' at the edge of the energy bands of extended states as in Fig. 15.1 rather than as in Fig. 15.2. If this is so then there is a true forbidden energy gap which may be only a little smaller than the mobility gap, defined again as the gap between two cut-off energies dividing the localized and extended states.

There will, of course, be some states in the gap but these are thought to be due to 'dangling' bonds produced by micro-voids and disruptions of the 'regular' disorder or to impurities. As for the element semiconductors, it is important to be sure that the material being studied is not microcrystalline. This is not always easy because many of these glasses tend to crystallize.

The conductivity of semiconductors of this type follows an exponential law of the form $\sigma = \sigma_0 \exp(-E_g/2\boldsymbol{k}T)$ for a range of values of σ extending over 4–6 orders of magnitude. Variations are found at low temperatures but, apart from this, there is clear indication of thermally activated conduction. The quantity E_g is approximately equal to the optically measured gap but not exactly so. This is compatible with the model suggested and indicates that the Fermi level for intrinsic material is near the middle of the gap (because of the $2kT$). Room temperature conductivities vary greatly but are usually of the order of $10^{-3}\ \Omega^{-1}\ \mathrm{cm}^{-1}$ falling to about $10^{-8}\ \Omega^{-1}\ \mathrm{cm}^{-1}$ at 150 °K. Mobilities (mainly μ_H) also vary greatly but are in the range 0.1–0.01 $\mathrm{cm}^2\ \mathrm{V}^{-1}\ \mathrm{s}^{-1}$ at room temperature falling, but not very fast (less than a factor of 10 between 300 °K and 150 °K), at low temperatures. They do not appear to be very sensitive to the addition of chemical impurities. Alloying usually decreases the conductivity and mobility. The electrical behaviour of glassy semiconductors has been discussed by H. Fritzsche (ref. as on p. 493).

15.7.1 Optical properties

The optical properties of these materials have also been extensively studied (see J. Tauc, ref. on p. 492). There appears to be more regularity, and less difference between samples, than for the element amorphous semiconductors. This is almost certainly due to the difficulty of preparing films of the latter free from imperfections. The form of absorption found for many semiconducting glasses is shown in Fig. 15.4 which gives the absorption of As_2S_3 as measured by F. Kosek and J.

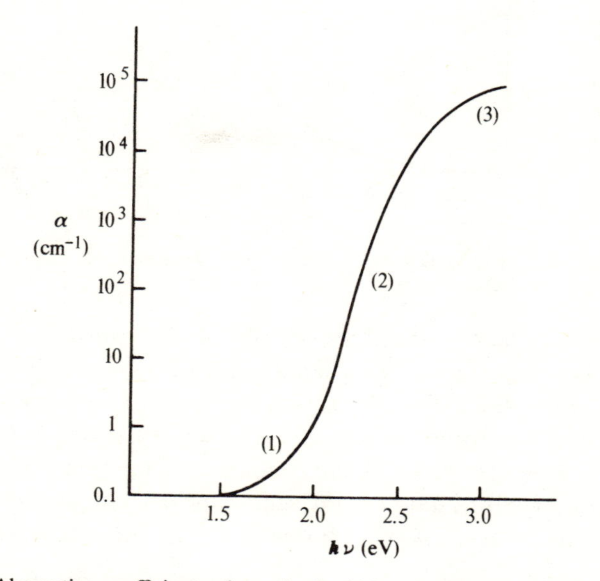

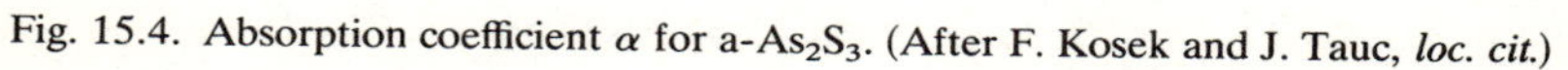
Fig. 15.4. Absorption coefficient α for a-As_2S_3. (After F. Kosek and J. Tauc, *loc. cit.*)

Tauc.[25] The absorption curve may be divided into three regions: (1) a long-wave tail which sometimes obeys Urbach's rule (see § 15.4.1); (2) another exponential region which may cover four to five orders of magnitude and for which the slope is largely independent of temperature (except at high temperatures); and (3) a region of high absorption which obeys a power law $\alpha = (h\nu - E_g^o)^r/h\nu$, from which we may define an 'optical energy gap' E_g^o. The value of r is frequently 2, although other values including 3 have been observed. For As_2S_3, if we plot the high-absorption region as $(h\nu\alpha)^{\frac{1}{2}}$ against $h\nu$ as in Fig. 15.5 we obtain a straight line showing that $r = 2$. This is similar to the variation for an indirect transition in a crystalline semiconductor.

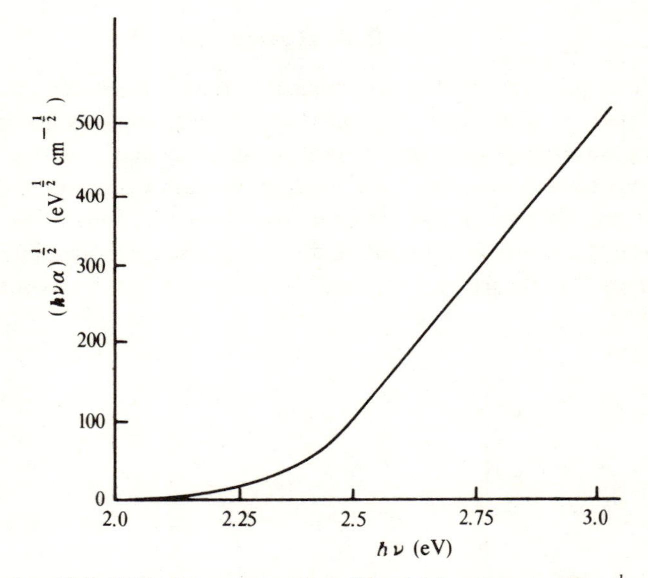

Fig. 15.5. High-absorption region of Fig. 15.4 replotted as $(h\nu\alpha)^{\frac{1}{2}}$.

The interpretation of the three regions of absorption in terms of the theoretical concepts we have outlined in § 15.2 has been discussed by J. Tauc (ref. on p. 492). The absorption in region (3) is considered to be due to transitions from the top of the extended states in the valence band to the extended states in the conduction band. Since we no longer have to conserve the momentum vector $\hbar\boldsymbol{k}$, an integration has to be performed over the density of states as for an indirect transition in a

[25] *Czech. J. Phys.* (1970) B**20**, 94.

crystal (see § 10.5.3). Since the densities of states $N_c(E)$, $N_v(E)$ are proportional to $(E-E_c)^{\frac{1}{2}}$ and $(E_v-E)^{\frac{1}{2}}$ we obtain the quadratic variation as for an indirect transition. Variation of the index r would indicate a different variation of $N(E)$ or an optically forbidden transition. By extrapolating the linear portion of the plot, as in Fig. 15.4, one may obtain the 'optical' energy gap E_g^o which corresponds to the difference of the quantities E_c and E_v shown in Fig. 15.1.

The region (2) is supposed to be due to transitions from the extended states at the top of the valence band to the localized states at the bottom of the conduction band, and also to transitions from the localized states at the top of the valence band to extended states in the conduction band. If the density of states in the tail $N(E)$ has the form $N(E) = B \exp(-E/E_t)$, where B and E_t are constants, then we get $\alpha = C \exp(-h\nu/E_t)$ provided E_t is the same for both tails, otherwise we get a sum of two exponentials. Tauc (*loc. cit.*) has analysed absorption spectra from a great many glasses and found this kind of law to be obeyed over a large range of absorption coefficients. In order to account quantitatively for the observed absorption, densities of states of the order of $10^{20}\ \mathrm{cm^{-3}\ eV^{-1}}$ are required.

The interpretation of the long-wave tail (region (1)) is not so simple. It is tempting to ascribe it to transitions between states in the two tails but this, as Tauc (*loc. cit.*) has shown, leads to difficulties. Moreover it is found to vary a great deal with method of preparation while the absorption in the other two regions varies very little. It is therefore almost certainly due to impurities or defects.

We have already discussed briefly in § 15.3 the absorption spectra of these materials due to lattice vibrations. These give information basically about the vibrational levels of the molecular groups making up the glass. Generally a few characteristic frequencies can be identified and associated with vibrations of particular chemical bonds. For example, the *restrahl* absorption in glasses although not so sharp as for crystals, occurs at the same place in the spectrum.[26]

15.7.2 Current instability and switching

A new phenomenon which has appeared in these glassy semiconductors is a particular type of current instability. This has been observed in a wide range of materials including the chalcogenide glasses, but rather more markedly in more complex glasses with room temperature

[26] D. L. Mitchell, S. G. Bishop and P. C. Taylor, *J. Non-Cryst. Solids* (1972) **10**, 231.

conductivities in the range 10^{-5}–10^{-8} Ω^{-1} cm^{-1}. When the field in the specimen (usually about 1 μm thick) exceeds about 10^6 V m^{-1} the conductivity rises dramatically to the order of 1–10 Ω^{-1} cm^{-1}. This is not an electrical breakdown of the usual kind since the current may now be reduced, as a linear function of the voltage, until a critical current is reached at which the normal current–voltage characteristic is restored. Moreover the polarity may be reversed. This is illustrated in Fig. 15.6. It is this current switching phenomenon that has led to the greatly increased interest in glassy semiconductors. The phenomenon was discovered by S. R. Ovshinsky[27] and has stimulated a great deal of device development, since it can be used as the basis for switches, memory devices, etc., having a very short switching time between the two states, of the order of 10^{-10} s. Although the phenomenon has been observed in a great variety of materials, glasses (mainly chalcogenide alloys) have been used for its exploitation. These include glasses such as $Te_{50}As_{30}Si_{10}Ge_{10}$ and other equally complex compounds. A review of these applications has been given by H. Fritzsche[28] who has also dis-

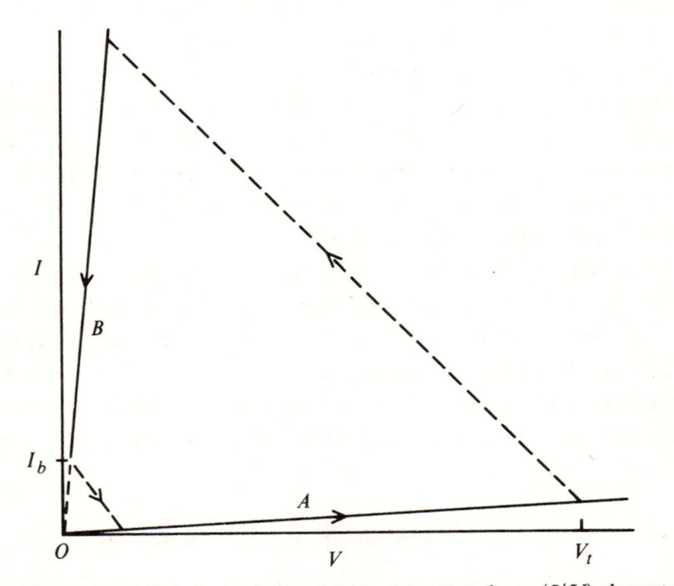

Fig. 15.6. Illustrating switch from high-resistance current–voltage (I/V) characteristic A to low-resistance characteristic B at critical voltage V_t and back at critical current I_b.

[27] *Phys. Rev. Lett* (1968) **21**, 1450.

[28] *Electronic and Structural Properties of Amorphous Semiconductors*, ed. P. G. Le Comber and J. Mort (Academic Press, 1973), p. 557; *Amorphous and Liquid Semiconductors*, ed. J. Tauc (Plenum Press, 1974).

cussed the physics of the phenomenon. Various proposals have been made to account for the switching of the current but no general agreement has yet been reached on the details of the mechanism involved. Indeed it is possible that several different processes occur, such as local thermal effects causing melting or change of crystal structure. Here is indeed a fascinating effect for which application has outstripped theory and which remains to be clarified.

15.8 Conclusion

It will be clear from our necessarily brief discussion of amorphous semiconductors that although a great deal has been achieved in understanding the basic physical mechanisms which give them their distinctive properties there are still many outstanding theoretical and practical problems to be solved before our knowledge of them compares with what we know of crystalline high-mobility semiconductors. On the theoretical side there are still alternatives to be resolved and a better understanding needed of some of the problems associated with electronic states in disordered systems. On the practical side there are formidable problems of preparation and purification to be overcome before we have material that we are sure is quite characteristic of the substance under examination. The progress made with a-Si and a-Ge shows that a step forward in technique leads to new developments both in basic physics and in application (see § 15.4). These problems present a challenge but also provide plenty of scope for new ideas and the hope for unforeseen and exciting new developments.

That amorphous materials will soon replace the single crystals now in general use in micro-electronics and opto-electronics seems most unlikely, but new uses have emerged for amorphous materials, as we have seen. This stimulus, as well as the fundamental interest in the subject, should ensure that rapid progress in reaching a full understanding of these materials will continue.

Bibliography

The following gives a short and very incomplete list of books available on the physics of semiconductors. They have been chosen to cover the range of topics dealt with, and most of them have been referred to in the text.

1 Serials and recurrent publications

Progress in Semiconductors ed. A. F. Gibson, P. Angrain, R. E. Burgess and (later) F. A. Kröger (Heywood, 1956–67); *Semiconductors and Semimetals*, ed. R. K Willardson and A. C. Beer (Academic Press, 1966–); *Solid State Physics*, ed. F. Seitz, D. Turnbull and (later) H. Ehrenreich (Academic Press, 1955–); see also proceedings of biennial international conferences on physics of semiconductors (1952–1976).

2 Theoretical solid state physics

Electrons and Phonons, J. M. Ziman (Oxford University Press, 1960); *Principles of the Theory of Solids*, J. M. Ziman (Cambridge University Press, 2nd edition, 1972); *Theoretical Solid State Physics*, W. Jones and N. H. March (Wiley Interscience, 1972); *Theoretical Solid State Physics*, A. Haug (Pergamon Press, 1972); *Insulators, Semiconductors and Metals*, J. C. Slater (McGraw-Hill, 1967); *Wave Mechanics of Crystalline Solids*, R. A. Smith (Chapman and Hall, 2nd edition, 1969) (referred to in text as *W.M.C.S.*).

3 General

Physics of Semiconductors, A. F. Ioffe (Infosearch, 1960); *Introduction to Semiconductor Physics*, R. B. Adler, A. C. Smith and R. L. Longini

(Wiley, 1964); *Physics of Semiconductors*, J. L. Moll (McGraw-Hill, 1964); *Solid State and Semiconductor Physics*, J. P. McKelvey (Harper and Row, 1966); *Solid State Electronics*, S. Wang (McGraw-Hill, 1966); *Physics of Electronic Conduction in Solids*, F. J. Blatt (McGraw-Hill, 1968); *Physics of Semiconductor Devices*, S. M. Sze (Wiley Interscience, 1969); *Semiconductors*, H. F. Wolf (Wiley Interscience, 1971); *An Introduction to Solid State Physics and its Applications*, R. J. Elliott and A. F. Gibson (Macmillan, 1974).

4 Impurities and imperfections

Deep Impurities in Semiconductors, A. G. Milnes (Wiley, 1973); *Lattice Defects in Semiconductors*, ed. F. A. Huntley (Inst. of Phys. Conf. Series, 1974); *Point Defects in Solids*, ed. J. H. Crawford and L. M. Slifkin (Plenum Press, 1975–); *Theory of Defects in Solids*, A. M. Stoneham (Clarendon Press, 1975).

5 Ion implantation and radiation damage

Ion Implantation in Semiconductors, J. W. Mayer, L. Eriksson and J. A. Davies (Academic Press, 1970); *Radiation Effects in Semiconductors*, ed. J. W. Corbett and G. D. Watkins (Gordon and Breach, 1971); *Ion Implantation*, G. Dearnaley, J. H. Freeman, R. S. Nelson and J. Stephen (North Holland, 1973); *Ion Implantation of Semiconductors*, G. Carter and W. A. Grant (Edward Arnold, 1976).

6 Distribution of electrons between levels

Semiconductor Statistics, J. S. Blakemore (Pergamon Press, 1962).

7 Transport properties

Electronic Structure in Solids, ed. E. D. Haidemenakis (Plenum Press, 1969); *The Thermoelectric Properties of Semiconductors*, ed. V. A. Kutasov (Consultants Bureau, N.Y., 1964); *Semiconductor Surfaces*, A. Many, Y. Goldstein and N. B. Grover (North Holland, 1965); *Electronic Properties of Semiconductor Surfaces*, D. R. Frankl (Pergamon Press, 1967); *Metal–Semiconductor Contacts*, ed. M. Pepper (Inst. of Phys. Conf. Series, 1974); *Physics of Solids in High Magnetic Fields*, ed. E. D.

Haidemenakis (Plenum Press, 1969); *The Electron Theory of Heavily Doped Semiconductors*, V. L. Bonch-Bruevich (Elsevier, 1966); *Heavily Doped Semiconductors*, V. I. Fistul (Plenum Press, 1969); *Tunneling Phenomena in Solids*, ed. E. Burstein and S. Lundquist (Plenum Press, 1969); also supplements to *Solid State Physics* (see § 1).

8 Optical properties

Infra-red Physics, J. Houghton and S. D. Smith (Oxford University Press, 1966); *Optical Properties and Band Structure of Semiconductors*, D. L. Greenaway and G. Harbeke (Pergamon Press, 1968); *Optical Properties of Solids*, ed. E. D. Haidemenakis (Gordon and Breach, 1970); *Optical Processes in Semiconductors*, J. I. Pankove (Dover Publ., 1975); *Optical Properties of Solids*, ed. F. Abeles (North Holland, 1972); *Semiconductor Opto-electronics*, T. S. Moss, G. J. Burrell and B. Ellis (Butterworths, 1973); *Optical Properties of Solids – New Developments*, ed. B. O. Seraphin (North Holland, 1976).

9 Band structure

Electronic Energy Bands in Solids, L. Pincherle (Macdonald, 1971); *Bonds and Bands in Semiconductors*, J. C. Phillips (Academic Press, 1973); also J. C. Slater (see § 2).

10 Materials preparation

Ultra-Purification of Semiconductors, M. S. Brooks and J. K. Kennedy (Macmillan, 1962); *The Art and Science of Growing Crystals*, ed. J. J. Gilman (Wiley, 1963); *Crystal Growth*, ed. W. Bardsley, D. J. T. Heurle and J. B. Mullin (North Holland, 1973–); *Electronic Materials*, ed. N. B. Hannay and U. Colombo (Plenum Press, 1973); *Characterization of Semiconducting Materials*, P. F. Kane and G. B. Larrabee (McGraw-Hill, 1975); *Semiconductor Measurements and Instrumentation*, W. R. Runyan (McGraw-Hill, 1975).

11 Materials

Materials used in Semiconductor Devices, ed. C. A. Hogarth (Wiley Interscience, 1965); *II–VI Semiconducting Compounds*, ed. D. G. Thomas (Benjamin, 1967); *Semiconductors and Semimetals*, see § 1;

Handbook of Electronic Materials, Vols. 2–7 (IFI Plenum, 1971–72); *Electronic Materials*, see § 10; *Physics of Semimetals and Narrow-Gap Semiconductors*, ed. D. L. Carter and R. T. Bates (Pergamon Press, 1971); *Refractory Semiconductor Materials*, Y. V. Shmartsev, Y. A. Valov and A. S. Borshchevskii (Consultants Bureau, N.Y., 1966); *Organic Semiconductors*, F. Gutman and L. E. Lyons (Wiley, 1967).

12 Special topics

Polarons and Excitons, ed. C. G. Kuper and G. D. Whitfield (Oliver and Boyd, 1962); *Polarons in Ionic Crystals and Polar Semiconductors*, ed. J. T. Devreese (North Holland, 1972); *Polaritons*, ed. E. Burstein and F. de Martini (Pergamon Press, 1974); *The Physics and Technology of Semiconductor Light Emitters and Detectors*, ed. A. Frova (North Holland, 1973); *Laser Physics*, M. Sargent, M. O. Scully and W. E. Lamb (Addison-Wesley, 1974); *Laser Applications to Optics and Spectroscopy*, ed. S. F. Jacobs *et al.* (Addison-Wesley, 1975); *Very High Resolution Spectroscopy*, ed. R. A. Smith (Academic Press, 1976); *Solids under Pressure*, ed. W. Paul and D. M. Warschauer (McGraw-Hill, 1963).

13 Amorphous and low-mobility semiconductors

Quantum Theory of Amorphous Semiconductors, A. I. Gabanov (Consultants Bureau, N.Y. 1965); *Amorphous Semiconductors*, D. Adler (Butterworths, 1971); *Electronic Processes in Non-Crystalline Materials*, N. F. Mott and E. A. Davis (Oxford University Press, 1971); *Electronic and Structural Properties of Amorphous Semiconductors*, ed. P. G. Le Comber and J. Mort (Academic Press, 1973); *Amorphous and Liquid Semiconductors*, ed. J. Tauc (Plenum Press, 1974); see also proceedings of biennial international conferences on amorphous semiconductors.

Author index

Subject index